Volumes already published

Volume 1: Trees I (1986)
Volume 2: Crops I (1986)
Volume 3: Potato (1987)
Volume 4: Medicinal and Aromatic Plants I (1988)
Volume 5: Trees II (1989)
Volume 6: Crops II (1988)
Volume 7: Medicinal and Aromatic Plants II (1989)
Volume 8: Plant Protoplasts and Genetic Engineering I (1989)
Volume 9: Plant Protoplasts and Genetic Engineering II (1989)
Volume 10: Legumes and Oilseed Crops I (1990)
Volume 11: Somaclonal Variation in Crop Improvement I (1990)
Volume 12: Haploids in Crop Improvement I (1990)
Volume 13: Wheat (1990)
Volume 14: Rice (1991)
Volume 15: Medicinal and Aromatic Plants III (1991)
Volume 16: Trees III (1991)
Volume 17: High-Tech and Micropropagation I (1991)
Volume 18: High-Tech and Micropropagation II (1992)
Volume 19: High-Tech and Micropropagation III (1992)
Volume 20: High-Tech and micropropagation IV (1992)
Volume 21: Medicinal and Aromatic Plants IV (1993)
Volume 22: Plant Protoplasts and Genetic Engineering III (1993)
Volume 23: Plant Protoplasts and Genetic Engineering IV (1993)
Volume 24: Medicinal and Aromatic Plants V (1993)
Volume 25: Maize (1994)
Volume 26: Medicinal and Aromatic Plants VI (1994)
Volume 27: Somatic Hybridization in Crop Improvement I (1994)
Volume 28: Medicinal and Aromatic Plants VII (1994)
Volume 29: Plant Protoplasts and Genetic Engineering V (1994)
Volume 30: Somatic Embryogenesis and Synthetic Seed I (1995)
Volume 31: Somatic Embryogenesis and Synthetic Seed II (1995)
Volume 32: Cryopreservation of Plant Germplasm I (1995)
Volume 33: Medicinal and Aromatic Plants VIII (1995)
Volume 34: Plant Protoplasts and Genetic Engineering VI (1995)
Volume 35: Trees IV (1996)
Volume 36: Somaclonal Variation in Crop Improvement II (1996)
Volume 37: Medicinal and Aromatic Plants IX (1996)
Volume 38: Plant Protoplasts and Genetic Engineering VII (1996)
Volume 39: High-Tech and Micropropagation V (1997)
Volume 40: High-Tech and Micropropagation VI (1997)
Volume 41: Medicinal and Aromatic Plants X (1998)
Volume 42: Cotton (1998)
Volume 43: Medicinal and Aromatic Plants XI (1999)

Volumes in preparation

Volume 44: Transgenic Trees (1999)
Volume 45: Transgenic Medicinal Plants (1999)
Volume 46: Transgenic Crops I (1999)
Volume 47: Transgenic Crops II

Biotechnology in Agriculture and Forestry 43

Medicinal and Aromatic Plants XI

Edited by Y.P.S. Bajaj

With 191 Figures, 4 in Colour and 64 Tables

Springer

Professor Dr. Y.P.S. Bajaj
A-137
New Friends Colony
New Delhi 110065, India

ISSN 0934-943-X
ISBN 3-540-63911-X Springer-Verlag Berlin Heidelberg New York

Library of Congress Cataloging-in-Publication Data. Medicinal and aromatic plants. (Biotechnology in agriculture and forestry; 4-). Includes bibliographies and index. 1. Medicinal plants – Biotechnology. 2. Aromatic plants – Biotechnology. 3. Plant cell culture. 4. Materia medica, Vegetable. I. Bajaj, Y.P.S., 1936– . II. Series. TP248.27.P55M43 1998 660.62 88-3059. ISBN 3-540-63911-X

Printed in Germany

Cover design: *design & production*, Heidelberg

Typesetting: Best-set Typesetter Ltd., Hong Kong

SPIN: 10569886 31/3136 – 5 4 3 2 1 0 – Printed on acid-free paper

Dedicated to
Mannat Bakshi

Preface

This series of books on biotechnology of *Medicinal and Aromatic Plants* is being compiled to provide a survey of the literature focusing on recent information and the state of the art in tissue culture and the in vitro production of secondary metabolites. This book, *Medicinal and Aromatic Plants XI*, like the previously ten volumes published in 1988, 1989, 1991, 1993, 1994, 1995, 1996, and 1998, is unique in its approach. It comprises 24 chapters dealing with the distribution, importance, conventional propagation, micropropagation, tissue culture studies and the production of important medicinal and pharmaceutical compounds in various species of *Anagallis*, *Azadirachta*, *Centranthus*, *Costus*, *Cuphea*, *Dioscorea*, *Drosera*, *Fagara*, *Frangula*, *Hyacinthus*, *Hypericum*, *Jamesoniella*, *Karwinskia*, *Lactarius*, *Lactuca*, *Marribium*, *Ornithopus*, *Petroselinum*, *Phellodendron*, *Physochlaina*, *Solanum*, *Solidago*, and *Zanthoxylum*. This book is tailored to the needs of advanced students, teachers, and research scientists in the field of pharmacy, plant tissue culture, phytochemistry, biomedical engineering, and plant biotechnology in general.

New Delhi, September 1998

Professor Dr Y.P.S. Bajaj
Series Editor

Contents

I *Anagallis arvensis* L. (Common Pimpernel): In Vitro Culture, Regeneration of Plants, and the Formation of Anagalline and Other Saponins
Y.P.S. Bajaj (With 2 Figures)

1 Introduction 1
2 Tissue Culture Studies 1
3 Formation of Saponins 8
4 Summary 9
References 9

II *Azadirachta indica* A. Juss. (Neem Tree): In Vitro Culture, Micropropagation, and the Production of Azadirachtin and Other Secondary Metabolites
E.J. Allan, T. Stuchbury, and A.J. Mordue (Luntz) (With 14 Figures)

1 General Account 11
2 In Vitro Approaches 20
3 Conclusions and Prospects 35
References 36

III *Centranthus* Species: In Vitro Culture and the Production of Valepotriates and Other Secondary Metabolites
P. Christen (With 4 Figures)

1 Introduction 42
2 In Vitro Culture Studies 44
3 Summary and Conclusion 53
References 54

IV *Costus speciosus* (Koenig) J.E. Smith: In Vitro Cultures, Micropropagation, and the Production of Diosgenin and Other Phytosteroids
G. Indrayanto, W. Utami, and A. Syahrani (With 10 Figures)

1 General Account 57
2 In Vitro Culture Studies 63
3 Conclusions 73

4 Protocols 74
References 75

V *Cuphea* Species: In Vitro Culture Studies, and the Production of Medium-Chain Fatty Acids
E.I.S. Floh, W. Handro, and I. Rita (With 5 Figures)

1 General Account 78
2 Economic Importance and Fatty Acids 79
3 Micropropagation 80
4 Summary and Conclusion 83
References 83

VI *Dioscorea balcanica* Košanin and *D. caucasica* Lipsky: In Vitro Culture and Production of Diosgenin
L. Ćulafić, K. Šavikin-Fodulović, D. Grubišić, and M. Nešković (With 5 Figures)

1 Introduction 85
2 In Vitro Culture Studies 86
3 Production of Diosgenin and Phytosterols 96
4 Summary and Prospects 101
References 102

VII *Drosera* Species (Sundew): In Vitro Culture and the Production of Plumbagin and other Secondary Metabolites
J. Šamaj, A. Blehová, M. Repčák, M. Ovečka, and M. Bobák (With 9 Figures)

1 Introduction 105
2 In Vitro Approaches 115
3 Biosynthesis and Extraction of Medicinal Compounds 125
4 Summary and Conclusions 129
References 130

VIII *Fagara zanthoxyloides* Lam. (Rutaceae): In Vitro Culture and the Production of Benzophenanthridine and Furoquinoline Alkaloids
E. Couillerot, C. Caron, C. Trentesaux, J.C. Chenieux and J.C. Audran (With 13 Figures)

1 Introduction 136
2 Chemical Constituents and Their Pharmacological Properties 138
3 In Vitro Culture Studies 145
4 Suggested Protocols 152
5 Summary and Conclusions 153
References 154

IX *Frangula* Species: In Vitro Culture and the Production of Anthraquinones
L. Sajc, N. Kovacević, D. Grubišić, and G. Vunjak-Novaković (With 7 Figures)

1 General Account 157
2 Analysis of Anthraquinones 160
3 In Vitro Culture Studies 162
4 Conclusion 171
References 173

X *Hyacinthus orientalis* L.: In Vitro Culture and the Production of Anthocyanin and Other Secondary Metabolites
K. Hosokawa (With 11 Figures)

1 Introduction 177
2 In Vitro Approaches 181
3 Summary and Conclusions 194
References 195

XI *Hypericum patulum*: In Vitro Culture and Production of Xanthones and Other Secondary Metabolites
K. Ishigoro, H. Oku, and K. Isoi (With 8 Figures)

1 Introduction 199
2 In Vitro Approaches 200
3 Antibacterial Activity of Xanthones 207
4 Discussion and Conclusions 208
References 210

XII *Jamesoniella autumnalis* (Liverwort): Culture and Production of Metabolites
K.P. Adam (With 5 Figures)

1 General Account 213
2 Chemical Constituents 215
3 In Vitro Culture Studies 217
4 Summary 219
5 Protocol for the Establishment of Aseptic Cultures of Folious Liverworts 220
References 221

XIII *Karwinskia* Species (Buckthorn): In Vitro Culture, Micropropagation, and the Production of Toxin Anthracenone
D. Lišková, A. Lux, A. Piñeyro-López, and R. Luján-Rangel

1 Introduction 223
2 In Vitro Approaches 228
3 Summary 238
References 239

XIV *Lactarius* Species (Mushrooms): In Vitro Culture and the Production of Sesquiterpenes and Other Secondary Metabolites
L.S. Trifonov and P. Chakravarty (With 7 Figures)

1 Introduction ... 243
2 In Vitro Cultures of *Lactarius* spp. ... 244
3 Secondary Metabolites Isolated from *Lactarius* Species ... 248
4 Summary and Conclusions ... 256
References ... 257

XV *Lactuca virosa* L. (Bitter Lettuce): In Vitro Culture and Production of Sesquiterpene Lactones
A. Stojakowska, J. Malarz, and W. Kisiel (With 6 Figures)

1 General Account ... 261
2 In Vitro Culture Studies ... 264
3 Summary and Concluding Remarks ... 271
References ... 272

XVI *Marrubium vulgare* (White Horehound): In Vitro Culture and the Production of Diterpene Marrubiin and Other Secondary Metabolites
W. Knöss (With 9 Figures)

1 Introduction ... 274
2 In Vitro Culture Studies ... 275
3 Summary and Conclusion ... 287
References ... 287

XVII *Menispermum* Species (Moonseed Vines): In Vitro Culture, and the Production of Dauricine and Other Secondary Metabolites
Y. Sugimoto (With 18 Figures)

1 General Account ... 290
2 In Vitro Studies ... 294
3 Summary and Conclusion ... 307
References ... 308

XVIII *Ornithopus sativus* Brot. (Serradella): In Vitro Culture, Phytochemical Studies, and Biotransformation
A. Kolbe, G. Krauss, and B. Schneider (With 9 Figures)

1 Introduction ... 310
2 In Vitro Culture ... 311
3 Phytochemical Studies ... 312
4 Biotransformation ... 316
5 Summary and Conclusions ... 321
References ... 322

XIX *Petroselinum crispum* (Mill.) Nyman (Parsley): In Vitro Culture, Production and Metabolism of Volatile Constituents
A.A. Gbolade and G.B. Lockwood (With 4 Figures)

1 Introduction ... 324
2 Constituents ... 324
3 In Vitro Culture Studies ... 326
4 Summary and Conclusion ... 333
5 Protocol ... 334
References ... 334

XX *Phellodendron amurense* (Cork Tree): In Vitro Culture, Micropropagation, and the Production of Berberine
Y.G. Park and M.S. Choi (With 8 Figures)

1 Introduction ... 337
2 In Vitro Approaches ... 339
3 Conclusion and Prospects ... 348
References ... 348

XXI *Physochlaina* Species: In Vitro Culture and the Production of Physochlaine and Other Tropane Alkaloids
N. Gorinova, A. Atanassov, and M. Welcheva (With 7 Figures)

1 General Account ... 350
2 Techniques ... 356
3 Tropane Alkaloids of *P. orientalis* Cell Suspension Culture ... 357
4 Summary and Conclusion ... 362
References ... 362

XXII *Solanum malacoxylon* Sendt. (*Solanum glaucophyllum* Desf.): In Vitro Culture and the Production of Sterols and Vitamin D_3 Metabolites
L.A. Manzocchi and M.L. Racchi (With 10 Figures)

1 Calcinogenic Plants, Vitamin D and Vitamin D-Like Substances ... 364
2 Calcitriol ... 368
3 In Vitro Approaches ... 372
4 Conclusions and Prospects ... 379
References ... 380

XXIII *Solidago* L. Species (Goldenrod): In Vitro Regeneration and Biologically Active Secondary Metabolites
L. Skrzypczak, M. Wesołowska, B. Thiem, and J. Budzianowski (With 6 Figures)

1 Distribution and Morphology of the Plants ... 384
2 Chemical Components and Their Significance ... 385
3 In Vitro Culture Studies ... 394
4 Summary ... 397
References ... 398

XXIV *Zanthoxylum* Species: In Vitro Culture, Aroma Emanation, and the Production of Secondary Metabolites
K. Imaizumi (With 7 Figures)

1 Introduction 404
2 In Vitro Approaches 407
3 Conclusions and Prospects 414
References 415

Index 417

List of Contributors

ADAM, K.P., FR 12.3, Pharmakognosie und Analytische Phytochemie, Universität des Saarlandes, 66041 Saarbrücken, Germany

ALLAN, E.J., Department of Agriculture, University of Aberdeen, 581 King Street, Aberdeen AB24 5UA, UK

ATANASSOV, A., Laboratory of Cell Genetics, Institute of Genetic Engineering, 2232 Kostinbrod – 2, Bulgaria

AUDRAN, J.C., Laboratoire de Biologie et Physiologie Végétales EA 2069, Université de Reims Champagne-Ardenne, UFR Sciences, BP 1039, 51687 Reims Cedex 2, France

BAJAJ, Y.P.S., Biotechnology in Agriculture and Forestry, A-137 New Friends Colony, New Delhi 110065, India

BLEHOVÁ, A., Department of Plant Physiology, Comenius University Bratislava, Mlynská dolina B-2, SK-842 15 Bratislava, Slovak Republic

BOBÁK, M., Department of Plant Physiology, Comenius University Bratislava, Mlynská dolina B-2, SK-842 15 Bratislava, Slovak Republic

BUDZIANOWSKI, J., Department of Pharmaceutical Botany, K. Marcinkowski University of Medical Sciences in Poznań, Św. Marii Magdaleny 14, 61-861 Poznań, Poland

CARON, C., Laboratoire de Pharmacognosie UPRES-A 6013, Université de Reims Champagne-Ardenne, UFR Pharmacie, BP 1039, 51687 Reims Cedex 2, France

CHAKRAVARTY, P., Canadian Forest Service, Northern Forestry Centre, 5320-122 St., Edmonton, Alberta, T6H 3S5, Canada

CHENIEUX, J.C., Laboratoire de Biologie Cellulaire et Biochimie Végétale, EA 1370, Université de Tours, UFR Pharmacie, 31 Avenue Monge, 37200 Tours, France

Choi, M.S., Department of Biotechnology, Forest Genetics Research Institute, Omokchon-dong, Kwonsun-ku, Suwon 441-350, Kyunggi-do, Korea

Christen, P., Laboratory of Pharmaceutical Analytical Chemistry, School of Pharmacy, University of Geneva, 20 Bd, d'Yvoy, 1211 Geneva 4, Switzerland

Couillerot, E., Laboratoire de Biologie et Physiologie Végétales EA 2069, Université de Reims Champagne-Ardenne, UFR Sciences, BP 1039, 51687 Reims Cedex 2, France

Ćulafić, L., University of Belgrade, Faculty of Biology, Institute of Botany, Takovska 43, 11000 Belgrade, Yugoslavia

Floh, E.I.S., Plant Cell Biology Laboratory, Department of Botany, Institute of Biosciences, University of São Paulo, CP 11461, 05421-970 São Paulo, Brazil

Gbolade, A.A., Department of Pharmacognosy, Faculty of Pharmacy, Obafemi Awolowo University, Ile-Ife, Nigeria

Gorinova, N., Laboratory of Enzymes, Institute of Genetic Engineering, 2232 Kostinbrod – 2, Bulgaria

Grubišić, D., Department of Plant Physiology, Faculty of Biology, Belgrade University, 29. novembra 142, 11000 Belgrade, Yugoslavia

Grubišić, D., University of Belgrade, Institute for Biological Research "S. Stanković", 29 novembra 142, 11060 Belgrade, Yugoslavia

Handro, W., Plant Cell Biology Laboratory, Department of Botany, Institute of Biosciences, University of São Paulo, CP 11461, 05421-970 São Paulo, Brazil

Hosokawa, K., Iwate Biotechnology Research Center, 22-174-4 Narita, Kitakami, Iwate 024- Japan. Present address: Tsukuba Medicinal Plant Research Station, National Institute of Health Sciences, 1 Hachimandai, Tsukuba, Ibaraki, 305-0843 Japan

Imaizumi, K., PL Gakuen Women's Junior College, 2172-1, Shindou, Tondabayashi, Osaka 584, Japan

Indrayanto, G., Laboratory of Pharmaceutical Biotechnology, Faculty of Pharmacy, Airlangga University, Jl. Dharmawangsa dalam, Surabaya 60286, Indonesia

ISHIGURO, K., School of Pharmaceutical Sciences, Mukogawa Women's University, Koshienkyuban-cho, Nishinomiya, 663-8179, Japan

ISOI, K., School of Pharmaceutical Sciences, Mukogawa Women's University, Koshienkyuban-cho, Nishinomiya, 663-8179, Japan

KISIEL, W., Department of Phytochemistry, Institute of Pharmacology, Polish Academy of Sciences, 12 Smętna Street, 31-343 Cracow, Poland

KNÖSS, W., Institute of Pharmaceutical Biology, University of Bonn, Nussallee 6, 53115 Bonn, Germany

KOLBE, A., Institut für Pflanzenbiochemie, Weinberg 3, 06120 Halle, Germany

KOVACEVIĆ, N., Department of Farmacognosy, Faculty of Pharmacy, Belgrade University, Vojvode Stepe 450, 11000 Belgrade, Yugoslavia

KRAUSS, G., UFZ – Umweltforschungszentrum Leipzig-Halle GmbH, Sektion Hydrogeologie, Hallesche Straße 44, 06246 Bad Lauchstädt, Germany

LIŠKOVÁ, D., Institute of Chemistry, Slovak Academy of Sciences, Dúbravská cesta 9, 84238 Bratislava, Slovakia

LOCKWOOD, G.B., School of Pharmacy and Pharmaceutical Sciences, University of Manchester, Manchester, M13 9PL, UK

LUJÁN-RANGEL, A., Departamento de Farmacología y Toxicología, Facultad de Médicina, U.A.N.L. Apdo. Postal 146, Col. del Valle, Monterrey, N.L. 66220, México

LUX, A., Department of Plant Physiology, Faculty of Natural Sciences, Comenius University, Mlynská dolina, 84215 Bratislava, Slovakia

MALARZ, J., Department of Phytochemistry, Institute of Pharmacology, Polish Academy of Sciences, 12 Smętna Street, 31-343 Cracow, Poland

MANZOCCHI, L.A., Istituto Biosintesi Vegetali, Consiglio Nazionale delle Ricerche, Via Bassini 15, 20133 Milan, Italy

MORDUE (LUNTZ), A.J., Department of Zoology, University of Aberdeen, Tillydrone Avenue, Aberdeen AB24 3TZ, UK

NEŠKOVIĆ, M., University of Belgrade, Faculty of Biology, Institute of Botany, Takovska 43, 11000 Belgrade, Yugoslavia

Oku, H., School of Pharmaceutical Sciences, Mukogawa Women's University, Koshienkyuban-cho, Nishinomiya, 663-8179, Japan

Ovečka, M., Institute of Botany, Slovak Academy of Sciences, Dúbravská cesta 14, SK-842 23 Bratislava, Slovak Republic

Park, Y.G., Department of Forestry, Kyungpook National University, 1370 Sakyup-dong, Buk-ku, Taegu 702-701, Korea

Piñeyro-López, A., Departamento de Farmacología y Toxicología, Facultad de Médicina, U.A.N.L. Apdo. Postal 146, Col. del Valle, Monterrey, N.L. 66220, México

Racchi, M.L., Genetics Unit, Institute of Silviculture, University of Florence, Via San Bonaventura 13, 50145 Florence, Italy

Repčák, M., Department of Experimental Botany and Genetics, Faculty of Natural Sciences, P.J. Šafárik University, Mánesova 23, SK-04167 Košice, Slovak Republic

Rita, I., Cenargen/Embrapa, CP 02372, 70849-970 Brasilia, Brazil

Sajc, L., Department of Analytical Chemistry, Faculty of Technology, Belgrade University, Karnegijeva 4, 11000 Belgrade, Yugoslavia

Šamaj, J., Institute of Plant Genetics, Slovak Academy of Sciences, Akademická 2, SK-950 07 Nitra, Slovak Republic

Šavikin-Fodulović, K., Institute for Medicinal Plant Research "Dr. Josif Pančić", T. Koš ćuš ka 1, 11000 Belgrade, Yugoslavia

Schneider, B., Max-Planck-Institut für Chemische Ökologie, Tatzendpromenade la, 07745 Jena, Germany

Skrzypczak, L., Department of Pharmaceutical Botany, K. Marcinkowski University of Medical Sciences in Poznań, Św. Marii Magdaleny 14, 61-861 Poznań, Poland

Stojakowska, A., Department of Phytochemistry, Institute of Pharmacology, Polish Academy of Sciences, 12 Smętna Street, 31-343 Cracow, Poland

Stuchbury, T., Department of Agriculture, University of Aberdeen, 581 King Street, Aberdeen AB24 5UA, UK

Sugimoto, Y., Arid Land Research Center, Tottori University, 1390 Hamasaka, Tottori 680-0001, Japan

Syahrani, A., Laboratory of Pharmaceutical Biotechnology, Faculty of Pharmacy, Airlangga University, Jl. Dharmawangsa dalam, Surabaya 60286, Indonesia

Thiem, B., Department of Pharmaceutical Botany, K. Marcinkowski University of Medical Sciences in Poznań, Św. Marii Magdaleny 14, 61-861 Poznań, Poland

Trentesaux, C., Laboratoire de Biochimie et Biologie Moléculaire, EA 2063 IFR 53, Université de Reims Champagne-Ardenne, UFR Pharmacie, 51 rue Cognacq-Jay, 51096 Reims Cedex, France

Trifonov, L.S., Chemistry Department, University of Alberta, Edmonton, Alberta, T6G 2G2, Canada

Utami, W., Laboratory of Pharmaceutical Biotechnology, Faculty of Pharmacy, Airlangga University, Jl. Dharmawangsa dalam, Surabaya 60286, Indonesia

Velcheva, M., Laboratory of Natural Products, Institute of Organic Chemistry with Center of Phytochemistry, Bulgarian Academy of Sciences, 1113 Sofia, Bulgaria

Vunjak-Novaković, G., Division of Health Sciences and Technology, Massachusetts Institute of Technology, Cambridge, Massachusetts 02139, USA

Wesołowska, M., Department of Pharmaceutical Botany, K. Marcinkowski University of Medical Sciences in Poznań, Św. Marii Magdaleny 14, 61-861 Poznań, Poland

I *Anagallis arvensis* L. (Common Pimpernel): In Vitro Culture, Regeneration of Plants, and the Formation of Anagalline and other Saponins

Y.P.S. Bajaj

1 Introduction

Anagallis arvensis L. (family Primulaceae) the common scarlet pimpernel, also called poor man's weather glass, because its flowers close before rain (Chevallier 1996), is a creeping annual herb with simple leaves and solitary axillary bisexual tubular flowers. It prefers waste land and untended sandy ground. This plant is a rich source of saponins (Table 1), and has been put to numerous uses. In India it is known by the common name jonkmari and used medicinally for gout, leprosy, epilepsy, and cerebral affections (Mathur 1965). In China it is a folklore remedy for snake bite and hydrophobia (Su Ching New Medical College 1977), while in Europe it is used as a diuretic (Stuart 1988). It has also been used for gallstones, liver cirrhosis, lung problems, urinary infection, kidney stones etc. The pattern of uses suggests a detoxifying action. It is also reported to stimulate the coughing up of mucus and help recovery from colds and flu (Chevallier 1996). It has antiviral activity against *Herpes simplex* virus type and polio virus (Amoros and Girre 1987), and fungicidal properties (Heitz et al. 1971).

This chapter is primarily based on published work on various in vitro culture studies on the establishment of callus cultures, somatic embryogenesis, and plant regeneration (Bajaj and Mäder 1974), and those of secondary metabolites, especially the saponins (Table 1) isolated from whole plants of *Anagallis arvensis*.

2 Tissue Culture Studies

2.1 Material and Methods

The seeds of *Anagallis arvensis* were surface sterilized with 1% chloramine T for about 15 min, washed thoroughly with distilled water, and germinated aseptically on moist filter papers in Petri dishes. The cotyledons, hypocotyl and stem segments, and leaves were excised from young seedlings and

Biotechnology in Agriculture and Forestry, A-137 New Friends Colony, New Delhi 110065, India

Biotechnology in Agriculture and Forestry, Vol. 43
Medicinal and Aromatic Plants XI (ed. by Y.P.S. Bajaj)

Table 1. Summary of studies on tissue culture, and the formation of saponins in *Anagallis arvensis*

Reference	Study
Tissue culture	
Bajaj and Mäder (1974)	Establishment of callus culture, differentiation of shoot buds, somatic embryos, and the regeneration of plants
Secondary metabolites	
Heitz et al. (1971)	Obtained novel triterpene anagalligenone B
Amoros and Girre (1987)	Elucidated structures of two antiviral triterpene saponins
Glombitza and Kurth (1987)	Five new saponins, i.e. anagalloside A,B,C, and desglucoanagalloside A,B
Mahato et al. (1991)	Elucidated structures of four new triterpenoid oligoglycosides
Shoji et al. (1994a)	Isolation of six saponins; anagallosaponin I–V, and methylanagallosaponin I
Shoji et al. (1994b)	Isolation of four oleanane glycosides, anagallosaponin VI–IX, and two artifact saponins apoanagallosaponin III and IV

mature plants and cultured on MS medium (Murashige and Skoog 1962), and also on WM (White 1943) containing various combinations and concentrations of IAA, 2,4-D, NAA, kinetin, zeatin, benzylamino purine, 6-(γ,γ-dimethylallylamino)-purine, coconut milk (CM) and casein hydrolysate (CH).

The pH of the medium was adjusted to 5.8, and it was then jelled with 0.8% agar. Unless mentioned otherwise the cultures were maintained for 14h in light from white fluorescent tubes (2000lx).

The callus tissue was initiated from 1-cm-long hypocotyl and stem segments grown on MS + 2,4-D (1mg/l) + kinetin (0.1mg/l) for 6 weeks, and later subcultured every 6 weeks on MS + 2,4-D (0.1mg/l) + kinetin (0.1mg/l) + CM (10%). The suspension cultures were raised from actively growing hypocotyl callus. About 250mg of the callus was used per flask (125ml) containing 30ml of the medium. The cultures were kept on a gyrorotary shaker, and observed periodically. For plating experiments the suspension cultures were filtered through cheese cloth, and plated on agar-solidified media in 5cm Petri dishes.

The fresh weight, dry weight and cell volume (of suspension cultures) were taken as the growth parameters. After the fresh weight of the callus tissue was taken, it was dried on a piece of aluminium foil in an oven at 70°C for 48h and the dry weight taken.

For histological studies, the callus tissue, hypocotyl segments, and leaf segments were fixed in formalin-acetic acid-alcohol (formalin 5ml, acetic acid 5ml, 70% ethyl alcohol 90ml), and the customary methods of dehydration and embedding were followed. Sections were cut 10–12μm thick, stained with Astrablue (3g Astrablue in 100ml methyl cellosolve, and 6g tartaric acid in 200ml distilled water), and photographed under a Zeiss photomicroscope.

2.2 Results and Discussion

Hypocotyl Segment Culture. The excised segments of hypocotyl and stem cultured on MS containing 2,4-D (1 mg/l) + kinetin (0.1 mg/l) started to enlarge within a week, and showed an overall increase in size. At the cut surface, the segments proliferated to form a callus. In about 4 weeks there was an all-round proliferation, and a mass of callus was formed. Proliferation was improved further if coconut milk (10%) was added to the medium (Fig. 1A). Replacement of 2,4-D with IAA (1 mg/l) induced the formation of roots (Fig. 1B). Within a week on MS + IAA (1 mg/l) + kinetin (1 mg/l) + casein hydrolysate (3000 mg/l) small papillae-like outgrowths made their appearance all around the surface of the segment. They continued to grow and eventually developed into buds and shoots (Fig. 1C). These shoots developed into plantlets (Fig. 1D) if allowed to remain on the segment, or if excised and subcultured separately.

The adventitious buds were observed to arise from the epidermal as well as from the subepidermal layers of the leaves and hypocotyl. The cells destined to become buds were distinguishable from others by being smaller in size, highly cytoplasmic and with prominent dark stained nuclei (Figs. 1G,H). Due to periclinal and anticlinal divisions, small protuberances developed all over the surface of the hypocotyl segments, which made the surface feel rough. These protuberances were somewhat round at first (Fig. 1H) with homogeneous cells. By further meristematic activity, adventitious buds were formed with a shoot apex surrounded by a pair of leaves (Fig. 1I).

Leaf Culture. The excised leaves when cultured on MS+ 2,4-D (0.05 mg/l) + kinetin (1 mg/l) + CM (10%) showed extensive proliferation to form a mass of compact and brownish callus, which differentiated into roots and shoots (Fig. 1F); or there was a direct formation of buds on the surface of leaves without callusing (Fig. 1E). After observing this morphogenetic behaviour, the leaves were grown on simple MS medium as well as on media containing various cytokinins, and the results are summarized in Table 2. Of the cytokinins used, 6-(γ,γ-dimethylallylamino)-purine proved to be the best for the frequency of cultures producing buds, and the number of buds per culture. The superiority of this cytokinin for growth has been earlier reported by the author (Bajaj and Dionne 1967). The regenerating capacity of the leaves also depended on their maturity and the age of the seedlings. Small leaves from comparatively young seedlings regenerated buds easily while older leaves grew much bigger in size and regenerated a few buds. Such a behaviour was also observed in *Torenia fournieri* (Bajaj 1972).

Callus Establishment and Growth. Segments of hypocotyl and stem grown on MS + 2,4-D (1 mg/l) + kinetin (0.1 mg/l) + CM (10%) formed a mass of callus in 4–8 weeks (Fig. 1A, 2J) by an all-round proliferation. This callus could be periodically subcultured on MS + 2,4-D (0.1 mg/l) + kinetin (0.1 mg/l) + CM (10%) (Fig. 1J) and maintained for over 2 years.

With a view to obtaining optimum growth of the callus tissue, the effect of various factors (pH, temperature, light and dark) was studied, and the results are presented in Table 3. Taking fresh and dry weight as the growth

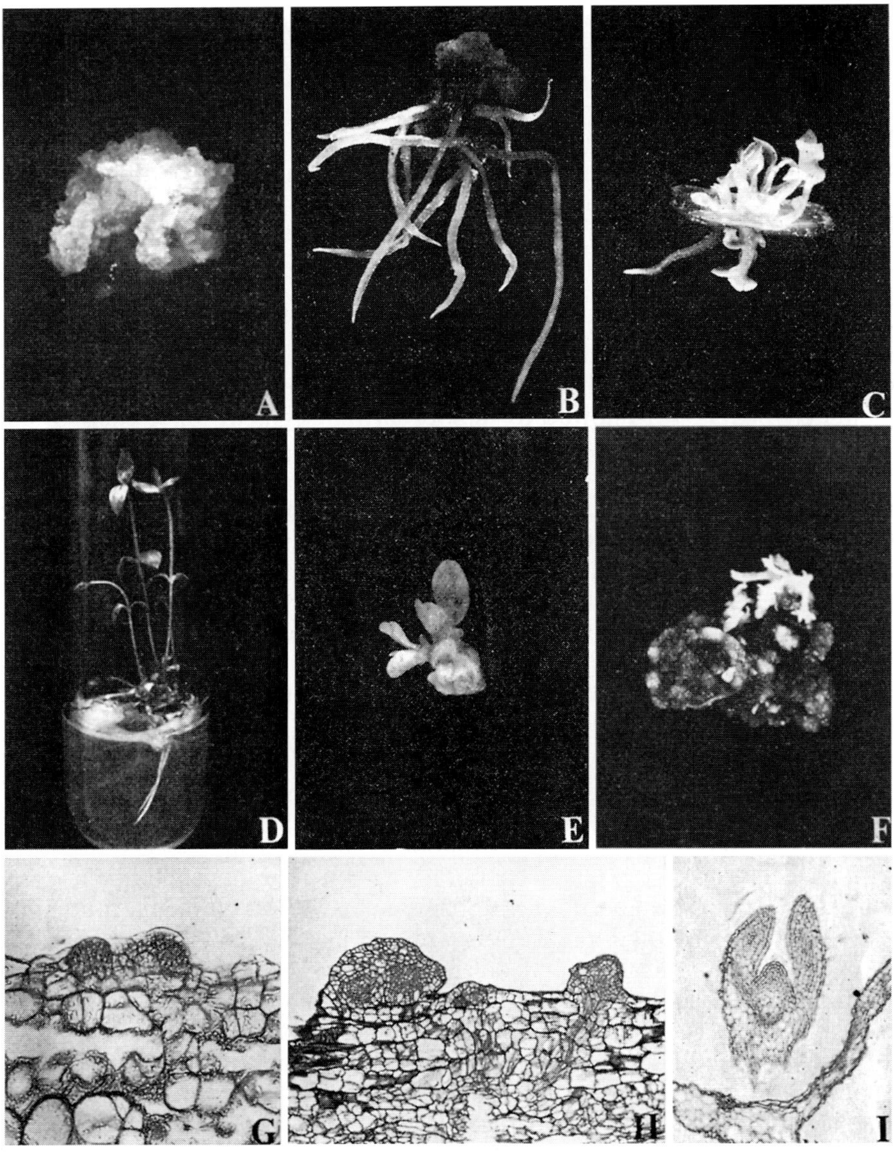

Fig. 1A–I. Growth and morphogenesis in the hypocotyl segments and excised leaves of *Anagallis arvensis*. **A** Four-week-old culture of a hypocotyl segment grown on MS + 2,4-D (1 mg/l) + kinetin (0.1 mg/l) + CM (10%), showing the formation of callus tissue. **B** Hypocotyl segment 2 weeks after culture on WM + IAA (1 mg/l) + CM (10%); note the formation of callus and roots. **C** Two-week-old culture of a hypocotyl segment grown on MS + IAA (1 mg/l) + kinetin (1 mg/l) + CH (3000 mg/l) showing the regeneration of root and shoots. **D** Same after 5 weeks, grown into plantlets. **E** Six-week-old culture of a young leaf showing the formation of shoots. **F** An excised leaf 12 weeks after culturing on MS + 2,4-D (0.05 mg/l) + kinetin (1 mg/l) + CM (10%); note the all-round formation of callus and buds. **G–I** Anatomy of various stages in the development of buds and shoots on the surface of leaf **G,I**, and hypocotyl segment **H**

Table 2. Effect of various cytokinins on growth and bud formation on excised leaves of *Anagallis arvensis*

Media	Growth responses and development	Buds per leaf	Time for bud induction, weeks
1. MS	Leaves turned yellow to colourless, and formed buds occasionally	Occasional	6
2. MS + IAA (0.5 mg/l)	" " "	Occasional	6
3. MS + IAA (0.5 mg/l) + kinetin (1 mg/l)	Leaves showed general enlargement, turned dark green, occasional callusing, but invariable bud formation	5–8	2–3
4. MS + IAA (0.5 mg/l) + 6-benzyl amino purine (1 mg/l)	" " "	5–8	6
5. MS + IAA (0.5 mg/l) + zeatin (1 mg/l)	" " "	6–9	2
6. MS + IAA (0.5 mg/l) + 6-(γ,γ-dimethyl-allylamino)-purine (1 mg/l)	" " "	8–12	2

Table 3. Effect of various factors on the growth of callus tissue cultures of *Anagallis arvensis* grown for 4 weeks on MS + 2,4-D (0.1 mg/l) + kinetin (0.1 mg/l) + CM (10%)

Factor		Fresh weight g	Dry weight mg
pH (temp. 23 °C, light 2000 lx)	5.0	1.68	111
	5.9	2.61	182
	7.0	2.07	144
Temperature (pH 5.8, light 2000 lx)	18 °C	0.67	53
	23 °C	2.18	118
	27 °C	2.69	127
	32 °C	0.69	43
Illumination (pH 5.8, temp. 23 °C)	dark	3.49	157
	(2000 lx)	2.88	186

parameters, optimum growth was obtained at pH 5.9, temperature 27 °C, and 4% sucrose as the carbon source. In complete darkness the callus was comparatively more friable, loose, white in colour, and had higher water content in light (2000 lx) grown cultures. However, growth as judged by dry weight was better under light conditions.

Suspension cultures. The friability, growth, colour and texture of the callus were strongly influenced by the media. On basal medium containing 2,4-D (0.2 mg/l) + kinetin (0.5 mg/l), the callus tissue dissociated into small clumps of cells and single cells within 4 days. The suspension obtained was white in colour, and within 2 weeks a vigorously growing, thick soup-like suspension of cells was produced (Fig. 2O). The shape and size of the cells was highly variable, however most of them were round or elongated. Occasionally on higher concentrations of 2,4-D (2 mg/l) with CM (10%) the cells showed anthocyanin pigmentation. When 2,4-D was replaced by IAA (2 mg/l), within a week round islets of meristemoids appeared on the surface of callus tissue, which differentiated into roots (Fig. 2P). If transferred to agar medium of the same composition, these roots continued to grow further.

Plating of Cell Suspensions. Since the dissociation of callus tissue into single cells and the growth of suspension was best on basal medium supplemented with 2,4-D (0.2 mg/l) + kinetin (0.5 mg/l), the suspensions from this medium were used for subsequent plating experiments. Two ml of the actively growing suspension (10-day-old), containing free cells and small clumps of cells, was plated on 5 ml of agar solidified MS + 2,4-D (0.1 mg/l) + kinetin (0.1 mg/l) + CM (10%). Within 4 days the cells started to divide, and later clumps of cells and cell aggregates were formed. After 6 weeks callus tissue spread all over the surface of the medium (Fig. 2N).

Differentiation of Callus, Somatic Embryogenesis, and Plant Regeneration. With a view to induce differentiation of callus, the effect of various concentrations and combinations of IAA and kinetin was studied. Organogenesis (Fig. 2K,L) was observed on a large variety of media. Addition of CM (10%)

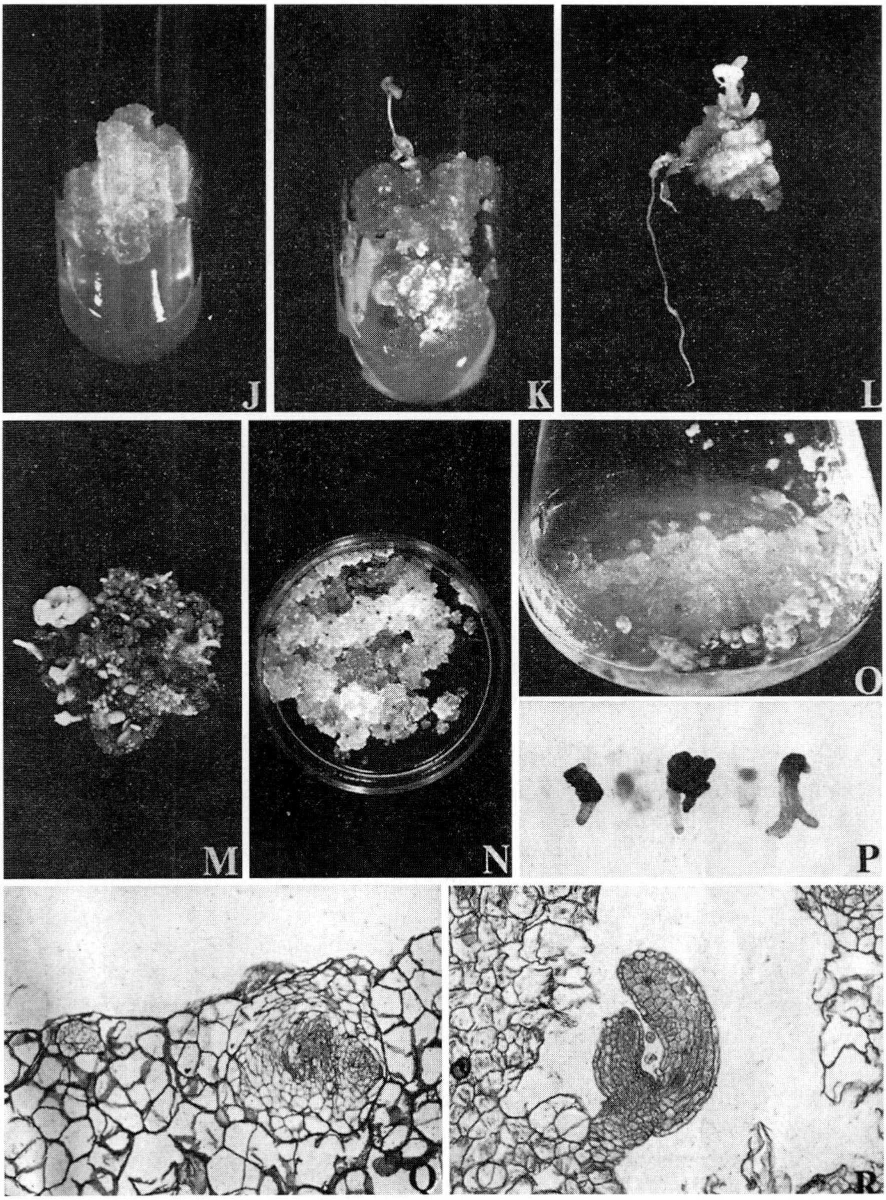

Fig. 2J–R. Organogenesis in callus tissue cultures of *Anagallis arvensis*. **J** Eight-week-old mass of callus tissue obtained from excised hypocotyl segment, and subcultured on MS + 2,4-D (0.1 mg/l) + kinetin (0.1 mg/l) + CM (10%). **K–L** Callus tissue showing differentiation into roots and shoots on MS + IAA (2 mg/l) + kinetin (0.2 mg/l). **M** Twelve-week-old callus culture showing an all-round differentiation of buds and shoots on MS + IAA (0.5 mg/l) + kinetin (1 mg/l) + CM (10%). **N** Suspension of callus cells 6 weeks after plating on agar solidified MS + 2,4-D (0.1 mg/l) + kinetin (0.1 mg/l) + CM (10%). **O** Four-week-old suspension of callus cells grown in liquid MS medium containing 2,4-D (0.2 mg/l) + kinetin (0.5 mg/l). **P** Two-week-old roots obtained from callus suspension cultures on WM + IAA (2 mg/l) + kinetin (0.5 mg/l) + CM (10%). **Q–R** Anatomy of differentiating callus tissue (from **M**); note the formation of "islets" of meristematic cells in **Q**, and the differentiation of a somatic embryo in **R**

to MS + IAA (1 mg/l) + kinetin (0.5 mg/l) greatly increased the percentage of cultures showing differentiation as well as the overall growth (Fig. 2M).

The callus tissue is composed of parenchymatous cells with scanty cytoplasm. In some places, the cells are densely cytoplasmic, small sized and form a nest of tissue or islet of meristem (Fig. 2Q). Callus exhibited two modes of morphogenesis, i.e. formation of shoot buds (Fig. 1G–I), and somatic embryogenesis (Fig. 2R). Somatic embryos at various stages of development were observed to form mostly from the superficial layer of the callus. The shoot buds elongated and eventually formed platlets.

3 Formation of Saponins

Numerous studies have been conducted on triterpenoids and their saponins in a number of plant species. These products now enjoy growing interest since their chemical extraction and purification have become easier. Their amphiphilic structures confer on them some well-known properties such as detergents, emulsives, and hemolytic and toxic substances (Henry 1989).

Various studies on saponins in *Anagallis arvensis* are summarized in Table 1, and anagalline, anagalligenine, anagalligenone, anagalloside, anagallasins etc. have been reported in various parts of the plants. Heitz et al. (1971) were the first to study secondary metabolites in *Anagallis arvensis* and reported the presence of anagalligene B, a new triterpene which is epoxy-13,28 ceto-16 oleananediol-3β, 23, 2a. They also described details of its isolation and structures. This compound manifests fungicidal properties. Amoros and Girre (1987) isolated two novel saponins, and on the basis of the spectral analysis, the structure proposed for one of the saponins was the 3-*O*-glucose (1–3 or 4) [arabinose 1 → 4 or 3]-glucose (1 → 2)-xyloside of 23-hydroxy protoprimulagenin A. The other saponin contained an additional glucose. These antiviral compounds were highly hemolytic, gave a copious lather when they were shaken with water and produced positive colour tests for triterpenoids.

Other studies included the isolation of five new triterpene saponins, i.e. anagalloside A,B,C, and desglucoanagalloside A and B. The structures of these compounds were elucidated by ^{1}H-, ^{13}C-NMR spectroscopy, FAB-MS, and the methylating and combined methylating-ethylating analysis methods (Glombitza and Kurth 1987). Later, four more new triterpenoid oligoglycosides were isolated (Mahato et al. 1991). Recently, Shoji et al. (1994a,b) have done extensive studies and elucidated structures of various anagallosaponins and their companion substances. Dried whole plants extracted with alcohol were analyzed, and the molecular formulae of various anagallosaponins found as colourless needles were given, i.e. anagallosaponin I ($C_{58}H_{96}O_{28}$), anagallosaponin II ($C_{54}H_{88}O_{25}$), anagallosaponin III ($C_{48}H_{78}O_{18}$), anagallosaponin IV ($C_{54}H_{88}O_{23}$), anagallosaponin V ($C_{54}H_{88}O_{23}$), anagallosaponin VI ($C_{46}H_{76}O_{17}$), anagallosaponin VII ($C_{52}H_{86}O_{2}$), and methylanagallosaponin I ($C_{59}H_{96}O_{28}$).

4 Summary

Anagallis arvensis is known for a variety of medicinal uses, such as a remedy for snake-bite, hydrophobia, epilepsy, leprosy, as a diuretic, and has antiviral and fungicidal activity. Various in vitro culture studies have been conducted, and the production of saponins reported in the plant.

In vitro cultures exhibited three modes of morphogenesis; (1) direct development of buds on excised explants, (2) indirect differentiation of shoot buds from callus, and (3) formation of somatic embryos in callus cultures.

Vigorously growing cell cultures capable of being periodically subcultured were established in liquid as well as on the agar-solidified MS medium supplemented with 2,4-D, kinetin and coconut milk. The callus induced from hypocotyl segments is white, soft, friable and fast growing, and has been subcultured over a period of 2 years without showing decline in growth. The optimum conditions for growth are pH 5.9, temperature 27°C, and 4% sucrose in the medium. The callus differentiated into roots, shoot buds, and somatic embryos.

The excised leaves, and segments of hypocotyl and stem regenerated shoot buds. Of the cytokinins used 6-(γ,γ-dimethylallylamino)-purine proved to be the best for the number of buds per culture. Anatomical studies revealed that buds arise from the epidermal and subepidermal layers of leaves and hypocotyl; these buds formed shoots which eventually developed into plants.

A number of saponins have been isolated from different parts of the plant by various workers. Some common ones are anagalline, anagalligenine, anagalligenone, anagalloside, anagallisins etc.

References

Amoros M, Girre RL (1987) Structure of two antiviral saponins from *Anagallis arvensis*. Phytochemistry 26:787–791

Bajaj YPS (1972) Effect of some growth regulators on bud formation by excised leaves of *Torenia fournieri*. Z Pflanzenphysiol 66:284–287

Bajaj YPS, Dionne LA (1967) Growth and development of potato callus in suspension cultures. Can J Bot 45:1927–1931

Bajaj YPS, Mäder M (1974) Growth and morphogenesis in tissue cultures of *Anagallis arvensis*. Physiol Plant 32:43–48

Chevallier A (1996) The encyclopedia of medicinal plants. Dorling Kindersley, London, pp 164–165

Glombitza KW, Kurth H (1987) Die Strukture einiger Triterpensaponine aus *Anagallis arvensis*. Planta Med 53:548–555

Heitz S, Billet D, Raulais D (1971) Triterpenes d'*Anagallis arvensis* L. (Primulacee). I. Structure de l'anagalligenone B. Bull Soc Chim Fr 6:2320–2324

Henry M (1989) *Saponaria officinalis* L.: in vitro culture and the production of triterpenoidal saponins. In: Bajaj YPS (ed) Biotechnology in agriculture and forestry, vol 7. Medicinal and aromatic plants II. Springer, Berlin Heidelberg New York, pp 431–442

Mahato SB, Sahu NP, Roy SK, Sen S (1991) Structure elucidation of four new triterpenoid oligoglycosides from *Anagallis arvensis*. Tetrahedron 47:5215–5230

Mathur RC (1965) Systematic botany. Agra Book Store, Agra 338 pp
Murashige T, Skoog F (1962) A revised medium for rapid growth and bio-assays with tobacco tissue cultures. Physiol Plant 15:473–497
Shoji N, Umeyama A, Yoshikawa K, Arihara S (1994a) Structures of anagallosaponins I–V and their companion substances from *Anagallis arvensis* L. Chem Pharm Bull (Tokyo) 42:1750–1755
Shoji N, Umeyama A, Yoshikawa K, Arihara S (1994b) Triterpenoid glycosides from *Anagallis arvensis*. Phytochemistry 37:1397–1402
Stuart M (1988) In: Nanba T (ed) The encyclopedia of herbs and herbalism. Enterprise, Tokyo, p 39
Su Ching New Medical College (1977) Dictionary of Chinese materia medica. Shanghai Science Technology Publishers, Shanghai
White PR (1943) A handbook of plant tissue culture. Cattel Press, Tempe, Arizona

II *Azadirachta indica* A. Juss. (Neem Tree): In Vitro Culture, Micropropagation, and the Production of Azadirachtin and Other Secondary Metabolites

E.J. Allan[1], T. Stuchbury[1], and A.J. Mordue(Luntz)[2]

1 General Account

Neem (*Azadirachta indica* A. Juss.) is a multi-purpose tree whose products have been used traditionally for centuries for insecticidal, antiseptic, contraceptive, antipyretic and antiparasitic purposes. In addition, it is used for reforestation and as a source of wood and provider of shade. The fruit produces oil which is used in soaps and detergents while other by-products are used for fertiliser and soil amendments. Over the last two decades, there has been a burgeoning of interest in neem, mainly due to the use of its extracts for insect control (Schmutterer et al. 1981; Schmutterer and Ascher 1984, 1987; National Research Council 1992). The recent book by Schmutterer (1995) provides a wealth of information on neem. Plant cell and tissue culture provides scope for improvement of the tree, an opportunity for understanding the biosynthesis of important neem metabolites and the possibility of standardised, year round production of biologically active materials.

1.1 Botany, Distribution, and Growth of Neem

Neem is a fast growing, evergreen tree (Fig. 1) belonging to the mahogany family (Meliaceae). The exact region of origin is not known although it is considered to be Mayanmar (Burma) and/or southern India (Schmutterer 1995). It is now widely distributed in the more arid, tropical and subtropical areas of the world, e.g. Asia (India, Bangladesh, Pakistan, Sri Lanka, Indonesia and Malaysia); in parts of the Middle East (including Iran, Qatar, Saudi Arabia and Yemen), East and West Africa (including the southern fringes of the Sahara), the Caribbean, Central and South America, Australia and the South Pacific islands (Benge 1989; Schmutterer 1995). Neem is now being cultivated as an economic forest crop in Central and South America (Jacobsen 1987). Associated with its wide distribution and traditional use is an array of common names such as Kohomba and Vampu (Sri Lanka), Dogon

[1] Department of Agriculture, University of Aberdeen, 581 King Street, Aberdeen AB24 5UA, UK

[2] Department of Zoology, University of Aberdeen, Tillydrone Avenue, Aberdeen AB24 3TZ, UK

Biotechnology in Agriculture and Forestry, Vol. 43
Medicinal and Aromatic Plants XI (ed. by Y.P.S. Bajaj)

Fig. 1. The neem tree, *Azadirachta indica* A. Juss

Yaro (Nigeria) and Nimbagaha (Singapore), noted by Koul et al. (1990) and Schmutterer (1995).

Neem grows well in many of the drier tropical and sub-tropical areas of the world where the rainfall is between 400–1200 mm per annum and, although it may grow where the rainfall is considerably greater, it tends to produce few flowers or fruits under these conditions. It grows best between 21–32 °C, but will tolerate temperatures up to 50 °C. It is well known for its resistance to drought and, although most leaves may be lost in extreme conditions, its deep tap-root ensures survival. It is also sensitive to frost (Radwanski 1977). The tree thrives up to about 800 m above sea level and is generally not associated with mountainous regions. The tree is tall and spreading with a short trunk (Fig. 1) and commonly reaches 15 m, and occasionally 25 m in height. The leaves are unpaired and pinnate and up to 40 cm in length with up to approximately 30 dark green lanceolate leaflets with toothed margins (Fig. 2). The flowers are small and white and carried in drooping panicles. In India and other Northern Hemisphere countries, fruits normally ripen over the period May to August and individual trees may have both flowers and ripening fruit at the same time. In Africa neem produces fruit twice a year (usually in March/April and July/August). Fruits are ellipsoidal drupes which turn from green to yellow-green to yellow-orange during ripening (Fig. 3). They consist of an exocarp (skin) and a fibrous mesocarp (pulp) enclosing a seed (Fig. 4) consisting of an endocarp (shell) containing normally one (but occasionally two or three) kernels which are surrounded by a testa (seed coat). Fruit production

Fig. 2. Neem leaf

Fig. 3. Immature neem fruits

can begin in 4–5 year old trees (Schmutterer 1995) and is well established in 10-year-old trees. Apparently trees may continue to bear seed for 200 years.

The tree can be propagated naturally via the seed, and seedlings may be readily grown in the nursery (Utz 1993). When transplanted in wet conditions some weed control may be necessary and in dry conditions removal of leaves to reduce water loss can be advantageous. Animals e.g. fruit bats, monkeys and birds help dispersal by eating the fruit and passage through the digestive system improves germination (Schmutterer 1995). Seeds lose viability within a few weeks of ripening under normal conditions (Ezumah 1986; National Research Council 1992) which may cause difficulty with respect to long-term storage for commercial propagation. This is further complicated by a debate

Fig. 4. Mature neem seeds

concerning whether the seed is recalcitrant, i.e. loses the ability to germinate on dehydration (Surendran et al. 1993); orthodox, i.e. tolerates drying (Gamene et al. 1996); or intermediate (Ezumah 1986). However, it appears that differences in habitat may determine the response of seeds to drying and this may be of particular relevance to neem. Some workers have found that neem seeds can be stored at low temperatures and remain viable for several months, especially after careful drying before storage (Radwanski and Wickens 1981; Roederer and Bellefontaine 1989; Eeswara et al. 1998).

Neem seeds deteriorate quickly if stored under moist conditions and may become infected with fungi which reduces seed quality. In particular, the spoilage organism *Aspergillus flavus* (Sinniah et al. 1983) produces aflatoxins which are acutely toxic, carcinogenic, mutagenic and teratogenic in animals (Medlock 1996). These toxins may contaminate neem products manufactured from seeds stored under adverse conditions and, although their carry over is much reduced in refined oils, they remain a concern for both human and animal health (Sinniah et al. 1983).

1.2 Uses of Neem

There is much literature concerning the properties and uses of the neem tree and its products (Pliske 1984; Koul et al. 1990; National Research Council 1992; Schmutterer 1995; Singh et al. 1996). For example, it is used as a woodsource, an animal and poultry feed and as a manure and fertiliser as well as in a range of traditional medicines and pesticides (Parmar and Ketkar 1993). Its wide array of phytochemicals (see Sect. 1.3) are responsible for its biological activity against a plethora of insects, micro-organisms, nematodes etc. although the tree itself is susceptible to a range of pests and pathogens (Boa 1995).

Neem trees are commonly grown along roads and in parks and gardens to provide shade and shelter throughout the year. Neem has also become extremely important in reforestation schemes. It has been planted to reduce soil erosion and provide shelter belts along the fringes of the Sahara and the severely deforested Majjca Valley (southern Sahel region of central Niger) where its drought tolerance and ability to tap water from considerable depths make it of particular value (National Research Council 1992). The economic, cultural and social dependence on neem is immense in areas such as Sub-Saharan Africa where it has become the most widely planted tree species (National Research Council 1992). The rapid growth rate of the tree and the high calorific value of its timber have also resulted in neem being planted as a source of fuel (Pliske 1984). The wood is moderately heavy, resembling mahogany, and is resistant to insect and fungal attack, making it suitable for building and furniture construction. In Sri Lanka, its durable nature makes it preferable for construction work and in particular for decorative wood carving (Ganesalingam 1987).

Neem leaves, seeds and oil have been used for centuries to protect people and stored agricultural products against insect attack. Schmutterer first noted, in 1959, that neem trees in Sudan were left untouched by locust swarms and since then, the trees have been shown to contain a variety of compounds which deter the feeding of many insects and disrupt their growth and development (Schmutterer 1995). Indeed, the effects of neem products on insects are well known (Schmutterer 1990; Mordue(Luntz) and Blackwell 1993), and are responsible for important commercial uses of neem.

Neem leaves, fruits, seeds, oil, root and bark are also used as components of traditional medicines (Siddiqui and Mitra 1945). Neem leaf infusions are used to treat digestive disorders, malaria (Udeinya 1993), and fevers while seed powders may be used to treat head lice (National Research Council 1992). In addition to insect control, neem oil is used to make soap and detergent, and as a component of toiletries and pharmaceuticals. Neem products have general antiseptic properties and neem oil has been used to treat a variety of skin infections, leprosy and parasitic conditions. Twigs and bark extracts are used to clean teeth and are apparently beneficial to the gums. Some neem products show promise in birth control and a commercial neem oil-based contraceptive is produced in India. The residual matter after extraction of oil, neem cake, is of value particularly to resource-limited farmers where it is used as a fertiliser (Parmar and Ketkar 1993), with apparent advantages of reducing nitrification and providing nematode control (Stoll 1987; Saxena 1990). Neem extracts are antimicrobial and recent research has shown that they are active against plant pathogens (Coventry and Allan 1997).

1.3 Chemical Constituents

Most of the active principles from neem fruit, as well as from seeds, twigs, stem and root bark are limonoid tetranortriterpenoids, although biologically active diterpenoids, triterpenoids, pentanortriterpenoids and a few nonterpenoidal

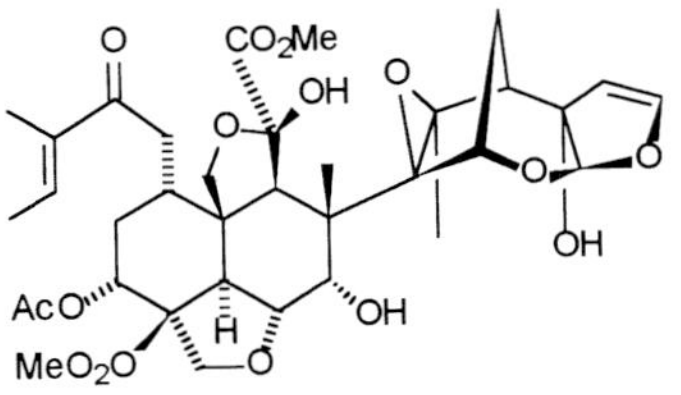

Azadirachtin

Fig. 5. Structure of azadirachtin. (Redrawn from Ley et al. 1993)

constituents are also present. To date more than 300 compounds have been isolated from various parts of the tree (Kumar et al. 1996) and most attention has been directed towards the major principle, azadirachtin (Fig. 5), which occurs mainly in the seed kernels. Azadirachtin was first isolated by Butterworth and Morgan (1968) and its structure took some years to determine due to the large number of oxygen atoms in the molecule which made molecular analysis difficult. Nakanishi and co-workers presented the first structural proposal (Zanno et al. 1975), the correct structure appeared in Kraus et al. (1985) and Broughton et al. (1986), and the full details were finally described by all groups in 1987 (Bilton et al. 1987; Kraus et al. 1987; Turner et al. 1987). The chemistry (Ley et al. 1993) and biological activity of azadirachtin (Mordue(Luntz) and Blackwell 1993) have been studied extensively. Azadirachtin has been shown to have extremely high antifeedancy to a wide range of insects with the Lepidoptera being the most sensitive (Blaney et al. 1990). The desert locust, *Schistocerca gregaria* (Orthoptera) also shows particular sensitivity to azadirachtin (Mordue(Luntz) and Nasiruddin 1992; Nasiruddin and Mordue(Luntz) 1994). Azadirachtin also causes strong growth inhibition, malformation and mortality when fed to insects and interrupts their reproduction (Mordue(Luntz) and Blackwell 1993). Marrangin (azadirachtin L) from seed kernels of *A. excelsa* shows higher biological activity than azadirachtin in the *Epilachna varivestis* bioassay (Ermel et al. 1991).

The precursors of the limonoids, which occur in the Meliaceae and other Rutales families (Buchanan and Halsall 1970; Dreyer 1984), are triterpenoids derived from the euphol or tirucallol skeleton (Fig. 6a). Meliantriol is one such example which is a strong antifeedant to the desert locust (Lavie and Jain 1967).

The tetranortriterpenoids are formed from 7 Δ euphol/tirucallol based precursors by oxidative degradation of the C-17 side chain, resulting in the loss of four carbon atoms. The resulting compounds can be divided into nine different basic structural groups i.e. the azadirone, amoorastatin (e.g. azadirachtanin), vepinin and vilasinin types (Fig. 6b–e); the D-seco gedunin types (Fig. 6f); the C-seco types nimbin, nimbolinin and salannin (Fig. 6g–i); and the azadirachtin group (Fig. 5).

The azadirone family consists of meliacins in which all four rings of the basic skeleton remain intact. Several have been isolated from neem oil, neem leaves and fruit, and most show low antifeedant or biological activity (e.g.

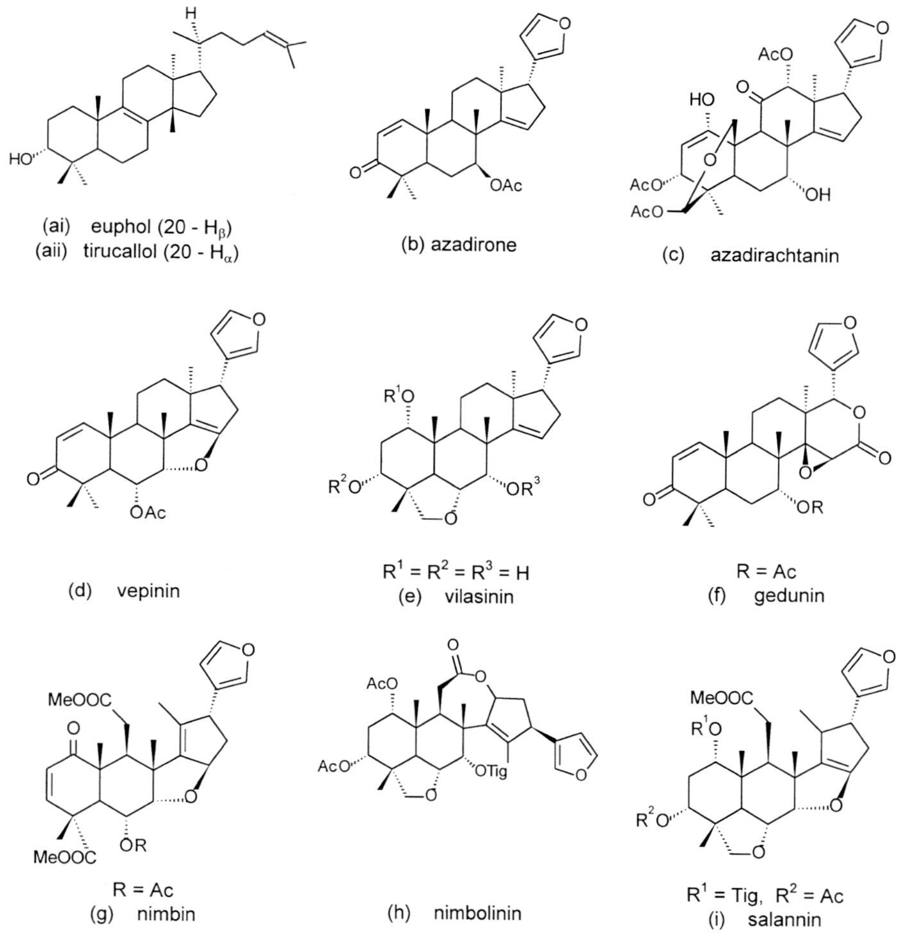

Fig. 6. Precursors and limonoid classes from neem: **a** euphol/tirucallol; **b** azadirone; **c** azadirachtanin; **d** vepinin; **e** vilasinin; **f** gedunin; **g** nimbin; **h** nimbolinin; **i** salannin. (Redrawn from Kraus 1995)

Lavie and Levy 1971; Schwinger et al. 1984). The more highly oxidised azadiradione is an active insect antifeedant and is also effective against a number of bacterial species (Schwinger et al. 1984; Siddiqui et al. 1990). Nimocinolide and isonimocinolide from green winter leaves possess larvicidal properties when applied to mosquito larvae, *Aedes aegypti* (Naqvi 1987).

The biological activity of the amoorastatin compounds and vepinin from neem seed oil, which are common in many Meliaceae species but which occur only rarely in fresh neem leaves, has not been assessed. The vilasinin group is widespread in the Meliaceae and is assumed to contain intermediates in the biosynthesis of the C-seco meliacins such as the nimbins, nimbolides, salannins and probably the azadirachtins. They occur in the leaves and seeds and several have good insect antifeedant activity. The gedunin group, formed from azadirone and its derivatives, have been isolated from neem seed oil, bark and seeds (Lavie et al. 1971; Kraus et al. 1981). Many show biological activity for example as antifeedants against termites or as antimalarial agents (Khalid et al. 1989; Ishida et al. 1992).

The C-seco meliacin group is the most important group of tetranortriterpenoids, containing a large number of biologically active compounds. Nimbin from neem seeds and leaves was one of the first constituents of neem to be isolated (Siddiqui 1942; Harris et al. 1968; Bokel et al. 1990). Nimbin has moderate insect antifeedant activity (Schwinger 1985; Blaney et al. 1990; Kraus et al. 1993a) whereas nimbolide has some antibacterial action (Rojanapo et al. 1985) and has been implicated as the principle cytotoxic component of neem seed insecticide preparations (Cohen et al. 1996).

Salannin and its derivatives from neem seeds and leaves show antifeedant effects against many insects e.g. *Earias insulana*, *Spodoptera littoralis*, *S.littura* (Meisner et al. 1981; Blaney et al. 1990; Govindachari et al. 1996; Aerts and Mordue(Luntz) 1997) although not against *S. frugiperda*, *Heliothis virescens* and *H. armigera* (Blaney et al. 1990). The generally higher antifeedant activity of the salannins compared with azadirone, gedunin or nimbin groups may well be related to the presence of the saturated A ring in the former (Kraus 1995).

Since 1987, three major groups of constituents related to azadirachtin have been isolated from neem and other Meliaceae species; the azadirachtols, azadirachtins and meliacarpins (equivalent to the azadirachtins A-L of Rembold et al. 1987a,b). The azadirachtols and meliacarpins show reasonable to good biological activity (Rembold 1989a,b; Mordue(Luntz) and Nasiruddin 1992; Ley et al. 1993).

Of the other non-terpenoidal constituents of neem, the water soluble arabinogalactan from the fruit pulp (Sen et al. 1993) and the polysaccharides (arabinofucoglucanes and fucogalactoglucoarabinanes) from neem bark possess anti-inflammatory and antitumour activity (Fujiwara et al. 1982, 1984; Shimizu et al. 1985). Hydrocarbons from neem leaves possess larvicidal activity against mosquitoes (Chavan 1984; Siddiqui et al. 1988). Isofraxidin and other coumarins from winter twigs of neem are insecticidal against *E. varivestis* (Kraus et al. 1993a,b). Sulphur containing compounds with antifungal and insect larvicidal effects have been found in fresh neem leaf and seed volatiles (Pant et al. 1986; Balandrin et al. 1988). Finally, the fatty acids from neem oil (e.g. palmitic, stearic, oleic, linoleic, linolenic, arachidic acids and their esters) may be important in control of aphids and mites.

1.4 Extraction of Products

Neem seeds are usually harvested by collection from the ground. The seeds must be depulped immediately and dried rapidly to prevent fungal growth and aflatoxin production. In most semi-arid tropical regions this is achieved by drying in the sun for up to 12 h and then in the shade for several days to reduce the moisture content to approximately 7%. Rapid drying is more difficult to achieve in more humid regions. The hard shell or endocarp makes up approximately 50% of the weight of the seed and may be removed by hand or by use of a mill or thresher. The method of preparation and uses of neem aqueous extract, oil, kernel powder and pressed cake have been described (Stoll 1987). Neem oil may be expressed from neem seed kernels by grinding and pressing or by using power-operated oil presses. The seed kernels yield about 45% neem oil and 55% neem cake.

Powdered neem seed, kernels or neem cake can be used as insecticides although they may need to be diluted with inert material before use. Aqueous extracts of these same materials may also be used since, although azadirachtin itself is only slightly soluble in water, other components present in the seeds increase its solubility (Hellpap and Dreyer 1995). Aqueous extracts of neem seeds may be prepared and used locally but the active ingredients tend to decompose rapidly.

Azadirachtin and other active ingredients can be extracted efficiently from neem kernels using polar solvents in which they are very soluble. Such extractions are usually preceded by extraction with non-polar solvents such as hexane or petroleum ether to remove non-polar neem oil constituents, although some products retain a considerable neem oil content. Procedures of this type can yield extracts with a high azadirachtin content which is relatively stable.

1.5 Commercial Products

The highest concentrations of active ingredients are found in the seed kernel and most commercial preparations are therefore derived from seeds. Neem-based products have recently been registered for use on both non-food and food crops in the USA (Ermel and Kleeberg 1995) and a large number of neem products are now available for insect control (Table 1).

It is generally considered that neem-based insecticides are environmentally safe, have minimal impact on non-target organisms and are non-toxic to mammals. However, it is now recognised that minor constituents, such as nimbolide, which are found at different concentrations in different neem

Table 1. Commercial neem products

Commercial product	Producer
Margosan O, Neemix, Neemazad, Azatin, Turplex and Align	Thermo Trilogy Corp, USA
NeemAzal	Trifolio M GmbH, Germany
Fortune Aza	Fortune Biotech, USA
Neemicide	Rym Exports, India
Bioneem, Neemesis & Safer's ENI	Ringer Corp, USA
Wellgro and RD-Repelin	India Tobacco Co, India
Neemark	West Coast Herbochem, India
Neemguard	Gharda Chemicals, India

products may have cytotoxic effects (Cohen et al. 1996). The development of neem-based products, and their associated patenting, by Western companies has caused some environmental and political concern (Pearce 1993; Kleiner 1995). The use of neem products however is dependent on the availability of seeds of good quality and several companies have now initiated seed collection schemes. The use of neem as a source of non-edible oil has been important in the development of its use for insect control. In the late 1970s the Indian Government introduced incentives for the solvent-extraction industry to use non-edible oilseeds for soap and shampoo manufacture rather than edible oilseeds which were in short supply (Immaraju 1998). Consequently, there was an increase in neem oil production and there are 141 neem seed collection and 70 oil production centres in India. Fortune Biotech Ltd. has commissioned the world's largest azadirachtin extraction facility, based in Hyderabad, India, which is capable of processing over 10000MT neem seeds per annum (Immaraju 1998). In addition to azadirachtin, all the by-products have valuable commercial uses which should contribute to the overall success of the process e.g. husks are used for soil amendment, neem cake as a fertilizer and a bionematocide, neem oil for soap and detergent manufacture as well as for insecticidal and medicinal uses (Parmar and Ketkar 1993; Immaraju 1998).

2 In Vitro Approaches

Numerous investigations have been conducted on various aspects of in vitro culture of neem, and the literature is summarized in Table 2. Two important applications of tissue culture with respect to neem were suggested by Schulz (1984): (1) to achieve organogenesis and plant regeneration because of the relatively rapid loss of the germinability of the seed, and (2) to stimulate the biotechnological production of the plant constituents which have biological effects on insects. In addition, regeneration through micropropagation systems, to multiply plants to produce clones, would allow rapid production of a uniform population of high azadirachtin-producing neem trees. This would be beneficial since the azadirachtin content of neem seeds varies from tree to tree and appears to be both ecotype and environment dependent (Ermel et al. 1987; Singh 1987; Eeswara 1996).

2.1 Explant Type and Sterilization

Aseptic cultures of neem have been established from most tissue types including anthers, bark, cotyledons, leaves and stems with the surface sterilization procedure being dependent on the parent plant. Western workers tend to use either aseptic seedlings (van der Esch et al. 1993) or laboratory grown plants (Kearney et al. 1994; Wewetzer 1997) since seeds can be easily transported to their laboratories (although the supply of germinable seeds can be problematic). Laboratory grown materials such as cotyledons and leaves are usually

Table 2. Chronological summary of various in vitro/micropropagation studies conducted on neem

Authors	Objective	Explant used	Observation/remarks
Rangaswamy and Promila (1972)	Callus production	Embryos and seedlings	Callus was produced from excised embryos and decapitated seedlings
Sanyal et al. (1983)	Effect of differentiation on nimbin and β-sitosterol content of neem callus and regenerated plantlets	Cotyledon callus and regenerated plantlets	Differentiated plantlets contained more nimbin but less β-sitosterol than callus
Schulz (1984)	Callus production and differentiation to produce shoots	Leaf callus	Conditions for the production and growth of callus. Shoots but no roots were produced from callus
Narayan and Jaiswal (1985)	Shoot and root formation from leaf callus	Leaf callus	Callus gave rise to shoots on BAP which produced roots when treated with NAA
Sarkar and Datta (1986)	β-sitosterol production by neem	Cotyledon callus	Effect of cholesterol on β-sitosterol production was studied
Sarkar et al. (1988)	Effect of glycine on nimbin and β-sitosterol production	Cotyledon callus	Nimbin biosynthesis was increased and β-sitosterol synthesis decreased by glycine
Subba Rao et al. (1988)	Callus production	Cotyledons, leaves, stems	Cotyledons were best source of callus
Naina et al. (1989)	Transformation of neem by *Agrobacterium tumefaciens*	Tumours, shoots and shoot callus	Transformed callus, shoots and plantlets were autotrophic
Ramesh and Padhya (1990)	Shoot formation from leaf discs	Leaves	Shoots could be produced directly or via callus. Rooting occurred
Gautam et al. (1993)	Root and shoot formation on callus from anthers	Anthers	Rooted plantlets obtained from callus
Shrikhande et al. (1993)	Optimisation of medium for callus and somatic embryo production	Cotyledon callus	Indirect somatic embryogenesis led to healthy plants
Subramani et al. (1993)	Shoot production from nodes, directly from leaf discs and from leaf callus	Nodes and leaves	Plants raised by axillary bud proliferation could be transferred to field
Allan et al. (1994)	Identification of azadirachtin	Leaf callus	Identification by a range of chromatographic methods
van der Esch et al. (1994)	Azadirachtin production in suspension cultures	Callus	Azadirachtin content and growth kinetics of suspension cultures are given
Kearney et al. (1994)	Determination of antifeedancy of neem cultures against desert locust	Callus and cell suspension cultures	Feeding was suppressed by extracts of seed, leaf, callus, suspension and shoot cultures
Wewetzer and Schultz (1994)	Regeneration of plantlets from neem cultures	Shoot-tip, leaves, bark, stem	Regeneration via callus was achieved from all explant types
Mordue(Luntz) et al. (1995)	Characterisation of production of biologically active compounds in neem cultures during growth	Callus and suspension cultures	Antifeedant compounds were present in callus and in cells and medium of suspension cultures

Table 2. *Continued*

Authors	Objective	Explant used	Observation/remarks
Akula and Drew (1996)	Somatic embryogenesis from various tissues	Roots and nodes and leaf callus	Embryoids developed into complete plantlets
Mordue(Luntz) et al. (1996)	Azadirachtin content of neem seeds and tissue cultures during growth	Callus and suspension cultures	No correlation between neem content in seeds and callus of same trees
Joarder et al. (1996)	Development of micropropagation methods	Nodes	Cultures were established on MS followed by MS + cytokinin
Joshi and Thengane (1996)	Regeneration from vegetative buds and somatic embryogenesis	Buds and cotyledons	Hormone concentrations for optimal regeneration and embryogenesis investigated
Drew (1996)	Multiplication and rooting of nodal segments	Nodes	Nodes were grown on MS without hormones and rooted by exposure to IBA for 4 days
Eeswara (1996)	Initiation and maintenance of hairy root cultures	Shoots and stems	Hairy roots produced azadirachtin, nimbin and salannin
Eeswara et al. (1997a)	Determination of nimbin, salannin and azadirachtin contents of callus and seeds	Leaf callus	No correlation between azadirachtin, nimbin and salannin yields of callus and seeds
Eeswara et al. (1997b)	Improvement of methods for micropropagation	Leaves	Standard methods for shoot intitiation, multiplication and rooting were developed
Jarvis et al. (1997)	Identification of azadirachtin	Suspension cultures from apical tips	Other tetranorterpenoids were also identified
Wewetzer (1997)	Azadirachtin production by callus	Leaf and bark callus	Differentiation was not needed for azadirachtin production

surface sterilized using a simple procedure with hypochlorite [approximately 10% (v/v) for 15–20min], often using proprietary disinfectants. On the other hand, researchers with access to naturally grown material, use a much more rigorous procedure where materials are washed in water and detergent (often as separate steps) followed by mercuric chloride treatment [0.05–0.15% (w/v)]. A study in Sri Lanka (Eeswara 1996) showed that leaves taken from laboratory grown plants, compared with leaves from wild-grown trees were, perhaps not surprisingly, easier to surface sterilize, requiring only a hypochlorite treatment. In this study, it was also found that leaves from naturally growing trees could not be surface sterilized using the methods described in the literature (Narayan and Jaiswal 1985; Ramesh and Padhya 1990; Subramani et al. 1993). A method which resulted in effective, although not 100% disinfection, with the explants going on to produce callus was developed (Eeswara 1996). This involved leaves being washed thoroughly in tap water, followed by mild liquid soap (Teepol C.I.C., Sri Lanka) and rinsed in sterile distilled water before being treated with 0.1% (w/v) mercuric chloride for 5min followed by 10% (v/v) of a proprietary sodium hypochlorite treatment (Bleech, Care Products, Sri Lanka) for 10min. Subsequently, it was also shown that Bleech could be replaced with 10% (v/v) sodium hypochlorite (Eeswara 1996). It was also interesting to note that in these studies treatment of the explants with an ethanolic wash prior to surface sterilization resulted in tissue necrosis and reduced callus initiation.

2.2 Callus Initiation and Growth

Several workers have obtained *A. indica* callus. Rangaswamy and Promila (1972) showed that neem exhibited "morphogenetic plasticity" in that decapitated seedlings, and sometimes intact seedlings, could form callus even on distilled water, which would subsequently differentiate to form leafy shoots. Such spontaneous production of callus is not always found and several groups have investigated the effect of medium composition on callus initiation (see Table 3). In examining these papers it appears that there is no consistency of results with repetitions of published work failing to provide the same results in other laboratories. An example of this can be seen for callus initiation in the contrasting results of Schulz (1984) and Subba Rao et al. (1988). The former obtained callus on MS medium (Murashige and Skoog 1962) but not on B5 medium (Gamborg et al. 1968), irrespective of the plant growth regulators used, while Subba Rao et al. (1988) found both media equally effective! Such differences could result from both the morphogenetic plasticity of neem and the different behaviour of explants derived from different ecological regions. There is no doubt that further detailed investigations are required. It is also evident that although there has been much work on neem tissue culture, many reports are not in the public domain. The formation of The International

Table 3. Media used for optimal callus initiation in *Azadirachta indica* A. Juss

Reference	Explant	Basal medium	Plant growth regulators ($mg\,l^{-1}$)
Sarkar and Datta (1983) Sarkar et al. (1988)	Cotyledons	MS	1 NAA + 0.5 Kn
Schulz (1984)	Leaves and petioles	MS	0.1 IAA alone
Narayan and Jaiswal (1985)	Leaves	MS	0.5 2,4-D[a] + 0.05 BA
Subba Rao et al. (1988)	Leaves, cotyledons and stems	B5 or MS	2 or 5 2,4-D + 1 IAA
Ramesh and Padhya (1988)	Leaves and cotyledons	Wood and Braun	0.43 Kn + 0.7 IAA
Ramesh and Padhya (1990)	Leaves	Wood and Braun	0.82 IBA alone 0.9 Kn + 0.82 IBA
Gautam et al. (1987, 1993)	Anthers	Nitsch's	1.75 IAA + 0.2 BA
Subramani et al. (1993)	Leaves	MS	3–5 BA alone or 3–5 Kn
Kearney et al. (1994)	Leaves	MS	4 IBA + 1 BA
Wewetzer (1996)	Leaves and bark	MS	0.2 IAA + 0.1 BA
Islam et al. (1996)	Cotyledons	MS	NAA
Nirmalakumari et al. (1996)	Cotyledons and leaves	MS	2 2,4-D + 0.5 Kn
	Bark		4 2,4-D + 0.5 Kn
	Stem		2 BA + 0.5 IAA

[a] 2,4-D: 2,4-dichlorophenoxyacetic acid.

Neem Network (Thomsen and Souvannavong 1994) should alleviate this problem, for example, in a recent publication, Parveen et al. (1995) have reviewed the work undertaken at the Arid Forest Research Institute in Jodhpur, India.

Table 3 shows that different workers have been successful in obtaining callus using a range of explants, basal media and growth regulators. Studies concerning the effect of explant type on callus initiation have been carried out by Subba Rao et al. (1988) and Nirmalakumari et al. (1996). The former workers showed that cotyledon explants produced callus most readily followed by young and old leaves, and finally stems. We use the same medium, Maintenance Medium (MM), for both callus initiation and prolonged subculture and have achieved success with explants originating from different provenances. This medium consists of MS basal salts with 3% (w/v) sucrose, solidified with 0.175% (w/v) Phytagel (Sigma, UK) and supplemented with the plant growth regulators, indole-3-butyric acid (IBA) 19.7 μM ($4\,mg\,l^{-1}$) and N^6-benzyladenine (BA) 4.44 μM ($1\,mg\,l^{-1}$). Successful callus initiation has been obtained from leaf material, originating from several different countries on this medium e.g. Ghana (Kearney et al. 1994), Niger (Mordue(Luntz) et al. 1995) and different ecological zones of Sri Lanka (Eeswara et al. 1997a) with many lines being maintained for over 4 years. Leaves are surface sterilized, and after thorough rinsing, are cut into sections of approximately $0.5 \times 1.5\,cm$ to remove areas where disinfectant may have impregnated and to provide explants that are easier to manipulate. Explants are incubated, in complete darkness at 25 °C and callus initiation can be observed routinely within 3–4 weeks, particularly at areas where the explant has been wounded (Fig. 7). Callus can normally be removed from the explant within 2 months and often earlier.

Dark/light conditions influence callus initiation (Guglielmini and Cei 1995). Leaf explants of young trees from Niger and Ghana were incubated on MM at 25 °C either in the light (16 h light and 8 h darkness) or in complete darkness, with the former requiring longer for cell proliferation (7 weeks) and a reduction in both the amount and quality of the callus. Thus, in dark conditions the callus was of good quality i.e. friable, pale yellow-white in colour, while in the light the callus grew poorly and produced polyphenols. After induction on MM, neem callus generally shows varying degrees of differentiation as evident by areas of varying hardness and colour with organogenesis rarely observed. However, the quality of callus improves with subculture when selected for friability and colour.

Tissue culture of trees is often hampered by polyphenol production and this is occasionally the case with neem. In such cases, addition of $300\,mg\,l^{-1}$ activated charcoal can improve the growth of neem callus (Guglielmini and Cei 1995). Callus cultivated on MM without activated charcoal showed reduced growth which was associated with browning of the callus and the growth medium (Fig. 8). Callus which was grown on medium supplemented with activated charcoal showed improved growth and morphology with the callus becoming pale white/yellow in colour with no discolouration of the growth medium. Others have used polyvinylpyrrolidone (PVP) for the same purpose (Gautam et al. 1993).

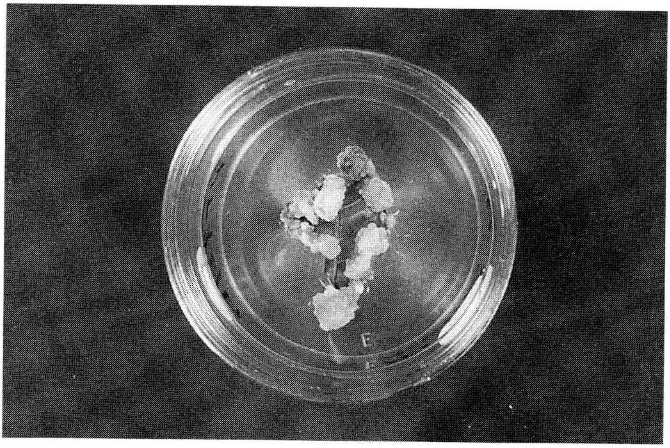

Fig. 7. *Azadirachta indica* (A. Juss.) callus initiated from a leaf explant

The effect of various parameters on *A. indica* callus growth on MM has been studied (Petty 1993). Over the range 21–35 °C, temperature did not have any dramatic effect on callus growth although the maximum biomass slightly decreased at higher temperatures (Fig. 9), e.g. at 25 °C a maximum biomass of 234 g was achieved while it decreased to 195 g at 35 °C. Different sugars had quite marked effects on callus growth at 25 °C; whilst 3% (w/v) glucose, fructose and sucrose all supported good callus growth, lactose did not (Fig. 10). Kinetic analyses of this data showed that the doubling times (t_d) for the first three media were also markedly different with a very fast specific growth rate (μ = 0.099 days, t_d = 6.70 days) being obtained on fructose while growth was slower but more prolonged on sucrose (μ = 0.045 days, t_d = 15.43 days). In addition, higher biomass was obtained throughout growth on the medium containing sucrose. Varying the levels of sucrose also had a marked effect on growth, at concentrations of 1% and 2% (v/v) exponential growth ceased after approx. 20 days while the exponential phase was more prolonged at higher concentrations (Fig. 11).

2.3 Callus Differentiation/Micropropagation

Neem is conventionally propagated through seed. Trees propagated from cuttings normally lack the deep tap root of seed-grown trees and are therefore less resistant to drought and strong winds. It is also possible to propagate neem from suckers, which are produced readily, especially in dry conditions. In order to stimulate timber production, neem can be pollarded and coppiced, and often regrows exceptionally rapidly after such treatment. Micropropagation, however, has a number of advantages.

Narayan and Jaiswal (1985) investigated neem callus differentiation on MS medium containing different concentrations of BA and 1-napthaleneacetic

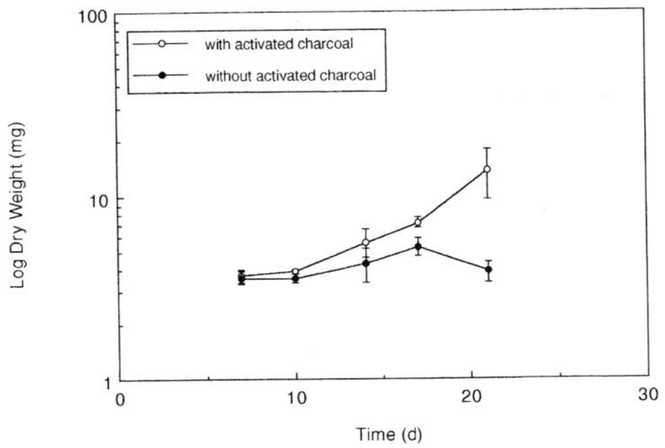

8

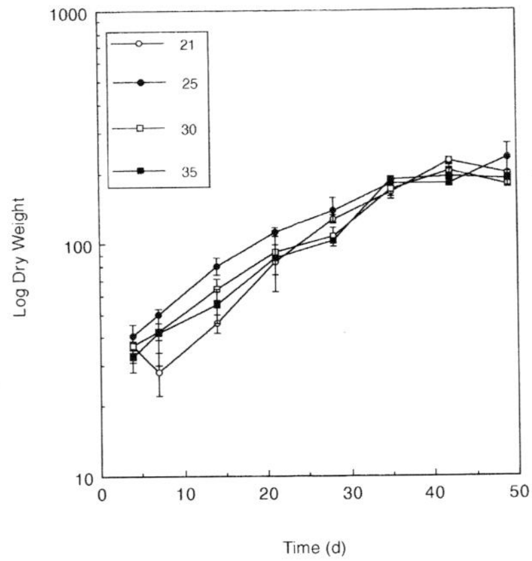

9

Fig. 8. Growth, as measured by dry weight, of *Azadirachta indica* callus on Maintenance Medium (Kearney et al. 1994) with and without supplementation with 300 mg l^{-1} activated charcoal. (Redrawn from Guglielmini and Cei 1995)

Fig. 9. The effect of temperature (21 °C, 25 °C, 30 °C and 35 °C) on the growth of *Azadirachta indica* callus on maintenance medium (Kearney et al. 1994). (Redrawn from data of Petty 1993)

acid (NAA). The presence of the auxin NAA alone at concentrations between 0.05 and 1 mg l^{-1} resulted in root production which was optimal at 0.5 mg l^{-1}. On the other hand, concentrations of the cytokinin, BA (0.05–1 mg l^{-1}), induced differentiation to produce shoot buds which subsequently grew into shoots. Addition of NAA at 0.05 mg l^{-1} enhanced shoot formation. Shoot development was also observed when *A. indica* callus was grown on MS, in the dark, with 0.1 mg l^{-1} BA and 0.1 mg l^{-1} NAA (Wewetzer and Schulz 1994).

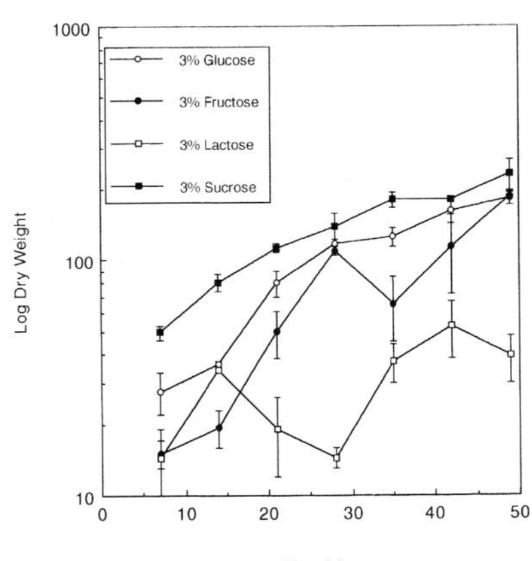

10

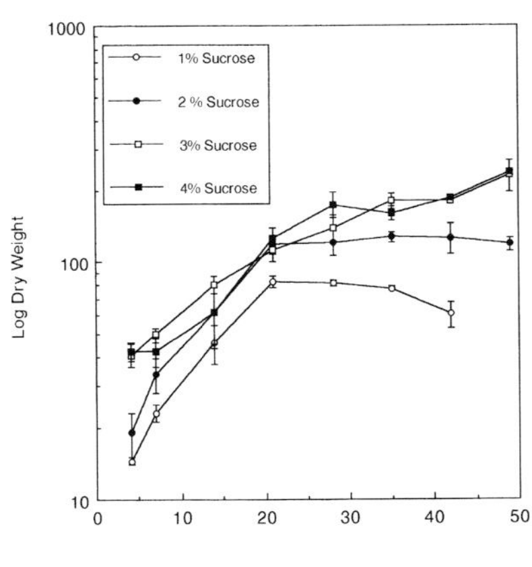

11

Fig. 10. The effect of different sugars on the growth of *Azadirachta indica* callus on maintenance medium (Kearney et al. 1994) at 25 °C. (Redrawn from data of Petty 1993)

Fig. 11. The effect of different concentrations of sucrose on the growth of *Azadirachta indica* callus on maintenance medium (Kearney et al. 1994) at 25 °C. (Redrawn from data of Petty 1993)

Subramani et al. (1993) observed that callus subcultured onto MS medium containing $1\,mg\,l^{-1}$ BA, $0.8\,mg\,l^{-1}$ kinetin (Kn) and $6\,mg\,l^{-1}$ adenine sulphate also initiated shoots. The development of shoots and roots in anther-derived callus of *A. indica* has been studied (Gautam et al. 1993). Again, MS medium with a relatively high concentration of cytokinin favoured shoot development (i.e. $4.4\,\mu M$ BA and $0.53\,\mu M$ NAA). In contrast to these studies, and again perhaps reflecting genotype variation, Subba Rao et al. (1988) found that neem callus failed to differentiate on medium containing Indole-3-acetic acid (IAA), Kn or BA alone or in combination.

2.3.1 *Shoot Multiplication*

Micropropagation from neem leaf explants has been reported by several researchers (Ramesh and Padhya 1990; Kearney 1993; Subramani et al. 1993; Eeswara et al. 1997b; Wewetzer 1997). Ramesh and Padhya (1990) studied the production of adventitious shoot buds from leaf discs on Wood and Braun medium supplemented with Kn, BA and adenine sulphate. They reported that although there was no growth on basal medium, addition of Kn or BA alone resulted in callus production while addition of Kn and BA together ($4\,\mu M$ each) resulted in production of up to 10–12 shoots per explant in 4 weeks. Similar results were found using nodal segments and leaf discs of *A. indica* on MS media supplemented with BA and Kn with the optimal type and concentrations of plant growth regulators being $1\,mg\,l^{-1}$ BA and $0.5\,mg\,l^{-1}$ Kn (Subramani et al. 1993). Kearney (1993) also reported that shoots could be induced directly on leaf explants by incubation on MS medium containing $0.9\,mg\,l^{-1}$ BA. Incubation on BA ($1\,mg\,l^{-1}$) and Kn ($0.8\,mg\,l^{-1}$) was also found to enhance adventitious bud formation and this could be further improved by addition of $6\,mg\,l^{-1}$ adenine sulphate (Subramani et al. 1993). Ramesh and Padhya (1990) used $10\,\mu M$ gibberellic acid (GA_3) and $1\,\mu M$ Kn to obtain further shoot development. In a study comparing differentiation of neem bark and leaf explants on different media (including MS) shoot differentiation only occurred on Nitsch's medium and this only occurred at low frequencies (Wewetzer 1997). The effect of IAA, IBA and NAA on root induction has also been studied (Ramesh and Padhya 1990; Subramani et al. 1993) and the auxin, IAA was found to be suitable for inducing roots, with $6\,\mu M$ giving optimum results (Ramesh and Padhya 1990). Shoots from anther-derived callus also produced roots when cultured on MS with the auxin, IBA (Gautam et al. 1993).

In later studies, Eeswara (1996), failed to obtain shoot formation in any of the conditions described above. Subsequently, Eeswara et al. (1997b) reported that neem leaf explants from different geographical locations in Sri Lanka, Ghana and Niger could be induced to produce shoots and that these could be successfully multiplied, rooted and grown in soil. They tested several media and culture conditions for efficiency of shoot production from leaves and reported that shoot initiation was best achieved in darkness on MS medium supplemented with the same plant growth regulators as used by Subramani

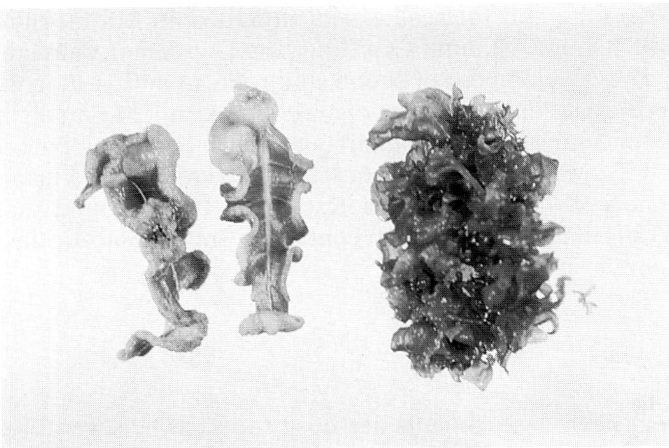

Fig. 12. Effect of dark (*left*) and light (16 h light: 8 h dark) (*right*) on shoot and callus production by leaf explants from neem trees originating in Ghana. Explants were incubated on Murashige and Skoog medium supplemented with N^6-benzyladenine (1 mg l^{-1}), kinetin (0.8 mg l^{-1}) and adenine sulphate (6 mg l^{-1})

et al. (1993) i.e. BA (1 mg l^{-1}), Kn (0.8 mg l^{-1}) and adenine sulphate (6 mg l^{-1}). No leaf explants produced shoots when incubated on a range of media in a 16 h light : 8 h dark regime (Fig. 12). When plantlets were allowed to continue to grow on the same medium, extensive vitrification resulted. Plantlets grew much better on a medium containing BA, Kn and adenine sulphate at concentrations one tenth of those used for shoot initiation (Fig. 13). Using these methods, 80 shoots from each leaf explant were produced in 32 weeks. Eeswara et al. (1997b) observed that rooting was induced in more than 50% of plantlets by treatment with 1 mg l^{-1} IBA. This rooting procedure is consistent with that of Joarder et al. (1996), Nirmalakumari et al. (1996) and Drew (1996), but contrasts with Narayan and Jaiswal (1985) and Subramani et al. (1993) who reported no rooting on medium containing IBA. Eeswara et al. (1997b) found that shoot initiation did not occur when leaf explants were incubated on the initiation medium in the light (2960 lx) and suggested that the fundamental difference between their work and that of others (Ramesh and Padhya 1990; Subramani et al. 1993) resulted from the different light intensities used, since the latter used subdued lighting (1000 lx).

Methods for rapid vegetative multiplication of neem from nodes or apical or axillary buds and which may be useful for the production of high azadirachtin-yielding genotypes, has been developed. In the work described by Drew (1996) neem shoots were established in culture from nodal bud explants of glasshouse grown plants. Apically dominant shoots were grown on hormone-free MS medium and multiplied by dissection into nodal sections. Actively growing axillary shoots quickly developed good root systems when

Fig. 13. Micropropagated neem shoots grown on Murashige and Skoog basal medium with 3% sucrose supplemented with N^6-benzyladenine (0.1 $mg\,l^{-1}$), kinetin (0.08 $mg\,l^{-1}$) and adenine sulphate (0.6 $mg\,l^{-1}$)

placed for 4 days on MS medium containing 10 μM IBA. Plants from eight countries were successfully established in the field using this system and produced seed within 2 years. Joarder et al. (1996) reported that the growth of similar nodal explants was improved by transferring to MS medium containing BA (1.5 $mg\,l^{-1}$) after growth for 2 weeks on MS alone. Joshi and Thengane (1996) induced multiple shoots from excised axillary and terminal buds by incubation on MS medium containing cytokinins. Although BA (0.5–5 $mg\,l^{-1}$) resulted in the greatest number of explants producing shoots, addition of BA and Kn gave a greater number of shoots per explant.

2.3.2 Somatic Embryogenesis

Somatic embryogenesis can be used to produce a large number of clonal plantlets directly in culture in a process where root and shoot formation occur simultaneously in an apparently co-ordinated manner. This may offer advantages for neem in terms of a method for rapid production of "elite" plantlets or, if the somatic embryos could be encapsulated to form artificial seeds (see Bajaj 1995) hence circumventing the problems of the recalcitrance of the natural seed. Immature cotyledons were used to test the effect of different plant growth regulators and carbohydrate on embryo induction, maturation and germination using a semi-solid MS medium (Shrikhande et al. 1993). The best induction response was found with BA (1 $mg\,l^{-1}$) while the number of normal embryos decreased with increasing sucrose concentration, with 5% sucrose giving the best results. Sixty to seventy percent germinated on MS medium containing IAA ($0.5\,mg\,l^{-1}$), BA ($2\,mg\,l^{-1}$) and 5% sucrose. Transfer of germinated embryos to half strength MS with 2% sucrose allowed for the development of complete plantlets. These had well developed shoots and roots and produced good plants which transferred well to soil. High frequency somatic embryogenesis has also been obtained from immature cotyledon explants in MS medium containing different concentrations of NAA although only low levels of shoot development occurred on transfer to cytokinin-supplemented medium (Islam et al. 1996). Joshi and Thengane (1996) induced embryos directly from cotyledonary explants or via callus and reported that the type of auxin was important in determining which process occurred. Akula and Drew (1996), in a detailed study using four different *A. indica* genotypes, also found efficient induction of somatic embryos, either directly from root or nodal cuttings or indirectly from leaf explants after a callus phase.

2.4 Production of Metabolites

More than 300 compounds have been isolated from various parts of the neem tree (Kumar et al. 1996). Some important ones obtained in callus cultures are shown in Figs. 5 and 6, and are discussed here.

2.4.1 Azadirachtin

The opportunity of using plant cell culture to provide a reliable and economic source of azadirachtin (van der Esch et al. 1993) is attractive for several reasons. Firstly, an additional supply may help alleviate the anticipated world-wide shortage of neem material which is likely to occur if its actual use in insect control keeps pace with the increase in research interest (Saxena 1989). Pure azadirachtin is very expensive (£54 per mg, Sigma Chemical Co. 1997), it cannot be synthesised chemically and has to be purified, using costly and sophisticated methods, from large quantities of seasonally produced seed. Also, an advantage of using a sterile, controlled production system is that it avoids the risks of microbial spoilage and especially mycotoxin contamination which is sometimes found in naturally occurring seed. Tissue culture methods should also give a more consistent product quality and yield, since the azadirachtin content of seeds is highly variable and can be decreased by heat or light (Jarvis et al. 1997), and may offer the opportunity for increasing yields and producing novel compounds.

Different groups have found varying azadirachtin yields in neem callus cultures: e.g. 0.5–8 $\mu g\,g^{-1}$ dry weight in callus derived from aseptic seedlings from Togo (van der Esch et al. 1996) while higher yields, ranging from 4–189 μg azadirachtin g^{-1} dry weight were detected in callus derived from wild trees in Sri Lanka (Eeswara 1996; Eeswara et al. 1997a). Investigations with various media (Wewetzer 1997) indicated that, in neem callus derived from seed from Nicaragua and Togo, the highest yield of 64 μg azadirachtin g^{-1} dry weight, as analysed by HPLC, was obtained on White's medium containing

$15\,g\,l^{-1}$ sucrose. Azadirachtin production was generally three times higher on media supplemented with 1.5% rather that 3% sucrose (Wewetzer 1996). Allan et al. (1994) undertook analyses of azadirachtin in callus cultures of Ghanaian origin using a range of chromatographic methods i.e. HPLC, TLC and supercritical fluid chromatography (SFC) and found azadirachtin yields of $7\,\mu g\,g^{-1}$ dry weight. These varying reports on yields may reflect differences in the explant type and indeed genotype. However, in a study with limited numbers of trees, no correlation was found between the amounts of azadirachtin, nimbin and salannin produced by callus and the amounts present in seeds collected from the same parental trees (Eeswara et al. 1997a). Therefore, it is likely that the differences may be associated with the environmental parameters used for callus cultivation by the different laboratories as well as the different extraction and detection techniques used for quantifying azadirachtin.

Comparison of growth and product formation in neem suspension cultures in an MS based medium at 28 °C in the light or dark showed that greatest biomass production was obtained in complete darkness (maximum biomass 16 mg dry weight ml^{-1} on day 24) with this being reduced by approximately 50% on a light/dark cycle of 16/8 h (van der Esch et al. 1993, 1994). Interestingly, both cultures produced azadirachtin, as determined by HPLC analysis, with the culture grown in the light showing a higher yield ($0.66\,\mu g\,mg^{-1}$ dry weight on day 12 compared with $0.19\,\mu g\,mg^{-1}$ on day 19). Further work by this group (van der Esch et al. 1994, 1996), indicated that in addition to azadirachtin, nimbin and salannin were also produced in suspension cultures. They also showed that azadirachtin production could be enhanced by elicitation using a chitosan preparation. In this study, azadirachtin yields were increased up to 0.25% of dry biomass and elicitation stimulated azadirachtin production during early exponential growth at a time when it was not normally produced. Kearney et al. (1994) showed that the production of insect antifeedant compounds in neem suspension cultures grown in the dark was typical of secondary metabolites, being produced at the end of the exponential and beginning of the stationary growth phases. This is in contrast to the results of van der Esch et al. (1993) where product formation was growth associated in dark-grown cultures.

It is becoming evident that differentiation may influence azadirachtin yields. Thus, neem suspension cultures grown in the light, which presumably increased the state of differentiation, had higher azadirachtin yields than in the dark (van der Esch et al. 1993) while greater yields have also been found in callus that had a hard texture and a tendency towards organogenesis (Eeswara 1996). Although these results are somewhat preliminary, they appear to be typical of undifferentiated tissue cultures which are renowned for suffering from both biochemical and genetic instability. To overcome some of these problems, the use of *Agrobacterium rhizogenes* to produce transformed root cultures of neem has been investigated (Eeswara 1996). Neem root cultures were obtained by infecting leaves and shoot cultures with *A. rhizogenes* and after a period of co-cultivation, the bacteria were eliminated from the explants by frequent subculture onto antibiotic-containing media (Eeswara 1996). Four weeks after infection, roots appeared and after further growth, these were removed from the explant and with time, cultivated in medium without antibiotic or plant growth regulators. These roots were analysed using HPLC and found to contain azadirachtin, nimbin and salannin at levels of 0.069, 0.0076 and 0.0093 $mg\,g^{-1}$ dry weight tissue respectively. The results for azadirachtin were confirmed using SFC by Professor Morgan's group at Keele University and the yield was greater, by a factor of ten, than results obtained from good quality, undifferentiated callus (Allan et al. 1994). Since then, these hairy root cultures have continued to produce azadirachtin, in addition to other biologically active compounds in a stable manner (unpubl. results).

In terms of cell screening and selection to enhance product formation, it is crucial that accurate assessments of yields are obtained. The extraction and detection of azadirachtin in tissue cultures, which usually contain very small amounts of material, is particularly difficult. The undeniable proof that azadirachtin is present in neem tissue cultures was confirmed by ^{1}H NMR and mass spectra from suspension culture extracts (Jarvis et al. 1997). The quantification of azadirachtin will become even more important as research develops to enhance yields since subpopulations of productive cells will have to be identified and selected, as it is the proportion of these and their rate of secondary metabolite accumulation which is likely to be the key to increasing yields. Although azadirachtin yields are well below those required for commercial production systems, tissue cultures do produce the highly complex compound azadirachtin, and do provide the potential for increasing yields.

2.4.2 Nimbin

Various workers (Sanyal et al. 1983; Sarkar and Datta 1986; Sarkar et al. 1988) have investigated the relationship between differentiation in neem tissue cultures and biosynthesis of nimbin, glycine and β-sitosterol (a steroidal drug product). These workers have found that undifferentiated callus tissue, derived from mature cotyledons, has high concentrations of β-sitosterol and low concentrations of nimbin and glycine while the reverse occurs in redifferentiated tissues and normal cotyledons. Interestingly, the amount of β-sitosterol in callus could be increased by addition of cholesterol (a precursor of β-sitosterol) to the growth medium (Sarkar and Datta 1986). Addition of exogenous glycine to callus resulted in an increase in the nimbin content of the callus and a decrease in the β-sitosterol concentration. Sarkar (Sanyal) et al. (1988) therefore suggested that glycine triggered the biosynthesis of nimbin, blocking the pathway to β-sitosterol. A study of the effect of different plant growth regulators on callus initiation and growth (Ramesh and Padhya 1988), confirmed that the levels of nimbin in tissue cultures were associated with the differentiation process. Although the production of antifeedant compounds by neem callus has been shown to be stable over a 4-year period (Kearney 1994; Eeswara 1996), the production of nimbin has been shown to decline with time (Sanyal et al. 1983).

2.4.3 *Flavonoids*

Ramesh and Padhya (1996) presented a detailed paper on the production of flavonoids in neem tissue cultures at the World Neem Conference held in India in 1993. They found that flavonoid production was related to the explant type, in that the highest amounts of the flavonoids, kaempferol and quercetin were obtained from callus tissue derived from flowers (as in the whole plant). These compounds were accumulated in suspension cultures during the stationary phase of growth. Subsequently, following studies on the effect of sucrose and nitrogen concentration, a medium which stimulated biomass formation and secondary metabolite production was developed. Feeding studies showed that addition of phenylalanine (flavonoid precursor) could also be used to enhance both biomass and product formation. They selected for high yielding cell lines using p-fluorophenylalanine and this led to a stable, threefold increase in production.

2.5 Activity of Neem Tissue Cultures Against Insects

The assessment of biological activity in neem tissue cultures is of crucial importance. The combination of chemical analysis and bioassay provides advantages and allows the possibility of screening for a range of bioactive compounds. Indeed, insects may be more selective and have lower detection limits than that offered by chemical analyses. Neem callus, suspension cultures (Fig. 14) and micropropagated shoots have been screened for insect antifeedancy using the desert locust, *Schistocerca gregaria* (Kearney et al. 1994) which is particularly sensitive to azadirachtin at the behavioural level and may detect

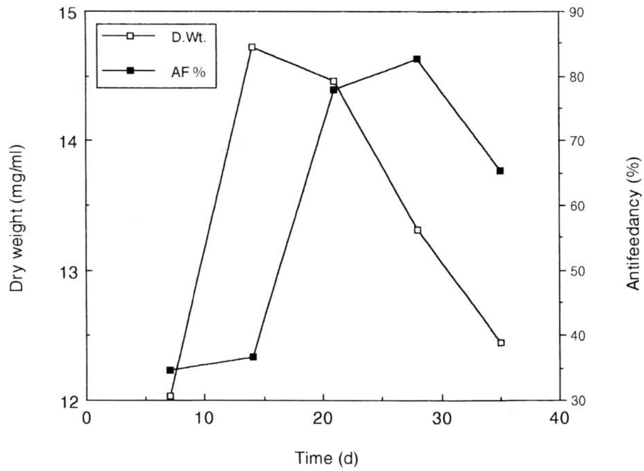

Fig. 14. Dry weight (*D.Wt.*) growth measurements and percentage antifeedancy (*AF%*) of *Azadirachta indica* suspension culture grown in Murashige and Skoog basal medium with 3% sucrose, $4\,mg\,ml^{-1}$ indole-3-butyric acid and $1\,mg\,l^{-1}$ N^6-benzyladenine

down to nanogram amounts. No-choice and choice feeding bioassays have been used to assay plant and cell culture extracts, using 3-day-old fifth instar nymphs of *S. gregaria*. The insects are kept separately and their state of hunger is standardized prior to the bioassay. They are given access to glass fibre discs pretreated with sucrose and the appropriate extract. Feeding is compared with that on control discs treated previously with sucrose only. The percentage antifeedancy is calculated according to the methods of Kearney et al. (1994) and Mordue(Luntz) et al. (1995). Feeding suppression was obtained in all of the neem derived tissue culture samples whereas none was found in the controls using carrot callus. Dose response curves using callus from different provenances have been investigated by the Aberdeen University group and all have shown good linear relationships with different lines from different areas having different effective doses. For example, a 6-week-old callus culture line derived from Niger was shown to produce a good dose response relationship with an effective dose (ED_{50}) of 1.74 mg dry weight callus ml^{-1} extract (1 µg "azadirachtin equivalents" g^{-1}) according to Mordue(Luntz) et al. (1995) while callus of the same age, derived from a tree of Ghanaian origin had an ED_{50} of 0.14 mg dry weight callus ml^{-1} extract (7 µg "azadirachtin equivalents" g^{-1}) according to Eeswara (1996). Such dose response curves have been compared with pure azadirachtin and have been found to be parallel, indicating that most of the antifeedant action is due to azadirachtin and its closely related analogues (Mordue(Luntz) et al. 1996; Wickramananda 1997). In addition, however, there may be other compounds which act as phagostimulants and/or deterrents which mask the full effect of azadirachtin (Eeswara 1996) and there may also be novel antifeedant compounds (Allan et al. 1994). Detailed screening of a Ghanaian neem callus line has also been undertaken with fractions of extracts being assessed both by different chromatographic methods and insect antifeedancy (Allan et al. 1994). Results clearly showed the presence of azadirachtin in addition to other novel compounds with biological activity (Allan et al. 1994). Mordue(Luntz) et al. (1995) confirmed and further developed the work of Kearney et al. (1994) showing that neem callus derived from different provenances showed high levels of insect antifeedancy. Another bioassay to test insect growth regulatory effects using the milkweed bug (*Oncopeltus fasciatus*) has also been developed for neem tissue cultures and shows equal promise to antifeedancy (Mordue(Luntz) et al. 1995; Wickramananda 1997).

2.6 Genetic Transformation

Genetic transformation of trees offers great potential (see Bajaj 1999). Neem can be transformed with *Agrobacterium tumefaciens* which is an established vector for genetically engineering plants (Naina et al. 1989). These authors transformed *in vitro* seedlings of neem and transformed plantlets were regenerated from infected regions of the seedlings as well as from the tumour callus which was formed. Eeswara (1996) also obtained hairy root cultures from aseptic seedlings using *A. rhizogenes* (see Sect. 2.4.1).

3 Conclusions and Prospects

The traditional uses of the neem tree and its products have progressed and developed so that the tree is now promoted in tropical and sub-tropical areas for reforestation programmes, as a source of wood, fertiliser, cattle feed, shade, oil, medicinals and pesticides. Neem extracts and the major limonoid azadirachtin in particular, are now being studied as environmentally sound insecticides for world-wide use and recently several commercial neem products have been registered for use on both non-food and food crops.

Major advances have been achieved in micropropagation, somatic embryogenesis and metabolite production in neem. Micropropagation is likely to be important in satisfying increasing demands for the tree, especially as its seeds lose viability rapidly after ripening. Several research groups have established micropropagation of neem from leaf explants and nodes. The recalcitrance of the natural seed may also be overcome by somatic embryogenesis with methods now established. In terms of metabolite production from neem cell cultures, several groups are working on the insect antifeedant and growth regulator, azadirachtin while others have identified different target compounds with commercial potential. Reported azadirachtin yields from cultures vary between different research groups and it is unknown whether this reflects environmental or genetic differences between cultures. There is no doubt, however, that azadirachtin yields in callus, suspension and hairy root cultures are presently much lower than those required for commercial exploitation. Despite this, it is likely that the increasing commercial demand for neem extracts will stimulate research in this area.

Acknowledgements. The authors gratefully thank their many students and collaborators who have, in different ways, contributed to this paper.

References

Aerts R, Mordue(Luntz) AJ (1997) Feeding deterrence and toxicity of neem triterpenoids. J Chem Ecol 23:2117–2132

Akula D, Drew R (1996) Somatic embryogenesis of *Azadirachta indica* (A. Juss.) and HPLC analysis of in vitro azadirachtin production. In: Abstr Vth Int Neem Conf, Gatton, Australia, pp 23

Allan EJ, Eeswara JP, Johnson S, Mordue(Luntz) AJ, Morgan ED, Stuchbury T (1994) The production of azadirachtin by in vitro tissue cultures of neem, *Azadirachta indica*. Pestic Sci 42:147–152

Bajaj YPS (ed) (1995) Biotechnology in agriculture and forestry, vol 30. Somatic embryogenesis and synthetic seed I. Springer, Berlin Heidelberg New York

Bajaj YPS (ed) (1999) Biotechnology in agriculture and forestry, vol 44. Transgenic trees. Springer, Berlin Heidelberg New York

Balandrin MF, Lee SM, Klocke JA (1988) Biologically active volatile organosulfur compounds from seeds of the neem tree, *Azadirachta indica* (Meliaceae). J Agric Food Chem 36:1048–1054

Benge MD (1989) Cultivation and propagation of the neem tree. In: Jacobsen M (ed) The neem tree. Focus on phytochemistry and pesticides, vol 1. CRC Press, Boca Raton, pp 1–18

Bilton JN, Broughton HB, Jones PS, Ley SV, Lidert Z, Morgan ED, Rzepa HS, Sheppard RN, Slawin AMZ, Williams DJ (1987) An X-ray crystallographic, mass spectroscopic, and NMR study of the limonoid insect antifeedant azadirachtin and related derivatives. Tetrahedron 43:2805–2815

Blaney WM, Simmonds MSJ, Ley WV, Anderson JC, Toogood PL (1990) Antifeedant effects of azadirachtin and structurally related compounds on lepidopterous larvae. Entomol Exp Appl 55:149–160

Boa ER (1995) A guide to the identification of diseases and pests of neem (*Azadirachta indica*). FAO, Bangkok

Bokel M, Cramer R, Gutzeit H, Reeb S, Kraus W (1990) Tetranortriterpenoids related to nimbin and nimbolide from *Azadirachta indica* A. Juss (Meliaceae). Tetrahedron 46:775–782

Broughton HB, Ley SV, Slawin AMZ, Williams DJ, Morgan ED (1986) X-ray crystallographic structure determination of ditigloyldihydro-azadirachtin and reassignment of the structure of the limonoid insect antifeedant azadirachtin. J Chem Soc Chem Commun: 46–47

Buchanan JGSTC, Halsall TG (1970) The conversion of turreanthin and turreanthin A into simple meliacins by a route involving an oxidative rearrangement of probable biogenetic importance. J Chem Soc (C): 2280–2284

Butterworth JH, Morgan ED (1968) Isolation of a substance that suppresses feeding in locusts. J Chem Soc Chem Comm 1:23–24

Chavan SR (1984) Chemistry of alkanes separated from leaves of *Azadirachta indica* and their larvicidal/insecticidal activity against mosquitoes. In: Schmutterer H, Ascher KRS (eds) Natural pesticides from the neem tree and other tropical plants. Proc 2nd Int Neem Conf, GTZ, Eschborn, pp 59–65

Cohen E, Quistad GB, Jefferies PR, Casida JE (1996) Nimbolide is the principal cytotoxic component of neem-seed insecticide preparations. Pestic Sci 48:135–140

Coventry E, Allan EJ (1997) The effects of neem based products on bacterial and fungal growth. In: Kleeberg H, Zeibitz CPW (eds) Practice orientated results on use and production of neem ingredients and pheromones. Proc 5th Neem Workshop. Druck and Graphic, Giessen, pp 237–242

Drew RA (1996) Clonal propagation of neem by tissue culture. In: Singh RP, Chari MS, Raheja AK, Kraus W (eds) Neem and the environment, vol 2. Science Publishers, Lebanon, New Hampshire, pp 999–1005

Dreyer DL (1984) Biogenetic relationships of degraded triterpenes in the Rutales. In: Nes WD, Fuller G, Tasi L-S (eds) Isopentenoids in plants, biochemistry and function. Dekker, New York, pp 247–265

Eeswara JP (1996) The production of the insect antifeedant azadirachtin and related metabolites by plant cell and tissue culture of neem (*Azadirachta indica*). PhD Thesis, University of Aberdeen, UK

Eeswara JP, Mordue(Luntz) AJM, Stuchbury T, Allan EJ (1997a) Azadirachtin, salannin and nimbin contents of seeds and callus of neem, *Azadirachta indica* A. Juss. In: Kleeberg H, Zeibitz CPW (eds) Practice orientated results on use and production of neem ingredients and pheromones. Proc 5th Neem Workshop. Druck and Graphic, Giessen, pp 251–262

Eeswara JP, Stuchbury T, Allan EJ, Mordue AJ (1997b) A standard procedure for the micropropagation of the neem tree (*Azadirachta indica* A Juss). Plant Cell Rep 17:215–219

Eeswara JP, Allan EJ, Powell AA (1998) The influence of stage of seed maturity, moisture content and storage temperature on the survival of neem (*Azadirachta indica*) seed in storage. Seed Sci Technol (in press)

Ermel K, Kleeberg H (1995) Commercial products, their standardisation and problems of quality control. In: Schmutterer H (ed) The neem tree. VCH, Weinheim, pp 375–384

Ermel K, Pahlich E, Schmutterer H (1987) Azadirachtin content of neem kernels from all over the world and its dependence on temperature, air, moisture and light. In: Schmutterer H, Ascher KRS (eds) Natural pesticides from the neem tree (*Azadirachta indica* A. Juss) and other tropical plants. Proc 3rd Int Neem Conf, GTZ, Eschborn, pp 185–194

Ermel K, Kalinowski H-O, Schmutterer H (1991) Isolierung und Charakterisierung von Marrangin, einer neuen, die Insektenmetamorphose störenden Substanz aus Samenkernen des Marrangobaumes *Azadirachta excelsa* (Jack). J Appl Entomol 112:512–519

Ezumah BS (1986) Germination and storage of neem (*Azadirachta indica* A Juss). Seed Sci Technol 14:593–600

Fujiwara T, Takeda T, Ogihara Y, Shimizu M, Nomura T, Tomita Y (1982) Studies on the structure of polysaccharides from the bark of *Melia azadirachta*. Chem Pharm Bull (Tokyo) 30:4025–4030

Fujiwara T, Sugishita EY, Takeda T, Ogihara Y, Shimizu M, Nomura T, Tomita Y (1984) Further studies on the structure of polysaccharides from the bark of *Melia azadirachta*. Chem Pharm Bull (Tokyo) 32:1385–1391

Gamborg OL, Miller RA, Ojima K (1968) Nutrient requirements of suspension cultures of soybean root cells. Exp Cell Res 50:151–158

Gamene CS, Kraak HL, Van Piljen JG, De Vos CHR (1996) Storage behaviour of neem (*Azadirachta indica*) seeds from Burkina Faso. Seed Sci Technol 24:441–448

Ganesalingam VK (1987) Use of the neem plant in Sri Lanka at the farmers level. In: Schmutterer H, Ascher KRS (eds) Natural pesticides from the neem tree and other tropical plants. Proc 3rd Int Neem Conf, GTZ, Eschborn, pp 95–100

Gautam VK, Nanda K, Gupta SC (1987) Morphogenic responses of in vitro cultured anthers of *Azadirachta indica* A. Juss. – a medicinal tree. Proc 14th Int Bot Congr, Berlin, p 123

Gautam VK, Nanda K, Gupta SC (1993) Development of shoots and roots in anther-derived callus of *Azadirachta indica* A. Juss. – a medicinal tree. Plant Cell Tissue Org Cult 34:13–18

Govindachari TR, Narasimhan NS, Suresh G, Partho PD, Gopalakrishnan G (1996) Insect antifeedant and growth-regulating activities of salannin and other C-seco limonoids from neem oil in relation to azadirachtin. J Chem Ecol 22(8):1453–1461

Guglielmini E, Cei S (1995) Tissue culture of neem. ERASMUS Project Rep, University of Aberdeen, UK

Harris M, Henderson R, McCrindle R, Overton KH, Turner DW (1968) Tetranortriterpenoids. – VIII. The constituents and stereochemistry of nimbin. Tetrahedron 24:1517–1523

Hellpap C, Dreyer M (1995) Neem products for pest management, practical results of neem application against arthropod pests, and probability of development of resistance. In: Schmutterer H (ed) The neem tree *Azadirachta indica* A. Juss and other meliaceous plants. VCH, Weinheim, pp 367–375

Immaraju JA (1998) The commercial use of azadirachtin and its integration into viable pest control programmes. Pestic Sci (in press)

Ishida M, Serit M, Nakata K, Juneja LR, Kim M, Takahashi S (1992) Several antifeedants from neem oil, *Azadirachta indica* A. Juss., against *Reticulitermes speratus* Kolbe (Isoptera: Rhinotermitidae). Biosci Biotechnol Biochem 56:1835–1838

Islam R, Joarder N, Joarder OI, Hossain M, Rahman SM, Naderuzzaman ATM (1996) Somatic embryogenesis and organogenesis in neem (*Azadirachta indica* A. Juss.) In: Singh RP, Chari MS, Raheja AK, Kraus W (eds) Neem and the environment, vol 2. Science Publishers, Lebanon, New Hampshire, pp 955–960

Jacobsen M (1987) Neem research and cultivation in the Western Hemisphere. In: Schmutterer H, Ascher KRS (eds) Proc 3rd Int Neem Conf, GTZ, Eschborn, pp 33–44

Jarvis AP, Morgan ED, Van der Esch SA, Vitali F, Ley SV, Pape A (1997) Identification of azadirachtin in tissue cultured cells of neem (*Azadirachta indica*). J Nat Prod 10:95–98

Joarder OI, Naderuzzaman ATM, Islam R, Hossain M, Joarder N, Biswas BK (1996) Micropropagation of neem through axillary bud culture. In: Singh RP, Chari MS, Raheja AK, Kraus W (eds) Neem and the environment, vol 2. Science Publishers, Lebanon, New Hampshire, pp 961–966

Joshi MS, Thengane SR (1996) Potential application of in vitro methods for propagation of neem (*Azadirachta indica* A. Juss.). In: Singh RP, Chari MS, Raheja AK, Kraus W (eds) Neem and the environment, vol 2. Science Publishers, Lebanon, New Hampshire, pp 967–973

Kearney M-L (1993) Plant cell and tissue culture of *Azadirachta indica* for the production of antifeedant compounds. MSc Thesis, University of Aberdeen, UK

Kearney M-L, Allan EJ, Hooker JE, Mordue(Luntz) AJ (1994) Antifeedant effects of in vitro culture extracts of the neem tree, *Azadirachta indica* against the desert locust (*Schistocerca gregaria* (Forskål)). Plant Cell Tissue Org Cult 37:67–71

Khalid SA, Duddeck H, Gonzales-Sierra M (1989) Isolation and characterization of an antimalarial agent of the neem tree *Azadirachta indica*. J Nat Prod 52:922–927

Kleiner K (1995) Pesticide tree ends up in court. New Sci 147:7
Koul O, Isman MB, Ketkar CM (1990) Properties and uses of neem, *Azadirachta indica*. Can J Bot 68:1–11
Kraus W (1995) Azadirachtin and other triterpenoids. In: Schmutterer H (ed) The neem tree *Azadirachta indica* A. Juss and other meliaceous plants. VCH, Weinheim, pp 35–73
Kraus W, Cramer R, Sawitzki G (1981) Tetranortriterpenoids from the seeds of *Azadirachta indica*. Phytochemistry 20:117–120
Kraus W, Bokel M, Klenk A, Pöhnl H (1985) The structure of azadirachtin and 22,23-dihydro-23β-methoxyazadirachtin. Tetrahedron Lett 26:6435–6438
Kraus W, Bokel M, Bruhn A, Cramer R, Klaiber I, Klenck A, Nagl G, Pöhnl H, Sadlo H, Vogler B (1987) Structure determination by NMR of azadirachtin and related compounds from *Azadirachta indica* (Meliaceae). Tetrahedron 43:2817–2830
Kraus W, Koll-Weber M, Maile R, Wunder T (1993a) Biologically active constituents of tropical and subtropical plants. 8th European Symposium on Organic Chemistry, Barcelona, p 448 (Abstr)
Kraus W, Maile R, Vogler B (1993b) Nematocidal constituents of *Azadirachta indica* seed cake. Phyt Soc Eur Int Symp Phytochem plants used in traditional medicine, Lausanne, p 59 (Abstr)
Kumar CSSR, Srinivas M, Yakkundi S (1996) Limonoids from the seeds of *Azadirachta indica*. Phytochemistry 43:451–455
Lavie D, Jain MK (1967) Tetranortriterpenoids from *Melia azadirachta* L. J Chem Soc Chem Commun: 278–280
Lavie D, Levy EC (1971) Meliane-meliacin relationship. Tetrahedron 27:3941–3947
Lavie D, Levy EC, Jain MK (1971) Limonoids of biogenetic interest from *Melia azadirachta* L. Tetrahedron 27:3927–3939
Ley SV, Denholm AA, Wood A (1993) The chemistry of azadirachtin. Nat Prod Rep: 109–157
Medlock VFP (1996) Fungal toxins in foods and feeds. Soc Gen Micro Q 23(3):71–73
Meisner J, Ascher KRS, Aly R, Warthen JD Jr (1981) Response of *Spodoptera littoralis* (Boisd.) and *Earias insulana* (Boisd.) larvae to azadirachtin and salannin. Phytoparasitica 9:27–32
Mordue(Luntz) AJ, Blackwell A (1993) Azadirachtin: an update. J Insect Physiol 39:903–924
Mordue(Luntz) AJ, Nasiruddin M (1992) Azadirachtin treatment and host plant selection. In: Menken SBJ, Visser JA, Harrewijn P (eds) Proc 8th Int Symp, Insect-Plant Relationships. Kluwer, Dordrecht, pp 176–178
Mordue(Luntz) AJ, Zounos A, Wickramananda IR, Allan EJ (1995) Neem tissue culture and the production of insect antifeedant and growth regulatory compounds. BCPC Symp Proc No 63, Integrated crop protection towards sustainability. BCPC, Farnham, pp 187–194
Mordue(Luntz) AJ, Eeswara JP, Wickramananda IR, Johnson S, Morgan ED, Stuchbury T, Allan EJ (1996) Neem metabolites with insect antifeedancy and growth regulatory properties are produced in tissue cultures. Vth Int Neem Conf, Gatton, Australia, p 26 (Abstr)
Murashige T, Skoog F (1962) A revised medium for rapid growth and bio-assays with tobacco cultures. Physiol Plant 15:473–497
Naina NS, Gupta PK, Mascarenhas AF (1989) Genetic transformation and regeneration of transgenic neem (*Azadirachta indica*) plants using *Agrobacterium tumefaciens*. Curr Sci 58:184–187
Naqvi SNH (1987) Biological evaluation of fresh neem extracts and some neem components, with reference to abnormalities and esterase activity in insects. In: Schmutterer H, Ascher KRS (eds). Natural pesticides from the neem tree and other tropical plants. Proc 3rd Int Neem Conf, GTZ, Eschborn, pp 315–330
Narayan P, Jaiswal VS (1985) Plantlet regeneration from leaflet callus of *Azadirachta indica* Juss. J Tree Sci 4:65–68
Nasiruddin M, Mordue(Luntz) AJ (1994) The protection of barley seedlings from attack by *Schistocerca gregaria* using azadirachtin and related analogues. Entomol Exp Appl 70:247–252
National Research Council (1992) Neem – a tree for solving global problems. National Academy Press, Washington, DC
Nirmalakumari A, Ramaswamy NM, Sree Rangaswamy SR (1996) Tissue culture studies in neem (*Azadirachta indica* A. Juss.). In: Singh RP, Chari MS, Raheja AK, Kraus W (eds) Neem and the environment, vol 2. Science Publishers, Lebanon, New Hampshire, pp 982–992

Pant N, Garg HS, Madhusudanan KP, Bhakuni DS (1986) Sulfurous compounds from *Azadirachta indica* leaves. Fitoterapia 57:302–304

Parmar BS, Ketkar CM (1993) Commercialization. In: Randhawa NS, Parmar BS (eds) Neem research and development. Society of Pesticides, India, pp 270–283

Parveen RKK, Vyas CS, Tomar UK (1995) Tissue culture studies on neem (*Azadirachta indica* A. Juss.). Neem Newslett Int Neem Network 2(3):15–17

Pearce F (1993) Pesticide patent angers Indian farmers. New Sci 140:7

Petty L (1993) Tissue culture of neem (*Azadirachta indica*) and the detection of insect antifeedant compounds in culture. M Sc Thesis, University of Aberdeen, UK

Pliske TE (1984) The establishment of neem plantations in the American tropics. In: Schmutterer H, Ascher KRS (eds) Natural pesticides from the neem tree (*Azadirachta indica* A. Juss) and other tropical plants. Proc 2nd Inter Neem Conf, GTZ, Eschborn, pp 521–526

Radwanski S (1977) Neem tree. I. Commercial potential, characteristics and distribution. World Crops Livest 29:65–66

Radwanski S, Wickens GE (1981) Vegetative fallows and potential value of the neem tree (*Azadirachta indica*) in the tropics. Econ Bot 35:398–414

Ramesh K, Padhya MA (1988) Isolation of nimbin from *Azadirachta indica* leaves and its callus cultures. Indian Drugs 25(12):526–527

Ramesh K, Padhya MA (1990) In vitro propagation of neem, *Azadirachta indica* (A. Juss), from leaf discs. Indian J Exp Biol 28:932–935

Ramesh K, Padhya MA (1996) De novo synthesis of secondary metabolites in cultured cells of neem. In: Singh RP, Chari MS, Rahejo AK, Kraus W (eds) Neem and environment, vol 1. Science Publishers, Lebanon, New Hampshire, pp 173–184

Rangaswamy NS, Promila (1972) Morphogenesis of the adult embryo of *Azadirachta indica* A Juss. Z Pflanzenphysiol 67:377–379

Rembold H (1989a) Isomeric azadirachtins and their mode of action. In: Jacobson M (ed) 1988 Focus on phytochemical pesticides, vol 1. The neem tree. CRC Press, Boca Raton, pp 47–86

Rembold H (1989b) Azadirachtins, their structure and mode of action. In: Arnason JT, Philogene BJR, Morand P (eds) Insecticides of plant origin. ACS Symp Ser 387, American Chemical Society, Washington, DC, pp 150–163

Rembold H, Forster H, Czoppelt CH (1987a) Structure and biological activity of azadirachtins A and B. In: Schmutterer H, Ascher KRS (eds) Natural pesticides from the neem tree and other tropical plants. Proc 3rd Int Neem Conf, GTZ, Eschborn, pp 149–160

Rembold H, Forster H, Sonnenbichler J (1987b) Structure of azadirachtin B. Z Naturforsch 42c:4–6

Roederer Y, Bellefontaine R (1989) Can neem seeds be expected to keep their germinative capacity for several years after collection? For Genet Resour Inf 17:30–33

Rojanapo W, Suwanno S, Somjaree R, Glinsukon T, Thebtaranonth Y (1985) Mutagenic and antibacterial activity testing in nimbolide and nimbic acid. J Sci Soc Thailand 11:177–181

Sanyal M, Dass A, Banerjee M, Datta PC (1981) In vitro hormone induced chemical and histological differentiation in stem callus of neem *Azadirachta indica* A. Juss. Indian J Exp Biol 19:1067–1068

Sanyal M, Sreeparna T, Datta PC (1983) Nimbin and β-sitosterol in cotyledon and cultured tissues of *Azadirachta indica*. Indian Drugs 81:479–481

Sarkar MS, Datta PC (1986) Biosynthesis of β-sitosterol in vitro culture of *Azadirachta indica* cotyledon tissues. Indian Drugs 23(8):478–479

Sarkar (Sanyal) MS, Mukherji A, Datta PC (1988) Glycine on in vitro biosynthesis of nimbin and β-sitosterol in tissues of *Azadirachta indica*. Curr Sci 57(1):40–41

Saxena RC (1989) Insecticides from Neem. In: Arnason JT, Philogene BJR, Morand P (eds) Insecticides of plant origin. ACS Symp Ser 387, American Chemical Society, Washington, DC, pp 110–135

Saxena RC (1990) Neem as a source of natural insecticides – an update. Proc Symp Botanical Pesticides in IPM, Indian Soc Tobacco Science, Rajahmundry, pp 1–24

Schmutterer H (1990) Properties and potential of natural pesticides from the neem tree *Azadirachta indica*. Annu Rev Entomol 35:271–297

Schmutterer H (ed) (1995) The neem tree VCH, Weinheim

Schmutterer H, Ascher KRS (eds) (1984) Natural pesticides from the neem tree (*Azadirachta indica* A. Juss) and other tropical plants. Proc 2nd Int Neem Conf, GTZ, Eschborn

Schmutterer H, Ascher KRS (eds) (1987) Natural pesticides from the neem tree (*Azadirachta indica* A. Juss) and other tropical plants. Proc 3rd Int Neem Conf, GTZ, Eschborn

Schmutterer H, Ascher KRS, Rembold H (eds) (1981) Natural pesticides from the neem tree (*Azadirachta indica* A. Juss) and other tropical plants. Proc 1st Int Neem Conf, GTZ, Eschborn

Schulz FA (1984) Tissue culture of *Azadirachta indica*. In: Schmutterer H, Ascher KRS (eds) Natural pesticides from the neem tree and other tropical plants. Proc 2nd Int Neem Conf, GTZ, Eschborn, pp 539–542

Schwinger M (1985) Über die fraßabschreckende Wirkung von Meliaceen-Inhaltsstoffen auf *Epilachna variavestis* (Muls.) und andere Insekten: Methoden – Versuchstechniken – Ergebnisse. PhD Thesis, University of Hohenheim, Germany

Schwinger M, Ehhammer B, Kraus W (1984) Methodology of the *Epilachna variavestis* bioassay of antifeedants demonstrated with some compounds from *Azadirachta indica* and *Melia azedarach*. In: Schmutterer H, Ascher KRS (eds) Natural pesticides from the neem tree and other tropical plants. Proc 2nd Int Neem Conf, GTZ, Eschborn, pp 181–198

Sen AK, Das AK, Banerji N (1993) A water soluble arabinogalactan from the fruit pulp of *Azadirachta indica*. Indian J Chem 32:862–866

Shimizu M, Sudo T, Nomura T (1985) China tree bark extracts with antineoplastic action. Patent (Switz) CH 650404 A. Chem Abstr 103(22):183551t

Shrikhande M, Thengane SR, Mascarenhas AF (1993) Somatic embryogenesis and plant regeneration in *Azadirachta indica* A. Juss. In Vitro Cell Dev Biol 29P:38–42

Siddiqui S (1942) A note on the isolation of three new bitter principles from the nim oil. Curr Sci 11:278–279

Siddiqui S, Mitra C (1945) Utilization of nim oil and its bitter constituents (nimbidin series) in the pharmaceutical industry. J Sci Ind Res 4:5–10

Siddiqui S, Mahmood T, Siddiqui BS, Faizi S (1988) Non-terpenoidal constituents from *Azadirachta indica*. Planta Med 54:457–459

Siddiqui S, Siddiqui BS, Ghiasuddin, Faizi S (1990) Triterpenoids from kernel of *Azadirachta indica*. Proc Pak Acad Sci 27:333–348

Singh RP (1987) Comparison of antifeedant efficacy and extract yields from different parts and ecotypes of neem (*Azadirachta indica* A. Juss) trees. In: Schmutterer H, Ascher KRS (eds) Natural pesticides from the neem tree (*Azadirachta indica* A. Juss) and other tropical plants. Proc 3rd Int Neem Conf, GTZ, Eschborn, pp 185–194

Singh RP, Chari MS, Raheja AK, Kraus W (eds) (1996) Neem and the environment. Science Publishers, Lebanon, New Hampshire

Sinniah D, Varghese G, Baskaran G, Koo SH (1983) Fungal flora of neem (*Azadirachta indica*) seeds and neem oil toxicity. Malays Appl Biol 12:1–4

Stoll G (1987) Methods of crop and storage protection. In: Natural crop protection based on local farm resources in the tropics and subtropics. Margraf, Weikersheim, pp 95–98

Subba Rao MV, Rao YV, Saideswara Rao YS, Manga V (1988) Induction and growth of callus in *Azadirachta indica* Juss. Crop Improv 15:203–205

Subramani J, Nair CAR, Gopinathan MC, Pillai SS (1993) Large scale production of neem through in vitro methods. In: Kleeberg H (ed) Practice oriented results on use and production of neem ingredients and pheromones. Proc 2nd Workshop Druck and Graphic, Giessen, pp 137–143

Surendran C, Rai RSV, Sivagnanam K, Shanmugham M, Kumaran K, Regupathi A, Vanangamudi K, Vimala S (1993) Tree improvement and seed management in Indian neem. In: Read MD, French JH (eds) Genetic improvement of neem and strategies for the future. Proc Int Consultation at Kasetart University, Thailand, pp 19–29

Thomsen A, Souvannavong O (1994) The international neem network. Forest Genetic Resources no 22. FAO, Rome

Turner CJ, Tempesta MS, Taylor RB, Zagorski MG, Termini JC, Schroeder DR, Nakanishi K (1987) An NMR spectroscopic study of azadirachtin and its trimethylether. Tetrahedron 43:2789–2803

Udeinya IJ (1993) Antimalarial activity of Nigerian neem leaves. Trans R Soc Trop Med Hyg 87:471

Utz V (1993) Geographic distribution, cultivation and propagation of neem. In: Kleeberg H (ed) Practice orientated results on use and production of neem ingredients and pheromones. Proceedings of the 5th Neem Workshop. Druck and Graphic, Giessen, pp 13–18

Van der Esch SA, Gianacovo G, Maccioni O, Vitali F (1993) Preliminary results on the production of azadirachtin by plant tissue culture of *Azadirachta indica*. G Bot Ital 127:927–928

Van der Esch SA, Giagnacovo G, Maccioni O, Vitali F (1994) Plant tissue culture of *Azadirachta indica* (A. Juss). In: Kleeberg H (ed) Practice orientated results on use and production of neem ingredients and pheromones. Proc 3rd Neem Workshop. Druck and Graphic, Giessen, pp 125–131

Van der Esch SA, Giagnacovo G, Maccioni O, Vitali F (1996) Production of the natural insecticide azadirachtin in suspension cultures of *Azadirachta indica* (A. Juss). Abstr Phytochem Soc Eur Meeting, Rolduc, The Netherlands, "Future Trends in Phytochemistry

Wewetzer A (1996) Etablierung von in vitro Kulturen aud *Azadirachta indica* (A. Juss) und Untersuchungen zu deren Azadirachtinproduktion. PhD Thesis, Humboldt University, Berlin

Wewtezer A (1997) Tissue culture of *Azadirachta indica* (A. Juss) and its potential for the production of azadirachtin. In: Kleeberg H, Zeibitz CP W (eds) Practice orientated results on use and production of neem ingredients and pheromones. Druck and Graphic, Giessen, pp 243–249

Wewetzer A, Schulz FA (1994) Plant regeneration from tissue culture of *Azadirachta indica* (Juss). 8th Int Congr Plant tissue and cell culture S1-28, p 12 (Abstr)

Wickramananda IR (1997) Insect antifeedant, growth regulatory and insecticidal properties of tissue cultures and plant extracts of neem (*Azadirachta indica* A. Juss). MPhil Thesis, University of Aberdeen, UK

Zanno PR, Miura I, Nakanishi K, Elder DL (1975) Structure of the insect phagorepellent azadirachtin. Application of PFRT/CWD carbon-13 nuclear magnetic resonance. J Am Chem Soc 97:1975–1977

III *Centranthus* Species: In Vitro Culture and the Production of Valepotriates and Other Secondary Metabolites

P. Christen

1 Introduction

1.1 The Plant

The family Valerianaceae comprises 13 genera and about 360 species (Evans 1996). Genera include *Valeriana* (over 200 spp.), *Valerianella* (80 spp.), *Patrinia* (20 spp.) and *Centranthus* (12 spp.). The Valerianaceae are annual to perennial herbs with opposite leaves and flowers which are usually zygomorphic.

Whereas numerous investigations on the genus *Valeriana* have been carried out (Houghton 1988), the genus *Centranthus* has been considerably less studied. According to the Flora europea (Tutin et al. 1976), the plants belonging to this genus are glabrous, usually glaucous, annual or rhizomatous perennial herbs with erect, usually unbranched flowering stems. The flowers are hermaphrodite or unisexual and form dense inflorescences. The corolla is made of five usually unequal lobes and the tube is cylindrical or infundibuliform, gibbous near the middle, or spurred near the base and with an internal longitudinal membrane from the insertion of the spur to the mouth. *Centranthus ruber* (L.) DC (synonym *Valeriana rubra* L.) is the common red valerian. It is 30–80 cm high, ascending, usually unbranched. The leaves are mostly lanceolate to ovate, obtuse to acuminate, the uppermost sometimes irregularly dentate and amplexicaul. The corolla may be red, pink or white. The species grows on walls and in rocky places around the Mediterranean basin and in Portugal. It is also frequently cultivated as an ornamental plant and is widely naturalized. There are two subspecies: ssp. *ruber* and spp. *sibthorpii*. The morphology and the flowers of *C. ruber* (L.) DC can be seen in Fig. 1.

1.2 Chemical Constituents and their Medicinal Importance

The roots of *Valeriana officinalis* have a long history of use as a sedative medicine in Europe, but the first report on a pharmacological investigation of

University of Geneva, Laboratory of Pharmaceutical Analytical Chemistry, 20, Bd. d'Yvoy, CH-1211 Geneva 4, Switzerland

Biotechnology in Agriculture and Forestry, Vol. 43
Medicinal and Aromatic Plants XI (ed. by Y.P.S. Bajaj)

Fig. 1. Flower heads of mature naturally occurring *Centranthus ruber* DC. (Photographed by Christen in the Botanical Garden, Geneva, Switzerland, June 1997)

C. ruber was only published in 1963 (Paris and Moury 1963). The authors demonstrated that the roots present an activity which lies between sedation and tranquilization.

It is generally accepted that valepotriates are the compounds responsible for the sedative activity of the Valerianaceae. However, other groups of secondary metabolites existing in the plants, like the volatile cyclopentane-sesquiterpenes and the small amount of alkaloids present in some species, may contribute to the pharmacological action. The scientific basis for using valerian as a mild sedative has not been elucidated. Reports on the activity of aqueous extracts have again raised the question as to whether some other groups of compounds contribute to the pharmacological effects (Leathwood et al. 1982; Leathwood and Chauffard 1985). However, the chemical investigation of the Valerianaceae has concentrated on the two major groups of constituents, the sesquiterpenes of the volatile oil and the valepotriates.

Valepotriates are esterified iridoid-monoterpenes discovered by Thies and Funke (1966) in *Valeriana wallichii*. As indicated by the name valepotriate

(*Val*eriana-*epo*xy-*tri*ester), these compounds are triesters of otherwise unstable polyhydroxycyclopenta-(c)-pyrans with acetic, isovaleric, β-methylvaleric, α-isovaleroxy-isovaleric, α-isovaleroxy-isovaleric, β-acetoxy-isovaleric, β-acetoxy-β-methylvaleric and β-hydroxyisovaleric acids (Thies et al. 1981). These compounds are divided into two classes: the monoene and the diene derivatives. A comprehensive review on the valepotriates isolated from the Valerianaceae has been published by Houghton (1988). With regard to their structures, valepotriates are very sensitive to temperature, acids and alkali. Baldrinal is a degradation product of valtrate and acevaltrate whereas homobaldrinal is a degradation product of isobaldrinal. The structures of some valepotriates frequently encountered in *C. ruber* are shown in Fig. 2.

Another group of secondary plant products, the essential oils, is also present in the roots of different Valerianaceae and consist of a wide variety of components: terpenes, esters, ethers, aldehydes, ketones, alcohols and acids (Gränicher et al. 1995a). Some of the oil components were suggested to have sedative properties, e.g. bornyl isovalerate, some kesso compounds, valeranone.

Valepotriates and essential oil are not only present in the subterranean parts of the plants but also in the aerial parts (Funke and Friedrich 1975; Hölzl and Jurcic 1975). The amount of valepotriates varies widely between species and genera. Becker et al. (1983) reported the highest amount (14.5% w/w) in the fresh roots of *V. thalictroides* Graebn. The underground parts contain normally higher amounts of valepotriates than the other parts of the plant. Mannetstätter et al. (1967) first reported the presence of valepotriates in the roots of *C. ruber* but no essential oil was detected. A survey of various valepotriates, isolated from *Valeriana* species and from *C. ruber*, has been published by Hazelhoff (1984).

2 In Vitro Culture Studies

Despite the fact that the active constituents, responsible for the sedative activity, are not known with precision and that the scientific basis for the use of valerian and related species has not been completely elucidated, in vitro cell and tissue cultures of the Valerianaceae have generated a growing interest. A comprehensive review on this topic has been published by Becker and Chavadej (1988). Numerous registered preparations containing extracts of *Valeriana edulis*, *V. officinalis* and *V. wallichii* are on the market in Europe, mainly in Germany, in the Netherlands and in Switzerland. These extracts are obtained from plants grown in the field and, therefore, their supply is limited and subject to climatic variations. That is why cell and tissue cultures may represent an interesting alternative to field-grown plants.

Diene valepotriates

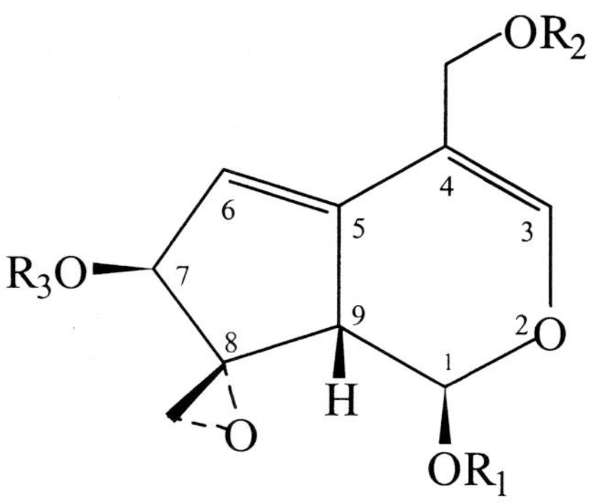

	R_1	R_2	R_3
VAL	IV	Ac	IV
IVAL	IV	IV	Ac
DIA-VAL	IV	Ac	Ac
HVAL	IV	Ac	MV

Monoene valepotriates

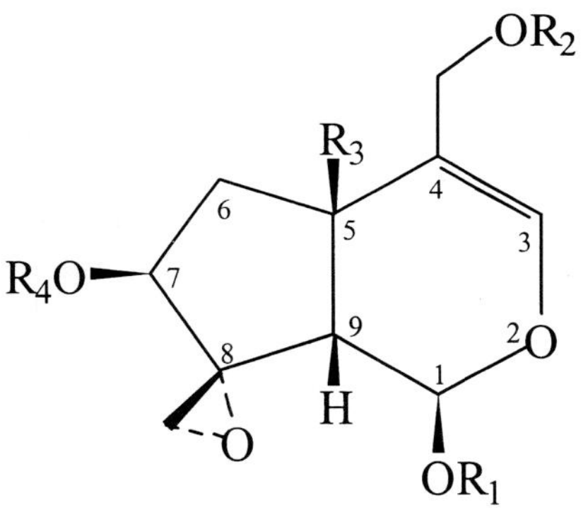

	R_1	R_2	R_3	R_4
DI	IV	IV	H	Ac
IVHD	IV	IV-IV	OH	Ac

Abbreviations :

Ac = acetyl
IV = isovaleryl
IV-IV = α - isovaleryloxy - isovaleryl
MV = β - methyl - valeryl

Fig. 2. Valepotriate structures frequently encountered in *Centranthus ruber*. (Gränicher et al. 1995c)

2.1 Callus and Suspension Cultures and Secondary Metabolites

The first report on the in vitro production of valepotriates was published by Becker et al. (1977) and concerned the establishment of callus culture of *V. wallichii* DC. It was demonstrated that the valepotriate content is qualitatively similar to that of the roots of the parent plants. A review on the formation of sesquiterpenes and valepotriates by in vitro culture of various *Valeriana* species has been published recently (Gränicher and Christen 1995).

Becker and Schrall (1980) reported on the tissue cultures of nine species belonging to the Valerianaceae, among them *C. ruber*. It was the first report on the biosynthesis of valepotriates by in vitro callus culture of the red valerian. Calli were initiated from seeds and maintained on agar MS medium (Murashige and Skoog 1962) supplemented with 5 mg/l 1-naphthylacetic acid (NAA) and 1 mg/l kinetin (kin). They were subcultured every 3 to 4 weeks and were kept in 12 h/day light at 22 ± 2 °C, as described by Schrall and Becker (1979). Calli produced valtrate, acevaltrate, isovaltrate and didrovaltrate, the first two compounds being in higher quantity. The culture failed to produce IVHD-valtrate. A few years later, Violon et al. (1983a) reported on the callus cultures of *Valeriana officinalis*, *Valerianella locusta* and *C. ruber* with the aim of determining the production of valepotriates under various culture conditions. The callus culture of *C. ruber* was optimized and the following medium was selected: solid (10 g/l agar) B5 medium (Gamborg 1970), containing 10 g/l sucrose, 2 mg/l glycine and 1 mg/l NAA. Furthermore, the production of callus culture of *C. ruber* was compared with a root organ culture of the same species obtained on solid B5 medium supplemented with 2 mg/l glycine, 1 mg/l NAA and 0.2 mg/l kin. The cultures were kept at 28 °C in the dark. Valtrate/isovaltrate, acevaltrate and didrovaltrate were quantified in the differentiated and undifferentiated tissues as well as in the growth medium. The valepotriate content of the root culture was about 17 times higher (4.2% dry wt.) than in the callus culture (0.25% dry wt.). Furthermore, no valepotriates were detected in the culture medium, indicating that the valepotriates accumulate in the tissues and are not excreted extracellularly. These results demonstrated that light does not seem to be an indispensable condition for valepotriate production, at least in this species. However, it was not clear whether the influence of the plant hormone composition was exerted on the production of valepotriates or whether the degree of valepotriate production was a consequence of differentiation.

Violon et al. (1983b) performed a microscopical study on *V. officinalis* and *C. macrosiphon* in order to localize valepotriates in the tissues. The plants were grown from seeds whereas callus cultures of both species and root organ cultures of *C. macrosiphon* were grown on solid media, differing in plant hormone content (Violon et al. 1983a). The roots of *V. officinalis* and *C. macrosiphon* contained 0.25 and 3.50% dry wt. of valtrate/isovaltrate, respectively and the in vitro cultures of *V. officinalis* contained 2.5% dry wt. of the same compounds. Callus and root organ cultures of *C. macrosiphon* contained 0.1 and 5% dry wt. of valtrate/isovaltrate, respectively. The microscopic analysis revealed the presence of a large quantity of lipid droplets of different sizes in the in vivo roots, as well as in the root organ cultures. These droplets were found to a much greater extent in the *C. macrosiphon* root than in the *V. officinalis* roots and were not confined to the hypodermis and root cortex but distributed all over the parenchyma, from the cortex to the vascular bundles. Furthermore, cultures with high valepotriate content showed more lipid droplets than the callus cultures with lower valepotriate content. The droplets were isolated by differential centrifugation and extracted by dichloromethane. The extract was analyzed by thin layer chromatography and compared with an extract of the intact *C. macrosiphon* in vitro root cultures. The valepotriate pattern was the same in both cases. Furthermore, the volatile oil, obtained by steam distillation from the roots of *C. macrosiphon* as well as from the in vitro cultures, contained only lower fatty acids, mainly isovaleric acid. The authors explained the presence of the essential oil and the valepotriates in the same lipid droplets by showing their chemical, lipophilic and biosynthetic relationship.

In order to study the influence of plant growth regulators, Violon et al. (1984a) set up callus cultures of *C. macrosiphon*, *C. ruber* and *V. officinalis* on solid B5 medium containing 10 g/l sucrose and 2 mg/l glycine. Two auxins, NAA and 2,4-dichlorphenoxyacetic acid (2,4-D) and one cytokinin, kinetin were tested in four media, always at a concentration of 1 mg/l as follows: medium I, NAA; medium II, NAA + 2,4-D + kin; medium III, NAA + 2,4-D; and medium IV, NAA + kin. The calli, initiated from sterile seedlings, were subcultured every 5 weeks and incubated in the dark at 26 °C. All callus cultures grew well on the four media tested, except those of *C. ruber* which did not grow on media II and III. The growth, differentiation level and valepotriate content were evaluated for 1.5 years. Furthermore, the valepotriate production was studied with the callus culture of *C. macrosiphon* during one growth period between two subcultures. These results were compared with those obtained with control plants grown in a greenhouse.

At the first subcultivation passage, the cultures of *C. ruber* showed formation of calli, shoots and mostly roots. The valepotriate content, calculated as the sum of valtrate/isovaltrate, acevaltrate and didrovaltrate, was stable and high (2%) in comparison with the roots of control plants (1.2%). After 50 weeks of culture on medium IV, root differentiation disappeared gradually, leaving well-growing calli with a lower valepotriate content.

The cultures of *C. macrosiphon* grown on media I and IV, which did not contain 2,4-D, showed many roots and shoot formation. Shoots disappeared promptly and the cultures became clearly root differentiated cultures with a high valepotriate content (5–8%). On media II and III, well-growing calli were formed and maintained thereafter. However, valepotriate content was much lower than on media I and IV and tended to decrease regularly during further subcultures (from 1 to 0.1%). The growth and the valepotriate production in *C. macrosiphon* were investigated in detail during one growth period (5 weeks) on the four different media. Media I, II and III were equivalent for the increase of biomass (fresh wt.), whereas the amount of dry wt was higher on the media without 2.4-D (I and IV). It was found that the valepotriate yield was much higher in the root differentiated tissue cultures (media I and IV) than in the callus cultures (media II and III). A stable, high valepotriate content was maintained during 14 weeks in differentiated root organ cultures.

Little is known about the volatile constituents from the genus *Centranthus*. A comparative study of the essential oils of in vivo and in vitro grown *V. officinalis* and *C. macrosiphon* was undertaken by Violon et al. (1984b). The establishment of the callus cultures of both plants and the optimization of the growth media were carried out as described by Violon et al. (1984a). The essential oils obtained from the roots of *C. macrosiphon*, grown in vivo and analysed by GC-MS, contain only lower fatty acids with large amounts of isovaleric acid, valeric acid and a third unidentified compound which could represent a thermal degradation of valepotriates. Furthermore, the analyses of the in vitro cultures (callus and root differentiated tissue cultures) present profiles similar to that of the roots grown in vivo.

In view of the different results reported above, it appears that a possible way of increasing yields of valepotriates by in vitro cultures would be to allow for a differentiation of tissues by establishing root cultures. In this context, the production of 'hairy root' by infection with *Agrobacterium rhizogenes* offers an attractive area for experimentation.

2.2 Hairy Root Cultures

2.2.1 Establishment of Hairy Root Cultures

Bacterial Strain. The supervirulent *Agrobacterium rhizogenes* strain R1601 (Pythoud et al. 1987) was used for the induction of hairy roots. The strain was grown on YMB solid medium (Hooykaas et al. 1977) supplemented with 100 mg/l kanamycin in the dark at 26 °C, as described by Gränicher et al. (1992).

Plant Material. Seeds of *Centranthus ruber*, obtained from Jelitto Staudensamen (Schwarmstedt, Germany), were surface sterilized for 15 min in a 2% sodium hypochlorite solution supplemented with two drops of Triton X-100 (Fluka, Switzerland), rinsed three times with sterile distilled water and left for germination on sterile wet filter paper at 25 °C in the light (Osram-L-Fluora 77R; 700 lx). Two-week-old plantlets were again sterilised in the same solution, rinsed and transferred on solid MS medium supplemented with 2% sucrose (MS-2 solid medium). Seeds of the same origin were sown in the field under normal conditions and 8-month-old untransformed roots were harvested for the purpose of comparison.

Transformation with Agrobacterium rhizogenes. Stems of eight-week-old plantlets were inoculated with *A. rhizogenes* strain R1601 and kept in the light at 26 °C. Five to 6 weeks after infection, hairy roots appeared at the infection sites. The tips of the roots were cut off and transferred to MS-2 solid medium, containing 0.25 g/l cefotaxime and 1 g/l ampicilline (Sigma, USA) to eliminate the excess of bacteria. The axenic hairy roots thus obtained were maintained in hormone-free B5 liquid medium supplemented with 3% sucrose (B5-3) and subcultured at 4 week intervals. All cultures were incubated in darkness on a rotary shaker at 80 rpm.

2.2.2 *Opine Assay*

Agrobacterium rhizogenes R1601 is a modified A4 strain (pTVK 291; Pythoud et al. 1987) leading to the production of agropine and mannopine. These opines are specifically found in *Agrobacterium*-mediated transformed tissues (crown gall and hairy roots). To prove the transformation, the opines were extracted according to the method described by Petit et al. (1983) and identified by paper electrophoresis (Otten and Schilperoort 1978). The electropherogram was visualized with an alkaline silver nitrate reagent (Trevelyan et al. 1950). Silver nitrate-positive compounds were identified through comparison with authentic agropine and mannopine.

2.2.3 *Preparation of N,N-Dimethylmorpholinium Iodide (DMI)*

DMI was prepared according to the method described by Schölly (1989). The purity was checked by NMR and mass spectrometry.

2.2.4 *Extraction, Identification, and Quantification of Valepotriates*

The powdered, lyophilized hairy roots were extracted essentially as described by Gränicher et al. (1992). Liquid media and roots of untransformed 8-month-old plants were extracted using the same procedure. Valtrate (VAL), isovaltrate (IVAL), 7-desisovaleroyl-7-acetylvaltrate (DIA-VAL), 7-homovaltrate (HVAL), didrovaltrate (DI) and isovaleroxyhydroxydidrovaltrate (IVHD) were identified by comparing their physical constants and spectral data with those of authentic samples. These had been isolated, in our laboratory, from a dichloromethane extract of *Centranthus ruber* untransformed roots, as described previously (Gränicher et al. 1995c). The valepotriates were quantified by HPLC using the simultaneous determination of the monoene and diene derivatives in a single HPLC run (Gränicher et al. 1994). Analyses were performed on a stainless steel Nucleosil C-18 column (25 × 0.4 cm; 5 µm; Macherey-Nagel, Germany) fitted with a guard column (3 × 0.4 cm; 5 µm). An isocratic methanol-water mixture (69–31) was used as the mobile phase at a flow rate of 0.7 ml/min for 10 min then at 1.4 ml/min for 30 min. The detection was performed at 208 nm and ethylbenzene was used as the internal standard. Two calibration curves were established, with

VAL for the diene-type valepotriates and DI for the monoene-type valepotriates.

2.2.5 *Optimization of the Culture Medium*

The effects of 12 different culture media, which had been supplemented with 2–7% sucrose, on the growth and on the content of six valepotriates were examined after 40 days and compared with the roots of 8-month-old non-transformed plants grown in the field (Fig. 3). The media tested were one-quarter, one-half and full-strength B5 (1/4 B5, 1/2 B5 and B5, respectively), one-half and full-strength Heller (1/2 H and H), one-half and full strength MS (1/2 MS and MS), and one-half and full-strength McCown's woody plant (1/2 WP and WP) (Lloyd and McCown 1980). All media were hormone-free and adjusted to pH 5.9 before autoclaving.

In all the media tested, the transformed roots produced a spectrum of valepotriates which qualitatively mirrored that of untranformed roots. The fastest growth (4.2 g fresh wt) was observed in 1/2 B5 and B5, both media being supplemented with 3% sucrose, but the latter medium led to a poor valepotriate content (0.6% dry wt). The highest concentration of valepotriates was observed in 1/2 and 1/4 strength B5-3 media and reached 3.0% dry wt. This value was in the same range as the valepotriate content of 8-month-old untransformed roots (3.4% dry wt.). VAL and HVAL were the main valepotriates produced by the hairy roots and by the untransformed roots. Half strength B5 medium supplemented with 3% sucrose was particularly favourable for the biosynthesis of DIA-VAL (0.5% dry wt.), whereas this compound was detected in a small amount only in the untransformed roots. The IVAL content was unaffected by the various culture media tested and remained very low. Heller, WP and MS media were suitable neither for hairy root growth, nor for valepotriate biosynthesis.

2.2.6 *Addition of DMI*

Förster and Becker (1987) demonstrated that the use of plant bioregulators, added at concentrations of 0.01 to 0.02 mmol during early exponential growth, may significantly improve terpenoid secondary metabolite production in cell cultures. The effect of feeding DMI on growth and valepotriate content was investigated over a period of 40 days. A sterile aqueous solution of DMI at various concentrations (0.5–50 ppm) was added to the hairy root cultures on day 0 of cultivation. Table 1 shows that all the concentrations of DMI reduced significantly the valepotriate content, as well as the hairy root growth.

It is noteworthy that the decrease of the valepotriate content is related to the increase of the DMI concentration. These results are not in line with those reported by Förster and Becker (1987) for the cell suspension cultures of *Valeriana wallichii* and *Fedia cornucopiae*.

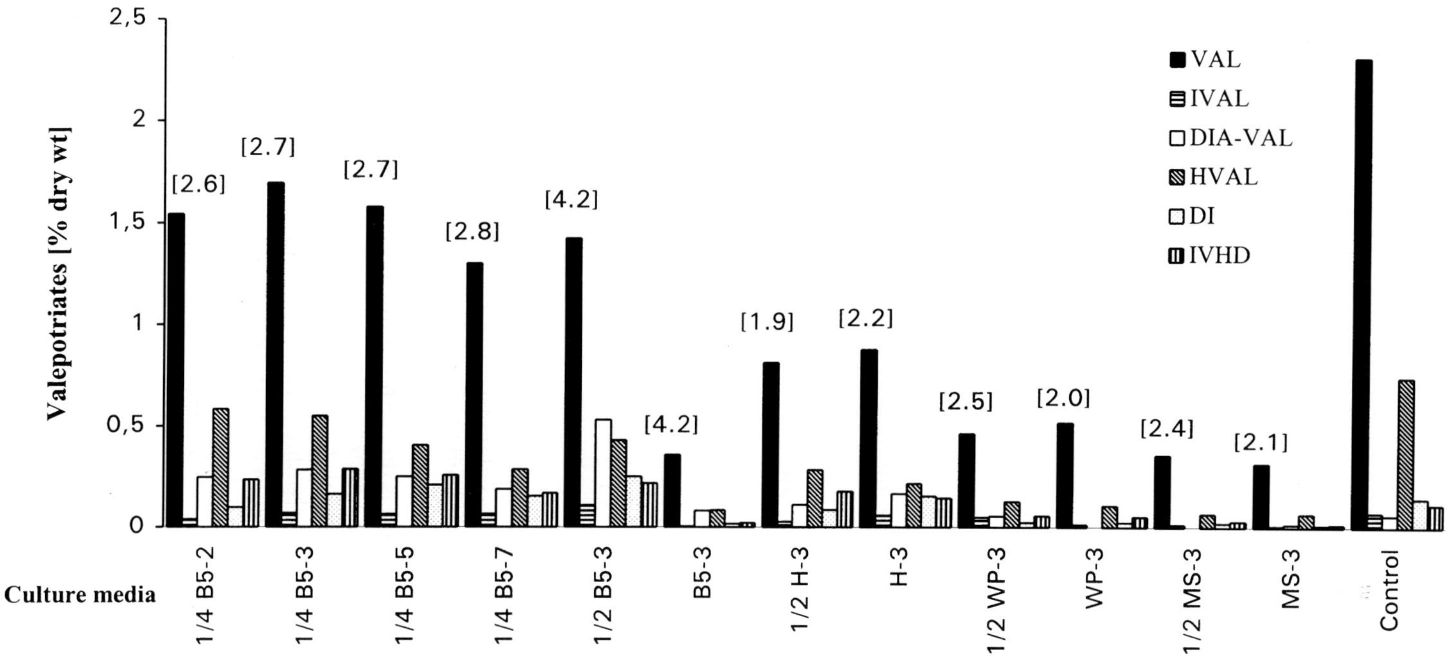

Fig. 3. VAL, IVAL, DIA-VAL, HVAL, DI and IVHD contents of *Centranthus ruber* hairy roots cultured in different liquid media for 40 days. Untransformed plants (control) were grown in the field during 8 months. *Numbers in brackets* indicate the fresh weight (g). (Gränicher et al. 1995c)

Table 1. Fresh weight and valepotriate content of *Centranthus ruber* hairy roots, cultured for 40 days in 1/2 B5-3 liquid medium supplemented with different concentrations of DMI. Valepotriate content is the sum of VAL, IVAL, DIA-VAL, HVAL, DI and IVHD. Each value is the mean of three separate determinations. (Gränicher et al. 1995c)

DMI content (ppm)	Valepotriate content (% dry wt.)	Fresh weight (g)
Control (no DMI added)	3.0	4.2
0.5	1.0	3.7
1	0.8	3.4
2.5	0.6	2.3
5	0.4	2.2
10	0.3	2.1
25	0.3	2.2
50	0.3	2.1

2.2.7 *Time-Course Study*

In a further experiment, the kinetics of root growth and valepotriate production, in 1/2 B5 medium supplemented with 3% sucrose, was investigated for 50 days (Fig. 4).

Three flasks were harvested at 5-day intervals, and growth and valepotriate production were measured. During a 45-day culture period, the fresh weight of the hairy roots increased from an original inoculum of ca. 50 mg to reach 4.2 g (growth index: 84). The maximum biomass was reached after 35 days of culture. On a mg/g dry wt basis, the overall valepotriate content reached 31.4 mg/g, representing a mean accumulation rate of 0.7 $mg^{-1}g^{-1}day^{-1}$. Between day 10 and day 25, the growth rate was maximum and reached 80% of the final fresh wt. VAL, DIA-VAL and HVAL were the major constituents of the valepotriate mixture synthetized by the hairy roots. The VAL content increased rapidly and regularly between the 10th and the 45th day of culture to reach 1.5% dry wt. DIA-VAL and HVAL contents increased continually between the 20th and the 40th day of culture to reach a constant level of about 0.5 and 0.4% dry wt., respectively. DI, IVHD and IVAL were also detected but their levels remained low. The lack of accumulation of these valepotriates may be due to other biosynthetic pathways competing for common precursors or to a very low activity of one or more of the relevant biosynthetic enzymes. The medium was also examined for the presence of valepotriates. During the first 45 days of culture, neither valepotriates nor their decomposition products were detected in the culture medium. This suggests that all valepotriates were retained within the tissues. After 45 days of culture, the hairy roots turned brown and died, and the valepotriates were released into the culture medium.

The results reported in this study differ significantly from those obtained with the hairy root cultures of *Valeriana officinalis* var. *sambucifolia*. In the

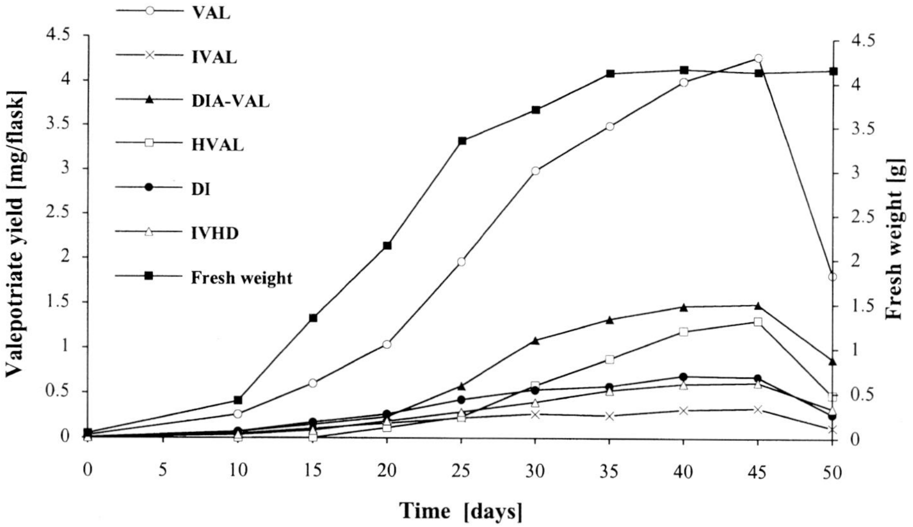

Fig. 4. Time-course study of growth and valepotriate production in *Centranthus ruber* hairy roots cultured in 1/2 B5-3 liquid medium. (Gränicher et al. 1995c)

latter case, the transformation by *A. rhizogenes* strain R1601 led to important modifications of the metabolic pathways. The spectrum of valepotriates was qualitatively and quantitatively modified. VAL was the most abundant valepotriate in the hairy roots, whereas it was IVAL in the untransformed roots. IVHD, a minor compound in the untransformed roots, reached 2.3% dry wt. in the hairy roots. DIA-VAL and a new iridoid compound, named valdiate, were detected in the hairy roots only (Gränicher et al. 1995b). On the other hand, the hairy roots of *Centranthus ruber* biosynthesized a spectrum of valepotriates which mirrored quantitatively and qualitatively that of the untransformed roots. The ratio VAL/IVAL and the monoene valepotriate content were not affected by the transformation. However, some similarities between the transformed roots of both species have to be reported. The best yield in valepotriates was recorded, in both species, with hairy roots cultured either in 1/4 B5 or in 1/2 B5 liquid media, the lowest yield occurring in MS liquid medium. Furthermore, 1/2 or 1/4 strength B5 liquid medium supplemented with 2 to 3% sucrose is the most appropriate for valepotriate biosynthesis in hairy root cultures of both species.

3 Summary and Conclusion

In vitro culture was applied to *Centranthus ruber* and *C. macrosiphon*. It appears that, in relation to production, aging of the culture is important only because it correlates with differentiation changes. Valepotriate production is inhibited or suppressed only if differentiation is simultaneously suppressed by the plant hormone content of the medium. Tissue cultures also demonstrated that the valepotriates and the essential oil are confined to lipid droplets which are located in the roots and distributed all over the parenchyma.

The *Agrobacterium*-mediated transformation of *C. ruber* is used for rapidly obtaining hairy roots with valepotriates which qualitatively and quantitatively mirror those of the parent plants. By optimization of the culture medium, it was possible to obtain the fastest growth (4.2 g fresh wt.), in 1/2 B5 medium supplemented with 3% sucrose, and the highest production of valepotriate (3% dry wt.), in 1/4 and 1/2 B5 medium with 3% sucrose. The valepotriates are confined within the tissues and are not excreted extracellularly. The addition of DMI as a plant bioregulator to the culture medium led to a quite surprising result. Compared with the control, all the DMI concentrations added significantly reduced the hairy root growth, as well as the valepotriate content.

In conclusion, it appears that *Centranthus* hairy root and callus lines present an interesting field of investigation. Presently, it is not clear whether callus cultures are able to synthesize the key enzymes which are involved in the valepotriate biosynthesis or whether the corresponding genes are missing.

References

Becker H, Chavadej S (1988) Valepotriates: production by plant cell cultures. In: Bajaj YPS (ed) Biotechnology in agriculture and forestry, vol 4. Medicinal and aromatic plants I. Springer, Berlin Heidelberg New York, pp 294–309

Becker H, Schrall R (1980) Valepotriates in tissue cultures of nine different Valerianaceae species in comparison to literature data of the intact plants. J Nat Prod 43:721–723

Becker H, Schrall R, Hartmann W (1977) Calluskulturen von Baldrian-Arten, 1. Mitt. Anlage einer Calluskultur von *Valeriana wallichii* DC. und erste analytische Untersuchungen. Arch Pharm (Weinheim) 310:481–484

Becker H, Chavadej S, Weberling F (1983) Valepotriates in *Valeriana thalictroides*. Planta Med 49:64

Evans WC (1996) Trease and Evans'pharmacognosy, 14th edn. Saunders, London, pp 50, 323–325

Förster W, Becker H (1987) Effects of plant bioregulators on the production of iridoid derived terpenoids in *Valeriana wallichii* and *Fedia cornucopiae* cell suspension cultures. Z Naturforsch [c] 42:33–40

Funke ED, Friedrich H (1975) Valepotriate in oberirdischen Organen weiterer Arten der Valerianaceen. Planta Med 28:215–224

Gamborg OL (1970) The effects of amino-acids and ammonium on the growth of plant cells in suspension culture. Plant Physiol 45:372–375

Gränicher F, Christen P (1995) *Valeriana* species: in vitro culture and the formation of sesquiterpenes and valepotriates. In: Bajaj YPS (ed) Biotechnology in agriculture and forestry, vol 33. Medicinal and aromatic plants VIII. Springer, Berlin Heidelberg New York, pp 442–458

Gränicher F, Christen P, Kapetanidis I (1992) High-yield production of valepotriates by hairy root cultures of *Valeriana officinalis* L. var. *sambucifolia* Mikan. Plant Cell Rep 11:339–342

Gränicher F, Christen P, Vuagnat P (1994) Rapid high performance liquid chromatographic quantification of valepotriates in hairy root cultures of *Valeriana officinalis* L. var. *sambucifolia*. Phytochem Anal 5:297–301

Gränicher F, Christen P, Kapetanidis I (1995a) Essential oils from normal and hairy roots of *Valeriana officinalis* var. *sambucifolia*. Phytochemistry 40:1421–1424

Gränicher F, Christen P, Kamalaprija P, Burger U (1995b) An iridoid diester from *Valeriana officinalis* var. *sambucifolia* hairy roots. Phytochemistry 38:103–105

Gränicher F, Christen, P, Kapetanidis I (1995c) Production of valepotriates by hairy root cultures of *Centranthus ruber* DC. Plant Cell Rep 14:294–298

Hazelhoff B (1984) Phytochemical and pharmacological aspects of valerian compounds. PhD Thesis, University of Groningen

Hölzl J, Jurcic K (1975) Valepotriates in the leaves of *Valeriana jatamansii*. Planta Med 27:133–139

Hooykaas PJJ, Klapwijk PM, Nuti MP, Schilperoort RA, Rörsch A (1977) Transfer of the *Agrobacterium tumefaciens* Ti plasmid to avirulent *Agrobacteria* and to *Rhizobium* ex planta. J Gen Microbiol 98:477–484

Houghton PJ (1988) The biological activity of Valerian and related plants. J Ethnopharmacol 22:121–142

Leathwood PD, Chauffard F (1985) Aqueous extract of valerian reduces latency to fall asleep in man. Planta Med 51:144–148

Leathwood PD, Chauffard F, Heck E, Munoz-Box R (1982) Aqueous extract of valerian root (*Valeriana officinalis* L.) improves sleep quality in man. Pharmacol Biochem Behav 17:65–71

Lloyd G, McCown B (1980) Commercially-feasible micropropagation of mountain laurel *Kalmia latifolia*, by use of shoot-tip culture. Proc Int Plant Soc 30:421–427

Mannetstätter E, Gerlach H, Poethke W (1967) Über die Inhaltsstoffe von Valerianaceen. 1. Mitt.: Der Nachweis einiger Inhaltsstoffe von *Kentranthus ruber* D.C. Pharm Zentralhalle 106:797–804

Murashige T, Skoog F (1962) A revised medium for rapid growth and bio-assays with tobacco tissue cultures. Physiol Plant 15:473–497

Otten LABM, Schilperoort RA (1978) A rapid micro scale method for the detection of lysopine and nopaline dehydrogenase activities. Biochim Biophys Acta 527:497–500

Paris R, Moury J (1963) Sur l'action sédative du Centranthe (*Centranthus ruber* DC). Therapie 18:1187–1192

Petit A, David C, Dahl GA, Ellis JG, Guyon P, Casse-Delbert F, Tempé J (1983) Further extension of the opine concept: plasmids in *Agrobacterium rhizogenes* cooperate for opine degradation. Mol Gen Genet 190:204–214

Pythoud F, Sinkar VP, Nester EW, Gordon MP (1987) Increased virulence of *Agrobacterium rhizogenes* conferred by the vir region of pTiBo542: application to genetic engineering of poplar. Biotechnology 5:1323–1327

Schölly T (1989) Culture in vitro de *Drosera anglica*. Etude phytochimique de *Drosera rotundifolia*. PhD Thesis, University of Geneva

Schrall R, Becker H (1979) Valepotriate in Callus- und Suspensionskulturen von verschidenen Valarianaceen. Herba Hung 18:75–83

Thies PW, Funke S (1966) Nachweis und Isolierung von sedativ wirksamen Isovaleriansäureestern aus Wurzel und Rhizomen von verschiedenen Valeriana- und Kentranthus-arten. Tetrahedron Lett 1155–1162

Thies PW, Finner E, David S (1981) XIV. Konstitutive Zuordnung der Acyloxy-substituofenten in Valepotriaten via C-13 NMR-Spektroskopie. Planta Med 41:15–20

Trevelyan WE, Procter DP, Harrison JS (1950) Detection of sugars on paper chromatograms. Nature 166:444–445

Tutin TG, Heywood VH, Burges NA, Moore DM, Valentine DH, Walters SM, Webb DA (1976) (eds) Flora europea, vol 4, Cambridge University Press, Cambridge

Violon C, Van Cauwenbergh N, Vercruysse A (1983a) Valepotriate content in different in vitro cultures of Valerianaceae and characterization of *Valeriana officinalis* L. callus during a growth period. Pharm Weekbl [Sci] 5:205–209

Violon C, Dekegel D, Vercruysse A (1983b) Microscopical study of valepotriates in lipid droplets of various tissues from valerian plants. Plant Cell Rep 2:300–303

Violon C, Dekegel D, Vercruysse A (1984a) Relation between valepotriate content and differentiation level in various tissues from Valerianofeae. J Nat Prod 47:934–940

Violon C, Sonck W, Vercruysse A (1984b) Comparative study of the essential oils of in vivo and in vitro grown *Valeriana officinalis* L. and *Centranthus macrosiphon* Boiss. by coupled gas chromatography-mass spectrometry. J Chromatogr 288:474–478

IV *Costus speciosus* (Koenig) J.E. Smith: In Vitro Cultures, Micropropagation, and the Production of Diosgenin and other Phytosteroids

G. INDRAYANTO, W. UTAMI, and A. SYAHRANI

1 General Account

1.1 The Plant

Costus speciosus (Koenig) J.E. Smith (Cs; syn. *Costus loureiri* Horan.; *Amomum arboreum* Lour.; *Banksia speciosa* Koenig) is a herbaceous plant of the Zingiberaceae family, growing throughout South and Southeast Asia (Inoue et al. 1995; Wijayakusuma 1992). This plant has been used for various medicinal purposes in that region. Koenig described that Cs cultivated in Malaya (in 1778) was used for medicinal purposes (cited by Burkill 1966). The plant has various local names, e.g.: Pacing (Java); Tabar-tabar (Sumatra); Lingkuas (Celebes); Setawar (Malay); Zhiang liu tou (China); Mia do (Vietnam); Co trong bon (Thailand) and Spiral ginger (English) (de Padua et al. 1981; WHO 1990; Wijayakusuma 1992).

It is an erect herbaceous plant about 1–2 m high and 1.5 cm in diameter. The leaves are spirally arranged oblong, acuminate, subsessile, soflty pubescent on the lower surface and about 30 cm long. The inflorescence is located at the apex of a normal leafy stem, and in the dense terminal spike, flowers are solitary, with a corolla which is white or tinged with purple, with red bracts. The rhizomes are stout and creep horizontally (Fig. 1A–D). The fruits are few to numerous, capsule, ellipsoid, finely sericeous-pubescent, red, 1–2 cm, and crowned by narrow calyx that is shorter than or as long as the fruit itself. There is a thick pericarpium and the black seed is about 3 mm in diameter, see Fig. 2 (Backer and van der Brink 1968; WHO 1990; de Padua et al. 1981). Our experience in Surabaya showed that only Cs plants that have purple flowers produced seeds. Cs can be easily propagated through rhizome and stem cuttings (Sarin et al. 1974; Nagendra and Abraham 1981).

Cs showed intraspesific chromosomal races, namely, diploidy ($2n = 2x = 18$), triploidy ($2n = 3x = 27$), and tetraploidy ($2n = 4x = 36$). A few cells in the diploids and tetraploids contained the triploid number of chromosomes, suggesting inconstancy in chromosome complements in somatic tissues. The diploids were distinguished from the triploids and tetraploids on account of their smaller leaves (Nagendra and Abraham 1981).

Laboratory of Pharmaceutical Biotechnology, Faculty of Pharmacy Airlangga University, Jl. Dharmawangsa dalam, Surabaya 60286, Indonesia

Biotechnology in Agriculture and Forestry, Vol. 43
Medicinal and Aromatic Plants XI (ed. by Y.P.S. Bajaj)

Fig. 1. **A** *Costus speciosus* general appearance of the plants. **B** Inflorescence at apex of normal leafy stems, **C** rhizomes, **D** solitary flower, with white corolla and red bracts

The rhizome of Cs is edible, and one-third of its dry weight is starch. As a food it is fairly nutritious, but it contains a large proportion of fiber. Boiled rhizomes are mucilaginous, feebly astringent but non aromatic (Burkill 1966).

1.2 Phytosteroids Content

According to Hegnauer (1986) the genus *Costus* is rich in sterols and steroidal saponins. Das Gupta and Pandey (1970) first reported the presence of diosgenin in the rhizome of Cs. Indrayanto et al. (1994) reported that the

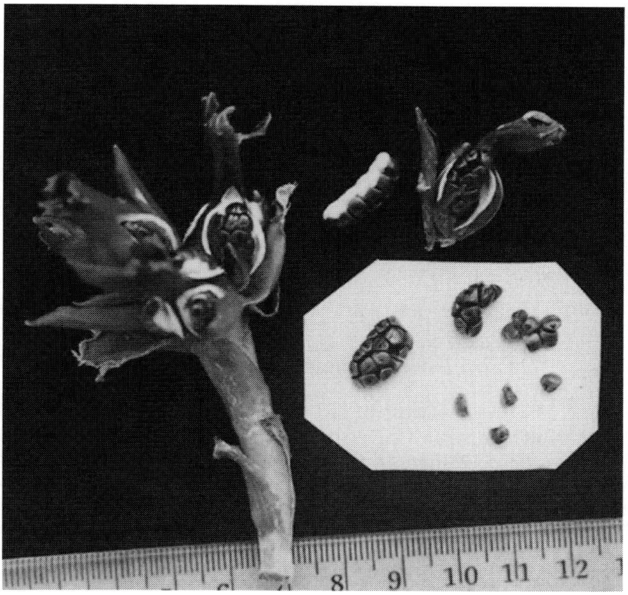

Fig. 2. Fruit and seeds of *Costus speciosus* (with purple flowers)

diosgenin content in roots, stem, flowers and leaves of Cs was very low, compared with the content in the rhizomes (Fig. 3A). Diosgenin content in the rhizomes was about 0.1–2% DW (dry weight) whilst its content in the seeds was 1–3% DW (Sarin et al. 1974; Sudiarto et al. 1985; Abdulah 1987; Lambert et al. 1988). The content of diosgenin of the rhizomes of Cs growing in the campus of Airlangga University was 0.3–0.7% DW (Setiawan 1993). According to Sudiarto et al. (1985), cultivation of the plants per hectare yielded 15–30 tons of fresh rhizomes and 1–15 tons of fresh seeds, whereas Singh et al. (1980) reported 40 tons of fresh rhizomes and 700 kg of fresh seeds from 1 ha of plants.

Eight spirostanol and furostanol glycosides were isolated and identified from the rhizome of Cs, whilst the seeds contained four spirostanol and five furostanol glycosides (Tschesche and Pandey 1978; Singh and Thakur 1982a,b; Inoue et al. 1995). Gupta et al. (1983) isolated four saponins from the roots of cs. The chemical structures of Cs saponins are presented in Table 1 and Fig. 4A.

It was already known that spirostanol glycosides were formed from furostanol glycosides during postharvest treatment and storage. In Cs plants, protogracillin is converted into gracillin by a β-glucosidase that accumulates in the rhizomes (Inoue et al. 1996). Inoue and Ebizuka (1996) described the purification and characterization of the enzyme involved (furostanol glycoside 26-O-β-glucosidase = F26G). The purified F26G was dimeric with a native apparent molecular weight of 110000, daltons and consisted of subunits of 54000 and 58000. This enzyme was highly specific for cleavage of the C-26-bound glucose moiety of furostanol glycoside, showing K_m for protogracillin of 50 μM. In rhizomes of Cs, which are known to accumulate furostanol and

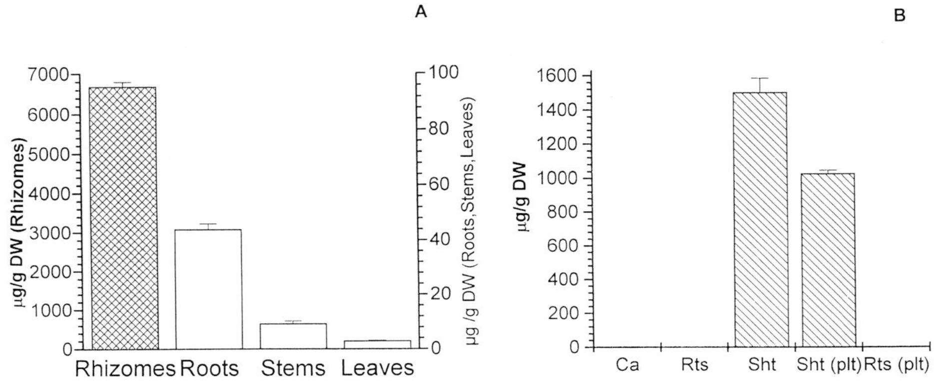

Fig. 3. A Diosgenin content of *Costus speciosus* (Cs) plants (with white flowers) growing in the campus of Airlangga University. **B** Diosgenin content of the in vitro cultures of Cs. *Ca* callus cultures, *Rts* root cultures, *Sht* shoot cultures, *Sht (plt)* shoot leaves of the plantlet cultures; *Rts (plt)* roots of the plantlet cultures. (Data from Indrayanto et al. 1994)

Table 1. Saponins of *Costus speciosus*

Type of saponins	Roots	Rhizomes	Seeds
Spirostanol	Prosapogenin A of dioscin **(1)**	Prosapogenin B of dioscin **(2)**	Prosapogenin A of dioscin **(1)**
	Prosapogenin B of dioscin **(2)**	Saponin S[a] **(5)**	Prosapogenin B of dioscin **(2)**
	Dioscin **(3)**	Dioscin **(3)**	Dioscin **(3)**
	Gracillin **(4)**	Gracillin **(4)**	Gracillin **(4)**
Furostanol		Methyl protogracillin **(6)**	Costusoside-I **(9)**
		Methyl protodioscin **(7)**	Costusoside-J **(10)**
		Protogracillin **(8)**	Protodioscin **(11)**
		Saponin F1[b] **(12)**	Saponin F2[c] **(13)**
			Methyl protodioscin **(7)**
Sterols	β-Sitosterol-β-D-glucoside **(14)**	β-Sitosterol-β-D-glucoside **(14)**	β-Sitosterol-β-D-glucoside **(14)**

Numbers in parentheses refer to the structures of the compounds presented in Fig. 4.
[a] Saponin S: Diosgenin 3-O-β-glucopyranosyl)(1→3)-β-glucopyranoside.
[b] Saponin F1: 26-O-β-glucopyranosyl-(25R)-furost-5-ene-3β,22ξ,26 triol.
[c] Saponin F2: 3-O-[α-L-rhamnopyranosyl(1→2)-β-D-glucopyranosyl],26-O-[β-D-glucopyranosyl]-22α-methoxy-(25R)-furost-5-en-3β,26-diol.

spirostanol glycosides, F26G activity was observed (62.6 nkat/mg protein), while other parts of the plants, which do not accumulate the glycosides, did not show any detectable F26G activity. Whole parts of the in vitro cultured plantlets showed this glucosidase activity (36.5, 47.7 and 16.4 nkat/mg protein for roots, stems and leaves, respectively, Inoue et al. 1996).

β-Sitosterol-β-D-glucoside was also isolated from the roots, rhizomes and seeds (Tschesche and Pandey 1978; Singh and Thakur 1982b; Gupta et al.

1983). Susilo (1992) reported that rhizomes that were grown in Surabaya contained about 0.02–0.03% DW of sterols, mostly as stigmasterol and β-sitosterol. The structures of the sterols are presented in Fig. 4B.

1.3 Traditional Therapeutic Uses, Bioactivity, and Metabolism

A daily dose of 10 to 20 g in the form of decoction of the rhizomes of Cs is used in the treatment of fever, urodynia, biliuria, rheumatism, lumbago and neuralgia, whilst the juice of pounded fresh tops of young branches is instilled in the ear for otitis (WHO 1990). The rhizomes can be used also for nephritic edema, urinary tract infection, eczema, urticaria (Hong Kong Chinese Medical Research Institute 1991), worms, dyspepsia, and skin diseases (Bandara et al. 1988). In the Philippines, powdered rhizomes are applied to wounds, bruises (to reduce swelling) and also to mumps and cancerous swellings. The leaves are applied externally for sore throat (de Padua 1981). In Indonesia, Cs is used for treatment of nephritic edema, ascites, dysuria, liver-cirrhosis and pertussis (Wijayakusuma 1992).

Mosihuzzaman et al. (1994) reported that the freeze-dried juices of the rhizomes of Cs plants, grown in the eastern Himalayan belt, have a significant hypoglycemic effect in rats (after glucose loading). According to Singh et al. (1972) the saponins of Cs showed an oestrogenic activity in rats.

Thewles et al. (1993)) showed that biliary cholesterol output in rats was stimulated over three fold by feeding diosgenin for 5 days, whereas biliary outputs of phospholipid and bile salts were not changed by diosgenin feeding. Hemolytic and antifungal activities of diosgenyl mono glycoside (Takechi and Tanaka 1991) and diosgenyl diglycosides (Takechi et al. 1992) have been reported. Recently, the same authors also described the activities of diosgenyl-β-D-xylopyranosyl-(1->6)-β-D-glucopyranoside as haemolytic and antifungal (Takechi et al. 1997), whilst diosgenin did not show these bioactivities. The rhizome also contained an antifungal principle that was identified as a methyl ester of para-coumaric acid (Bandara et al. 1988).

The metabolism of diosgenin and its glycosides in ruminants were extensively studied by Miles et al. (1995). Firstly the saponins were rapidly hydrolyzed, then the double bond at C-5 was reduced. Next, they were epimerited to smilagenin, and finally metabolized into epismilagenin (Fig. 5). This last metabolite precipitated as glucoronide calcium salts to form crystalloid deposits. Recently Meagher (1996) reported that dosing of a diosgenin-derived saponins mixture from Cs to lamb, revealed that ruminal conversion to episimilagenin was relatively low. This author also described the availability of three distinct zones of metabolic activity in the gastrointestinal tract of the lamb dosed with saponins extract from Cs. In the rumen, diosgenin was converted to tigogenin, smilagenin, smilagenone and epismilagenin; the duodenum, the jejunum and ileum formed an absorption zone where there was variation in levels of both free and conjugated sapogenins; sapogenins were concentrated and epimerized in the caecum and large intestine, in a manner similar to metabolism in the rumen.

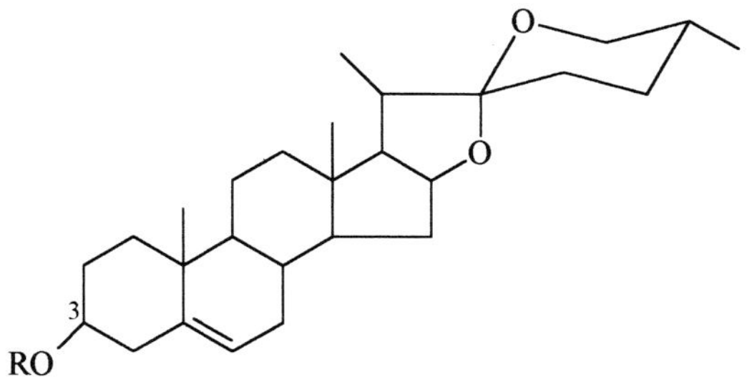

Compound	R
Diosgenin	H
1	Rha - ^{2} Glc -
2	Rha - ^{4} Glc -
3	Rha ^{4}, Rha ^{2} > Glc -
4	Glc ^{3}, Rha ^{2} > Glc -
5	Glc - ^{3} Glc -

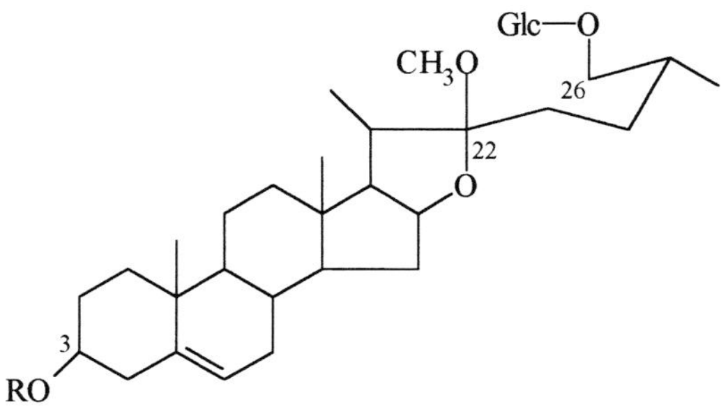

Compound	R
6	Glc ^{3}, Rha ^{2} > Glc -
7	Rha ^{4}, Rha ^{2} > Glc -
9	Rha ^{4}, Glc-^{2}Rha ^{2} > Glc -

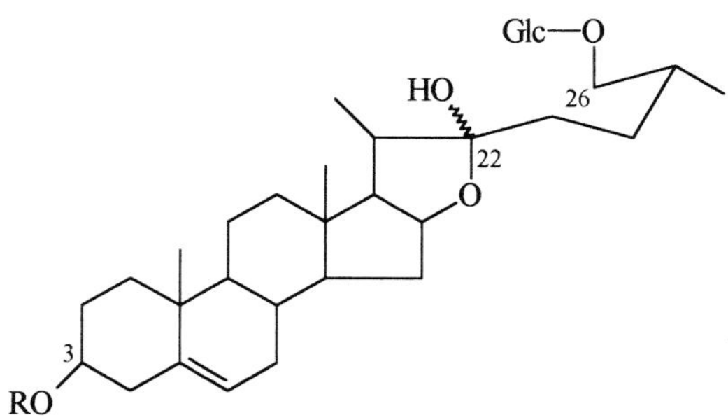

Compound	R
8	Glc ^{3}, Rha ^{2} > Glc -
10	Rha ^{4}, Glc-^{2}Rha ^{2} > Glc -
11	Rha ^{4}, Rha ^{2} > Glc -
12	H
13	Rha - ^{2} Glc -

A

Fig. 4. **A** The chemical structure of the saponins. The names of the compounds 1–13 are presented in Table 1. **B** The chemical structure of the sterols and squalene (19). The name of the compounds are, β-sitosterol-β-D-glucoside (14), cholesterol (15); campesterol (16); β-sitosterol (17); stigmasterol (18); and lanosterol (20)

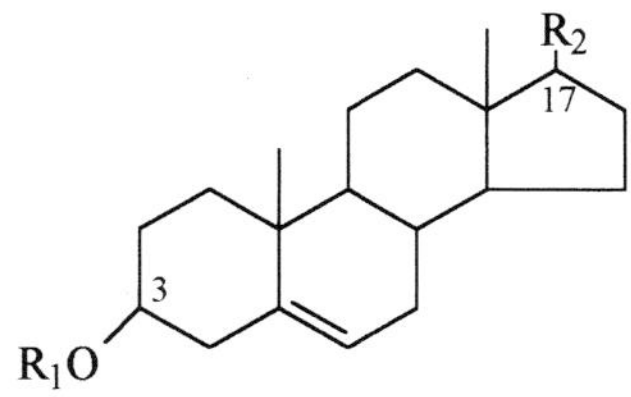

Compound	R1	R2
14	Glc -	
15	H	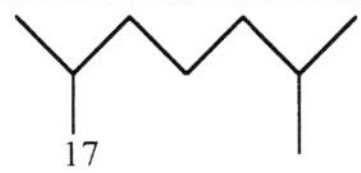
16	H	
17	H	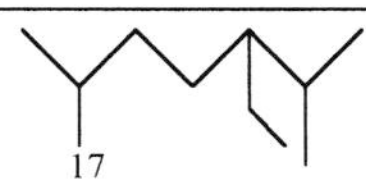
18	H	

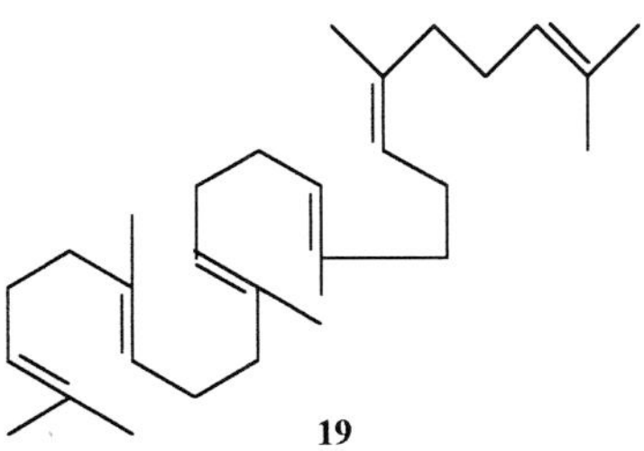

19

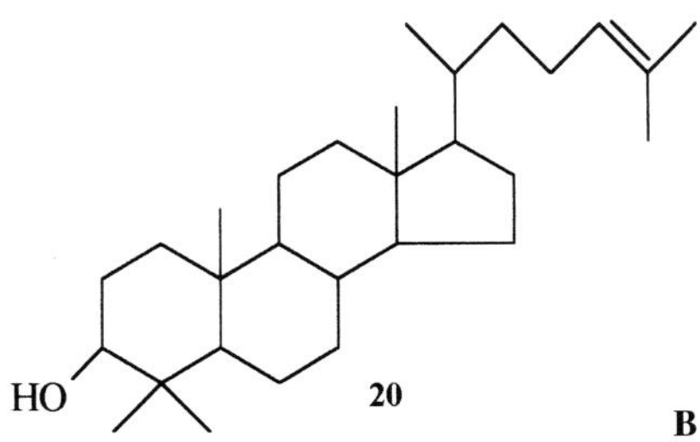

20

B

Fig. 4. *Continued*

2 In Vitro Culture Studies

Several studies have been conducted on various aspects of in vitro cultures of Cs. A summary of these works is presented in Table 2. Although a number of sapogenin steroids and sterols have been isolated, identified and assayed from the in vitro cultures of Cs, steroid glycosides have not yet been reported.

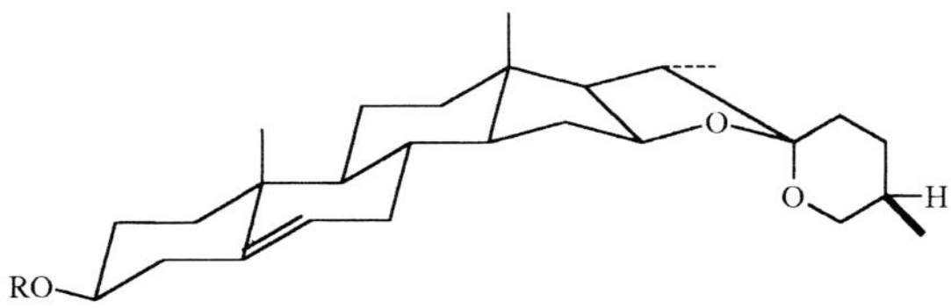

Compound	R
Diosgenin glycosides	Sugars
Diosgenin	H

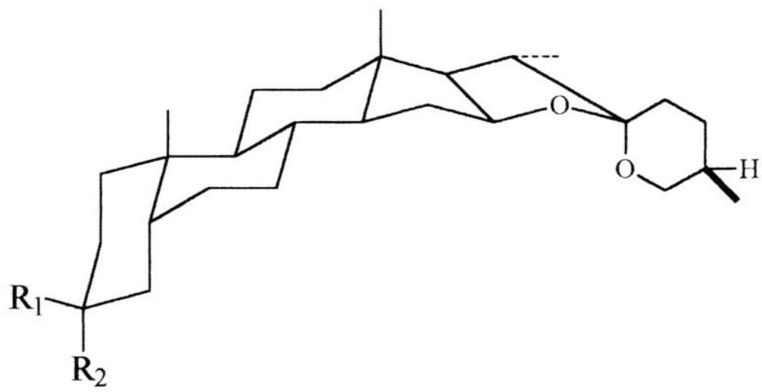

Compound	R_1	R_2
Smilagenin	OH	H
Epismilagenin	H	OH
Smilagenone	O	
Epismilagenin β-D-glucoronide	H	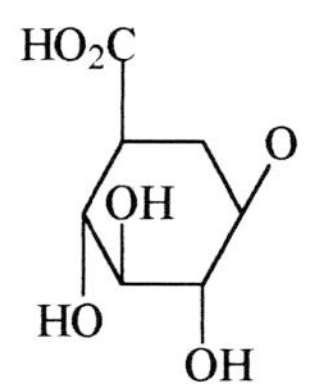

Fig. 5. Chemical structure of the ruminal metabolites of diosgenin

2.1 Establishment of Tissue Cultures, Micropropagation, and Regeneration

In our laboratory, the shoot cultures were first initiated from some clusters of Cs (codes F1, F2, F3, F7 and F8) plants growing in the campus of Airlangga University (Setiawan 1993). The in vitro cultures (Fig. 6A-C, 8A) were initiated from young stems (shoots), using MS medium (Murashige and Skoog 1962) modified with the addition of 2 mg/l BAP (medium B_2 for shoot cultures); 5 mg/l IBA (medium I_5 for root cultures); 1 mg/l NAA and 0.2 mg/l

Table 2. Summary of the in vitro studies conducted on *Costus speciosus*

Authors	Culture system used	Observations/remarks
Rathore and Khanna (1979)	Callus cultures	Isolation and identification of sapogenins and sterols
Pal and Sharma (1982)[a]	Callus cultures	Callus induction
Indrayanto (1983)	Suspension cultures	Isolation and identification of sterols and squalene
Chaturvedi et al. (1984)	Shoot cultures	Clonal multiplication
Jain and Chaturvedi (1985)	Callus cultures	Organogenetic differentiation
Setiawan (1993)	Callus and organ cultures	Quantitative determination of diosgenin and sterols
Indrayanto et al. (1994)	Callus and organ cultures	Differentiation diosgenin accumulation
Chairunnisa (1994)	Shoot cultures	Effect of casein on diosgenin content
Nuryati (1994)	Shoot cultures	Effect of sucrose on diosgenin content
Emiliawati (1994)	Shoot cultures	Effect of calcium on diosgenin content
Roy and Pal (1995)	Callus cultures	Morphogenetic differentiation
Utami et al. (1997)	Callus and organ cultures	Effect of elicitors on phytosteroids content
Kustiawan (1997)	Shoot cultures	Effect of salicylic acid on diosgenin content
Wartono (1997)	Shoot cultures	Effect of salicylic acid on diosgenin content

[a] Cited from George et al. (1987).

kinetin (medium $N_1K_{0.2}$ for plantlet cultures)and 0.2 mg/l kinetin and 1 mg/l 2,4-D (medium $K_{0.2}D_1$ for callus cultures. Three percent sucrose and 0.7% agar were added in all above media. In her work, Chairunnisa (1994) added 2–3 g casein hydrolysate to the media B_2 and found that the GI (growth index) increased significantly. Recently, Utami et al. (1997) has initiated root cultures from roots of the shoot cultures of Cs (code F8) by using medium MS modified by the addition of 2 mg/l IBA (medium I_2). She also initiated suspension cultures by using MS medium modified with the addition of 2 mg/l kinetin and 0.5 mg/l 2,4-D (medium $K_2D_{0.5}$). The growth rates of the in vitro cultures of Cs (codes F1 and F8) are presented in Fig. 7. This figure shows that the organ cultures of Cs have a relatively low growth index (GI). Callus was also induced from seed embryos of Cs by using a combination of 3 mg/l NAA and 2 mg/l kinetin (George et al. 1987).

Almost 90% of the Cs shoot cultures that used the basal (lower part) of shoots as inoculating explants, cultivated on medium B_2, could grow roots after 6–7 weeks of cultivation. Recently Wartono (1997) and Kustiawan (1997) initiated shoot cultures from the top parts of shoots (F1 and F8), but these new shoot cultures (F8-1 and F1-1) did not show root formation after 4–5 weeks. Simultaneous shoot and root formation were observed when the basal shoots were inoculated on medium $N_1K_{0.2}$ (plantlet cultures), but the rate of shoot

Fig. 6. Root (**A** on medium I_2), callus (**B** on medium $K_{0.2}K_1$) and cell suspension (**C** on medium $K_2D_{0.5}$) cultures of *Costus speciosus*, code F8. All cultures were maintained in continuous light (1500 lx) at 25 ± 1 °C

formation was lower compared with the shoot cultures. About 75% of the plantlets (code F3; 8 week old cultures) cultivated on medium B_2 survived and grew well after transplanting to common trays containing sterilized humus. After 4 weeks of acclimatization, the small plants were ready to be transferred to soil (Fig. 8B–C). These results showed that micropropagation of Cs can be achieved by using only one step of cultivation (without transferring the shoots to a rooting medium). Our recent unpublished results using shoot cultures (code F7) showed that by lowering the hormone BAP from 2 mg/l (medium B_2) to 0.4 mg/l (medium $B_{0.4}$), an almost fourfold increase in GI was achieved. Chaturvedi et al. (1984) reported on the clonal propagation of Cs. For shoot proliferation, they used a modified SH medium (Schenk and Hildebrandt 1972), supplemented with 0.5 mg/l BAP or 1 mg/l kinetin, 1 mg/l IAA and 15 mg/l adenine sulfate (AdS). A shoot tip regenerated about 10 adventitious

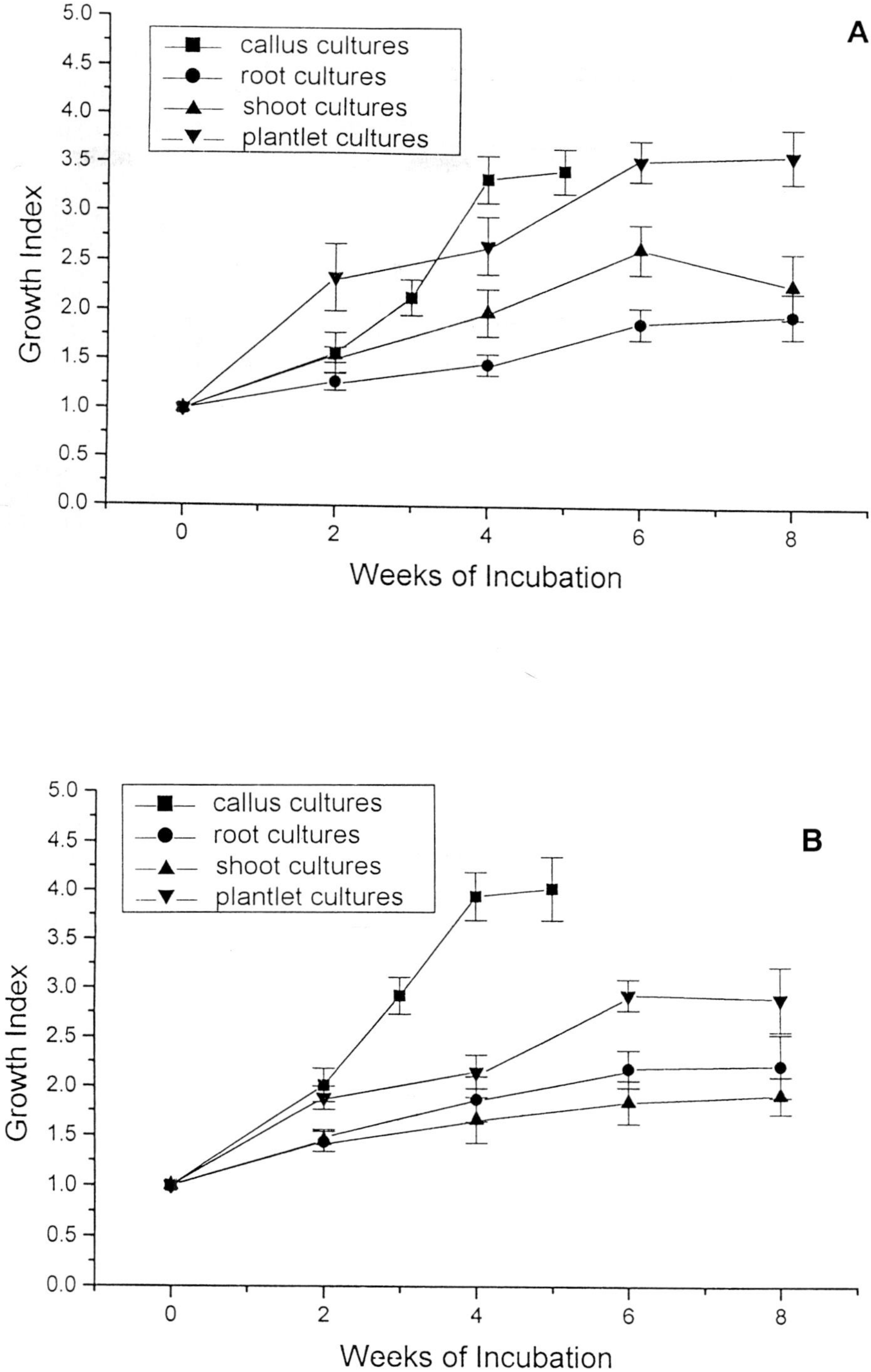

Fig. 7. Growth curves of the in vitro cultures of *Costus speciosus*, code F1 **A** and F8 **B** on medium B_2 (1500 lx at 25 ± 1 °C). (Data from Setiawan 1993)

Fig. 8. A Shoot cultures of *Costus speciosus* (Cs) (F8; 6 week old cultures) on medium B_2. **B** Young plants of Cs (code F3), 3 weeks after transplanting into sterilized humus for acclimatization. **C** Cs plants (F3) 2 weeks after transferring to soil

shoots after 60 days. The excised shoots rooted 100% in SH basal medium supplemented with 1 mg/l IAA.

Jain and Chaturvedi (1985) reported the formation of shoots from calli by using an SH basal medium supplemented with 2 mg/l IAA, 15 mg/l AdS, 200 mg/l malt extract and 0.25 mg/l BAP. The shoots were then rooted on medium with the addition of 3 mg/l IAA. After initiation of three different types of callus (soft, nodular and hard) from two tetraploid populations of Cs, Roy and Pal (1995) reported that morphogenetic differentiation was observed only from the hard type of callus. They also reported the histological evidence to clarify the mode of shoot differentiation. The differentiation was achieved by sequential lowering of the growth hormone regulators, abrupt withdrawal of the growth hormone regulators (combinations of IAA and BAP or NAA and BAP) or by addition of cytokinin and lowering of auxin in the medium.

2.2 Phytosteroid Content of the In Vitro Cultures

Rathore and Khanna (1979) described the identification (by TLC, GLC and IR) of steroidal sapogenins (diosgenin and tigogenin) and sterols (stigmasterol and lanosterol) from the callus cultures of Cs. Indrayanto (1983) reported that two cell lines of suspension cultures contained cholesterol (trace), campesterol (trace), sitosterol and stigmasterol, whilst diosgenin was not detected (identified by TLC, GLC and GC-MS). Squalene (identified by GLC, GC-MS), a precursor of phytosteroid, was also identified in these cultures.

Table 3. The diosgenin and sterols content of the in vitro cultures of *Costus speciosus* (mean ± SD, n = 4)[a]

In vitro cultures	Code	Medium	Free sterol (μg/g DW)	Bound sterol (μg/g DW)	Diosgenin (μg/g DW)
Callus	F1	$K_{0.2}D_1$	201 ± 13.1	87.5 ± 7.6	–
	F2		243.9 ± 10.5	112.4 ± 7.6	–
	F8		300.4 ± 7.2	57.2 ± 1.6	–
Roots	F1	I_5	132.8 ± 4.4	129.0 ± 6.8	–
	F2		157.8 ± 6.2	144.4 ± 6.6	–
	F8		190.3 ± 9.1	68.5 ± 0.9	–
Shoots	F1	B_2	131.4 ± 7.4	236.2 ± 1.2	2493.7 ± 64.2
	F2		119.0 ± 4.4	330.3 ± 12.1	1640.6 ± 75.0
	F8		123.7 ± 3.9	88.6 ± 4.9	1502.0 ± 88.1
Shoot leaves of plantlet cultures	F1	$N_1K_{0.2}$	142.7 ± 11.3	256.2 ± 4.4	1357.6 ± 32.2
	F2		149.5 ± 8.8	261.6 ± 16.4	1171.2 ± 27.1
	F8		120.7 ± 8.1	115.6 ± 7.9	1028.3 ± 17.3
Roots of the plantlet cultures	F1		172.8 ± 9.9	126.7 ± 4.4	–
	F2		116.6 ± 2.9	98.9 ± 6.3	–
	F8		138.3 ± 10.1	69.1 ± 2.7	–

[a] Data cited from Setiawan (1993).

From the various types of in vitro cultures of Cs that were grown in our laboratory, only shoot cultures and shoot-leaves of the plantlet cultures accumulated diosgenin (identified by TLC and GC-MS). In callus cultures, root cultures and roots of the plantlet cultures, diosgenin was not detected (Fig. 3B). The relatively high diosgenin content in shoot cultures and shoot-leaves of the plantlet cultures compared with the leaves and stems of the in vivo plants may be due to the lack of an accumulating organ (rhizome). This result showed that the active site of sapogenin steroid biosynthesis was the shoot and leaves (Indrayanto et al. 1994).

Setiawan (1993) reported that in all in vitro cultures of Cs, free and bound sterols were detected. The diosgenin and sterols content of several cell lines of in vitro cultures that were grown in our laboratory are presented in the Table 3 (data cited from Setiawan 1993). In the suspension cultures of Cs (code F8) the main free sterol and bound sterol (sterol glycosides) was sitosterol, whilst diosgenin was not detected. In this study the structures of the sterol glycosides were not investigated. The maximum sterol content was achieved in 12 days of cultivation. The characterization of this suspension cultures over a culture period of 14 days, together with the time courses of the sterol accumulation, are presented in Fig. 9 (Muhammad 1998).

2.3 Quantification of the Phytosteroids

Diosgenin, sterol glycosides (bound sterols) and free sterols were extracted from the dried biomass of the in vitro cultures using a previously published method (Indrayanto et al. 1993). The diosgenin and sterols content can be

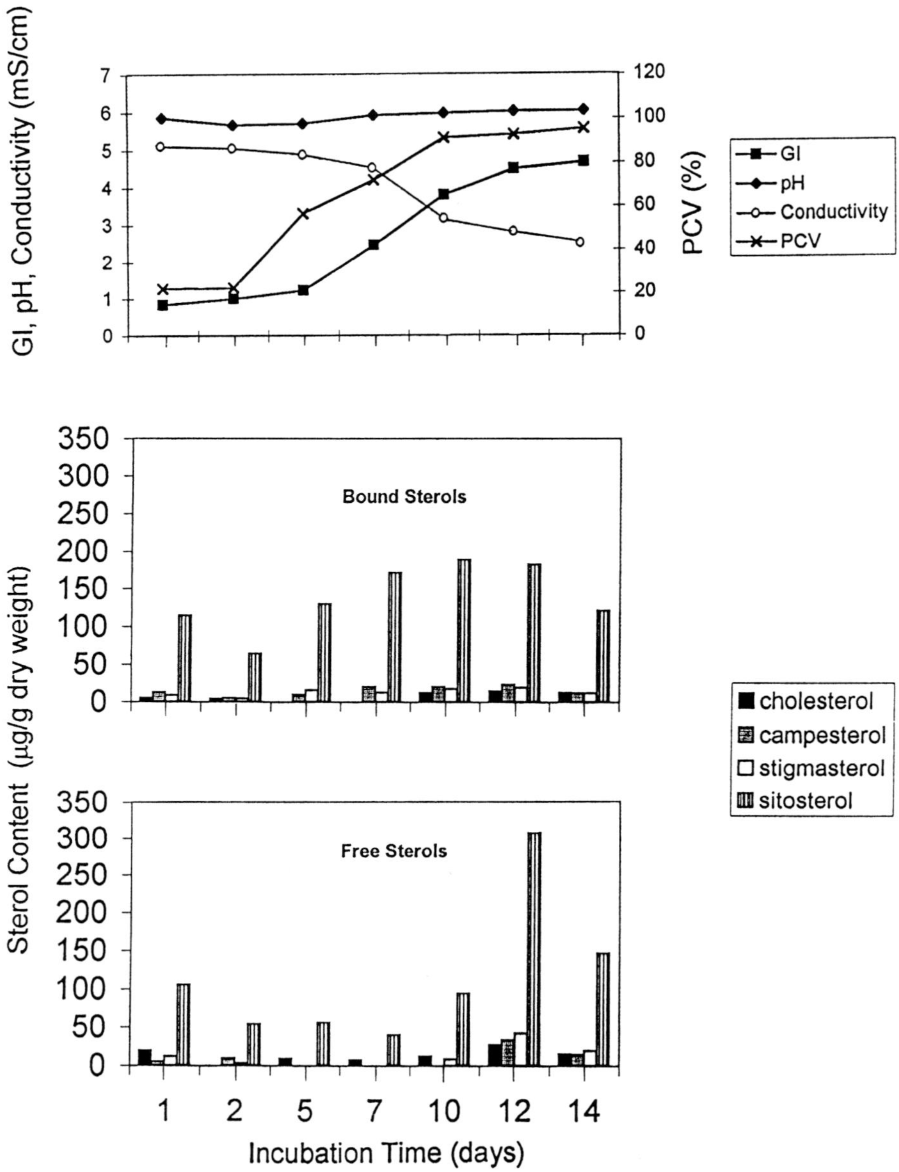

Fig. 9. Characterization (*upper panel*) of suspension cultures of *Costus speciosus* (code F8) in medium K_2 $D_{0.5}$ and the time courses of sterol accumulation (*middle* and *lower panels*). (Data from Muhammad 1998)

determined simultaneously by densitometry using Kieselgel $60F_{254}$ precoated plates (E. Merck) and n-hexane: ethyl acetate (3:1) as the eluent. The diosgenin (yellow-orange spot; $R_f = 0.36$) and sterols (violet; $R_f = 0.45$) were detected by anisaldehyde-H_2SO_4 reagent (ca. 10 min, 100 °C). Quantification was done by measuring the maximum absorbance reflectance at 427 nm (diosgenin) and 397 nm (sterols). Diosgenin and cholesterol (Sigma) were used as external standards. By this method, linearity was achieved over a concentration of 0.01 to 12 µg/spot (for diosgenin; n = 10; r = 0.0993) and 0.4 to 10 µg/spot (for cholesterol; n = 8; r = 0.997); the detection limit (DL) was 0.004 µg/spot (diosgenin) and 0.016 µg/spot (cholesterol); the quantification limit (QL) was 0.012 µg/spot and 0.053 µg/spot (cholesterol); accuracy with the standard addition method was 98.86 ± 0.65% (diosgenin; mean ± SD; n = 5) and 98.47 ± 0.76% (cholesterol; mean ± SD, n = 5). In this case, the sterols content in the biomass was calculated as cholesterol (Setiawan 1993; Indrayanto et al. 1994). To determine each individual sterols such as cholesterol, campesterol, stigmasterol, sitosterol etc., HPLC or GLC methods were essential (Indrayanto et al. 1998). Susilo (1992) reported the determination of diosgenin and content in the rhizomes of various *Costus* spp. by an HPLC method. He used a Partisil ODS 3 (Whatman) column and a mixture of methanol: distilled water (95:5) was used as the mobile phase. The injection volume was 20 µl, and the separation was performed at 40 °C. Detection was performed by using UV 205 nm. By this method, linearity was achieved from 2.5 to 199 µg/ml (diosgenin; n = 7; r = 0.999) and from 5.0 to 200 µg/ml (cholesterol; n = 6; r = 0.999); DL were 0.041 µg/peak (diosgenin) and 0.032 µg/peak (cholesterol). When the biomass also contains the C_{25} epimer of diosgenin (yamogenin), an HPLC method according to Wu and Wu (1991) is recommended. They used a Zorbax-ODS column (DuPont) and a mixture of MeOH: water (94:6) as the mobile phase. Separation was performed at O °C. For detection, a refractive index detector RI 401 (Waters) was used.

Miles et al. (1994) and Meagher (1996) developed a selected ion monitoring (SIM) method using a GC-MS system equipped with a 20 × 0.25 mm i.d. HP-1 methyl silicone capillary column (Hewlett Packard) for quantitative analysis of sapogenins and sterols. For the internal standard she used 3-deoxysarsapogenin. The GC column was temperature programmed from 200 °C (0.5 min hold) to 250 °C at 35 °C/min, and then to 285 °C at 2 °C/min (10 min hold). Total ion and a series of selected ions of m/z 139, 396, 282 (for spirost-5-ene-3-ol) and 368, 370, 386, 394, 396, 398 and 414 (for sterols) mode of acquisitions were undertaken.

2.5 Effect of Some (Bio) Chemicals on the Growth and Phytosteroids Content of the In Vitro Cultures

Although addition of 2–3 g/l casein hydrolysate could increase the GI of shoot cultures of Cs (code F3) significantly, its diosgenin content remained relatively unchanged (Chairunnisa 1994). Depletion and reduction of the sucrose concentration in the media did not affect the growth rate and diosgenin content of the shoot cultures F3. By using 60 g/l sucrose in the medium, its diosgenin content was increased by almost 2.5-fold (Nuryati 1994). The diosgenin content of the same cultures was also increased almost 1.6-fold by using modified MS medium without calcium ions, whilst its GI was relatively unchanged (Emiliawati 1994). The increase of other sapogenin (hecogenin) and alkaloid contents by depletion of calcium ions in the medium has already been reported (Sierra 1991; Indrayanto et al. 1993).

Recently, Utami et al. (1997) studied the effect of using a relatively high concentration of some metal ions (2.5 mg/l $CuSO_4.5H_2O$; 2.5 mg/l $CoCl_2.6H_2O$; 2.2 g/l $CaCl_2.2H_2O$; 1.85 g/l $MgSO_4.7H_2O$; 223 mg/l $MnSO_4.H_2O$; and 4 g/l $SrCl_2.6H_2O$) in media on the growth rate and diosgenin content of the shoot cultures (code F8). Figure 10 shows that the growth rate of shoot cultures on media containing a high concentration of metal ions (Ca, Mg, Sr and Mn) were relatively low compared with cultures on modified MS medium, supplemented with 1 mg/l BAP and 2 g/l casein hydrolysate (medium B_1C_2). Shoot cultures on media with high levels of Cu^{2+} and Co^{2+} ions did not show an inhibition of their GI. Increase in the diosgenin content was observed in shoot cultures cultivated on media with a high concentration of Cu^{2+} and Sr^{2+} ions, whilst on other media the cultures showed an inhibition of its formation. The stimulation of other steroidal sapogenin formations by the use of relatively high concentrations of metals ions (5 mM Mg^{2+} and 2.5 mg/l Cu^{2+} ions) in media was also observed in the callus cultures of *Agave amaniensis* (Indrayanto et al. 1996). Schübel et al. (1989) reported increased raucaffricine content in *Rauwolfia* cells by using relatively high concentrations (10 mM) of $MgSO_4$ and $CaCl_2$ in media. The suspension cultures of Cs (code F8) cultivated on medium $K_2D_{0.5}$ did not produce diosgenin (Indrayanto et al. 1994). Interestingly, by increasing the amount of micronutrient $CoCl_2.6H_2O$ by 100-fold (to 2.5 mg/l) in the medium, diosgenin formation could be stimulated ($413.0 \pm 45.3\,\mu g/g$ DW, n = 4). The induction of diosgenin formation ($459.0 \pm 52.8\,\mu g/g$ DW, n = 4) in root cultures of Cs (code F8) was also observed by the addition of 100 mg/l dried mycelium of *Rhizopus stolonifer* L. in medium I_2. The root cultures did not produce diosgenin when cultivated on the medium without addition of the fungus. Stimulation of diosgenin formation in the suspension cultures of *Dioscorea deltoidea* by the addition of fungal mycelia of *Rhizopus arrhizus* and other fungi was previously reported by Rokem et al. (1985). The result of the described studies showed that the formation of some steroidal sapogenins (diosgenin, hecogenin) could be stimulated by the addition of certain abiotic or biotic elicitors in media.

Significantly increased diosgenin formation was also reported by Wartono (1997) and Kustiawan (1997) by the addition of 100 mg/l salicylic acid (SA) in medium B_2. For this work, shoot cultures of Cs (code F8-1 and F1-1) were used. They also showed that, by using the top part of the shoots as the inoculating explants, reduction of the diosgenin content in the shoot cultures occurred. It was already known that exogenous application of SA could affect the physiological process in plant cells (Dombrowski and Alfermann 1993). The inducer effect of SA for different metabolic pathways in *Catharanthus roseus* cell suspension cultures were reported by Andayani and Verpoorte (1996).

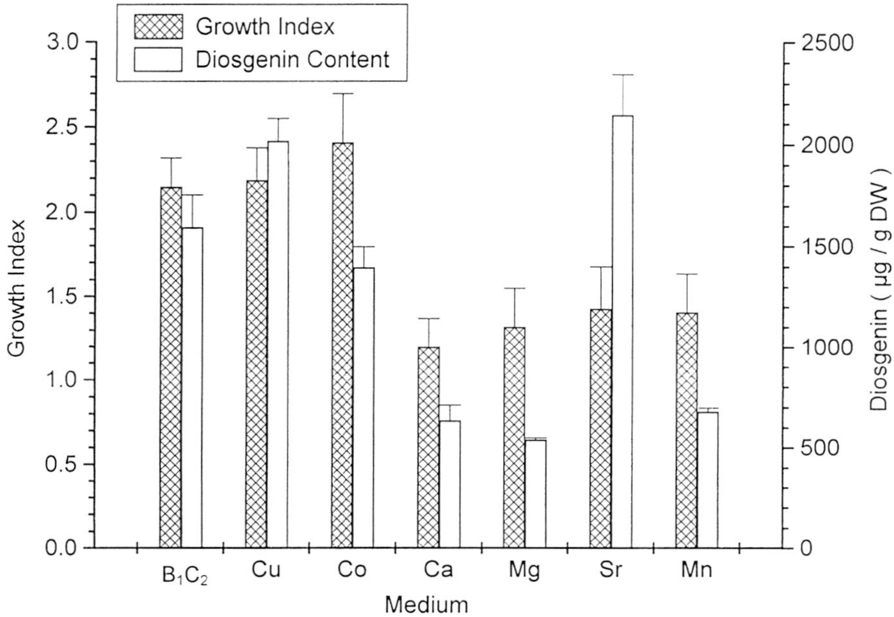

Fig. 10. Effect of using high concentrations of metal ions on the growth and diosgenin content of shoot cultures of *Costus speciosus* (F8), cultivated on medium B_1C_2. All cultures were maintained in continuous light (1500 lx) and 25 ± 1 °C. The concentrations of certain salts in the medium B_1 C_2 were increased to 2.5 mg/l $CuSO_4.5H_2O$ (medium Cu), 2.5 mg/l $CoCl_2.6H_2O$ (medium Co), 2.2 g/l $CaCl_2.2H_2O$ (medium Ca), 1.85 g/l $MgSO_4.7H_2O$ (medium Mg), 4 g/l $SrCl_2.6H_2O$ (medium Sr) and 223 mg/l $MnSO_4.H_2O$ (medium Mn). Other components of the treatment media were same as medium B_1C_2. Values are mean ± SD (n = 4–5). (Data from Utami et al. 1997).

3 Conclusions

Regeneration of organs (shoot, root and plantlet), and callus cultures of *Costus speciosus* were initiated from young stems using modified MS medium supplemented with 1–2 mg/l BAP (shoot cultures), 2–5 mg/l IBA (root cultures), 1 mg/l NAA + 0.2 mg/l kinetin (plantlet cultures) and 0.2 mg/l kinetin + 1 mg/l NAA or 2 mg/l + 0.5 mg/l 2,4-D (callus and suspension cultures).

One-step clonal propagation of Cs could be achieved by using modified MS medium with the addition of 2 mg/l BAP. After 6–7 weeks of cultivation, the individual plantlets could be transferred directly to sterilized humus for acclimatization.

Diosgenin accumulation was only observed in shoot cultures and shoot leaves of the plantlet cultures, whilst sterols were detected in all organ and callus cultures. Squalene, a precursor of phytosteroid was also detected in suspension cultures of Cs.

Diosgenin production could be increased by using a relatively high concentration of sucrose, Cu^{2+} and Sr^{2+} ions in media. Addition of dried mycelium of *Rhizopus stolonifer*, salicylic acid and increasing by 100-fold the micronutrient Co^{2+} ions stimulated the diosgenin formation in the callus and root cultures.

4 Protocols

4.1 Establishment of the In Vitro Cultures

Young stems (shoots), isolated from the mother plant, were cut (ca. 5 cm lengths), then washed with distilled water, ethanol (95%), and surfaced sterilized with 1–2% sodium hypochlorite for 5–15 min. The sterilized shoots were cut into 1–2 cm lengths and placed on modified MS medium. The hormone compositions used for initiating various in vitro cultures are described in Section 2.1. The cultures were maintained in continuous light (1500–2000 lx) at 25 ± 1 °C. Established in vitro cultures were subcultured for 4–6 weeks (shoot, plantlet and root cultures) and 2–3 weeks (callus cultures).

Suspension cultures can be initiated by transferring ca. 5 g soft friable calli into 50 ml MS medium modified with the addition of 2 mg/l kinetin and 0.5 mg/l 2,4-D (medium $K_2D_{0.5}$), and 3% sucrose in 250–300 ml Erlenmeyer flasks. Shake at about 100 rpm on a gyratory shaker with the same conditions as for callus cultures. Subculture every 7 days.

The growth rate was expressed as a growth index (GI = ratio of the final fresh weight and inoculated fresh weight) and % PCV (packed cell volume, for suspension cultures).

4.2 Micropropagation

About four to five lower parts of shoots with one to two axillary buds (that were cut about 2–3 cm long from the stock cultures) were transferred to 25 ml medium B_2 or $B_{0.4}$/flask. After about 7–8 weeks of cultivation (1500–2000 lx, 25 ± 1 °C), the plantlets (in vitro plants that have roots, shoot and leaves) can be separated individually and directly transferred to sterilized humus for acclimatization (3–4 weeks).

4.3 Quantitative Analysis of Sterols and Diosgenin by the Densitometric Method

An accurate weight of 1 g of powdered dried biomass was extracted 3–4 times, using a vortex-mixer (15 min) with 7.5 ml chloroform for each. All the extracts were combined and evaporated under N_2 to dryness. This extract contains free sterols, squalene and triterpene.

The residue was hydrolyzed with 2N HCl (2 h, 100 °C), then neutralized with NaOH 10N, the bound sterols (from sterol glycosides) and diosgenin were extracted 4 times with 10 ml chloroform, using a vortex mixer (15 min). The extracts were combined and evaporated to dryness in N_2.

Diosgenin and sterols (total) can be determined simultaneously using a densitometric method. Kieselgel 60 precoated plates (E. Merck) are used as the stationary phase. As eluent, a mixture of n-hexan: ethyl acetate (3:1) or chloroform: ethyl acetate (4:1) can be used. After visualizing with H_2SO_4 – anisaldehyde reagent (100 °C, 5–10 min), quantification was performed by measuring the absorbance reflectance of the analyte spots at 427 nm (diosgenin) and 397 nm (sterols). Diosgenin (Sigma) and cholesterol (Sigma) were used as the external standard. With this method, the total sterols in the biomass are calculated as cholesterol. It is recommended that the presence of the sapogenin(s) and sterols in the biomass is confirmed by GC-MS analysis (according to Carle 1979) by comparison with authentic standards.

For analyzing individual sterols, triterpene, diosgenin or other sapogenins (if any) an HPLC, GLC or SIM-GC MS system are recommended. The methods have been described previously (Meagher 1996; Indrayanto et al. 1998).

Acknowledgments. The authors are grateful to Professor A.W. Alfermann (Düsseldorf, Germany), Professor A.L. Wilkins (Hamilton, New Zealand) for providing some useful data and references; Ms. U. Chasanah, Ms. M. Ervina and Mr. D. Wartono (Faculty of Pharmacy, Airlangga University) for helping to carry out some experiments described here. This work was supporterd in part by a grant from the Department of Education and Culture Republic of Indonesia, Jakarta (No. 020/PPIPD/DPM/1996). Special thanks to Aloysius Indrayanto, SMU St. Louis, Surabaya for help with typing.

References

Abdullah A (1987) Plasma nutfah tanaman penghasil bahan dasar obat Kontrasepsi. Edisi Khusus Littro (III) 3:28–31

Andayani R, Verpoorte R (1996) Salicylic acid and hydrogen peroxide as inducers for different metabolic pathways in *Catharanthus roseus* cell suspension cultures. Poster presented at 44th Annu Congr Society for Medicinal Plant Research and a Joint Meeting with the Czech Biotechnology Society Prague, Czech Republic, 3–7 Sept 1996

Backer CA, van der Brink (1968) Flora of Java, vol 3. Wolters Noordhoff, Groningen, pp 74–76

Bandara BMR, Hewage CM, Karunaratne V, Adikaram NK (1988) Methyl ester of para-coumaric acid: antifungal principle of the rhizome of *Costus speciosus*. Planta Med 54:477–478

Burkill IH (1966) A dictionary of the economic products of the Malay penninsula, vol 1. Ministry of Agriculture and Cooperatives, Kuala Lumpur, Malaysia, pp 680–683

Carle R (1979) Untersuchungen zur Steroidalkaloid- und -sapogeninführung in Pflanzen und Zellkukturen der Gattung *Solanum* L., PhD Thesis, University of Tübingen

Chairunnisa (1994) Pengaruh kasein hidrolisat terhadap kandungan diosgenin kultur tunsa *Costus speciosus*. BSci Thesis, Faculty of Pharmacy, Airlangga University, Surabaya, Indonesia

Chaturvedi HC, Misra P, Jain M (1984) Proliferation of shoot tips and clonal multiplication of *Costus speciosus* in long term culture. Plant Sci Lett 35:67–71

Das Gupta B, Pandey VB (1970) A new Indian source of diosgenin. Experentia 26:475

De Padua LS, Lugod GC, Pancho JV (1981) Handbook on the Philippine medicinal plants, vol 3. The Documentation and Information Section Office of the Director of Research, University of The Philippine, Los Banos, p 49

Dombrowski K, Alfermann AW (1993) Salicylsäure – das Universalpharmakon der Human-und Phytomedizin. Pharm Unserer Zeit 22:275–285

Emiliawati S (1994) Pengaruh ion kalsium terhadap kandungan diosgenin pada kultur tunas *Costus speciosus* (Koen) Sm. BSci Thesis, Faculty of Pharmacy, Airlangga University, Surabaya, Indonesia

George EF, Puttock DJM, George HJ (1987) Plant culture media, vol 1. Formulation and uses. Exegetics, Edington, pp 58–59

Gupta MM, Shukla YN, Lal RN (1983) Investigation of *Costus* IV, chemical studies of saponins of *Costus speciosus* roots. Planta Med 48:64

Hegnauer R (1986) Chemotaxonomie der Pflanzen, vol 7. Birkhäuser, Basel, pp 788–789

Hong Kong Chinese Medical Research Institute (1991) Chinese medical herbs of Hong Kong, vol 1. Hong Kong Chinese Medical Research Institute, Hong Kong, p 196

Indrayanto G (1983) Steroide und Triterpene in Zellkulturen. PhD Thesis, Univrsity of Tübingen

Indrayanto G, Rahayu R, Rahman A, Noerani PE (1993) Effect of calcium, strontium and magnesium ions on the formation of phytosteroids in callus cultures of *Agave amaniensis*. Planta Med 59:97–98

Indrayanto G, Setiawan B, Cholies N (1994) Differential diosgenin accumulation in *Costus speciosus* and its tissue cultures. Planta Med 60:483–484.

Indrayanto G, Sondakh R, Syahrani A, Utami W (1998) *Solanum mammosum* L (Terong Susu): in vitro cultures and the production of secondary metabolites. In: Bajaj YPS (ed), Biotechnology in agriculture and forestry, vol 41. Medicinal and aromtic plants X. Springer Berlin Heidelberg New York, pp 394–414

Inoue K, Ebizuka Y (1996) Purification and characterization of furostanol glycoside 26-O-β-glucosidase from *Costus speciosus* rhizomes. FEBS Lett 378:157–160

Inoue K, Kobayashi S, Noguchi H, Sankawa U, ebizuka Y (1995) Spirostanol and furostanol glycosides of *Costus speciosus* (Koenig) Sm. Nat Med 49:336–339

Inoue K, Shimomura K, Kobayashi S, Sankawa U, Ebizuka Y (1996) Conversion of furostanol glycoside by β-glucosidase in *Costus speciosus*. Phytochemistry 41:725–727

Jain M, Chaturvedi HC (1985) Caulogenesis in rhizome callus of *Costus speciosus*. Planta Med 50:462–463

Kustiawan I (1997) Pengaruh pemberian asam salisilat terhadap kandungan fitosteroid dari kultur pucuk *Solanum mammosum* (sm-12) dan *Costus speciosus* (F-1). BSci Thesis, Faculty of Pharmacy, Airlangga University, Surabaya, Indonesia

Lambert N, Baccou JC, Sauvaire Y (1988) Screening for diosgenin in rhizome from 3 *Costus* species (*C. deistellii, C. igneus, C. lucanusianus*). Planta Med. 54:366–367

Meagher LP (1996) Aspects of the chemistry of some agricultural myco- and phytotoxins, PhD Thesis, University of Waikato, Hamilton, New Zealand

Miles CO, Wilkins AL, Erasmus GL, Kellerman TS, Coetzer JAW (1994) Photosensitivity in South Africa. VII. Chemical composition of biliary crystals from a sheep with experimentally induced geeldikkop. Onderstepoort J Vet Res 61:215–222

Miles CO, Wilkins AL, Kellerman TS, Smith BL, Flaoyen A (1995) Saponin metabolism and biliary crystal formation. In: Gartwaite L (ed) Toxicology and food safety research reports 1992–1995 AgResearch, Hamilton, New Zealand, pp 61–62

Mosihuzzaman M, Nahar N, Ali L, Rokeya B, Khan AK, Nur-E- Alam M, Nandi RP (1994) Hypoglycemic effects of three plants from eastern Himalayan belt. Diabetes Res 26:127–138

Muhammad R (1998) Profil pertumbuhan kultur suspensi sel *Costus speciosus* Koen (kode F8) dan pembentukan fitosteroidnya, BSci thesis, Faculty of Pharmacy, Airlangga University, Surabaya, Indonesia

Murashige T, Skoog F (1962) A revised medium for rapid growth and bio-assays with tobacco tissue cultures. Physiol Plant 15:473–497

Nagendra P, Abraham PZ (1981) Polyploidy and speciation in *Costus speciosus* (Koen) Sm. Curr Sci 50:26–28

Nuryati N (1994) Pengaruh beberapa konsentrasi sakarosa terhadap kandungan diosgenin pada kultur tunas *Costus speciosus* (Koen) Sm BSci Thesis, Faculty of Pharmacy, Airlangga University, Surabaya, Indonesia

Rathore AK, Khanna P (1979) Steroidal constituents of *Costus speciosus* (Koen) Sm. callus cultures. Planta Med 35:289–290

Rokem JS, Tal B, Goldberg I (1985) Methods of increasing diosgenin production by *Dioscorea deltoidea* cells in suspension cultures. J Nat Prod 48:210–222

Roy A, Pal A (1995) Callus cultures from zygotic embryos of *Costus speciosus* and their morphogenetic responses. J Plant Biochem Biotechnol 4:29–32

Sarin YK, Bedi KL, Atal CK (1974) *Costus speciosus* rhizome as source of diosgenin. Curr Sci 43:569–570

Schenk RU, Hildebrandt AC (1972) Medium and technique for induction and growth of monocotyledonous and dicotyledonous plant cell cultures. Can J Bot 50:199–204

Schübel H, Ruyter CH, Stöckigt J (1989) Improved production of raucaffricine by cultivated *Rauwolfia* cells. Phytochemistry 28:491–494

Setiawan B (1993) Inisiasi kultur *Costus* spp. dan analisis kandungan steroidnya. M Sci Thesis, Faculty of Graduate Studies, Airlangga University, Surabaya, Indonesia

Sierra MI (1991) Aspects of indole alkaloid accumulation in *Tabernaemontana* tissue cultures: differentiation, peroxidases and stability. PhD Thesis, University of Leiden

Singh SB, Thakur RS (1982a) Costusoside-I and costusoside-J, two new furostanol saponins from the seeds of *Costus speciosus*. Phytochemistry 21:911–915

Singh SB, Thakur RS (1982b) Saponins from the seed of *Costus speciosus*. J Nat Prod 45:667–670

Singh SB, Sanyal AK, Bhattacharya SK, Pandey VB (1972) Oestrogenic activity of saponins from *Costus speciosus* (Koen) Sm. Indian J Med Res 60:287–290

Singh SB, Gupta MM, Lal RN, Thakur RS (1980) *Costus speciosus* seeds as an additional source of diosgenin. Planta Med 38:185–186

Sudiarto, Chairani F, Rosita SM, Wahid P (1985) Perkembangan penelitian budidaya tanaman bahan baku pil kontrasepsi. J Litbang Pertanian 4:72–76

Susilo NE (1992) Penetapan kadar steroid dari beberapa rimpang *Costus* spp. secara densitometri dan kromatografi kinerja tinggi. BSci Thesis, Faculty of Pharmacy, Airlangga University, Surabaya, Indonesia

Takechi M, Tanaka Y (1991) Structure activity relationship of syntheic diosgenyl monoglycosides. Phytochemistry 30:2557–2558

Takechi M, Shimada S, Tanaka Y (1992) Structure activity relationship of synthetic diosgenyl diglycosides. Phytochemistry 31:3280–3281

Takechi M, Uno C, Tanaka Y (1997) Biological activity of synthetic triterpenoid and steroid β-D-xylopyranosyl-(1->6)-β-D-glucopyranosides. Phytochemistry 44:299–303

Thewles A, Parslow RA, Coleman R (1993) Effect of diosgenin on biliary cholesterol transport in the rat. Biochem J 291:793–798

Tschesche R, Pandey VB (1978) Steroidal saponins of *Costus speciosus*. Phytochemistry 17:1781–1782

Utami W, Indrayanto G, Santosa MH (1997) Pengaruh elisitor abiotik dan biotik pada kandungan diosgenin pada kultur tunas *Costus speciosus*. Laporan Penelitian Dasar, Airlangga Research Institute, Surabaya, Indonesia

Wartono D (1997) Pengaruh asam salisilat terhadap pertumbuhan dan kandungan fitostroid kultur tunas *Solanum laciniatum* (Sl-7) dan *Costus speciosus* (F-8). BSci Thesis, Faculty of Pharmacy, Airlangga University, Surabaya, Indonesia

WHO (1990) Medicinal plants in Viet Nam. Regional publications, Western Pacific no 3. WHO Regional Office for the Western Pacific, Manila and Institute of Materia Medica, Hanoi, P 61

Wijayakusuma HMH (1992) Tanaman berkhasiat obat Indonesia, Jilid ke 2. Pustaka Kartini, Jakarta, pp 101–103

Wu Y, Wu P (1991) Separation of C^{25} epimers of steroidal sapogenins by reversed phase high performance liquid chromatography at low temperature. Phytochem Anal 2:220–224

V *Cuphea* Species: In Vitro Culture Studies, and the Production of Medium-Chain Fatty Acids

E.I.S. FLOH[1], W. HANDRO[1], and I. RITA[2]

1 General Account

The genus *Cuphea* (Lythraceae) comprises about 260 species growing wild in temperate, tropical, and subtropical regions in the New World, from the eastern United States to southern Argentina. It comprises annual or perennial herbs and small shrubs. Many species are common in disturbed habitats, others are endemics with narrow distributions. Two centers of species diversification are recognized, southeastern Brazil and western Mexico. *Cuphea* species have characteristic tubular, zygomorphic flowers, and an unique seed dispersal mechanism. The genus is extremely diverse with respect to chromosome number, ranging from n = 6 to n = 54, n = 8 being the base number (Graham 1989a). The interest in *Cuphea* species deals with their unique seed lipid composition. This interest has increased from the 1960s, resulting in an extensive bibliography reviewed by Graham (1989b).

Agronomic studies on *Cuphea* have been carried out since the early 1980s (see Hirsinger and Knowles 1984; Hirsinger 1985; Graham 1989b); it is a wild plant which must be domesticated. Adaptation for economic production will depend on overcoming problems such as: (1) dormancy conditioned by the hard seed coat, (2) early seed scattering and shattering, (3) continuous flowering and seed production, and (4) production of viscous exudates from glandular hairs in the stem and floral tube. In order to become a commercial crop *Cuphea* must have a high seed yield, a high oil content with the desired chemical composition, and larger seeds which are easy to handle and harvest. However, there is significant variation for seed dormancy and weight, and for lauric and capric acid composition in several species (Knapp and Tagliani 1990; Thompson et al. 1990). Mass selection could reduce seed dormancy as described for some *Cuphea* species by Knapp (1990). On the other hand, such species show some advantages, such as ability to form stands naturally which aids agricultural cultivation, and adaptability to different climates; no specific photoperiods seem to be required to induce flowering. *Cuphea* appears easy to propagate vegetatively; Thompson (1984) propagated five different species through rooting stem tip cuttings, without hormone application.

[1] Plant Cell Biology Laboratory, Dept of Botany, Institute of Biosciences, University of São Paulo, CP 11461, São Paulo, SP, 05421-970, Brazil
[2] CENARGEN/EMBRAPA, CP 02372, Brasilia, 70849-970, Brazil

Biotechnology in Agriculture and Forestry, Vol. 43
Medicinal and Aromatic Plants XI (ed. by Y.P.S. Bajaj)

2 Economic Importance, and Fatty Acids

Until the 1950s, some *Cuphea* species were considered only as ornamentals (Bailey 1958) although several species were used as treatment for syphilis and fevers, as laxatives, to control dysentery and intestinal inflammations, hemorrhoids etc. (Kohne 1877; Hohene 1939; Martinez 1959; Von Reis and Lipp 1982). The ethanolic extract of *C. balsamona*, a species used in folk medicine to treat arterial hypertension, was demonstrated to induce hypotension in animals (Ericeira et al. 1985).

From the early 1960s, the study of oilseed of various *Cuphea* species (Graham and Kleiman 1985, 1987; Wolf et al. 1983; Graham 1989b) showed that in this genus an unparalleled diversity of fatty acid patterns occurs, with high levels of medium-chain fatty acids, such as caprylic, capric and lauric (Table 1). Lauric acid is used for manufacturing soaps, detergents, surfactants, lubricants etc., and its primary sources are coconut and palm kernel oils. Medium-chain triglycerides have potentially important applications in medical, nutritional and dietetic fields, which include: as a tool in the control of obesity; as a factor in the lowering and inhibition and/or limitation of cholesterol deposition in the tissues; for providing quick, high energy in both animals and humans, and treatment of childhood epilepsy (Babayan 1974, 1981; Bach and Babayan 1982; Thompson 1984).

As no plants other than palms have *Cuphea's* fatty acid composition, and in view of the increasing interest in medium-chain fatty acids, *Cuphea* may

Table 1. Diversity in fatty acid composition of oilseed from some *Cuphea* species (% of total fatty acid), according to Graham (1989b)

Species	Fatty acids									
	8:0	10:0	12:0	14:0	16:0	18:0	18:1	18:2	18:3	20:0
C. mimuloides[a]	–	–	25.0	10.0	20.0	3.0	10.0	30.0	–	–
C. utriculosa	–	0.2	0.8	0.5	25.0	3.5	26.4	30.7	–	–
C. decandra	–	1.3	0.6	0.3	29.6	2.4	20.6	44.7	–	–
C. fruticosa	–	tr	tr	0.1	16.8	0.5	12.8	67.2	–	–
C. setosa[a]	–	1.1	62.2	26.0	5.3	–	1.3	1.4	–	–
C. tetrapetala[a]	–	0.4	32.4	51.0	7.0	0.9	1.5	6.4	0.4	0.1
C. calophylla[a]	0.1	11.1	76.9	4.1	1.9	0.1	2.1	3.1	0.3	–
C. polymorpha[a]	tr	7.4	80.1	3.6	2.0	0.1	2.5	4.3	–	–
C. ferrisiae[a]	1.2	82.2	1.9	1.0	3.2	tr	2.7	4.4	6.4	–
C. wrightii[a]	–	29.4	53.9	5.1	2.3	0.4	3.1	4.6	–	–
C. vesiculigera[a]	tr	0.4	71.2	18.0	2.6	0.2	2.4	5.0	–	0.1
C. viscosissima[a]	9.1	75.5	3.0	1.3	3.1	0.3	1.9	4.7	0.5	0.3
C. koheneana[a]	0.1	91.6	1.5	0.6	1.3	0.3	1.1	3.1	0.2	0.1
C. aequipetala[a]	24.6	1.3	1.8	56.0	6.6	–	4.9	3.6	–	–
C. spectabilis[a]	0.6	0.2	1.2	2.7	15.7	1.2	26.0	22.2	30.8	–
C. pinetorum[a]	48.1	40.3	0.4	0.4	2.6	0.4	3.0	4.1	–	–
C. cyanea[a]	67.8	24.4	–	–	1.4	–	2.4	3.1	–	–

tr = Traces.

[a] Species with high level of medium-chain fatty acids.

become an economically viable crop because of the value of the unique oils it can provide.

Considering the number of *Cuphea* species, and the wide range of applications for their products, a significant increase will be expected in their economic importance as well as in research programs using these plants. In recent years, many studies have been made on the biosynthesis of medium-chain fatty acids using *Cuphea* species (e.g. Bafor et al. 1990; Bafor and Stymne 1992; Fuhrman and Heise 1993; Toepfer and Martini 1994; Jones et al. 1995). By using conventional techniques it is possible to select strains and mutants for medium-chain fatty acid production (Knapp and Tagliani 1991; Brandt and Knapp 1993). Nevertheless, most of the potential for tissue culture of *Cuphea* remains unexplored, especially regarding the improvement by selection of genotypes in vitro. In view of the occurrence of a great diversity in the chromosome number (6–54) in the genus *Cuphea* (Graham 1989a), the production of haploids would also be of interest.

Efforts are being made to transfer genes that regulate medium-chain fatty acids from *Cuphea* to already established oilseed crops. Recently, an acyl-ACP thioesterase cDNA from *C. hookeriana* was expressed in seeds of transgenic canola (*Brassica campestris*) resulting in a dramatic increase of caprylic and capric acid in a species that normally does not accumulate these two fatty acids (Dehesh et al. 1996). The contrary may also be envisaged, i.e. the use of recombinant DNA transfer techniques to modify some undesirable *Cuphea* characteristics. The apparent ease in propagation of *Cuphea* species through tissue culture could be relevant in programs to improve these plants.

3 Micropropagation

In 1986, Janick and Whipkey propagated *Cuphea wrightii* by culturing leaves, shoot tips and stem segments in a basal medium containing: MS salts, 0.1 $mg\,l^{-1}$ thiamine-HCL, 0.5 $mg\,l^{-1}$ pyridoxine-HCL, 100 $mg\,l^{-1}$ i-inositol, 0.5 $mg\,l^{-1}$ nicotinic acid, 2 $mg\,l^{-1}$ glycine, 1 $g\,l^{-1}$ casein hydrolysate, 3% sucrose and 0.8% agar. This medium was supplemented with different combinations of BA and 2,4-D (Table 2), and was able to induce callus, shoot proliferation or roots. Subculture of organogenic callus yielded more shoots, which could be isolated and rooted. Excised embryos from freshly harvested seeds of *C. wrightii* were also cultured in vitro; it was shown that 1 or 10 $mg\,l^{-1}$ BA in the basal medium doubles the germination (Janick and Whipkey 1986).

Rita and Floh (1995) reported the micropropagation of *C. ericoides*, a native species from Brazil. In this species, plant regeneration occurred on leaf and stem-derived callus under various combinations of auxins and cytokinins (Table 2). Vigorous calli were obtained when tissues were cultured in a basal medium with MS salts, vitamins (Nitsch 1969), 3% sucrose and several combinations of an auxin (IAA, NAA or 2,4-D) plus a cytokinin (BA or kinetin).

Table 2. Selected conditions and responses of different cultured tissues of *Cuphea* (from Janick and Whipkey 1986; Rita and Floh 1995; Millan et al. 1997). Data for *C. ingrata* are published for the first time

Material	Growth regulators ($mg\,l^{-1}$)	Response
C. wrightii		
Shoot tip	2,4-D (0.1) + BA (1)	Callus
	2,4-D (0.01) + BA (1)	Shoots
	2,4-D (0.1)	Roots
Leaf	2,4-D (0.01–0.1) + BA (0.1–10)	Callus
	2,4-D (0.0–0.01) + BA (1–10)	Shoots
	2,4 D (0.01)	Roots
Organogenetic callus	BA (0.1–1)	Shoots
Isolated shoot	BA (0.1–5)	Roots
C. ericoides		
Leaf	2,4-D (0.05–1) + BA (0.5)	Organogenic callus
Stem	2,4-D (1) + BA (0.5)	Organogenic callus
Leaf organogenic callus	IAA, NAA (0.5–2) + BA (0.5)	Shoots
Stem organogenic callus	IAA, NAA (2) + BA (0.5)	Shoots
Isolated shoots	IBA or NAA (0.1)	Roots
C. ingrata		
Leaf	NAA (0.05) + kinetin (2)	Shoots (direct)
Shoots	IBA or NAA (0.1–0.5)	Roots
C. lutea, C. lanceolata, C. leptopoda		
C. paucipetala, C. laminuligera		
Leaf	BA (2) + NAA (0.2)	Shoots

More than 200 cultures could be easily established, without problems of contamination, and with usual procedures of sterilization. Leaf explants developed calli with buds (Fig. 1), specially in medium with NAA (0.5–2 $mg\,l^{-1}$) or 2,4-D (0.05 $mg\,l^{-1}$), plus BA or kinetin (1 $mg\,l^{-1}$). When these organogenic calli were subcultured to medium containing NAA or IAA (0.5 $mg\,l^{-1}$) plus BA or kinetin (1 $mg\,l^{-1}$), profuse shoot regeneration occurred (Figs. 2 and 3). These organogenic calli can be subcultured indefinitely, producing more and more shoots.

Leaf and stem explants grew only in medium containing auxin plus cytokinin; BA was more effective than kinetin in inducing shoot regeneration. Direct organogenesis never occurred; buds and shoots arose from the mass of callus developed from the explant. Leaf explants showed higher shoot proliferation potential than stem explants. Regenerated shoots ca. 5 cm high can be isolated and put to root (Fig. 4) in basal medium without growth regulators. However, rooting was increased when the medium was supplied with 0.5 $mg\,l^{-1}$ IBA or NAA. Rooted shoots (Fig. 5) were planted into pots, transferred to a greenhouse or soil in the open air, where plants grew and flowered. More than 200 plants were obtained from 20 explants. Another Brazilian species, *C. ingrata* (unpubl.) regenerated shoots directly from leaf explants, in a medium with NAA (0.05 $mg\,l^{-1}$) plus kinetin (2 $mg\,l^{-1}$).

Recently, Millan et al. (1997) has reported the establishment of tissue cultures and regeneration for several *Cuphea* species (*C. lutea*, *C. lanceolata*,

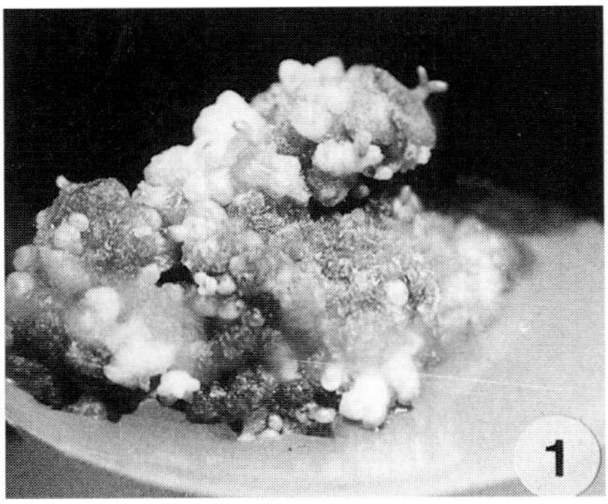

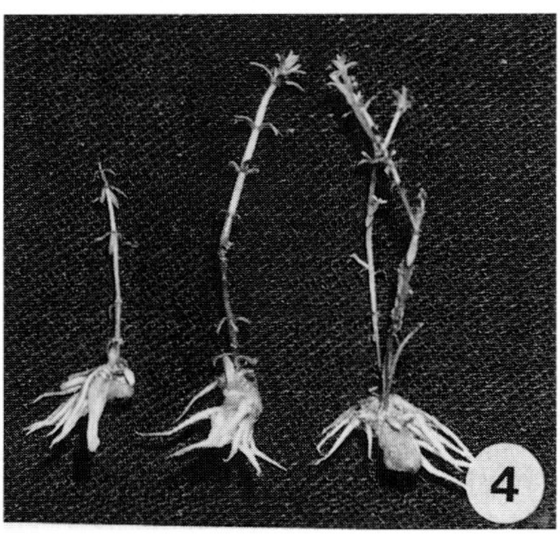

Fig. 1. Organogenic callus on leaf explant of *Cuphea ericoides* (40 days culture) in medium with $0.05\,mg\,l^{-1}$ 2,4-D + $0.5\,mg\,l^{-1}$ BA

Fig. 2. Organogenic calli 70 days after isolation from the explant and cultured in a medium with $2\,mg\,l^{-1}$ NAA + $0.5\,mg\,l^{-1}$ BA

Fig. 3. Shoots developed from a culture shown in Fig. 2 (90-day culture)

Fig. 4. Roots developed from shoots grown in a medium with $0.5\,mg\,l^{-1}$ IBA

Fig. 5. Regenerated plant in greenhouse, 8 weeks from in vitro transfer

C. leptopoda, *C. paucipetala*, *C. lanuligera*), using mainly leaf explants (Table 2). After 3–6 weeks of culture, in a medium supplemented with 8.8 μM BA plus 1 μM NAA, up to 12 shoots per explant can be regenerated. Shoot development was better in medium gelled with agar instead of Phytagel.

4 Summary and Conclusion

Cuphea species (Lythraceae) show an unparalleled diversity in fatty acids. In many species medium-chain fatty acids predominate, making them potentially economically viable crops because the value of their unique oil. Most of *Cuphea* species are wild plants to be domesticated, with constraints for their propagation. Tissue culture could provide an easy means to propagate these plants. Leaf, stem and shoot tip explants of *C. wrightii*, *C. ericoides* and *C. ingrata* when cultured in vitro in MS medium with auxins (2,4-D, IAA or NAA) plus a cytokinin (BA or kinetin) produced an organogenic callus, with multiple shoots which could be isolated for rooting, producing new plants that are able to flower. The use of recombinant DNA technology to modify some undesirable characteristics in *Cuphea* would be highly rewarding.

References

Babayan VK (1974) Modification of food to control fat intake. J Am Oil Chem Soc 51:260–264

Babayan VK (1981) Medium chain length fatty acid esters and their medical and nutritional applications. J Am Oil Chem Soc 58:49A–51A

Bach AC, Babayan VK (1982) Medium-chain triglycerides: an update. Am J Clin Nutr 36:950–962

Bafor M, Stymne S (1992) Substrate specificities of glycerol acylating enzymes from developing embryos of two *Cuphea* species. Phytochemistry 31:2973–2976

Bafor M, Jonsson L, Stobart AK, Stymne S (1990) Regulation of tryacylglycerol biosynthesis in embryos and microsomal preparations from the developing seeds of *Cuphea lanceolata*. Biochem J 272:31–38

Bailey LH (1958) The standard cyclopedia of horticulture, vol 1. Macmillan, New York

Brandt T, Knapp SJ (1993) Genetic selection at fatty acid and allozyma marker loci and meiosis within *Cuphea viscosissima x Cuphea lanceolata* populations. Crop Sci 33:1138–1143

Dehesh K, Jones A, Knutzon DS, Voelker TA (1996) The production of high levels of 8:0 and 10:0 fatty acids in transgenic canola by overexpression of Ch $FatB_2$, a thiosterase cDNA from *Cuphea hookeriana*. Plant J 9:167–172

Ericeira VR, Martins MMR, Souccar C, Lapa AJ (1985) Atividade farmacológica de extrato etanólico de "sete sangrias", *Cuphea balsamona* Chem & Sohltd. Cad Pesq São Luis 1:44–62

Fuhrman J, Heise KP (1993) Factors controlling medium-chain fatty acids synthesis from maturing *Cuphea* embryos. Z Naturforsch [c] 48:616–622

Graham AS (11989a) Chromosome number in *Cuphea* (Lythraceae): new counts and a summary. Am J Bot 76:1530–1540

Graham SA (1989b) *Cuphea*: a new plant source of medium-chain fatty acids. Crit Rev Food Sci Nutr 28(2):139–173

Graham SA, Kleiman R (1985) Fatty acid composition in *Cuphea* seed oils from Brazil and Nicaragua. J Am Oil Chem Soc 62:81–82

Graham SA, Kleiman R (1987) Seed lipids of the Lythraceae. Biochem Syst Ecol 15:433–439

Hirsinger F (1985) Agronomic potential and seed composition of *Cuphea*, an annual crop for lauric and capric seed oil. J Am Oil Chem Soc 62:76–80

Hirsinger F, Knowles PF (1984) Morphological and agronomic description of selected *Cuphea* germplasm. Econ Bot 38:439–451

Hohene FC (1939) Plantas e substâncias vegetais tóxicas e medicinais. Graphicars, São Paulo, Brasil

Janick J, Whipkey A (1986) In vitro propagation of *Cuphea wrightii*. Hortscience 21:135–137

Jones A, Maelor H, Voelker TA (1995) Palmitoyl-acyl carrier protein (ACP) thioesterase and the evolutionary origin of plant acyl-ACP thioesterases. Plant Cell 7:359–371

Knapp SJ (1990) Recurrent mass selection reduced seed dormancy in *Cuphea lanceolata* and *Cuphea laminuligera*. Plant Breeding 104:46–52

Knapp SJ, Tagliani LA (1990) Genetic variation for seed dormancy in *Cuphea laminuligera* and *Cuphea lanceolata*. Euphytica 47:65–70

Knapp SJ, Tagliani A (1991) Two medium chain fatty acid mutants of *Cuphea viscosissima*. Plant Breeding 106:338–341

Kohne BAA (1877) Lythraceae. In: Martius CFP, Eichler AW (eds) Flora Brasiliensis, vol 13(2). Frid Fleischer, Leipzig, pp 185–370

Martinez M (1959) Las plantas medicinales de México. Botas, Ciud. Mexico

Mathias R, Espinosa S, Roebbellen G (1990) A new embryo rescue procedure for interspecific hybridization. Plant Breeding 104:258–261

Millan S, Mitchell SM, Moscheni E, Lyon JE (1997) The establishment and regeneration of a range of *Cuphea* germplasm in vitro. Plant Cell Tissue Organ Cult 48:143–146

Nitsch JP (1969) Experimental androgenesis in *Nicotiana*. Phytomorphology 19:389–404

Rita I, Floh EIS (1995) Tissue culture and micropropagation of *Cuphea ericoides*, a potential source of medium-chain fatty acids. Plant Cell Tissue Organ Cult 40:187–189

Thompson AE (1984) *Cuphea* – a potential new crop. Hortscience 19:352–354

Thompson AE, Dierig DA, Knapp SJ, Kleiman R (1990) Variation in fatty acid content and seed weight in some lauric acid rich *Cuphea* spp. J Am Oil Chem Soc 67:611–617

Toepfer R, Martini N (1994) Molecular cloning of cDNAs or genes encoding proteins involved in de novo fatty acid biosynthesis in plants. J Plant Physiol 143:416–425

Von Reis S, Lipp FJJ (1982) New plant sources for drugs and foods from the New York Botanical Garden Herbarium. Harvard University Press, Cambridge, MA

Wolf RB, Graham SA, Kleiman R (1983) Fatty acid composition of *Cuphea* seed oils. J Am Oil Chem Soc 60:27–28

VI *Dioscorea balcanica* Košanin and *D. caucasica* Lipsky: In Vitro Culture and Production of Diosgenin

L. Ćulafić[1,2], K. Šavikin-Fodulović[3], D. Grubišić[1,3], and M. Nešković[1,2]

1 Introduction

1.1 Distribution, Morphology and Importance

Dioscorea is by far the largest genus in the family Dioscoreaceae, with about 600 species, grouped into 23 sections on the basis of morphologic and chemotaxonomic characters. According to Burkill (1960), the genus *Dioscorea* originated in East Asia. Certain species migrated westwards to Europe, and others to Japan and North America. *D. balcanica* and *D. caucasica* are the only wild *Dioscorea* species living in Europe nowadays. *D. balcanica* was discovered by Košanin (1914), in northern Albania, at 900–1200 m above sea level, while subsequently a few more localities were found in Montenegro and southwest Serbia (Košanin 1929; Lakušić 1966), and in Macedonia (Rizovski 1977). *D. balcanica* is an endemic, endangered species and therefore has been put under protection.

D. caucasica, discovered by Lipsky (1894), is the closest relative of *D. balcanica*. It is an endemic species in the region between the Caucasus and the Black Sea, and grows at the altitudes of up to 1600 m. It is also a protected species, included in the Red Book of the former USSR.

D. balcanica and *D. caucasica* are morphologically very similar. They are right-twining climbers with herbaceous stems, reaching a height of up to 2 m. Both are dioecious, female plants being larger than the males. The most conspicuous difference between the two species concerns the leaf size and shape: in *D. caucasica* all leaves are of equal size, longer than wide; in contrast, lower leaves in *D. balcanica* are larger, approximately of equal length and width, while the upper leaves are smaller, but more elongated and acute (Košanin 1929). Both species are perennial, due to the rhizomes that survive cold winters. The growth period is rather short, the plants flower from April to June, and bear fruits in September.

[1] Institute for Biological Research "S. Stanković", University of Belgrade, 29 novembra 142, 11060 Belgrade, Yugoslavia
[2] University of Belgrade, Faculty of Biology, Institute of Botany, Takovska 43, 11000 Belgrade, Yugoslavia
[3] Institute for Medicinal Plant Research "Dr. Josif Pančić", T. Košćuška 1, 11000 Belgrade, Yugoslavia

Biotechnology in Agriculture and Forestry, Vol. 43
Medicinal and Aromatic Plants XI (ed. by Y.P.S. Bajaj)

The family Dioscoreaceae embraces two unique groups of species, each one being of utmost importance for humans. The first comprises cultivated species, known as yams, or true yams; they produce edible tubers rich in starch, which are one of the main foodstuffs in tropical and subtropical regions of West Africa, Southeast Asia, China, Japan, the Pacific Islands and the Caribbean (Ammirato 1984).

The second group consists of wild *Dioscorea* species, which are a very important source of secondary metabolites, used in the pharmaceutical industry and medicine. Most of them produce steroid saponins, whose aglycons (sapogenins) are used as precursors in the synthesis of sex hormones, corticosteroids, fertility control compounds, anabolic agents, and cardiatonic glucosides (Van Staden and Fowlds 1992). Diosgenin is the most abundant sapogenin and the main component used for chemical or microbial conversion. The yearly production of steroid hormones at the end of the 1970s was about 700 tons. About 70% of that production was based on diosgenin as the initial raw material, about 19% used the diosgenin epimers, and only 1% derived the substance from *Solanum* species. The remaining 10% of steroid hormones was produced from cholesterol (Wagner 1980).

1.2 Conventional Practice for Obtaining Diosgenin

Van Staden and Fowlds (1992) compiled a list of nine medicinal *Dioscorea* species that produce on average 4–5% diosgenin on a dry weight basis, although some produced as much as 13%. Diosgenin has generally been extracted from the tubers of wild plants. Due to the ever increasing medical use of steroid products, the need for diosgenin has greatly expanded in the last two decades, such that this requirement can hardly be met by harvesting plants in nature. Attempts to cultivate several *Dioscorea* species (Martin and Gaskins 1968) for commercial diosgenin production do not seem to be economical. Besides, most of the wild species, being adapted to the high temperatures and humidity of the tropics, do not tolerate frost and have a long vegetation period, up to 10 months (Ammirato 1984).

It was, therefore, reasonable to pay more attention to the two wild species growing in Europe. *D. caucasica* was succesfully introduced into Moscow surroundings, while *D. balcanica* has grown in Belgrade Botanical Garden since 1919. First analyses showed that both species produce saponins (Kičenko and Panina 1965; Petričić et al. 1973).

2 In Vitro Cultures Studies

2.1 Survey of Work with *Dioscorea* Species

The need for improving the breeding of edible yams, as well as to providing more diosgenin-producing plant material, has already prompted research into unconventional propagation techniques. It was shown that many *Dioscorea* species are amenable to in vitro culture, while their cells and tissues preserve the capacity for diosgenin synthesis. Previous research has been reviewed by Ammirato (1984), Furmanova and Guzewska (1988), Van Staden and Fowlds (1992), and Ng (1992), Table 1 summarizes recent data not included in these articles.

2.2 Nutrient Media and Culture Conditions

Cultivation of organs and tissues of *D. balcanica* and *D. caucasica* in vitro has been done on media commonly used for many species, including other Dioscoreaceae. Basal medium (BM) contained mineral salts according to Murashige and Skoog (1962), sucrose (3%), agar (0.7%), and a mixture of vitamins, consisting of thiamine (2–10 $mg\,l^{-1}$), pyridoxine (1–2 $mg\,l^{-1}$), nicotinic acid (1–5 $mg\,l^{-1}$), and myo-inositol (80–100 $mg\,l^{-1}$). Various auxins and cytokinins were the only growth regulators examined. Callus formation was induced using high concentrations of 2,4-dichlorophenoxyacetic acid (2,4-D), such as 1–5 $mg\,l^{-1}$. For callus maintenance 2,4-D was lowered to 0.1–0.5 $mg\,l^{-1}$, or replaced by indole-3-acetic acid (IAA, 0.18–1 $mg\,l^{-1}$). Aerial tubers were formed on a medium containing BA 0.1 $mg\,l^{-1}$ and IBA 0.01–5 $mg\,l^{-1}$ (Grubišić et al. 1993). Somatic embryos were induced at 2,4-D 1 $mg\,l^{-1}$, and further developed on a regulator-free medium, or at a lower 2,4-D concentration (0.3 $mg\,l^{-1}$). Germinated embryos were raised to plantlets using a combination of 0.18 $mg\,l^{-1}$ IAA and 2.25 $mg\,l^{-1}$ BA (Ćulafić et al. 1994).

For obtaining callus culture in liquid media, about 500 mg of callus tissue was transferred to each 100 ml Erlenmeyer flask, containing 30 ml of the medium. The flasks were kept on an orbital shaker at 100 rpm, and subcultured every 28 days, diluting the suspension in a ratio 1:3.

The cultures were grown at 25 ± 2 °C, in darkness, or in light (warm white fluorescent tubes, fluence rate 47 $\mu mol\,cm^{-2}s^{-1}$, photoperiod 16 h).

2.3 Cultivation of Callus Tissues

In *D. balcanica* and *D. caucasica*, the callus cultures most often were initiated using parts excised from seedlings, such as nodal stem segments bearing axillary buds, apical segments, or young leaves (Ćulafić 1991). In addition, mature embryos, isolated from soaked seeds could be used as an alternative explant suitable for callus induction (Grubišić et al. 1991). The explants were cultured on the callus induction medium for 5 days, and then transferred to the medium for callus maintenance.

The calli obtained in culture were by no means homogeneous in structure. Šavikin-Fodulović et al. (1997) noted that an embryo-derived callus tissue of *D. balcanica*, which was initially green and compact, after about 2 months produced friable segments that differed in morphology and colour. The authors visually separated five tissue lines (A–E), which have retained their

Table 1. Summary of in vitro studies on *Dioscorea*[a]

Species	Purpose of study	Morphogenetic response	Secondary product	Reference
Dioscorea spp.[b]	Establishment of an in vitro germplasm collection	Slow growth of nodal cuttings, basal and aerial tubers	NS[c]	Malaurie et al. (1993)
D. abyssinica Hoch.	Photoperiod influence on microtuber production	Microtubers on nodal cuttings	NS	Jean and Cappadocia (1991)
	Effect of NAA, BA, and ABA on microtuber production	Microtubers on nodal cuttings	NS	Jean and Cappadocia (1992)
	Micropropagation, cytological study	Shoot multiplication, microtubers	NS	Lauzer et al. (1992)
D. alata L.	Photoperiod influence on microtuber production	Microtubers on nodal cuttings	NS	Jean and Cappadocia (1991)
	Effect of NAA, BA, and ABA on microtuber production	Microtubers on nodal cuttings	NS	Jean and Cappadocia (1992)
	Effects of cytokinins, NAA, ABA, and light on tuberization	Microtubers	NS	John et al. (1993)
	In vitro morphogenesis	Somatic embryos, plants	NS	Twyford and Mantell (1996)
D. balcanica Košanin	Vegetative propagation	Callus tissue, regenerated shoots	NS	Ćulafić (1991)
	Optimization of in vitro growth	Callus tissue, organogenesis, plants	NS	Grubišić et al. (1991)
	Micropropagation	Aerial microtubers	NS	Grubišić et al. (1993)
	Morphogenesis, cryopreservation	Somatic embryos, growth restoration after freezing/thawing	NS	Ćulafić et al. (1994)
	Diosgenin synthesis	Callus tissue, shoots, roots	Diosgenin	Šavikin et al. (1994)
	Morphogenesis, synthesis of diosgenin and phytosterols	Callus tissue, cell suspension, organogenesis	Diosgenin, sitosterol, stigmasterol, campesterol	Šavikin-Fodulović (1996)
	Selection of morphologically different callus lines	Callus, roots, shoots, somatic embryos, cell suspension	NS	Šavikin-Fodulović et al. (1997)
D. bulbifera L.	Tissue culture competence in leaves	Callus tissue	NS	Wernicke and Park (1993)

D. caucasica Lipsky	Optimization of in vitro growth	Callus tissue, organogenesis, plants	NS	Grubišić et al. (1991)
	Micropropagation	Aerial microtubers	NS	Grubišić et al. (1993)
	Morphogenesis, cryopreservation	Somatic embryos, growth restoration after freezing/thawing	NS	Ćulafić et al. (1994)
	Diosgenin synthesis	Callus tissue, shoots, roots	Diosgenin	Šavikin et al. (1994)
	Diosgenin synthesis	Callus tissue, roots	Diosgenin	Deliu et al. (1989a)
	Selection of callus lines, nutrient media	Callus tissue, roots	Diosgenin, sitosterol, stigmasterol, campesterol	Deliu et al. (1989b)
	Selection of cell lines, primary and secondary metabolites	Cell suspension cultures	Diosgenin, sitosterol, stigmasterol, campesterol	Deliu et al. (1992)
D. cayenensis	Responses to mineral media and growth regulators	Callus, shoots	NS	Mitchell et al. (1995)
D. cayenensis/ D. rotundata complex	Effect of growth regulators and genotype on meristem tip culture	Callus, shoots, roots	NS	Malaurie et al. (1995)
D. deltoidea Wall.	Effect of high sugar conc. on cell growth	Cell suspension	NS	Battat et al. (1989)
	Accumulation and location of steroid glucosides in cell strains	Cell suspension	Deltoside, protodioscine, their 25S-analogues	Butenko et al. (1992)
	Ultrastructure of cultured cells	Cell suspension	Electron dense material, probably steroid glucosides	Knyazkov et al. (1994)
	Furostanol glucosides, biomass production, cell viability	Continuous cell culture	Furostanol glucosides	Kandarakov et al. (1994)
	Cell growth and synthesis of steroid glucosides in closed continuous culture	Cell suspension	Steroid glucosides	Oreshnikov et al. (1994)
	Cryopreservation, effect of vitamin deficiency	Cell suspension	NS	Popov and Volkova (1994)

Table 1. *Continued*

Species	Purpose of study	Morphogenetic response	Secondary product	Reference
D. mangenotiana Miege	Micropropagation, cytological study	Shoot multiplication, microtubers	NS	Lauzer et al. (1992)
D. opposita Thunb.	Micropropagation of virus-free plants	Multiple buds on leaves, microtubers, plants	NS	Kohmura et al. (1995)
D. praehensilis	Effect of growth regulators and genotype on meristem tip culture	Callus, roots, shoots	NS	Malaurie et al. (1995)
D. rotundata Poir.	Vegetative propagation from axillary buds and shoot tips	Multiple shoots, plants	NS	Mix-Wagner (1994)
	Responses to mineral media and growth regulators	Callus, shoots	NS	Mitchell et al. (1995)
D. trifida L.	Obtaining virus-free plants through meristem culture	27% YMV-free plants	NS	Saleil et al. (1990)
	Responses to mineral media and growth regulators	Callus, shoots	NS	Mitchell et al. (1995)

[a] See also the reviews by Ammirato (1984), Furmanowa and Guzewska (1988), Van Staden and Fowlds (1992), and Ng (1992), for earlier papers not cited here.

[b] *D. abyssinica*, *D. alata*, *D. bulbifera*, *D. burkilliana*, *D. cayenensis-D. rotundata* complex, *D. cayanensis-D. rotundata* complex cv. Krengle × *D. praehensilis*, *D. dumetorum*, *D. esculenta*, *D. hirtiflora*, *D. mangenotiana*, *D. minutiflora*, *D. praehensilis*, *D. schimperana*, *D. togoensis*.

[c] NS = not studied.

characteristics for more than 3 years. On a medium containing $0.178\,mg\,l^{-1}$ IAA and $0.5\,mg\,l^{-1}$ BA, the line A produced only roots, while lines B and C produced both shoots and roots. The line D did not produce any organs on media that were tried, but was found suitable for studying the effects of precursors and inhibitors on sterol synthesis. The line E was undifferentiated on the same solid medium, but in liquid medium it became embryogenic. The callus lines also differed in anatomy, cell size, friability, and growth potential (Šavikin-Fodulović et al. 1997). It is not known if the lines originated from different cells of primary explants, or from subsequent variations. The variants were interesting, however, since their synthetic capacities were also shown to differ (see Sect. 3.2). In callus tissues of *D. caucasica*, Deliu et al. (1989b) sorted out 11 callus lines, which differed in colour and friability, and produced different amounts of diosgenin and phytosterols.

2.4 Callus Culture in Liquid Media

The five *D. balcanica* callus lines were tested for the ability to grow in liquid culture, in order to obtain a larger biomass yield, and/or to enable addition of various agents that could influence their synthetic capacity. Callus lines A, B, and C were not suitable for liquid culture, having remained constantly compact and organogenic. The line D was dispersed into small cell clumps, while the friable line E readily gave rise to a suspension of cells and small cell aggregates which displayed the greatest biomass yield (Šavikin-Fodulović et al., in prep.). The same line showed embryogenic potential upon transfer to appropriate media (see Sect. 2.6).

Deliu et al. (1992) selected 13 cell lines of *D. caucasica*, which differed in colour, size of cell aggregates, and growth, as well as in the capacity for synthesis of primary and secondary metabolites.

2.5 Organogenesis

The nodal and apical stem segments, which were cultured to induce callus growth, were also suitable for initiating propagation by multiple shoot production. Axillary shoots readily elongated on media with low auxin content, and low to high cytokinin concentrations (Ćulafić 1991; Grubišić et al. 1991). Certain *D. balcanica* and *D. caucasica* callus lines also produced adventitious buds, at IAA $0.135–0.178\,mg\,l^{-1}$ and BA $0.5\,mg\,l^{-1}$ (Grubišić et al. 1991; Šavikin et al. 1994). These shoots continued to branch and were not distinguishable from lateral shoot cultures in later subcultures (Fig. 1A). The shoots elongated in culture and produced well developed, green leaves. Roots were inducible by auxins, preferably by IBA treatment (Fig. 1B).

Fig. 1. Organogenesis and plant regeneration in *D. caucasica*. **A** Multiplication phase, shoot induction in a callus tissue. **B** Acclimation phase, plantlet from the greenhouse, ready for transfer to the field. (Grubišić et al. 1991)

2.6 Somatic Embryogenesis

Somatic embryogenesis has already been described in *D. floribunda* (Ammirato 1978), *D. rotundata* (Osifo 1988), *D. opposita* (Nagasawa and Finer 1989), and *D. alata* (Twyford and Mantell 1996). The callus tissues of *D. balcanica* and *D. caucasica* also displayed the capacity of producing somatic embryos (Ćulafić et al. 1994); essentially the same pattern of somatic embryogenesis was confirmed in selected callus line E (Šavikin-Fodulović 1996). Pieces of organogenic calli were grown for 4 weeks on a medium containing 2,4-D $1\,mg\,l^{-1}$, to induce formation of embryos (Ćulafić et al. 1994). At the end of that period, white transparent globular structures, later identified as embryogenic tissue, covered the callus surface. This tissue was separated from the non-embryogenic parts and cultured in an agitated liquid medium, hormone-free, or supplemented with 2,4-D $0.3–3\,mg\,l^{-1}$, where the tissue proliferated and disintegrated into proembryonic cell clumps of various sizes. Among them, bipolar embryonic structures were discernible, which gradually developed during subsequent subcultures into somatic embryos (Fig. 2). They were returned to solidified media and the number of embryos vs. cell clumps was scored. In *D. balcanica* the embryo percentage was highest at $0.3\,mg\,l^{-1}$ 2,4-D (about 60%), while in *D. caucasica* the highest number (75%) appeared when 2,4-D was omitted.

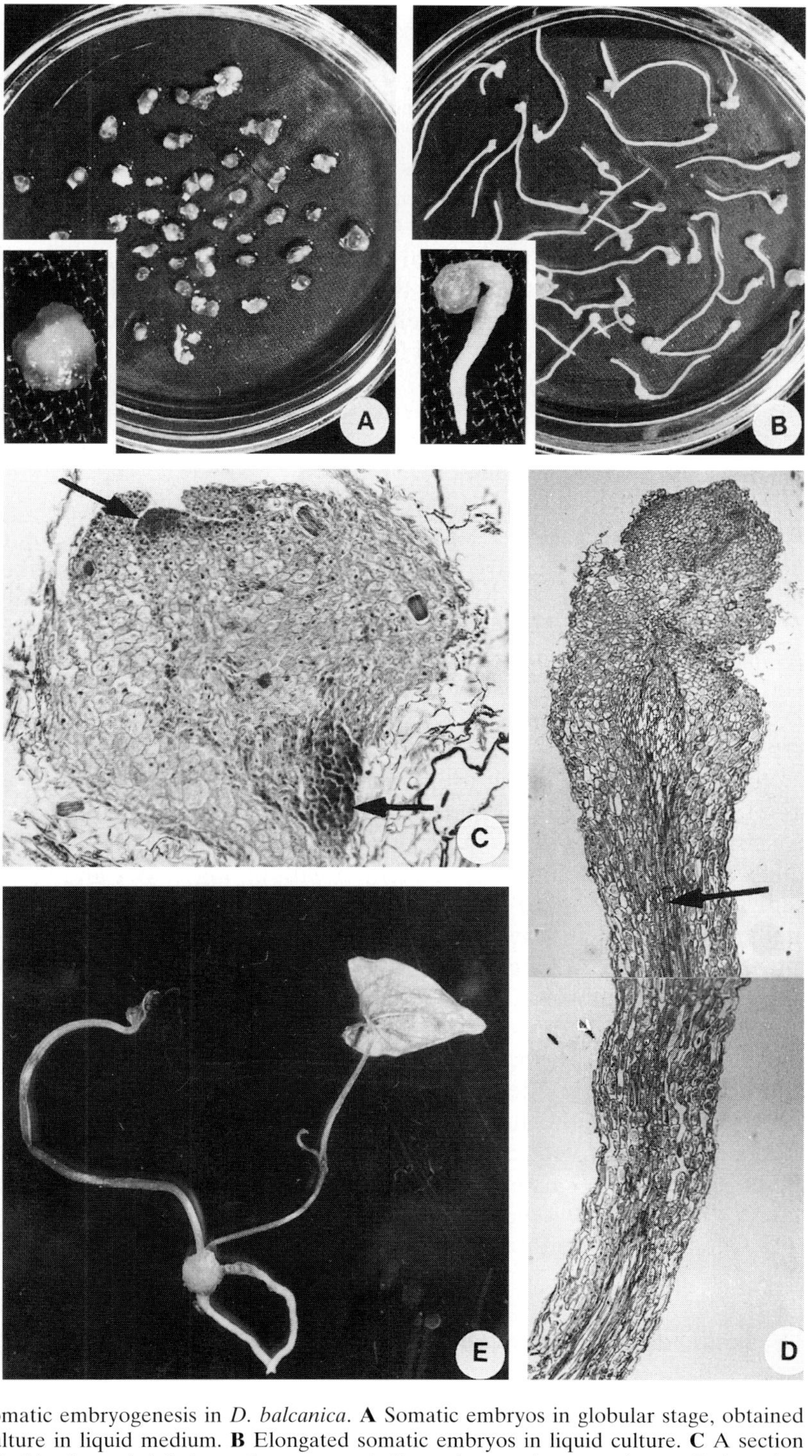

Fig. 2. Somatic embryogenesis in *D. balcanica*. **A** Somatic embryos in globular stage, obtained from a culture in liquid medium. **B** Elongated somatic embryos in liquid culture. **C** A section showing the histology of globular embryo; arrows indicate shoot and root meristems. **D** A section showing the histology of an elongated embryo; *arrow* points to the vascular initials. **E** A germinating somatic embryo. (Ćulafić et al. 1994)

Dioscorea somatic embryos had a rather peculiar shape. In the globular stage (Fig. 2A), shoot and root apical meristems were discernible in opposite poles of a bulb-like structure (Fig. 2B). In the next stage, embryos elongated (Fig. 2C) and a vascular tissue was formed, connecting the apex and the root (Fig. 2D). The medium routinely used for micropropagation, containing BA ($2.2\,mg\,l^{-1}$) and IAA ($0.18\,mg\,l^{-1}$) stimulated further growth of somatic embryos. Roots of somatic embryos in *D. caucasica* were branched, while those in *D. balcanica* were not, this being the only morphological difference between the two species. Plantlets that developed from somatic embryos (Fig. 2E) were transferred to soil and acclimated for 3 months in the greenhouse. The whole process takes almost a year.

2.7 Formation of Microtubers

Aerial tubers are reported to occur in nature in many *Dioscorea* species growing in tropical regions. In vitro tuber formation was described in *D. floribunda* (Sengupta et al. 1984), *D. bulbifera* (Forsyth and Van Staden 1984), *D. alata* and *D. bulbifera* (Mantell and Hugo 1989), *D. rotundata* (Ng 1992), *D. alata* and *D. abyssinica* (Jean and Cappadocia 1991, 1992), *D. abyssinica* and *D. mangenotiana* (Lauzer et al. 1992), and *D. alata* (John et al. 1993). *D. balcanica* and *D. caucasica* belong to the rhizomatous species and were not reported to form tubers in their natural habitats. Nevertheless, the occurrence of tubers was noted in culture, when apical stem segments were cut from aseptically grown seedlings and cultured (Grubišić et al. 1993). The segments, consisting of an apex with one or two subtending internodes, were put in test tubes, on a basal medium supplemented with $0.1\,mg\,l^{-1}$ BA and various concentrations of IBA (0.01–$5\,mg\,l^{-1}$). The tubers are shortened and thickened axillary sprouts, that grow out from one of the three axillary buds, instead of lateral branches. In a long day of 16 h light, the microtubers appeared within 8 weeks. A single tuber was formed initially on each stem base, but additional tubers developed later from other lateral buds, that otherwise remain quiescent and do not form branches (Fig. 3A). If the cuttings consisted of two internodes, the tubers developed on both nodes (Fig. 3B). With increasing IBA concentration, the number of microtubers also increased. At $5\,mg\,l^{-1}$ their number, compared with controls, was 100% higher in *D. balcanica*, and 20% higher in *D. caucasica* stems. The microtubers could be separated and germinated, to develop new plantlets. The tubers contained starch and may prove very valuable for vegetative propagation under natural conditions.

2.8 Acclimation of Regenerated Plantlets

Regenerated plantlets of both *D. balcanica* and *D. caucasica* could be obtained in three ways; (1) induction and rooting of shoots, (2) somatic embryogenesis, and (3) germination of microtubers. They were usually acclimated in a greenhouse for 3 months and then planted in the experimental field near Belgrade. About 90% of plants survived the transfer to the field and flowered for the first

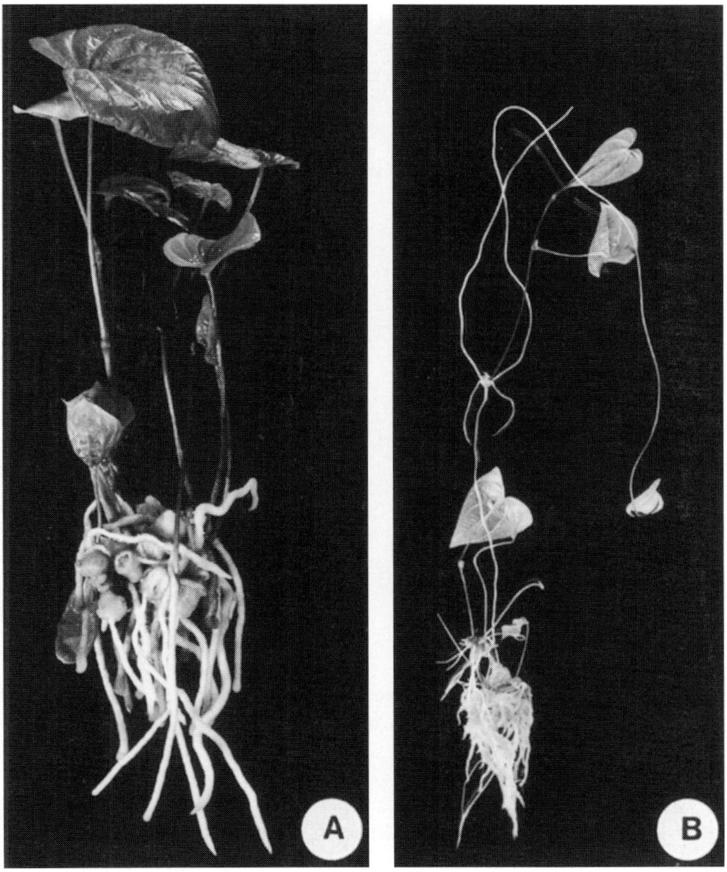

Fig. 3. Formation of aerial tubers in vitro in *D. balcanica*. **A** Microtubers at the stem base. **B** A microtuber in the leaf axil at the second node. (Grubišić et al. 1993)

time in the next season, which in our first experiments occurred in 1995. During 1996–1997, new shoots developed from the rhizomes and the plants reached a height of 2–3 m (Fig. 4A,B).

2.9 Cryopreservation

Organogenic calli of *D. caucasica* and *D. balcanica*, which were used for cryopreservation, were cultured for 10 to 45 weeks on a medium supplemented with 0.6 $mg l^{-1}$ 2,4-D and 0.5 $mg l^{-1}$ BA (Ćulafić et al. 1994). Pieces of callus tissue (1.5–3 mm in diameter) were cooled on ice and a cold culture medium with cryoprotectants was added. Two types of protective solution were chosen: 7% DMSO and 7% DMSO with 10% sucrose. Cell viability before freezing was 65–75%. The dense suspension was shaken for 5 min and poured into 1.2–2.0 ml plastic ampoules. They were kept at 4 °C for 1.0–1.5 h,

Fig. 4. Regenerated plants in the experimental field. **A** *D. balcanica* (*left*) and *D. caucasica* (*right*) lifted from the soil after 2 years. **B** A view of several *D. balcanica* plants in the experimental field. (Šavikin-Fodulović, unpubl.)

transferred to a freezing chamber, gradually cooled to −70 °C and finally immersed into liquid nitrogen. The ampoules were stored from several days up to seven months. They were thawed in a water bath at 40 °C, and transferred to an ice bath until the last ice crystals disappeared. The callus pieces were placed on filter paper disks and held for one day on the agarose surface, in darkness. After deep freezing and thawing, the average cell viability was 28% when 7% DMSO was used, and 15% with DMSO and sucrose. The restoration of growth of the organogenic callus was achieved after 4 to 6 weeks.

The plantlets of *D. balcanica* and *D. caucasica* were successfully regenerated from both organogenic and embryogenic calli that were frozen and thawed. The frozen samples have been deposited in the cryobank of the Timiryazev Institute of Plant Physiology (Moscow, Russia; Popov et al. 1995a,b).

3 Production of Diosgenin Phytosterols

Callus cultures, cell suspensions, somatic embryos, and regenerated shoots of *D. balcanica* and *D. caucasica* were extracted and analyzed for the presence of diosgenin and phytosterols (sitosterol, stigmasterol, and campesterol), depicted in Fig. 5. The steroid content in cultured tissues was compared with selected data obtained in the analyses of other *Dioscorea* species, known as good steroid producers. Likewise, steroid production in plants regenerated in culture and planted in the field were compared with wild plants collected at their natural habitats.

3.1 Methods of Analysis

Plant material for analysis was extracted and purified according to Drapeau et al. (1986). Quantification of diosgenin and phytosterols was done by gas chromatography, using a Hewlett-Packard Gas Chromatograph, model 5890 II, with a flame-ionizing detector. The external calibration method was used for quantification. The calibration curves were produced with standard diosgenin (Roth, Germany) and phytosterol (Fluka, Switzerland) solutions (Šavikin-Fodulović 1996). All experiments were repeated at least 3 times. The results are presented as diosgenin percentage of dry weight. Standard deviations of the mean were calculated using arc-sin transformed data.

3.2 Diosgenin and Phytosterol Contents in Cultured Cells, Tissues, and Organs

Callus strains, cell suspensions and organs of *D. balcanica* and *D. caucasica* were analysed on several occasions and the presence of diosgenin, sitosterol, stigmasterol and campesterol was confirmed. The representative data are shown in Table 2. It is clear that the content of all substances varied widely in different tissues. The experiment with five *D. balcanica* callus lines (A–E) indicates that a selection pressure is worth applying to establish lines that may be better producers than the original non-selected calli. The three organogenic lines A, B, and C produced less diosgenin than the lines D and E. In the two latter lines, the diosgenin content in darkness was remarkably increased. Light conditions also affected the qualitative composition of phytosterols. The dominant compound in light was sitosterol, while in darkness the amount of stigmasterol prevailed, particularly in lines D and E.

Eleven selected *D. caucasica* callus lines were reported as possessing different synthetic capacities (Deliu et al. 1989b). Also, in selected cell lines these authors (Deliu et al. 1992) found a very variable diosgenin content, from 0.04 to 0.74% (Table 2).

Most authors showed that the diosgenin content in calli and cells is lower than that found in plant parts in vivo. According to the data compiled by Furmanowa and Guzewska (1988) and Van Staden and Fowlds (1992), diosgenin content in cultured cells and tissues of several *Dioscorea* species is very variable, from less than 0.1% (*D. composita*, *D. bulbifera*) to 2.5% or more (*D. deltoidea*). The amounts of diosgenin found in *D. balcanica* and *D. caucasica* (Table 2) fall within the same range.

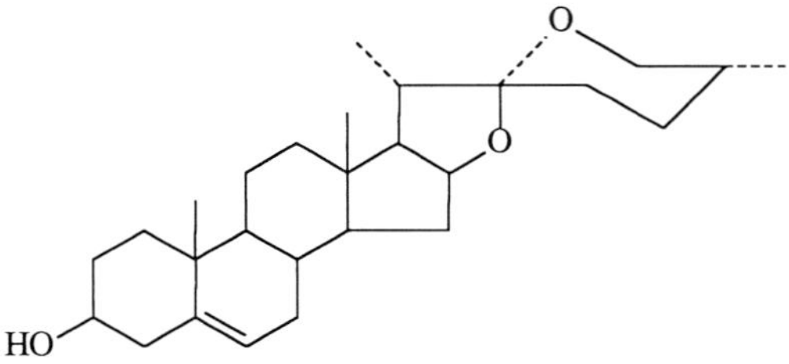

Diosgenin

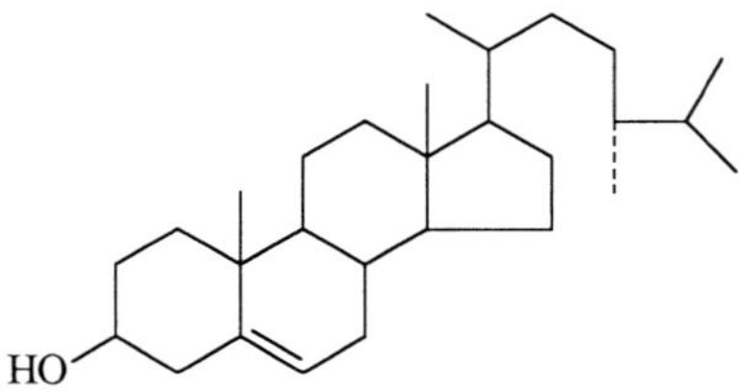

Campesterol

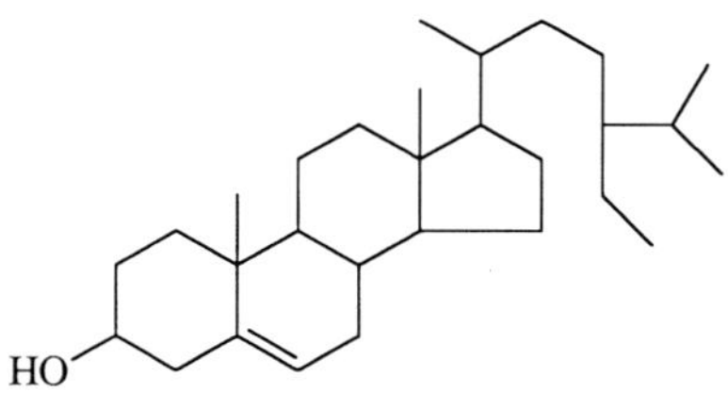

Sitosterol

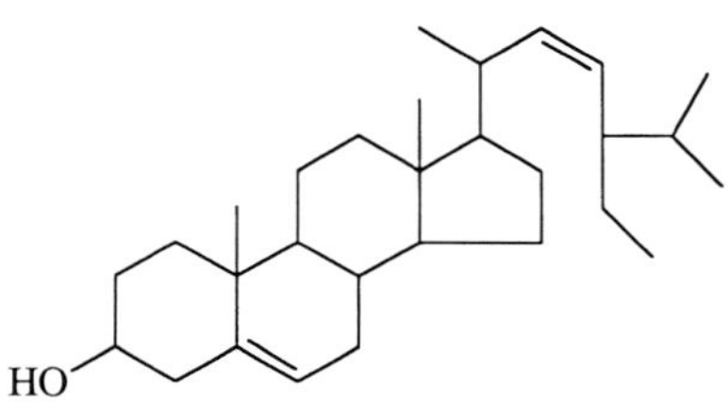

Stigmasterol

Fig. 5. Structural formulae of diosgenin, campesterol, sitosterol, and stigmasterol

Table 2. Diosgenin and phytosterol content (% on dry weight basis ± SD) in *D. balcanica* and *D. caucasica* tissues and organs in vitro

Plant species and material	Diosgenin	Sitosterol	Stigmasterol	Campesterol	Reference
D. balcanica					
Leaves	0.24 ± 0.02	–	–	–	Šavikin et al. (1994)
Roots	0.37 ± 0.02	–	–	–	
Callus	0.28 ± 0.02	–	–	–	
Cell suspension	0.41 ± 0.02	0.07 ± 0.01	0.16 ± 0.02	0.03 ± 0.01	Šavikin-Fodulović (1996)
Somatic embryos	0.16 ± 0.01	0.55 ± 0.03	0.09 ± 0.01	t.[a]	
Selected callus lines					
A, light	0.03 ± 0.01	0.02 ± 0.01	t.	0.01 ± 0.00	
darkness	0.05 ± 0.01	0.01 ± 0.00	0.13 ± 0.02	0.03 ± 0.01	
B, light	0.11 ± 0.02	0.06 ± 0.01	0.02 ± 0.00	0.01 ± 0.00	
darkness	0.22 ± 0.02	0.01 ± 0.00	0.06 ± 0.01	0.02 ± 0.01	
C, light	0.06 ± 0.01	0.04 ± 0.01	t.	0.01 ± 0.00	
darkness	0.05 ± 0.01	0.01 ± 0.00	0.05 ± 0.01	0.02 ± 0.01	
D, light	0.14 ± 0.02	0.08 ± 0.01	t.	0.02 ± 0.00	
darkness	0.52 ± 0.02	0.01 ± 0.00	0.3 ± 0.02	0.06 ± 0.01	
E, light	0.15 ± 0.01	0.10 ± 0.02	t.	0.02 ± 0.00	
darkness	0.58 ± 0.03	0.02 ± 0.01	0.22 ± 0.02	0.05 ± 0.01	
D. caucasica					
Leaves	0.23 ± 0.02	–	–	–	
Roots	0.61 ± 0.03	–	–	–	Šavikin et al. (1994)
Callus	0.20 ± 0.01	–	–	–	
Callus	0.40	–	–	–	Deliu et al. (1989a)
Cell lines					Deliu et al. (1992)
Non-selected	0.20 ± 0.1	–	–	–	
2	0.04 ± 0.01	–	–	–	
6	0.28 ± 0.02	–	–	–	
8	0.74 ± 0.03	–	–	–	
10	0.18 ± 0.01	–	–	–	
13	0.12 ± 0.00	–	–	–	
15	0.48 ± 0.02	–	–	–	
17	0.09 ± 0.00	–	–	–	
20	0.15 ± 0.00	–	–	–	
21	0.06 ± 0.00	–	–	–	
25	0.34 ± 0.01	–	–	–	
48	0.10 ± 0.00	–	–	–	
59	0.59 ± 0.03	–	–	–	
74	0.23 ± 0.01	–	–	–	

[a] t., trace; –, not studied.

In fact, higher values reported by some authors, were due to various manipulations of the nutrient media, devised to improve the diosgenin production. Deliu et al. (1989b) found in *D. caucasica* that the MS mineral solution, supplemented with adequate organic and hormonal substances, was superior for phytosterol production, than some other solutions. An improvement of diosgenin production in *D. deltoidea* cells was achieved by a particular ratio of ammonium to nitrate nitrogen (Tal et al. 1982), or by increasing the phosphate content (Rokem et al. 1985).

Sucrose and glucose(1–10%) were the most frequently used sugars for *Dioscorea* tissues. Although other sugars were suitable for cell suspension growth, diosgenin was produced only in the presence of sucrose (1.5%) in

D. deltoidea (Tal et al. 1982) and *D. caucasica* (Deliu et al. 1989a). In our experiments with *D. balcanica* and *D. caucasica* (Šavikin, unpublished) sucrose was added in concentrations from 0.5–8.0%, and the range of diosgenin content was between 0.12–0.19%, and 0.10–0.21%, respectively, with the highest value at 2–4% sucrose.

As for the growth regulators' effect, Marshall and Staba (1976) hold that they generally have a minor role in diosgenin production. Deliu et al. (1989b) tried several auxins and concluded that 0.5 $mg\,l^{-1}$ 2,4-D, in addition to MS medium produced the highest diosgenin content (3.66%) in *D. caucasica*, while IAA, NAA, and IBA did not stimulate, or even inhibited the diosgenin accumulation. In our experiments with callus tissues, 2,4-D was applied in concentrations from 1.1–12 $mg\,l^{-1}$. While the callus mass decreased approximately fourfold over that range of 2,4-D, the diosgenin content increased from 0.13–0.22% in *D. balcanica*, and 0.17–0.24% in *D. caucasica* (Šavikin et al. 1994). At the concentrations of BA in the range of 0.2–3.2 $mg\,l^{-1}$, the diosgenin content showed slight changes, not in correlation with BA increasing doses (Šavikin et al. 1994).

It was assumed that the addition of appropriate precursors might increase the synthesis of sterols. Kaul et al. (1969) found that 1 $mg\,ml^{-1}$ of the diosgenin precursor cholesterol increased the diosgenin content from 1% in the control, to 2.5%. We have checked the effect of 10–1000 $mg\,l^{-1}$ cholesterol in callus line D of *D. balcanica*; the lower concentrations (10 and 50 $mg\,l^{-1}$) doubled the diosgenin content in light (from 0.11 to 0.25%), while at higher concentrations this effect was not observed. In darkness, where the control level was anyway higher (0.52%), cholesterol even decreased the diosgenin content.

Rokem et al. (1985) found an increase in diosgenin production in *D. deltoidea* cells upon the addition of norflurazon, the inhibitor of carotenoid biosynthesis. Deliu et al. (1992) confirmed an inverse correlation between carotenoid and diosgenin biosynthesis, i.e. between primary and secondary metabolism. We studied the effect of 0.3–30 mM norflurazon on diosgenin and phytosterol synthesis in callus line D of *D. balcanica*. Norflurazon at 0.3–3 mM increased the diosgenin content about 2.5-fold in light. However, the simultaneous addition of 100 $mg\,l^{-1}$ cholesterol significantly increased the diosgenin level in darkness, which at 0.3 and 3 mM norflurazon reached values of 0.69 and 0.67%, respectively.

3.3 Diosgenin and Phytosterols Content in Regenerated Plants and their Vegetative Descendants

The plantlets of both *D. balcanica* and *D. caucasica*, that were regenerated in vitro through organogenesis or somatic embryogenesis, required an acclimation period of about 3 months in the greenhouse. Several plants were grown for a year in the greenhouse, and then their roots, rhizomes, and leaves were extracted to measure the diosgenin content (Table 3). Most acclimated plantlets were planted in the experimental field of the Institute for Medicinal Plant Research, on the plain about 20 km north of Belgrade. After 2 years, the

Table 3. Diosgenin content (% on dry weight basis ± SD) in *D. balcanica* and *D. caucasica* plants from different localities

Species and sample origin	Roots	Leaves	Rhizomes	Reference
D. balcanica				
Greenhouse	0.16 ± 0.01	0.29 ± 0.01	0.76 ± 0.04	Šavikin et al. (1994)
Experimental field	–	0.33 ± 0.01	2.93 ± 0.03	Šavikin-Fodulović (1996)
Botanical Garden	0.31 ± 0.01	0.42 ± 0.02	2.25 ± 0.07	
Montenegro	0.49 ± 0.02	0.39 ± 0.01	2.18 ± 0.08	
S.W. Serbia	0.36 ± 0.01	0.35 ± 0.01	1.98 ± 0.09	
D. caucasica				
Greenhouse	0.17 ± 0.01	0.41 ± 0.02	0.73 ± 0.02	Šavikin et al. (1994)

diosgenin content was analysed in rhizome and leaf extracts. The results were compared with the analyses of plants harvested at two natural habitats, in south-west Serbia and Montenegro, and the plants from Belgrade Botanical Garden (Table 3). As was expected, the rhizomes of all plants were the richest sources of diosgenin. Petričić et al. (1973) reported that the underground organs of *D. balcanica*, harvested in Montenegro contained about 1% diosgenin, and low amounts of a few other saponins, which were not identified. We found 1.98% diosgenin in the plants from the same region. However, the plants grown in the Botanical Garden, and particularly those in the experimental field, contained still higher amounts, 2.25 and 2.93%, respectively. This may be due to milder climatic conditions and to better agrotechniques, permitting more vigorous vegetative development. The plants cultivated in the greenhouse were intermediate between the in vitro cultures and plants in the soil.

4 Summary and Prospects

It has been shown that the two *Dioscorea* species, *D. balcanica* and *D. caucasica* share with many other members of the Dioscoreaceae the capacity for producing in vitro callus tissues, which may be continuously cultured in solid or liquid media. We have established protocols for three systems for plant in vitro propagation: shoot multiplication and rooting, somatic embryogenesis and microtuber formation. Since the calli and the regenerated plants also produce diosgenin and related steroid compounds, like the *Dioscorea* plants grown in nature, the projects for utilizing in vitro cultures as sources of these pharmaceuticals sound feasible.

It may be envisaged that further investigations could be carried out in two directions. First, the callus tissues or cells in suspension culture might be grown on a large scale to provide material for extraction and purification of the drugs. Our experiments indicate several approaches, that may possibly increase diosgenin production. These include (1) selecting diosgenin-rich cell lines, such as line D for callus and cell cultures, or line E, which also has an embryogenic potential; (2) culturing callus tissues and cell suspensions in darkness; (3) optimizing conditions for high diosgenin production in selected lines, by checking various combinations of nutrients in culture media; and (4) supplying precursors, such as cholesterol, or compounds like norflurazon, which could divert the metabolic pathways towards diosgenin synthesis. All these approaches have been poorly explored so far with *D. balcanica* and *D. caucasica* tissues and should be given much more attention.

The second line of research may be directed towards the micropropagation of plantlets, which could be grown in plantation fields. This could also be preceded by in vitro selection for high drug producers. Since the *D. balcanica* and *D. caucasica* grow in remote places at high altitudes, it is very important that they can adapt to agricultural soils in temperate regions, such as the surroundings of Belgrade. Since harvesting from nature is not permitted, field-grown plants from in vitro cultures could provide a stable source of plant material.

Acknowledgments. The authors would like to thank Drs. Nebojša Menković and Mihajlo Ristić, the Institute for Medicinal Plant Research "Dr. Josif Pančić", for their collaboration in analyses of secondary metabolites. Most of the research on *D. balcanica* and *D. caucasica* was supported through the grant #03E21, funded by the Ministry of Science and Technology of Serbia. The scientific collaboration with the "Timiriazev" Institute of Plant Physiology Moscow, Russia, is greatly acknowledged.

References

Ammirato PV (1978) Somatic embryogenesis and plantlet development in suspension cultures of the medicinal yam, *Dioscorea floribunda*. Am J Bot 65 (Suppl):27

Ammirato PV (1984) Yams. In: Ammirato PV, Evans DA, Sharp WR, Yamada Y (eds) Handbook of plant cell culture, vol 3. Crop species. Macmillan, New York, pp 327–354

Battat E, Rokem JS, Goldberg I (1989) Growth of *Dioscorea deltoidea* at high sugar concentrations. Plant Cell Rep 7:652–654

Burkill IH (1960) The organography and the evolution of Dioscoreaceae, the family of the Yams. J Linn Soc Bot 56:319–412

Butenko RG, Vorobyov AS, Nosov AM, Knyazkov IE (1992) Synthesis, accumulation, and location of steroid glucosides in cells of different strains of *D. deltoidea* Wall. Sov Plant Physiol 39:763–768

Ćulafić L (1991) Clonal micropropagation of *Dioscorea balcanica* – an endemic species of the Yugoslav flora. In: Butenko RG (ed) Biology of cultivated cells and plant biotechnology. Nauka, Moscow (in Russian)

Ćulafić L, Grubišić D, Vujičić R, Volkova LA, Popov AS (1994) Somatic embryo production in vitro in *Dioscorea caucasica* Lipsky and *Dioscorea balcanica* Košanin and cryopreservation of their organogenic callus tissues. Russ J Plant Physiol 41:821–826

Deliu C, Tamas M, Farcas A (1989a) *Dioscorea caucasica* Lipsky tissue culture. I. The action of some factors on callus development and diosgenin biosynthesis. 4th National Symposium on Plant Cell and Tissue Culture. Cluj-Napoca, pp 110–111

Deliu C, Farcas A, Tamas M (1989b) *Dioscorea caucasica* Lipsky tissue culture. II. Effect of hormonal balance on sapogenin and phytosterol biosynthesis in two callus strains. 4th Natl Symp on Plant cell and tissue culture, Cluj-Napoca, pp 111–113

Deliu C, Tamas M, Dobrota C, Munteanu-Deliu C (1992) Kinetics of cellular growth and diosgenin biosynthesis in *Dioscorea caucasica* cell suspensions, cultured in batch system. Plant Sci 85:99–105

Drapeau D, Sauvaire Y, Blanch HW, Wilke CR (1986) Improvement of diosgenin yield from *Dioscorea deltoidea* plant cell culture by use of a non-traditional hydrolysis method. Planta Med 6:474–478

Forsyth C, van Staden J (1984) Tuberization of *Dioscorea bulbifera* stem nodes in culture. J Plant Physiol 115:79–83

Furmanowa M, Guzewska J (1988) *Dioscorea*: in vitro culture and the micropropagation of diosgenin-containing species. In: Bajaj YPS (ed) Biotechnology in agriculture and forestry, vol 7. Medicinal and aromatic plants II. Springer, Berlin Heidelberg New York, pp 162–184

Grubišić D, Ćulafić L, Bojović-Cvetić D (1991) Regeneration of yam plants (*Dioscorea caucasica* Lipsky and *D. balcanica* Košanin) in in vitro culture. Fiziol Rast 38:1018–1022 (in Russian)

Grubišić D, Ćulafić L, Bojović-Cvetić D (1993) Development of aerial tubers in plants of relict species *Dioscorea balcanica* Košanin and *Dioscorea caucasica* Lipsky. Sov Plant Physiol 40:245–249

Jean M, Cappadocia M (1991) In vitro tuberization in *Dioscorea alata* L. "Brazo fuerte" and "Florido" and *D. abyssinica* Hoch. Plant Cell Tissue Organ Cult 26:147–152

Jean M, Cappadocia M (1992) Effects of some growth regulators on in vitro tuberization in *Dioscorea alata* L. "Brazo fuerte" and *D. abyssinica* Hoch. Plant Cell Rep 11:34–38

John JL, Courtney WH, Decoteau DR (1993) The influence of plant growth regulators and light on microtuber induction and formation in *Dioscorea alata* L. cultures. Plant Cell Tissue Organ Cult 34:245–252

Kandarakov OF, Vorobev AS, Nosov AM (1994) Biosynthetic characteristics of a *Dioscorea deltoidea* cell population grown in continuous culture. Russ J Plant Physiol 41:805–809

Kaul B, Stohs SJ, Staba EJ (1969) *Dioscorea* tissue culture. III. Influence of various factors on diosgenin production by *Dioscorea deltoidea* callus and suspension culture. Lloydia 32:347–359

Kičenko VI, Panina VV (1965) Diosgenin content in rhizomes of several *Dioscorea* species, introduced into Moscow surroundings. Rastit Resur 1:397–401 (in Russian)

Knyazkov IE, Lobakova EC, Nosov AM (1994) Steroid glucoside location and ultrastructure of *Dioscorea deltoidea* cells in culture. Russ J Plant Physiol 41:789–794

Kohmura H, Araki H, Imoto M (1995) Micropropagation of "Yamatoimo" Chinese yam (*Dioscorea opposita*) from immature leaves. Plant Cell Tissue Organ Cult 40:271–276

Košanin N (1914) *Dioscorea balcanica* Koš. n. sp. Österr Bot Z 64:37–39

Košanin N (1929) Nachträge zur *Dioscorea balcanica* Koš. Bull Inst Jard Bot Univ Beograd 1:199–203

Lakušić R (1966) Einige Resultate ökologischer Untersuchungen an der Art *Dioscorea balcanica* Košanin. Angew Pflanzensoziol 18:47–53

Lauzer D, Laublin G, Vincent G, Cappadocia M (1992) In vitro propagation and cytology of wild yams, *Dioscorea abyssinica* Hoch. and *D. mangenotiana* Miege. Plant Cell Tissue Organ Cult 28:215–223

Lipsky AK (1894) *Dioscorea caucasica* Lipsky. Zap Kiewskago Obš Estestw 13:143

Malaurie B, Pungu O, Dumont R, Trouslot MF (1993) The creation of an in vitro germplasm collection of yam (*Dioscorea* spp.) for genetic resources preservation. Euphytica 65:113–122

Malaurie B, Pungu O, Trouslot MF (1995) Effect of growth regulators concentrations on morphological development of meristem-tips in *Dioscorea cayenensis-D. rotundata* complex and *D. praehensilis*. Plant Cell Tissue Organ Cult 41:229–235

Mantell SH, Hugo SA (1989) Effects of photoperiod, mineral medium strength, inorganic ammonium, sucrose and cytokinin on root, shoot and microtuber development in shoot cultures of *Dioscorea alata* L. and *D. bulbifera* L. yams. Plant Cell Tissue Organ Cult 16:23–37

Marshall JG, Staba EJ (1976) Hormonal effects on diosgenin biosynthesis and growth in *Dioscorea deltoidea* tissue cultures. Phytochemistry 15:53–55

Martin FW, Gaskins MH (1968) Cultivation of the saponin-bearing *Dioscorea* species. US Dept Agric Prod Res Rep 103:3–19

Mitchell SA, Asemota HN, Ahmad MH (1995) Effects of explant source, culture medium strength and growth regulators on the in-vitro propagation of three Jamaican yams (*Dioscorea cayenensis*, *D. trifida* and *D. rotundata*). J Sci Food Agric 67:173–180

Mix-Wagner G (1994) In-vitro multiplication of white yam (*Dioscorea rotundata* Poir.) and taro (*Colocasia esculenta* L.) for the production of vegetative propagation stock. Plant Res Dev 40:24–36

Murashige T, Skoog F (1962) A revised medium for rapid growth and bio-assays with tobacco tissue cultures. Physiol Plant 15:473–497

Nagasawa A, Finer JJ (1989) Plant regeneration from embryogenic suspension cultures of Chinese yam (*Dioscorea opposita* Thunb.). Plant Sci 60:263–271

Ng SYC (1992) Micropropagation of white yam (*Dioscorea rotundata* Poir.). In: Bajaj YPS (ed) Biotechnology in agriculture and forestry, vol 19. High-Tech and micropropagation III. Springer, Berlin Heidelberg New York, pp 135–159

Oreshnikov AV, Nosov AM, Manakov MN (1994) Characterization of *Dioscorea deltoidea* cells grown in closed continuous culture. Russ J Plant Physiol 41:810–814

Osifo EO (1988) Somatic embryogenesis in *Dioscorea*. J Plant Physiol 133:378–380

Petričić J, Apostolski R, Bedenko D (1973) Saponine zweier in Jugoslawien vorkommender *Dioscoreazeen*. Acta Pharm Jugosl 23:45–47

Popov AS, Volkova LA (1994) Cryopreservation and some characteristics of *Dioscorea deltoidea* cell cultures in the vitamin-free medium. Russ J Plant Physiol 41:815–820

Popov AS, Volkova LA, Ćulafić L (1995a) Cryopreservation of in vitro plants and plant tissue germplasm. Bull Inst Jard Bot Univ Beograd 29:1–8 (in Russian)

Popov AS, Volkova LA, Butenko RG (1995b) Cryopreservation of germplasm of *Dioscorea deltoidea* (medicinal yam). In: Bajaj YPS (ed) Biotechnology in agriculture and forestry, vol 32. Cryopreservation of plant germplasm I. Springer, Berlin Heidelberg New York, pp 487–499

Rizovski R (1977) Beitrag zur Kenntnis die Verbreitung von *Dioscorea balcanica* Koš. Glas Repub Zavoda Zašt Prir 10:101–104

Rokem JS, Tal B, Goldberg I (1985) Methods for increasing diosgenin production by *Dioscorea* cells in suspension cultures. J Nat Prod 48:210–222

Saleil V, Degras L, Jonard R (1990) Obtention de plantes indemnes du virus de la mosaïque de l'igname (YMV) par culture in vitro chez l'igname américaine *Dioscorea trifida* L. Agronomie 10:605–615

Šavikin K, Grubišić D, Menković N, Ristić M, Ćulafić L (1994) Diosgenin content in calluses and organs of *Dioscorea balcanica* Košanin and *D. caucasica* Lipsky, regenerated in in vitro cultures. Rastit Resur 30:102–106 (in Russian)

Šavikin-Fodulović K (1996) Endemic relict species *Dioscorea balcanica* Košanin – a potential producer of diosgenin in in vitro cultures. PhD Thesis, University of Beograd (in Serbian)

Šavikin-Fodulović K, Grubišić D, Ćulafić L (1997) Selection of *Dioscorea balcanica* callus lines. Arch Biol Sci (Beograd) 49:105–112

Sengupta J, Mitra GC, Sharma AK (1984) Organogenesis and tuberization in cultures of *Dioscorea floribunda*. Plant Cell Tissue Organ Cult 33:325–331

Tal B, Gressel J, Goldberg I (1982) The effect of medium constituents on growth and diosgenin production by *Dioscorea deltoidea* cells grown in batch cultures. Planta Med 44:111–115

Twyford CT, Mantell SH (1996) Production of somatic embryos and plantlets from root cells of the greater yam. Plant Cell Tissue Organ Cult 46:17–26

Van Staden J, Fowlds DL (1992) Micropropagation of medicinal *Dioscorea* species. In: Bajaj YPS (ed) Biotechnology in agriculture and forestry 19. High-Tech and micropropagation III. Springer Berlin Heidelberg New York, pp 425–442

Wagner H (1980) Pharmazeutische Biologie. Drogen und ihre Inhaltsstoffe. Gustav Fischer-Verlag Stuttgart New York

Wernicke W, Park HY (1993) The apparent loss of tissue culture competence during leaf differentiation in yams (*Dioscorea bulbifera* L.). Plant Cell Tissue Organ Cult 34:101–105

VII *Drosera* Species (Sundew): In Vitro Culture and the Production of Plumbagin and Other Secondary Metabolites

J. Šamaj[1], A. Blehová[2], M. Repčák[3], M. Ovečka[4], and M. Bobák[2]

1 General Account

Carnivorous plants have attracted considerable attention from many biologists for several hundred years, not only because of their special nutritional requirements and ecological adaptations, but also because of their value as medicinal herbs (Lecoufle 1990). Among them, Charles Darwin studied insectivorous plants in detail and used them in his evolutionary studies (Darwin 1875). The genus *Drosera*, which consists of about 125 species (Culham and Gornall 1994), represents a really good example of plant evolution and functional adaptation. Importantly, extracts from numerous species of *Drosera* have been traditionally used for various medicinal purposes, especially as efficient agents against respiratory diseases (Table 1). These therapeutic effects are thought to correlate with the content of secondary metabolites, namely naphthoquinones which are synthesized and accumulated in various *Drosera* species (Zenk et al. 1969; Culham and Gornall 1994). Among naphthoquinones, the most intensively studied has been plumbagin, because of its broad medicinal and other effects (see Finnie and van Staden 1993). However, many *Drosera* species also contain the second of the two most abundant naphthoquinones of *Droseraceae*, called 7-methyljuglone. Besides this, *Drosera* plants synthesize several minor, but unique, naphthoquinones as well as a broad spectrum of flavonoids with practical or potential medicinal use. Plumbagin production in vivo and in vitro has been well described, especially for the two South-African species *D. capensis* and *D. natalensis*, by Crouch et al. (1990). Therefore, in this chapter attention is focused on new data on the production of 7-methyljuglone, plumbagin and minor naphthoquinones as well as flavonoids in some new *Drosera* species, e.g. *D. spathulata* and *D. rotundifolia*.

[1] Institute of Plant Genetics, Slovak Academy of Sciences, Akademická 2, SK-950 07 Nitra, Slovak Republic, Present address till end of April 1999: Botanisches Institute, Venusbergweg 22, D-53115 Bonn, Germany
[2] Department of Plant Physiology, Comenius University Bratislava, Mlynská dolina B-2, SK-842 15 Bratislava, Slovak Republic
[3] Department of Experimental Botany and Genetics, Faculty of Natural Sciences, P.J. Šafárik University, Mánesova 23, SK-04167 Košice, Slovak Republic
[4] Institute of Botany, Slovak Academy of Sciences, Dúbravská cesta 14, SK-842 23 Bratislava, Slovak Republic

Biotechnology in Agriculture and Forestry, Vol. 43
Medicinal and Aromatic Plants XI (ed. by Y.P.S. Bajaj)

Table 1. Medicinal and other properties of *Drosera* and naphthoquinones

Use	Reference
Against old age and arteriosclerosis	Grieve (1959)
Allelophatic	Harborne (1982)
Anti-asthma	Frenzer (1980)
Anti-cancer	Krishnaswamy and Purushothaman (1980); Chandrasekaran et al. (1982); Keller (1985); Kreher et al. (1990); Fujii et al. (1992)
Antifeedant	Kubo et al. (1980); Thomson (1987)
Antifertility, Abortifacient	Bhargava (1984); Bhargava and Dixit (1985)
Antifungal	Krishnaswamy and Purushothaman (1980) Békésiová (1997)
Antileprosy	Bokemo (1984)
Antimicrobial	Lloyd and Middlebrook (1994); Denoel (1949); van der Vijver and Lotter (1971); Heble et al. (1974); Ray and Majumdar (1976); Krishnaswamy and Purushothaman (1980); Kubo et al. (1980); Wurm et al. (1984); Didry et al. (1986); Lakshmi et al. (1987); Durga et al. (1990); Fujii et al. (1992)
Anti-mutagenic	Farr et al. (1985); Durga et al. (1992)
Anti-nematodes	Fetterer and Fleming (1991)
Anti-spasmodic	Juniper et al. (1989)
Anti-viral	Walt (1972)
Aphrodisiac	Grigson (1958)
Bronchial infection	Denoel (1949)
Cardiotonic	Itoigawa et al. (1991)
Cosmetic	Slack (1980)
Chitin synthethase inhibitor	Kubo et al. (1983); Mitchell and Smith (1988)
Immunomodulator	Kreher et al. (1990)
Inhibition of seed germination	Reynolds (1987)
Insecticidal	Kubo et al. (1980); Thomson (1987); Gujar and Mehrotra (1988); Joshi and Sehnal (1989)
Against tuberculosis, *Staphylococcus*	Heise and Steenken (1941); Lloyd and Middlebrook (1944); Denoel (1949)
Neisseria	Gundidza and Manwa (1990)
Against whooping cough	Grieve (1959); Weiss (1991); Ragazzi et al. (1993)

It is now well established that aseptic cultures of *Drosera* species represent a promising alternative source of material for medicinal use, since natural plants are endangered and are therefore protected by law in many countries (Bonnet et al. 1984a; Blehová et al. 1995; Wawrosch et al. 1996).

1.1 Plant Distribution and Characterization

Various *Drosera* species are distributed worldwide across tropical to tundra regions, except for deserts and Antarctica, and they are adapted to various climates with wet or seasonally wet open ground. Numerous *Drosera* species

Fig. 1. *Drosera capensis.* **a** Flowering plant. **b** Detail of thread-like trap with glandular hairs having pink drops of mucilage

are concentrated in Australia and New Zealand, while others can be found only in South Africa or in Europe. *Drosera* plants can withstand low as well as high temperatures between 4–40 °C. European species form winter buds during cold acclimation (Slack 1985; Kukulczanka 1991), while some Australian species form gemmae in response to the reduction in photoperiod and temperature (Albert et al. 1992). Gemmae represent special organs for vegetative reproduction which are functionally and even structurally analogous to seeds.

Drosera species have different morphological and anatomical adaptations for carnivory. Their trap consists of a great number of stalked glands which exude sticky droplets attracting insects (Fig. 1a,b). These droplets do not dry out in the sun, and because of this the name sundew originated for these plants. Secretory cells of these glands produce digestive hydrolytic and proteolytic enzymes (Lloyd 1942; Heslop-Harrison 1975), e.g. proteases and acid phosphatases (Clancy and Coffey 1977; Dexheimer 1978) which are activated and secreted extracellularly together with a mucilage after gland stimulation (Clancy and Coffey 1977; McNally et al. 1988). Recently, glucanase with anti-fungal properties has also been isolated from *Drosera* plants cultured in vitro after gelatine stimulation (Békésiová 1997). For example, in

D. rotundifolia, the mucilage is secreted by tentacles (there are no separate glands for enzyme secretion and mucilage production) and serves for prey capture and its digestion (McNally et al. 1988). Länger et al. (1995) analysed 52 *Drosera* species and found 14 different types of glandular hairs, including two-celled papillae, peltate scales, several types with unbranched, bi- or multiseriate stalks with two- or multicellular glands, and one type with a multiseriate stalk and a two-armed gland. The type of glandular hairs confirms the actual classification of the genus and facilitates the identification of species. Interestingly, vascularized glands are present only in Droseraceae (Albert et al. 1992).

1.1.1 Taxonomic Classification and Characterization of the Plant

The genus *Drosera* contains about 125 species (Culham and Gornall 1994) represented by annual and perennial terrestrial herbs. They have rhizomes, fibrous roots, or tubers with a vertical underground stolon. Stems are often very short or long and erect or climbing. Leaves are alternate or basal, usually forming a rosette, having petioles and glandular hairs (Fig. 1a,b). Flowers are bisexual, actinomorphic, usually with five sepals, five free petals, five stamens and three–five styles (rarely two or six) free or connate at the base. The ovary is superior and loculate. Placentation is parietal. Ovules are characterized by nucellar elongation and strongly enlarged cells of the nucellar epidermis. Capsules are dehiscent, two–six valved. Seeds are numerous, ellipsoidal, linear or winged. *Drosera* has a less complex testa compared with *Drosophyllum* or *Dionaea*. The operculate seed is composed of a non-lignified seed coat, starchy endosperm and small embryo (Boesewinkel 1989). Pollen grains remain united in permanent tetrads, which is accomplished by cross-wall adhesion, a unique structure in the angiosperms (Takahashi and Sohma 1982). Pollen tetrads behave as single units and larger pollen masses are thought to enhance self-pollination, because each flower in a *Drosera* inflorescence opens only for a few hours in the morning (Takahashi 1988).

1.1.2 Nutritional Requirements

In natural conditions, *Drosera* grows mostly on acidic substrates which are often deficient in nitrogen and phosphorus (Chandler and Anderson 1976; Stewart and Nilsen 1992). For example, *D. rotundifolia* is well adapted to low salt concentrations in the substrate. On the other hand, higher salt concentrations negatively influence plant growth, and this effect is more pronounced than insect starvation (Stewart and Nilsen 1992). The facultative or obligatory nature of insectivory and its relevance for plant nutrition has often been discussed for *Droseraceae* (Darwin 1878; Lloyd 1942; Chandler and Anderson 1976; Thum 1989a,b; Schulze and Schulze 1990; Stewart and Nilsen 1992). Insects as prey were proposed to be a valuable additional source of nitrogen. For example, an optimal growth of *D. binata* was achieved after application of insects to plants, while addition of inorganic nitrogen resulted in slower growth and flower inhibition (Chandler and Anderson 1976). Several studies have shown that the source of nitrogen is important for *Drosera* nutrition (Chandler and Anderson 1976; Small et al. 1977; Schulze and Schulze 1990; Stewart and Nilsen 1992). Usually, *Drosera* competes with the moss *Sphagnum* for light and nitrogen in their natural biotopes (Svensson 1995). Besides nitrogen, the level of phosphorus is also critical for optimal growth and development (Karlsson and Pate 1992b; Stewart and Nilsen 1992). It was proposed that captured insects can represent an alternative source of phosphorus for *Drosera* plants (Chandler and Anderson 1976; Stewart and Nilsen 1992).

1.2 Importance of the Plant

1.2.1 Evolutionary Impacts

The family Droseraceae, represents an ideal model for developmental studies of plant evolution (Darwin 1875), including developmental and physiological experiments that could address specific roles for phytohormones in macromorphological evolution (Albert et al. 1992). Moreover, the cosmopolitan genus *Drosera* with numerous species which are adapted to various environmental conditions is particularly interesting for studies of functional and structural adaptations (including insectivory nutrition and gemmae production) to various environments. Furthermore, this genus with pollen grains permanently organized in tetrads may exhibit an intermediate state in an evolutionary shift in pollination systems from cross-pollination to self-pollination (Takahashi 1988).

1.2.2 Medicinal and Other Uses

Numerous *Drosera* species are medicinal herbs containing valuable secondary metabolites. The most important are naphthoquinones, especially plumbagin and 7-methyljuglone (Zenk et al. 1969; Finnie and van Staden 1993; Culham and Gornall 1994) and flavonoids (Schölly and Kapetanidis 1989b, 1993; Budzianowski et al. 1993; Repčák and Galambosi 1996). Extracts from *Drosera* which contain naphthoquinones have medicinal and other valuable properties (Table 1). Importantly, naphthoquinones are thought to be responsible for therapeutic effects against respiratory diseases in humans, including asthma, bronchial infections, whooping cough and tuberculosis. Plumbagin has been the most studied in this respect. For example, this compoud can inhibit enzyme activity reacting with SH and NH_2 radicals. Its liposoluble properties are used against tuberculosis (Denoel 1949). Plumbagin also inhibits development of parasitic nematodes and insects. It was also shown that naphthoquinones have antiviral, antibacterial, antifungal and anticancer properties. Moreover, they have aphrodisiac, antispasmodic, antileprosy and antisclerotic effects on humans, and they are supposed to act in *Drosera*

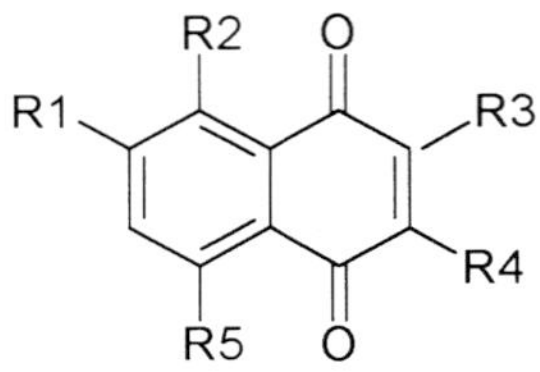

	R1	R2	R3	R4	R5	
[1]	Me	H	H	H	OH	**7-methyljuglone**
[2]	H	H	Me	H	OH	**plumbagin**
[3]	H	H	Me	OH	OH	**droserone**
[4]	OH	H	Me	OH	OH	**hydroxydroserone**
[7]	H	H	Me	OH	OGlc	**droserone-5-glucoside**
[8]	OH	H	Me	OH	OGlc	**hydroxydroserone-5-glucoside**

Fig. 2. Chemical structure of naphthoquinones and some glucosides isolated from *Drosera* species. Both 7-methyljuglone (5-hydroxy-7-methyl-1,4-naphthoquinone) [**1**] and plumbagin (5-hydroxy-2-methyl-1,4-naphthoquinone) [**2**] have been shown to be major naphthoquinones in the family Droseraceae. Apart from these two compounds, *Drosera* plants contain several minor, but often unique, naphthoquinones. Droserone (3,5-dihydroxy-2-methyl-1,4-naphthoquinone) [**3**] and hydroxydroserone (3,5,8-trihydroxy-2-methyl-1,4-naphthoquinone) [**4**] (Schölly and Kapetanidis 1989a,b) as well as glycosides like droserone-5-glucoside (3,5-dihydroxy-2-methyl-1,4-naphthoquinone-5-glucoside) [**7**] (Schölly and Kapetanidis 1989a), and hydroxydroserone-5-glucoside (3,5,8-trihydroxy-2-methyl-1,4-naphthoquinone-5-glucoside) [**8**] (Schölly and Kapetanidis 1989b) were isolated and characterized from various *Drosera* species

plants as protective chemicals having allelopathic, molluscidal, antimicrobial and antifeedant functions (Table 1). The chemical structure of *Drosera* naphthoquinones and naphthoquinone-glucosides is shown in Figs. 2 and 3.

Besides naphthoquinones, *Drosera* plants also contain other protective chemicals, namely flavonoids. They act as antioxidants, enzyme inhibitors, precursors of toxic substances, light screens and insect attractants in plants. The biological activity of these compounds depends on the presence of a free hydroxyl group. Flavonoids have been demonstrated to affect the activity of many mammalian and other cell systems in vitro. For example, kaempferol is a specific inhibitor of myosin light chain kinase and acts competitively with ATP. Quercetin and myricetin are potent inhibitors of glyoxylase enzymes which might be important in the regulation of cell division by detoxifying α-ketoaldehydes. Some of them significantly affect the function of the immune system and inflammatory cells (Middleton and Kandaswami 1993). Quercetin,

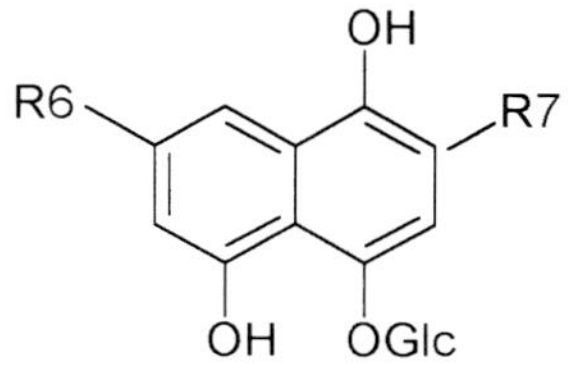

	R6	R7	
[5]	Me	H	**rossoliside**
[6]	H	Me	**2-methyl-hydrojuglone-4-glucoside**

Fig. 3. Chemical structure of rossoliside (7-methyl-hydrojuglone-4-glucoside) [**5**] (Vinkenborg et al. 1969, 1970, Sampara-Rumantir 1971) and 2-methyl-hydrojuglone-4-glucoside [**6**] (Kreher et al. 1990) isolated from *Drosera* species

kaempferol and myricetin can inhibit the release of rat mast cell histamine. Regarding medicinal use of flavonoids, quercetin can be used in anti-cancer prevention and therapy (Smith and Yang 1994). This compound has been shown to inhibit the mutagenic activity of benzopyrene, a representative polynuclear aromatic hydrocarbon carcinogen in bacterial mutagenicity studies (Middleton and Kandaswami 1993).

1.3 Conventional Practices for *Drosera* Propagation

Drosera plants are conventionally propagated by seeds or by leaf and root cuttings, and tubers (Finnie and van Staden 1993). Seeds require stratification by cold temperature (around 4 °C) before sowing in spring. Vegetative propagation by leaf cuttings is very effective and, therefore, widely used in practice (Lloyd 1942).

Drosera species are normally reproduced in nature by seeds, however, some Australian sundews (e.g. *D. dichrosepala* and *D. pygmaea*) also produce gemmae, the vegetative reproductive organs, on the specialized leaves (Albert et al. 1992). Gemmae represent terminal abscising propagules which are reproductively adequate as seeds, however, no fertilization event occurs during their genesis. In contrast to sexual reproduction from a zygote, gemmae originate from diploid somatic cells. This type of vegetative reproduction was briefly described previously by Slack (1980) and Lowrie (1989) which reported that gemmae consist of nutritive tissue and embryogenic leaf and root structures. Appearance of gemmae in place of trapping lamina is homeotic and likely represents a shift in cell differentiation rather than development of a de novo structure (Albert et al. 1992). Karlson and Pate (1992a) reported greater gemma production in several sundew species (8–52 gemmae per plant during the season) compared with seed formation (just eight mature seeds per plant during the season). These data indicate that the vegetative type of reproduction is more effective for plants under changing environmental conditions. Plants in their natural environment (where normal fertilization might be problematic) create gemmae instead of seeds, because it takes less time after a dramatic change of climatic conditions. Gemmae germinate very effectively (80–100%), and usually just within 7–14 days (Lowrie 1989). Plant development from gemma to mature plant takes only 70–90 days compared with mature seed which needs 120–150 days.

1.4 Production of Medicinal Compounds

Different *Drosera* species produce two major naphthoquinones – plumbagin and 7-methyljuglone in vivo (Table 2) as well as in vitro (Table 3). These two naphthoquinones are, however, not unique for *Drosera*; they are also produced by other species. For example, 7-methyljuglone is a predominant quinone in the Ebenaceae, and plumbagin is produced by Plumbaginaceae, Nepenthaceae and Dioncophyllaceae (Thomson 1987; Culham and Gornall 1994).

The presence of plumbagin and 7-methyljuglone was studied in different *Drosera* species (e.g. Zenk et al. 1969; Culham and Gornall 1994) as chemotaxonomic markers. Rarely, their level was also quantitatively estimated in plants. For example, Crouch et al. (1990) reported that the content of plumbagin in plants of *D. natalensis* and *D. capensis* growing in vivo represented 0.0025%, and 0.0048% of their fresh mass, respectively. On the other hand, Repčák and Galambosi (1994) found that the content of 7-methyljuglone in *D. rotundifolia* and *D. anglica* collected in Finland depends on ontogenetic development of leaves. They detected the highest levels of 7-methyljuglone in young leaves (2.7% of dry weight in *D. rotundifolia* and 2.1% in *D. anglica*) and flowers (2.7% in *D. rotundifolia*) compared with mature leaves (only 0.8–1.2% in *D. rotundifolia* and 0.6–1.6% in *D. anglica*). Moreover, red pigmentation of leaves was connected with low levels of this naphthoquinone (0.04–0.1% in both species). Recently, Krenn et al. (1995) estimated the content of plumbagin and 7-methyljuglone in a drug containing *D. peltata* and *D. madagascariensis*. The amount of plumbagin in two samples of *D. peltata* was 0.569% and 0.305% of dry weight. The amount of plumbagin in 11 samples of *D. madagascariensis* varied between 0.002 and 0.01% of dry weight, and that of 7-methyljuglone between 0.004 and 0.054%.

Importantly, *Drosera* plants also synthesize some minor naphthoquinones like droserone, hydroxydroserone and several naphthoquinone glucosides (Table 4) which could not be found in other plant species.

Drosera also contains a broad spectrum of flavonoids and flavonol-glucosides like quercetin, isoquercitin, kaempferol, astragalin, hyperin, gossypin and gossypitrin (see Sect. 3.2). Plants of *D. rotundifolia* and

Table 2. Occurrence of two major naphthoquinones isolated from in vivo cultured *Drosera* species

Plumbagin	Taxon [Reference][a]	7-Methyljuglone
D. andersoniana [8]		*D. adelae* [8]
D. auriculata [1,2,8]		*D. aliciae* [1,2,8]
D. binata [1,2,4,8]		*D. aliciae x capensis* [8]
D. bulbosa [8]		*D. anglica* [9]
D. capensis [1,4,7]		*D. auriculata* [8]
D. capillaris [2,4]		*D. burkeana* [1,2,8]
D. cistiflora [1,8]		*D. burmanii* [8]
D. dichotoma [1,2]		*D. capensis* [1,2,8]
D. erythrorhiza [2,8]		*D. cistiflora* [1,8]
D. indica [1,8]		*D. cuneifolia* [1,8]
D. intermedia [1,2,4,5,8]		*D. dielsiana* [9]
D. longifolia [1[b]]		*D. filiformis* [8]
D. lunata [1]		*D. hamiltonii* [1,8]
D. macrophylla [8]		*D. hilaris* [8]
D. madagascariensis [10]		*D. indica* [8]
D. microphylla [2,8]		*D. intermedia* [5,8]
D. modesta [8]		*D. longifolia* [8]
D. natalensis [7]		*D. longifolia x rotundifolia* [8]
D. peltata [6,8,10]		*D. madagascariensis* [1,8,10]
D. platypoda [8]		*D. ramentacea* [3]
D. prolifera [8]		*D. regia* [8]
D. pygmaea [9]		*D. rotundifolia* [5,8,9]
D. regia [8]		*D. spathulata* [1,2,8]
D. rotundifolia [2,4,8]		*D. stolonifera, ssp. stolonifera* [8]
D. slackii [8]		*D. tracii* [1,2]
D. stolonifera, ssp. rupicola [2,8]		*D. trinervia* [1,8]
D. venusta [8]		*D. villosa* [8]
D. villosa [8]		
D. whittakeri [1,2,4,8]		

[a] References: [1] Zenk et al. (1969); [2] Durand and Zenk (1974a); [3] Luckner and Luckner (1970); [4] Nahrstedt (1980); [5] Bonnet et al. (1984a); [6] Leclerq and Angenot (1984); [7] Crouch et al. (1990); [8] Culham and Gornall (1994); [9] Länger et al. (1995); [10] Krenn et al. (1995).
[b] Possible artifact by Culham and Gornall (1994).

Table 3. Occurrence of two major naphthoquinones isolated from in vitro cultured *Drosera* species

Plumbagin	7-Methyljuglone
D. capensis (Crouch et al. 1990)	*D. capensis* (Wawrosch et al. 1996)
D. intermedia (Bonnet et al. 1984a; Budzianowski et al. 1993)	*D. hilaris* (Wawrosch et al. 1996)
D. natalensis (Crouch et al. 1990)	*D. intermedia* (Bonnet et al. 1984a; Budzianowski et al. 1993)
	D. rotundifolia (Bonnet et al. 1984a; Wawrosch et al. 1996)
	D. spatulata (Budzianowski et al. 1993; Blehová et al. 1995; Budzianowski 1995)

Table 4. Occurrence of other minor naphthoquinones isolated from *Drosera* species

Naphthoquinone	Species	Type of culture	Reference
Droserone	*D. whittakeri*	In vivo	Thomson (1949)
	D. rotundifolia	In vivo	Schölly and Kapetanidis (1989b)
Hydroxydroserone	*D. whittakeri*	In vivo	Macbeth and Winzor (1935)
		In vivo	Macbeth et al. (1935)
	D. rotundifolia	In vivo	Schölly and Kapetanidis (1989b)
Biramentaceone[a]	*D. ramentacea*[c]	In vivo	Krishnamoorthy and Thompson (1969)
Droserone-5-glucoside	*D. rotundifolia*	In vivo	Schölly and Kapetanidis (1989a)
Rossoliside	*D. rotundifolia*	In vivo	Vinkenborg et al. (1969, 1970)
		In vivo	Sampara-Rumantir (1971)
		In vitro	Budzianowski (1996)
	D. spatulata	In vitro	Budzianowski (1995)
Hydroxydroserone-5-glucoside	*D. rotundifolia*	In vivo	Schölly and Kapetanidis (1989b)
2-Methylnaphtazarin-5--O-glucoside[b]	*D. rotundifolia*	In vitro	Budzainowski (1997)
	D. spatulata	In vitro	Budzainowski (1997)

[a] Artifact produced by extraction with organic solvents, by Thomson (1987).
[b] Artifact produced by extraction with methanol.
[c] Possible misidentification of *D. madagascariensis* as *D. ramentacea*, by Culham and Gornall (1994).

Table 5. Content of flavonol aglycones in leaves, stems, and flowers of *Drosera rotundifolia* and *D. anglica*. (After Repčák and Galambosi 1996)

	Leaf		Stem		Flower	
	%	SE	%	SE	%	SE
Drosera rotundifolia						
Quercetin	5.96	0.311	4.69	0.243	6.40	0.366
Kaempferol	0.15	0.007	0.07	0.003	0.45	0.039
Drosera anglica						
Quercetin	5.14	0.128	3.61	0.071	3.72	0.088
Kaempferol	0.11	0.004	0.07	0.003	0.82	0.047

% of compound in dry matter, SE = standard error.

D. anglica grown in vivo in Finland were analyzed for flavonol aglycone content by HPLC (Table 5). A high percentage of quercetin (3.6 to 6.4% of dry weight) was found in both sundew species when compared with kaempferol (only 0.07 to 0.45% of dry weight). Both flavonols are present in leaves, stems and flowers. The main accumulation of kaempferol is only found in flowers (Repčák and Galambosi 1996). Many of the above-mentioned naphthoquinones and flavonoids, isolated from *Drosera*, have practical or potential medicinal use (Table 1; Middleton and Kandaswami 1993; Smith and Yang 1994).

2 In Vitro Approaches

2.1 Establishment of Tissue Cultures and Micropropagation

As a consequence of intensive plant collection, *Drosera* species can be found relatively rarely in nature (Juniper et al. 1989). In many countries they are protected by law. Therefore, new biotechnological methods based on in vitro culture of plants and tissues are considered for plant propagation which ensures the necessary amount of pharmaceutically important raw material. Simultaneously, the naphthoquinone production from in vitro growing plants is evaluated for commercial use. Most previous research on tissue culture of sundew stressed nutritional requirements and plant micropropagation. In spite of much effort, and good results with some *Drosera* species, the main goal still remains the optimization of culture conditions (Simola 1978a,b; Kukulczanka and Czastka 1988; Kukulczanka 1991; Blehová et al. 1992; Anthony 1992), leading to higher production of naphthoquinones in vitro (Bonnet et al. 1984a, Blehová et al. 1995, Wawrosch et al. 1996).

Drosera species have high regeneration potential when cultured under appropriate in vitro conditions (Table 6). Up until now, about 12 *Drosera* species have been successfully cultured in vitro. Mostly seeds, leaf rosettes or isolated leaves, and occasionally also roots, flowers and gemmae have been used as explants for the establishment of tissue cultures. Different agents including commercial bleach, CaOCl, $HgCl_2$ or NaOCl can be applied for explant sterilization. Best results are usually achieved when NaOCl is used in low concentrations as a sterilant (Finnie and van Staden 1993). Sterilization of leaf explants originating from in vivo plants is often difficult because of mucilage which possesses various microorganisms (Anthony 1992). Contamination in such cultures is often very high. It is therefore much better to use in vitro growing plants as a source from which sterile leaf explants can be obtained directly (Blehová et al. 1992). *Drosera* explants were cultured on agar, MS medium (Murashige and Skoog 1962) or RM medium (Reinert and Mohr 1967) which was originally developed for orchids. In the case of MS medium, its different modifications and dilutions including one half, one fourth, one fifth, one sixth and one eighth strength MS media were used in order to reduce salt concentrations. It was shown by Simola (1978a,b) that *D. rotundifolia*, unlike other carnivorous species, is able to grow and reproduce in aseptic cultures using relative simple media supplemented with inorganic and organic nitrogen. Culture media can be supplemented with casein hydrolysate, grapevine exudate, coconut milk, activated charcoal, nitrogen in various inorganic and organic forms and with plant hormones. These hormonal supplements include mostly benzyladenine (BA) as the source of cytokinin and α-naphthalenacetic acid (NAA) as the source of auxin.

For example, Bonnet et al. (1984b) used BA (10^{-5}M) with or without NAA to induce bud and shoot regeneration from isolated leaf explants of *D. rotundifolia*. They obtained more than 20 buds per leaf explant. By contrast, only roots are formed in the precence of NAA (10^{-5}M) and low concentra-

Table 6. Survey of in vitro approaches with various *Drosera* species

Species	Explant	Medium	Reference
Drosera	Seeds	Agar	Schmid (1912)
D. intermedia	Seeds	–	Burger (1961)
	Leaves	RM	Kukulczanka (1991)
	Seeds	RM	Budzianowski et al. (1993); Budzianowski (1995)
D. pygmaea	Seeds	Agar	Harder (1964)
	Gemmae	MS	Šamaj et al. (this work)
D. aliciae	Seeds	RM	Small and Hendrikz (1974)
	Seeds	RM	Small et al. (1977)
D. rotundifolia	Seeds	–	Simola (1978a,b)
	Seeds	MS	Bonnet et al. (1984b)
	Seeds	RM	Kukulczanka and Czastka (1988)
	Seeds	RM	Budzianowski (1996, 1997)
	Leaves	RM	Kukulczanka (1991)
	Leaves	MS	Anthony (1992); Šamaj et al. (1994, 1995)
	Leaves	MS	Bobák et al. (1995)
	Shoots, roots	MS	Wawrosch et al. (1993)
	Shoots, leaves	MS	Wawrosch et al. (1996)
D. regia	Leaves	–	Janssens (1986)
D. natalensis	Leaves, shoots, roots, flowers	MS	Crouch and van Staden (1988)
	Leaves	MS	Crouch et al. (1990)
D. hilaris	Seeds	RM	Kukulczanka and Czastka (1988)
	Leaves	RM	Kukulczanka (1991)
	Shoots, roots	MS	Wawrosch et al. (1993)
	Shoots, leaves	MS	Wawrosch et al. (1996)
D. spathulata	Seeds	MS	Bobák et al. (1989, 1990); Blehová et al. (1990)
	Seeds	RM	Budzianowski et al. (1993); Budzianowski (1995, 1997)
	Leaves	MS	Blehová et al. (1992); Bobák et al. (1993)
	Leaves	MS	Bobák et al. (1994); Blehová et al. (1995)
D. capensis	Leaves	MS	Crouch et al. (1990); Anthony (1992)
	Shoots, roots	MS	Wawrosch et al. (1993)
	Shoots, leaves	MS	Wawrosch et al. (1996)
D. brevifolia	Leaves	RM	Kukulczanka (1991)
D. binata	Leaves	MS	Anthony (1992)
D. dielsiana	Shoots, roots	MS	Wawrosch et al. (1993)

MS medium (Murashige and Skoog 1962); RH medium (Reinert and Mohr 1967).

tions of BA. The neoformed buds have the potential to initiate roots a few days later which lead to the regeneration of complete plantlets. Crouch and van Staden (1988) observed about 20 regenerated buds per leaf explant of *D. natalensis* after 40–60 days in culture when they used one fifth strength MS media supplemented with both NAA and BA in low concentrations. These authors reported that up to 1500 plantlets could be obtained within 8 months from a single leaf explant. As another example, Kukulczanka and Czastka (1988) used a combination of $1\,\mathrm{mg\,l^{-1}}$ BA and $0.2\,\mathrm{mg\,l^{-1}}$ NAA to induce adventitious buds in serveral *Drosera* species and they reported the production of

Table 7. Influence of BA and NAA on somatic embryogenesis of *Drosera rotundifolia* isolated leaves after 60 days in culture. (After Bobák et al. 1995)

NAA [M]	BA [M]						
	0	10^{-8}	10^{-7}	10^{-6}	5×10^{-6}	10^{-5}	2×10^{-5}
0	E	S	S	C	C	C	C
10^{-8}	E	S	S	C	C	C	C
10^{-7}	S	S	S	C	C	C	C
10^{-6}	ΔN	ΔN	ΔN	ΔC	ΔC	ΔC	ΔC
5×10^{-6}	ΔN	ΔN	ΔN	ΔC	ΔC	ΔC	ΔC
10^{-5}	ΔN	ΔN	ΔN	ΔC	ΔC	ΔC	ΔC
2×10^{-5}	ΔN	ΔN	ΔN	ΔC	ΔC	ΔC	ΔC

E = somatic embryogenesis, S = shoot regeneration, C = callus formation, N = necrosis; Δ = red pigmentation of explants and calli induced by medium composition.

100 buds and rosettes from a single explant. Later, Crouch et al. (1990) studied both regeneration and plumbagin production in two South-African species *D. capensis* and *D. natalensis*. Shoot differentiation of *D. natalensis* requires higher levels of BA, while rooting of both species needs higher levels of NAA. Interestingly, Kukulczanka (1991) showed that germplasm of various *Drosera* species can be stored in vitro long-term using temperatures around 5 °C. She reported the production of 160000 plants per annum from one rosette and 5400 plants per annum from a single leaf explant. Moreover, Anthony (1992) claimed that *D. rotundifolia*, *D. capensis* and *D. binata* formed flowers in vitro when cultured on the half-strength MS medium without hormones or supplemented with NAA and BA in low concentrations. Regenerated shoots of *D. binata* rooted better when low supplements of both BA and NAA were used. Flowering in vitro was also described in our experiments with *D. spathulata* and *D. rotundifolia* (Blehová et al. 1990), and was seasonally dependent, expressing a maximum in April and May. It was connected with a lower production of fresh and dry raw material in vitro. Similar results were reported by Lowrie (1989) during the flowering period of *D. pygmaea* and other Australian sundew species growing under natural environmental conditions.

Using leaf explants of *D. rotundifolia*, Bobák et al. (1995) screened 49 variants of MS media supplemented with various concentrations of NAA and BA (Table 7). Somatic embryos and buds can also be grown from leaves without hormones (Fig. 4b) or just with a very small supplement of hormones like NAA and BA (Bonnet et al. 1984b; Kukulczanka and Czastka 1988). In spite of the fact that the number of regenerated plants is lower in this case, e.g. about seven regenerants per explant (Table 8), these plants have no structural and developmental abnormalities (Bobák et al. 1995), which are commonly observed when higher hormone concentrations are used. Recently, Wawrosch et al. (1996) successfully used other cytokinins like γ,γ-dimethylallylamino purine (2iP), zeatin and kinetin for micropropagation of *D. rotundifolia*, *D. hilaris* and *D. capensis*. The highest regeneration potential was achieved using 20 μM 2iP for all three species. *D. hilaris* produced about 50 shoots,

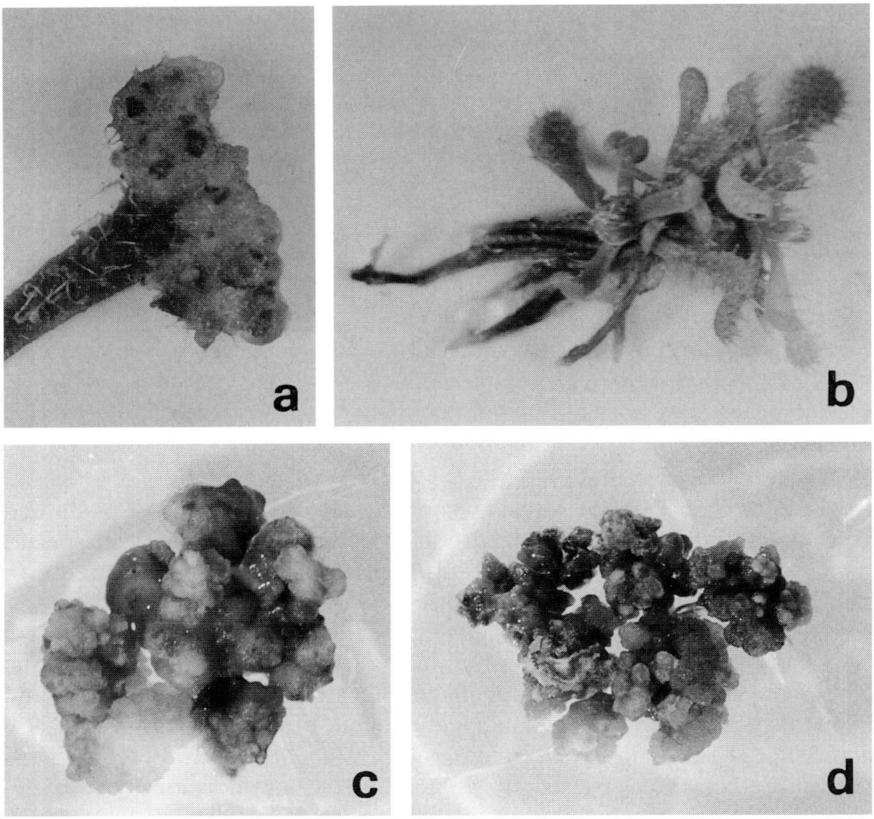

Fig. 4. Callus formation and plant regeneration in *Drosera rotundifolia* (**a,b**) and *Drosera spathulata* (**c,d**). **a** Nodular callus, regenerated from the base of petiole on MS medium supplemented with NAA and BA, possessing red spots. **b** Plantlets directly regenerated via somatic embryogenesis from the leaf explant on the MS medium without hormones after 60 days in culture. **c** Yellow callus, long-time cultured in the dark on MS medium supplemented with 2,4-D and KIN, having red-pigmented spots. **d** Green compact callus long-time cultured in the light (long-day conditions) on MS medium supplemented with 2,4-D and KIN

Table 8. Quantitative evalution of direct shoot regeneration after 30 days in culture (After Bobák et al. 1995). Mean number of shoots per leaf explant ± SD on solidified MS media with various concentrations of NAA and BA

NAA [M]	BA [M]		
	0	10^{-8}	10^{-7}
0	6.8 ± 1.5	4.1 ± 1.4	3.4 ± 1.2
10^{-8}	6.5 ± 1.4	1.4 ± 0.8	1.3 ± 0.9
10^{-7}	1.3 ± 1.0	1.6 ± 1.4	1.5 ± 1.0

NAA = α-naphathalene acetic acid; BA = benzyladenine.

while *D. rotundifolia* and *D. hilaris* produced just about 20 shoots from the single leaf explant.

Callus formation in vitro was observed very occasionally, particularly due to the high natural regeneration potential of *Drosera* species. Rarely, a small amount of compact green callus was observed during shoot regeneration (Finnie and van Staden 1993). Beebe (1980) studied callus induction in *Dionaea muscipula* (Droseraceae) cultured in vitro, and reported the low production of friable callus originating from cotyledons cultured on MS medium with the same amount of both NAA and BA, while complete plants were regenerated on the hormone-free medium. We have induced callus formation from isolated leaves of *Drosera rotundifolia* (Fig. 4a, Table 7) and *D. spathulata* (Fig. 4c,d). In *D. spathulata*, we studied the influence of various light conditions and hormones on callus induction and development during two-year experiments (Blehová et al. 1992). The best callus production was induced on the MS medium supplemented with 2,4-D and kinetin. Depending on physical conditions, red, green and yellow pigmented callus clones were obtained, with different consistencies (Fig. 4c,d). Even after 2 years of subculture, calli were able to regenerate complete plants after their transfer to the hormone-free medium. It was possible to regenerate 25–30 new plants from 1 g of these organogenic calli. Red pigmentation of some callus clones originates from anthocyanoplasts which are active in anthocyane biosynthesis (Bobák et al. 1994). Such red-pigmented callus also contains higher levels of 7-methyljuglone compared with the green and the yellow calli (Blehová et al. 1995).

Finally, we also tested gemma germination and plant production of *D. pygmaea* from gemmae during in vitro culture. It was possible to induce gemma germination (see also Fig. 9b,c,d) and subsequent plant regeneration using MS culture medium without growth regulators. Complete new plants regenerated from gemmae after 6 weeks.

Importantly, *D. rotundifolia* plants regenerated in vitro, can be successfully transplanted to nature where they quickly adapt and show vigorous growth. Using this approach, natural populations of endangered *Drosera* species may be restored (Wawrosch et al. 1993, 1996).

2.2 Structural Aspects of *Drosera* Regeneration In Vitro

The optimal formation of secondary metabolites is usually connected with structural requirements of differentiated cell types, tissues and plant organs (Šamaj et al. 1997). The differentiated organs are able to produce more secondary metabolites than nondifferentiated calli during in vitro culture. In some cases it is also connected with differentiation of specialized cells and tissues for product storage, e.g. laticifers in *Papaver* species (Galewsky and Nessler 1986). Subcellular studies, moreover, indicate that vacuoles are the most preferable storage compartments of secondary metabolites in differentiated as well as in undifferentiated cells (Nessler and Mahlberg 1979; Šamaj et al. 1990). Unfortunately, their genesis during cell differentiation, and especially during events accompaning production and storage of secondary metabolites, is poorly understood.

We have studied cytological aspects of direct and indirect regeneration of *D. spathulata* (Bobák et al. 1989, 1990, 1993, 1994; Blehová et al. 1990), and direct somatic embryogenesis of *D. rotundifolia* (Šamaj et al. 1994, 1995; Bobák et al. 1995) from isolated leaves as well as direct regeneration of *D. pygmaea* from gemmae.

2.2.1 Direct Organogenesis of Drosera spathulata

We screened different culture media to induce shoot organogenesis and callus from isolated leaves of *D. spathulata* (Bobák et al. 1989; Blehová et al. 1992). On the MS medium supplemented with NAA and BAP parenchymatic cells of leaf vascular bundles divide first. Clusters of dividing cells which form meristemoids can be observed in the vicinity of vascular bundles after 6 to 15 days in culture (Fig. 5a). Meristemoids arising from the leaves have a cylindrical shape and yellow-green or green coloration. They are composed of different cell types – meristematic, parenchymatic and protracheal. Protracheal cells actively participate in vascular system development. Shoot primordia can be observed clearly after 21 days in culture; they have apical meristems with leaf primordia in their apical parts as well as tracheal elements connecting them to the mother explant in their basal parts (Fig. 5b). Regenerated shoots are able to create roots and complete regenerated plants can be obtained subsequently.

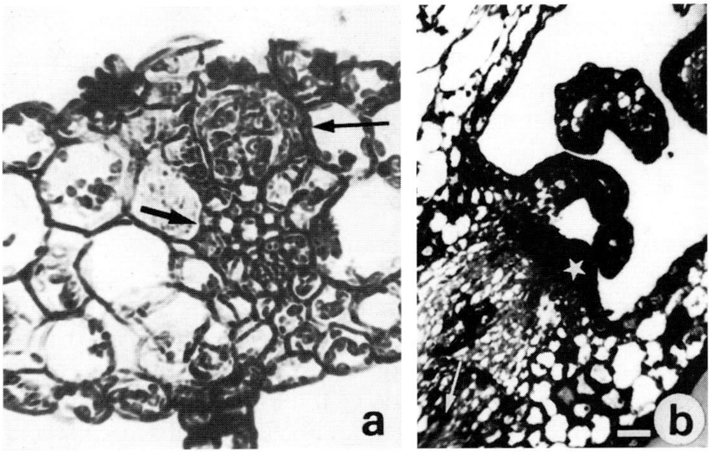

Fig. 5. Direct organogenesis from leaf explants of *Drosera spathulata* (Bobák et al. 1989). **a** Globular meristemoid (*thin arrow*) created from the bundle sheath cells surrounding the leaf vein (*thick arrow*). **b** Shoot primordium regenerated from a meristemoid, having an apical meristem (*asterisk*), leaf primordia and vascular bundles (*arrow*). Scale *bar* represents 30 μm for **a** and 90 μm for **b**

2.2.2 *Indirect Organogenesis of Drosera spathulata*

We obtained callus from isolated leaves on MS medium supplemented with 2,4-D and kinetin (Blehová et al. 1992). Shoot organogenesis was induced from this callus within 30 days after its transfer to hormone-free MS medium. Meristemoids, shoot buds and complete plants regenerated on this induction medium (Bobák et al. 1993). Clusters of meristematic cells were formed on the surface of calli within the first 10 days. Subsequently, globular and irregular meristemoids developed (Fig. 6a,b) which gave rise to the shoot primordia with apical meristems and leaf primordia. The development of shoot primordia takes approximately 30 days. By contrast, tracheal elements or organized vascular systems, which connect regenerated shoots with the inner callus part, originate from the inner meristematic centres of the callus. Rarely, embryo-like structures can be observed on the periphery of large meristematic clumps (Fig. 6c). Isolated regenerated shoots form roots, and thus complete plants.

2.2.3 *Direct Embryogenesis of Drosera rotundifolia*

Direct somatic embryogenesis was achieved from isolated leaves of *D. rotundifolia* (Bobák et al. 1995) on hormone-free or NAA-supplemented MS media (Table 6). Anatomical studies revealed that leaf epidermal cells divide within 48 h on the induction medium. These cells show the highest regeneration potential and represent cells from which somatic embryos are regenerated. First divisions of reactivated epidermal cells are usually anticlinal (Fig. 7a) or transverse followed by periclinal ones which lead to the formation of the epidermal cell clusters on the explant surface (Fig. 7b). Globular somatic embryos are formed from these clusters after 10 to 12 days in culture (Fig. 7c). New protodermis is developed on the surface of late globular embryos after 14 to 16 days in culture. Meristematic cells located in the inner somatic embryo part are highly cytoplasmic with small provacuoles and prominent nuclei and nucleoli. Mostly, the central vacuole is not differentiated and several small vacuoles contain dense deposits. Cells of epidermal origin usually separate somatic embryos from the mother explant tissue, especially during early stages of their development (Fig. 7c). After 18 to 20 days in culture, this coherent layer of cells is disrupted by meristematic cells, which give rise to the root primordia in the basal part of somatic embryos.

The development of somatic embryos is not synchronized. Globular-, heart- and cotyledonary-shaped somatic embryos can be observed after 15 days on the induction medium (e.g. Fig. 8a). After 20 days in culture, the shoot primordia are visible, which mostly have three leaf primordia (Fig. 8b). Subsequently, root primordia are formed. Further development of these organ primordia results in the regeneration of complete plants on the leaf explant surface. Plantlets regenerated via somatic embryogenesis usually have three to seven leaflets and primary roots with numerous root hairs (Fig. 8c). These young complete plants are easily separable from the mother explant

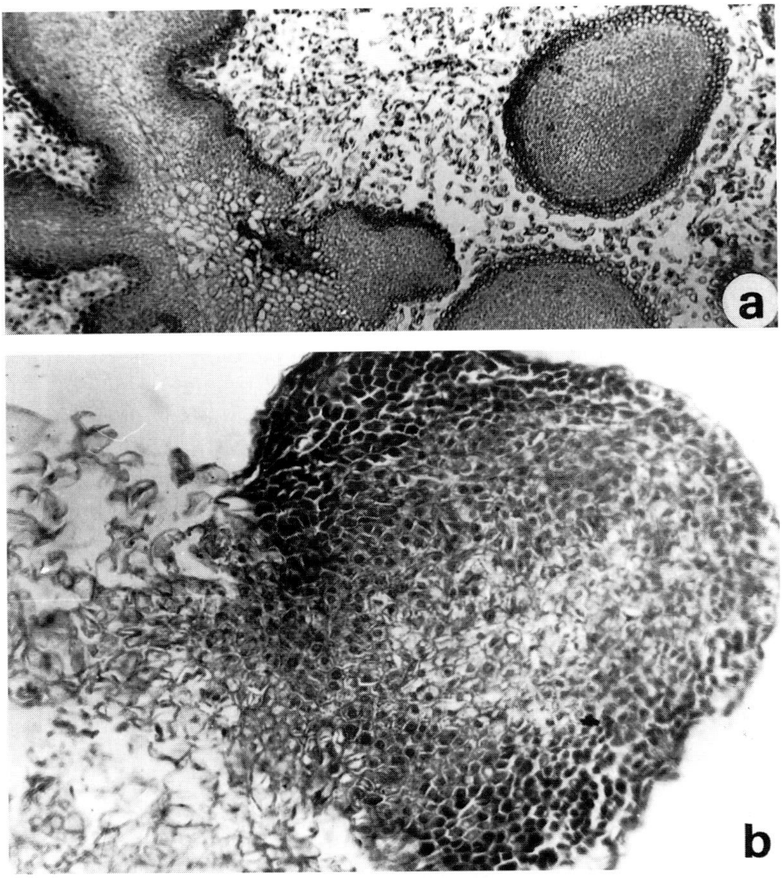

Fig. 6. Indirect organogenesis from callus of *Drosera spathulata* (Bobák et al. 1993). **a** Section through meristemoids of various shapes regenerated from organogenic callus. **b** Section through a globular meristemoid regenerated on the surface of the callus. **c** Morphology of embryo-like structure located on the callus surface. Scale *bar* represents 200 μm for **a**, and 100 μm for **b** and 10 μm for **c**

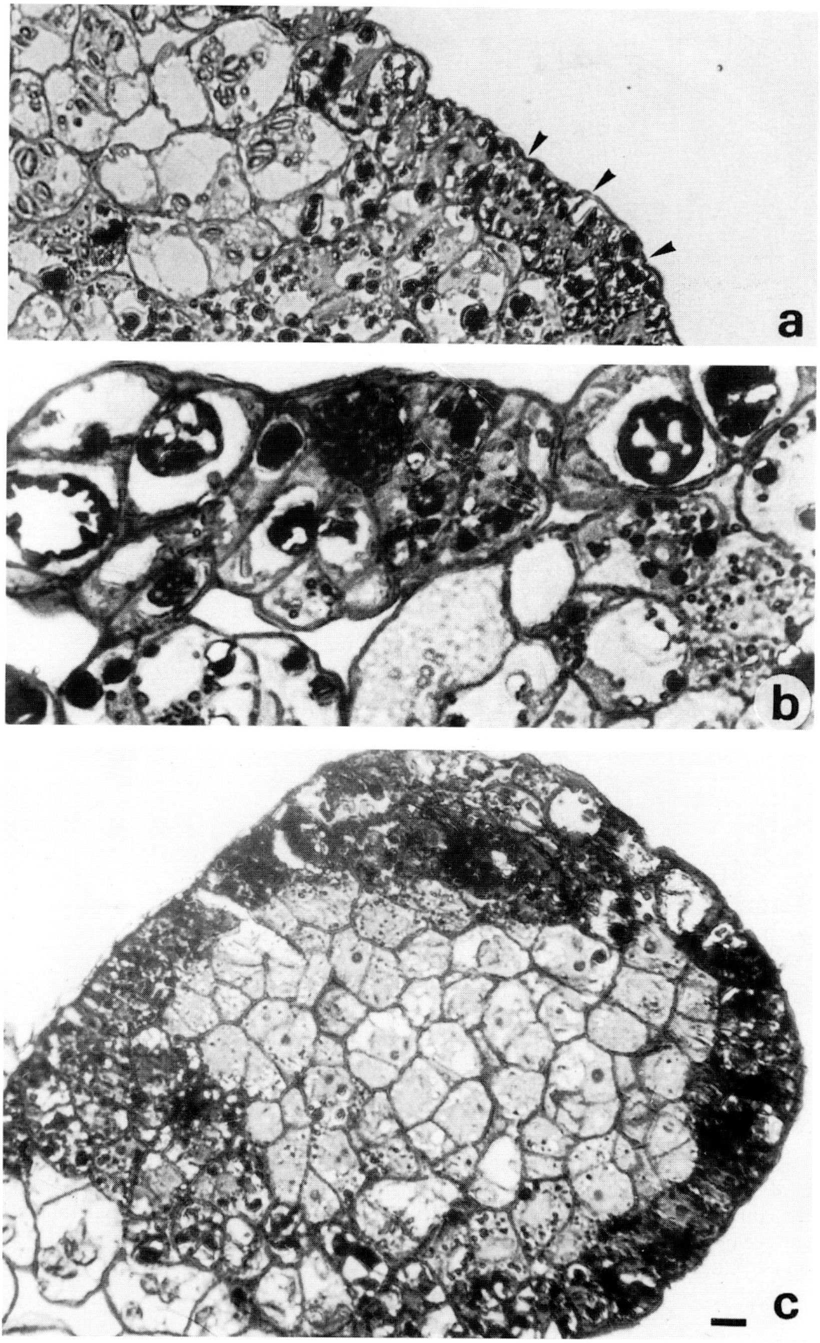

Fig. 7. Anatomy of direct somatic embryogenesis from leaves of *Drosera rotundifolia* (Bobák et al. 1995). **a** First anticlinal divisions (*arrowheads*) of reactivated epidermal cells. **b** Clusters of dividing epidermal cells. **c** Globular somatic embryo with meristematic cells (inner part) surrounded by cells of epidermal origin. Scale *bar* represents 10 μm for **a** and **c** and 5 μm for **b**

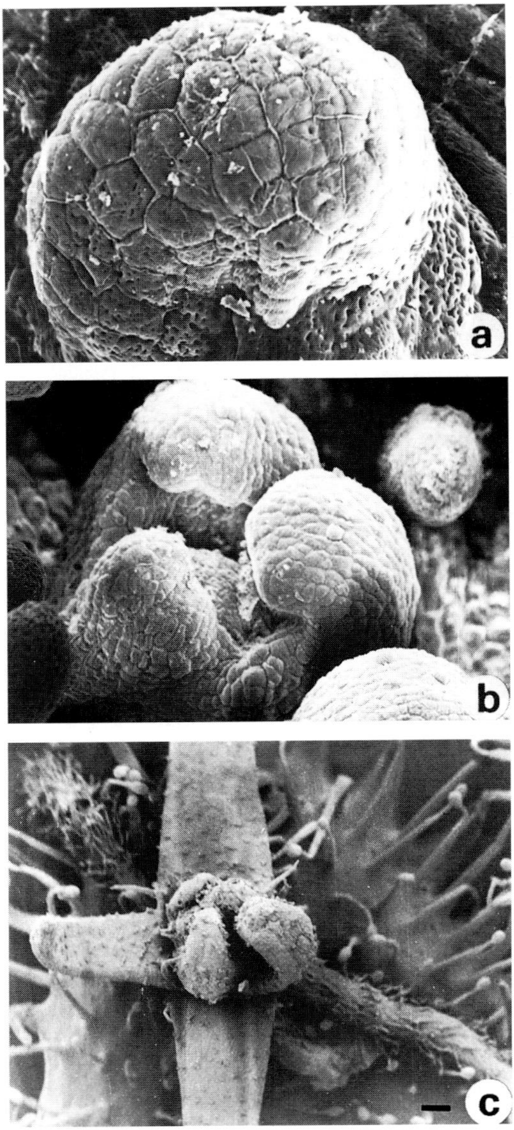

Fig. 8. Morphology of direct somatic embryogenesis from leaves of *Drosera rotundifolia* (Bobák et al. 1995). **a** Globular somatic embryo with new protodermis on its surface. **b** Shoot primordium with three leaf primordia. **c** Complete regenerated plantlet with leaf rosette and three roots on the surface of mother leaf explant after 38 days in culture. Scale *bar* represents 10 μm for **a**, 25 μm for **b** and 500 μm for **c**

tissue. They grow to maturity after their transfer to the medium without hormones.

We also studied the cell surface structures of proembryos and somatic embryos developing from isolated leaves of *Drosera rotundifolia* (Šamaj et al.

1994, 1995, 1997; Bobák et al. 1995). The first sign of direct somatic embryogenesis is visible when somatic proembryos emerge from the leaf epidermis (mostly in the basal part of tentacles) after 7 to 10 days on the induction medium. Detailed morphological studies revealed that the surface of proembryogenic cells are covered with extracellular matrix (ECM) which forms a fibrillar network. Such a fibrillar network is never seen on the surface of leaf explant epidermal cells. The network is gradually reduced to the coarse strands which link neighbor surface cells during subsequent differentiation of somatic embryos (after 10 to 12 days) and completely disappears when new protodermis differentiates around globular somatic embryos.

2.2.4 Gemma Germination and Regeneration In Vitro

Drosera pygmaea produces gemmae within the basal part of a stipule bud. A globular embryo-like structure is located within the basal part of the gemma ventral side. This meristematic structure is surrounded by storage tissue. Anatomical studies have shown that the embryo-like structure of a gemma has an apparent polar organization with two meristematic regions. These meristems are functional and give rise to the shoot and root primordia, and subsequently to the new regenerated plants during in vitro culture on the hormone-free MS medium (Fig. 9b–d). The embryo-like structure is surrounded by storage tissue resembling the endosperm of seed plants. These structures swell and appear pear-like after 3 days in culture (Fig. 9a). Subsequently, the organ primordia (shoot and root poles) are differentiated within these structures. Primary roots emerge first after 7 days in culture (Fig. 9b,d), and subsequently shoot primordia differentiate after 11 days (Fig. 9c). Shoot primordia have relatively wide primary apical meristems and two leaf primordia.

3 Biosynthesis and Extraction of Medicinal Compounds

Chemical analyses of *Drosera* plants have shown that they accumulate pharmacologically active compounds, namely naphthoquinones and flavonoids. Various naphthoquinones and flavonoids were isolated and identified from native as well as from in vitro cultured and micropropageted *Drosera* species (Zenk et al. 1969; Juniper et al. 1989; Budzianowski et al. 1993; Culham and Gornall 1994).

3.1 Naphthoquinones

Plumbagin and 7-methyljuglone are biosynthesized from tyrosine via the acetate-polymanolate pathway (Durand and Zenk 1971, 1974a,b, 1976). The incorporation of tyrosin into naphthoquinones is indirect. It is first broken

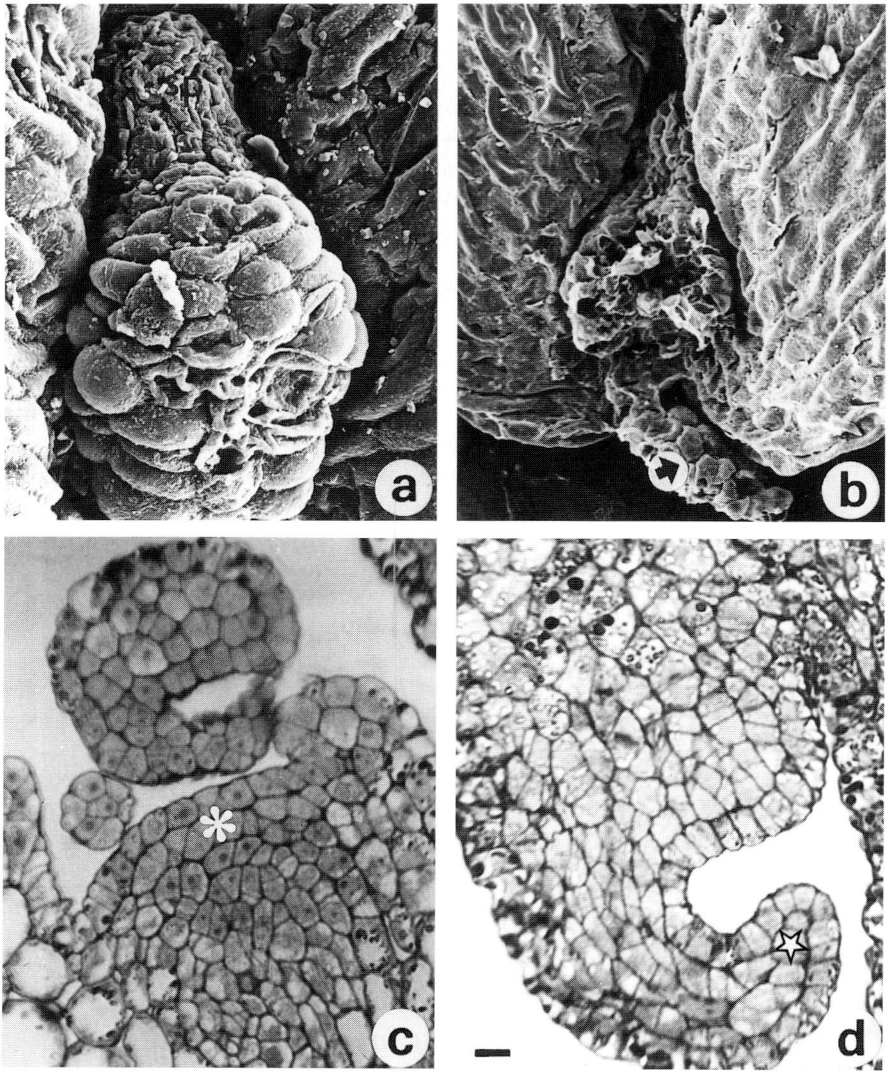

Fig. 9. Gemma germination of *Drosera pygmaea* during in vitro culture. **a** Embryo-like structure of gemma changes its shape from globular to pear-like after 3 days in culture. **b** Morphology of root primordia (*arrow*) initiation after 7 days in culture. **c** Anatomy of shoot primordium with apical meristem (*snowflake*) and leaf primordia after 11 days in culture. **d** Anatomy of root primordium (*star*) after 7 days in culture. Scale *bar* represents 10 μm for **a**, 20 μm for **b** and 30 μm for **c** and **d**

down to acetate via a ring-cleavage homogenate pathway, and subsequently incorporated through intermediates into plumbagin or closely related 7-methyljuglone.

Extraction by dry or 80% methanol, benzene, toluene, diethylether and steam distillation from fresh, dry or freeze dried plant material are used for naphthoquinone isolation (Bonnet et al. 1984a; Kreher et al. 1990; Culham and Gornall 1994; Budzianowski 1995; Blehová et al. 1995). For the separation, identification and quantification of naphthoquinones, thin layer chromatography on silica gel plates using toluene-HCOOH in the ratio 99:1 (Budzianowski 1995), gas chromatography with a Carbowax 20M capillary column (Blehová et al. 1995), GC-mass spectrometry (Bonnet et al. 1984a; Blehová et al. 1995), and high performance liquid chromatography (Crouch et al. 1990; Repčák and Galambosi 1994) can be used.

Various naphthoquinones and their glycosylated derivatives were isolated and identified in *Drosera* plants growing in vivo (Tables 2, 4, Figs. 2, 3). Additionally, the occurrence of volatile naphthoquinones like 7-methyljuglone and plumbagin in plants cultured in vitro was also described (Table 3). For example, plumbagin was detected in *D. intermedia* (Bonnet et al. 1984a; Budzianowski et al. 1993), *D. capensis* and *D. natalensis* (Crouch et al. 1990). While in the case of *D. intermedia*, this compound represented up to 9.86% of dry weight, in *D. capensis* and *D. natalensis* it was just about 0.001% of the fresh mass.

Later, it was shown that conjugated naphthoquinone rossoliside, which is present in plants of *D. spathulata* and *D. rotundifolia* cultured in vitro, can easily release 7-methyljuglone (Budzianowski 1995). The formation of some artifactual naphthoquinones was described by Budzianowski (1995, 1996, 1997) during extraction with methanol or in water containing plant extracts.

3.2 Flavonoids

Flavonoids represent one of the largest classes of plant phenolics. The basic skeleton of a flavonoid contains 15 carbons with two aromatic rings connected by a three-carbon bridge. This structure results from two separate biosynthetic pathways. The bridge and one aromatic ring are, like phenylpropane units, synthesized from phenylalanine produced in the shikimic acid pathway. The six carbons of the other aromatic ring originate from the three acetate units via the malonic acid pathway. The key enzyme for the formation of the flavonoid skeleton is chalcone synthase, which catalyses the stepwise condensation of three acetate units from malonyl-CoA with coumaroyl-CoA to the C 15 intermediate. Subsequent glycosylation of flavonoids is an important modification to increase their water solubility and for deposition in vacuoles (Taiz and Zeiger 1991; Heller and Forkmann 1993). Almost all the flavonoids possess either catechol or pyrogallol structures which appear to be characteristic for the genus *Drosera* (Ramachandran Nair et al. 1990).

Methanol extraction and preparative TLC on silica gel with toluene-formic acid (100:1) and chromatography on polyamide (water-*n*-butanol-acetone) were used for isolation, and subsequently UV and NMR spectroscopy was carried out to identify flavonoids in *Drosera* species (Budzianowski et al. 1993). Recently, Repčák and Galambosi (1996) used extraction by 75% methanol, hydrolysis of glycosides by 6% HCl and estimation of flavonol aglycones by HPLC.

Phytochemical research on *D. rotundifolia* documented the presence of flavonols, quercetin, kaempferol (Bienenfeld and Katzlmeier 1966), isoquercitrin (quercetin-3-glucoside), astragalin (kaempferol-3-glucoside), gossypin (gossypetin-8-glucoside) and gossypitrin (gossypetin-7-glucoside) (Schölly and Kapetanidis 1989b, 1993). Repčák and Galambosi (1996) quantified the content of qercetin and kaempferol in leaves, stems and flowers of *D. rotundifolia* and *D. anglica* growing in vivo in Finland (Table 5). From in vitro cultured *D. spathulata* and *D. intermedia*, quercitrin (quercetin-3-rhamnoside), hyperin (quercetin-3-galactoside), isoquercitrin, astragalin and myricetin-3-rhamnoside were identified (Budzianowski et al. 1993).

3.3 Production of 7-Methyljuglone In Vitro

The content of 7-methyljuglone from in vitro cultured *Drosera* plants varies depending on species and culture conditions. For example, Bonnet et al. (1984a) found a high content of 7-methyljuglone (6.1% of dry weight) in plants of *D. rotundifolia* cultured on the MS medium supplemented with 10^{-6}M NAA and 10^{-7}M BA, and just moderate amounts of this compound (0.69%) in *D. intermedia* plants cultured on the same medium supplemented with 10^{-7}M NAA and 10^{-5}M BA. Besides 7-methyljuglone, plumbagin was also found in *D. intermedia* at a very high level (9.86%). These values were up to 5 times higher compared with in vivo plants of the same species. Later, Budzianowski et al. (1993) and Budzianowski (1995) reported the occurrence of 7-methyljuglone in *D. spathulata* and of plumbagin and traces of 7-methyljuglone in *D. intermedia* plants from in vitro cultures. This author also showed that 7-methyljuglone can be easily released from rossoliside. Unfortunately, the quantitative data about naphthoquinone content are not presented in these excellent works.

We have shown for the first time that the content of 7-methyljuglone estimated in the callus cultures, shoots, roots and regenerated plants of *D. spathulata* is related to the stage of tissue and organ differentiation (Table 9). The highest level of 7-methyljuglone (1.6%) is found in complete plants regenerated from seeds. The content of this substance varies in callus cultures. The callus with intensive red pigment spots contains higher percentages of 7-methyljuglone. Red pigmentation can be associated with the production of 7-methyljuglone because slightly lower amounts of this compound can be found in the green or in the yellow-pigmented calli. Moreover, there are significant differences between calli growing in dark or in light, using medium with a lower concentration of 2,4-D. After a long period of culture (23 passages) the amount of 7-methyljuglone is reduced. The content of this naphthoquinone also varies significantly in the morphogenic callus cultured on half-strength MS medium without hormones. The level of 7-methyljuglone in shoots was regularly higher than in roots. Complete plants, regenerated from the callus, contain about 1.27% 7-methyljuglone, which is not much less than in plants growing directly from seeds. We concluded that the content of 7-methyljuglone in plant tissues decreases with the duration of their culture

Table 9. Occurrence and distribution of 7-methyljuglone in plants regenerated from seeds, calli, shoots, roots and regenerants of *Drosera spathulata* (% of dry mass). (After Blehová et al. 1995)

Plant material and culture conditions	Medium	Passage	Content of 7-methyljuglone	
			x	s
Complete differentiated plants – light-dark period	MS + E	4	1.655	0.119
Green callus – light-dark period	MS + CM + 2,4-D	12	0.057	0.038
Red callus – dark	MS + CM + 2,4-D	12	0.349	0.032
Yellow callus – dark	MS + CM + 2,4-D + KIN	23	traces	traces
Regenerated roots – light-dark period	$^1/_2$ MS	24	0.145	0.026
Regenerated shoots – light-dark period	$^1/_2$ MS	24	0.215	0.020
Complete regenerated plants – light-dark period	$^1/_2$ MS	25	1.275	0.471

MS = Murashige-Skoog medium; CM = coconut milk; E = grape vine exudate; 2,4-D = 2,4-D-dichlorophenoxyacetic acid; KIN = kinetin; $^1/_2$ MS = $^1/_2$ strength Murashige-Skoog medium; x = arithmetic mean; s = standard deviation.

(especially when strong auxins like 2,4-D are used), and depends on morphogenic factors, increasing with the level of cell differentiation.

Recently, Wawrosch et al. (1996) reported the production of 0.135% of 7-methyljuglone in *D. rotundifolia*, 0.142% in *D. hilaris* and 0.06% in *D. capensis* when plants were cultured on MS media supplemented with cytokinin 2iP.

4 Summary and Conclusions

Numerous *Drosera* species are found relatively rarely in nature as a consequence of intensive plant collection. In many countries they are protected by law. Therefore, new biotechnological methods are considered for plant propagation which would ensure the necessary amount of pharmaceutically important drugs to meet demands. Simultaneously, naphthoquinone production from in vitro growing plants should be evaluated for commercial use. For example, the production of plumbagin was shown to be higher in *Plumbago* roots compared with in vivo and in vitro grown plants of *Drosera capensis* and *Drosera natalensis*. Based on this finding in two South African species, it was concluded that the extraction of plumbagin from *Drosera* is not commercially feasible, especially using in vitro cultures (Finnie and van Staden 1993). On the other hand, Bonnet et al. (1984a) reported that the production of naphthoquinones in vitro can be 5 times higher than in vivo in two other species, *D. rotundifolia* and *D. intermedia* when cultured under optimal conditions. These results are of particular interest since *D. rotundifolia* represents probably the most pharmacologically important *Drosera* species (Weiss 1991). This species was traditionally collected as a major component of pharmaceutical drugs designed for medicinal purposes (Czygan 1989). However, now this protected species is replaced by *D. ramentacea*, *D. madagascariensis* and *D. peltata* which contain significantly less naphthoquinones (Wawrosch et al. 1993; Krenn et al. 1995). This problem can be resolved by searching for alternative sources of plants including in vitro production of naphthoquinones from the species naturally rich in these compounds, e.g. *D. rotundifolia*, *D. spathulata* and *D. intermedia*. Moreover, Caniato et al. (1989) reported that *D. capensis* is an interesting source of 7-methyljuglone, while *D. binata* var. *dichotoma* is a good natural source of plumbagin. The efficiency of the in vitro system can be further enhanced by the optimization of nutritional and physical conditions in the culture and by selection of plant clones which overproduce secondary metabolites. Taking into account the extremely high regeneration potential of *Drosera* during in vitro culture, accompanied by easy mass-production of plants (up to 160000 plants from single explant per annum), all this can result in the commercial use of aseptic *Drosera* plants for pharmaceutical and medicinal uses. Moreover, the use of hormone-free and diluted culture media can make the in vitro production of plants cheaper and so economically feasible.

Acknowledgements. The authors would like to thank Dr. Irene Lichtscheidl (Institute of Plant Physiology, University of Vienna, Austria) for providing Fig. 1, and Prof. Dieter Volkmann (Institute of Botany, University of Bonn, Germany) for critical reading of the manuscript.

References

Albert VA, Williams SE, Chase MW (1992) Carnivorous plants: phylogeny and structural evolution. Science 257:1491–1495

Anthony JL (1992) In vitro propagation of *Drosera* spp. HortScience 27:850

Beebe JD (1980) Morphogenetic responses of seedlings and adventitious buds of the carnivorous plant *Dionaea muscipula*. Bot Gaz 141:396–400

Békésiová I (1997) Identification of antifungal proteins in *Drosera rotundifolia*. Book of Abstracts, Molecular Biology for Agriculture, Institute of Plant Molecular Biology, Ceské Budejovice, 25–29 Aug, p 123

Bhargava SK (1984) Effect of plumbagin on reproductive function of male dog. Indian J Exp Biol 22:153–156

Bhargava SK, Dixit VP (1985) Antifertility effects of embelin and plumbagin in female rats. Plant Med Phytother 19:29–34

Bienenfeld W, Katzlmeier H (1966) Flavonoide aus *Drosera rotundifolia* L. Arch Pharm 299:598–602

Blehová A, Šomšáková V, Bobák M (1990) Anatomical studies of the development of new plants from the leaves of the sundew (*Drosera spathulata* L.) in vitro. Acta FRN Uni Comen; Physiol Plant 26:33–42

Blehová A, Erdelský K, Bobák M (1992) Cultivation of organ and callus culture of *Drosera spathulata* Labill. in vitro conditions. Acta FRN Uni Comen; Physiol Plant 27:93–102

Blehová A, Erdelský K, Repčák M, Garčár J (1995) Production and accumulation of 7-methyljuglone in callus and organ culture of *Drosera spathulata* Labill. Biologia 50:397–401

Bobák M, Blehová A, Erdelský K (1989) Histological and cytological study of early phases of organogenesis on excised leaves of *Drosera spathulata* L. cultivated in vitro. Biologia 44:785–792

Bobák M, Blehová A, Erdelský K, Cholvadová B, Šomšáková V (1990) Morphogenesis of the plastid apparatus in the process of organogenesis at the leaves of the sundew (*Drosera spathulata* L.) cultivated in vitro. Acta FRN Uni Comen – Physiol Plant 25:33–41

Bobák M, Blehová A, Šamaj J, Ovečka M, Krištín J (1993) Studies of organogenesis from the callus culture of the sundew (*Drosera spathulata* Labill.). J Plant Physiol 142:251–253

Bobák M, Blehová A, Lux A, Šamaj J, Krištín J, Gajdoš J (1994) Occurence, genesis and ultrastructure of anthocyanoplast-like structures in cells of callus culture of *Drosera spathulata* Labill. Acta Bot Hung 38:221–226

Bobák M, Blehová A, Krištín, J, Ovečka M, Šamaj J (1995) Direct plant regeneration from leaf explants of *Drosera rotundifolia* cultured in vitro. Plant Cell Tissue Org Cult 43:43–49

Boesewinkel FD (1989) Ovule and seed development in *Droseraceae*. Acta Bot Neerl 38:295–311

Bokemo W (1984) Les plantes antilepreuses de Kisangani (Haut Zaire). Bull Soc R Bot Belg 117:305–311

Bonnet M, Coumans M, Hofinger M, Ramaut JL, Gaspar T (1984a) High-performance gas chromatography of 1,4-naphthoquinomes from *Droseraceae*. Chromatographia 18:621–622

Bonnet M, Coumans M, Hofinger M, Ramaut JL, Gaspar Th (1984b) Vegetative multiplication in vitro of the sundew *Drosera rotundifolia*. Arch Int Physiol Biochem 92:16–17

Budzianowski J (1995) Naphthoquinones of *Drosera spathulata* from in vitro cultures. Phytochemistry 40:1145–1148

Budzianowski J (1996) Naphthohydroquinone glucosides of *Drosera rotundifolia* and *D. intermedia* from in vitro cultures. Phytochemistry 42:1145–1147

Budzianowski J (1997) 2-Methylnaphthazarin 5-*O*-glucoside from the methanol extracts of in vitro cultures of *Drosera* species. Phytochemistry 44:75–77

Budzianowski J, Skrypczak L, Kukulczanka K (1993) Phenolic compounds of *Drosera intermedia* and *Drosera spathulata* from in vitro cultures. Acta Hortic 330:277–280

Burger RM (1961) Axenic seedlings of *Drosera*. Nature 191:724–725

Caniato R, Filippini R, Cappelletti EM (1989) Naphthoquinone contents of cultivated *Drosera* species *Drosera binata*, *D. binata* var. *dichotoma*, and *D. capensis*. Int J Crude Drug Res 27:129–136

Chandler GE, Anderson JW (1976) Studies on the nutrition and growth of *Drosera* species with reference to the carnivorous habit. New Phytol 76:129–141

Chandrasekaran BR, Vijayendran KK, Purushothaman B, Nagarajan B (1982) New methods for urinary estimation of antitumor compounds echitamine and plumbagin. Indian J Biochem Biophys 19:148–149

Clancy FG, Coffey MD (1977) Acid phosphatase and protease release by the insectivorous plant *Drosera rotundifolia*. Can J Bot 55:480–488

Cline SD, Coscia CJ (1989) Ultrastuctural changes associated with accumulation and secretion of sanguinarine in *Papaver bracteatum* suspension culteres treated with fungal elicitor. Planta 178:303–314

Crouch IJ, van Staden J (1988) In vitro propagation of *Drosera natalensis*. S Afr J Bot 54:94–95

Crouch IJ, Finnie JF, van Staden J (1990) Studies on the isolation of plumbagin from in vitro and in vivo grown *Drosera* species. Plant Cell Tiss Org Cult 21:78–82

Culham A, Gornall RJ (1994) The taxonomic significance of naphthoquinones in the *Droseraceae*. Biochem System Ecol 22:507–515

Czygan FC (1989) Sonnentaukraut. In: Wichtl M (ed) Teedrogen. Wissenschaftliche Verlagsgesellschaft, Stuttgart, pp 462–465

Darwin C (1875) Insectivorous plants. John Murray, London

Darwin F (1878) Experiments on the nutrition of *Drosera rotundifolia*. J Linn Soc Bot 17:17–27

Denoel A (1949) Determination de l'activité des *Droseras* indigénes et de leurs teintures. J Pharm Belg 1–2:1–19

Dexheimer J (1978) Localization ultrastructurale des phosphatases acides (phosphomonoesterases acides) dans les cellules des glandes digestives du *Drosera capensis* L. pendant la synthese de mucilage. Rev Cytol Biol Veget 1:49–57

Didry N, Pinkas M, Dubreuil L (1986) Activité antibacterienne de naphtoquinones d'origine végétale. Ann Pharm Fr 44:73–78

Durand R, Zenk MH (1971) Biosynthesis of plumbagin (5-hydroxy-2-methyl-1,4-naphthoquinone) via the acetate pathway in higher plants. Tetrahedron Lett 32:3009–3012

Durand R, Zenk MH (1974a) Homogenisate ring-cleavage pathway in the biosynthesis of acetate-derived naphthoquinones of the *Droseraceae*. Phytochemistry 13:1483–1492

Durand R, Zenk MH (1974b) Enzymes of the homogenisate ring-cleavage pathway in cell suspension cultures of higher plants. FEBS Lett 39:218–220

Durand R, Zenk MH (1976) The biosynthesis of the naphthoquinone 7-D methyljuglone. Biochem Physiol Pflanz 169:213–217

Durga R, Sridhar P, Polasa H (1990) Effect of plumbagin on antibiotic resistance in bacteria. Ind J Med Res 91:18–20

Durga R, Sridhar P, Polasa H (1992) Antimutagenic activity of plumbagin in Ames *Salmonella-Typhimurium* test. Ind J Med Res 96:143–145

Eilert U, Constable F (1985) Ultrastructure of *Papaver somniferum* cells cultured in vitro and treatment with fungal homogenate elicited alkaloid production. Protoplasma 128:38–42

Farr SB, Natvig DO, Kogoma T (1985) Toxicity and mutagenicity of plumbagin and the induction of a possible new DNA repair pathway in *Escherichia coli*. J Bacteriol 164:1309–1314

Fetterer RH, Fleming MW (1991) Effects of plumbagin on development of the parasitic nematodes *Heamonchus contortus* and *Ascaris suum*. Comp Biochem Physiol 100:539–542

Finnie JF, van Staden J (1993) *Drosera* spp. (Sundew): micropropagation and the in vitro production of plumbagin. In: Bajaj YPS (ed) Biotechnology in agriculture and forestry 24. Medicinal and aromatic plants V. Springer, Berlin Heidelberg New York, pp 164–177

Frenzer R (1980) Sonnentau extract in Herbalpina. Carniv Plant Newsl 9:33–34

Fujii N, Yamashita Y, Arima Y, Nagashima M, Nakano H (1992) Induction of topoisomerase II-mediated DNA cleavage by the plant naphthoquinones plumbagin and shikonin. Antimicrob Agents Chemother 36:2589–2594

Galewsky S, Nessler GL (1986) Synthesis of morphinane alkaloids during opium poppy somatic embryogenesis. Plant Sci 45:215–222

Grieve M (1959) A modern herbal. Jonathan Cape, London

Grigson G (1958) The Englishman's flora. Phoenix, London

Gujar GT, Mehrotra KN (1988) Toxicity and morphogenetic effects of plumbagin on *Dyscerdus koenigii* F. (Het, Pyrrhocoridae). J Appl Entomol 105:466–470

Gundidza M, Manwa G (1990) Activity of chloroform extracts from *Plumbago zeylanica* against *Neisseria gonorrhoeae*. Fitoterapia 61:47–49

Harborne JB (1982) Introduction to ecological biochemistry, 2nd edn. Academic Press, London

Harder R (1964) Vegetative entwicklung und blütenbildung in axenischen kulturen der insektivore *Drosera pygmaea* Dc. Planta 63:316–325

Heble MR, Narayanaswamy S, Chadha MS (1974) Tissue differentiation and plumbagin synthesis in variant cell strains of *Plumbago zeylanica* L. in vitro. Plant Sci Lett 2:405–409

Heise FH, Steenken W (1941) Growth and virulence of tubercle bacilli. Am Rev Tuberc 44:635–636

Heller W, Forkmann G (1993) Biosynthesis of flavonoids. In: Harborne JB (ed) The flavonoids: advances in research since 1986. Chapman & Hall, London, pp 499–535

Heslop-Harrison Y (1975) Enzyme release in carnivorous plants. In: Dingle JT, Dean RT (eds) Lysosomes in biology and pathology. North-Holland, Amsterdam, pp 525–578

Itoigawa M, Takeya K, Furukawa H (1991) Cardiotonic action of plumbagin on guinea-pig papillary muscle. Planta Med 57:317–319

Janssens J (1986) In vitro propagation of sundew, *Drosera regia* Stephens. Meded Fac Landbowwet Rijksuniv Gent 51:61–66

Joshi NK, Sehnal F (1989) Inhibition of ecdysteroid production by plumbagin in *Dyscercus cingulatus*. J Insect Physiol 35:737–741

Juniper BE, Robins RJ, Joel DM (1989) The carnivorous plants. Academic Press, London

Karlsson PS, Pate JS (1992a) Resource allocation to asexual gemma production and sexual reproduction in south-western Australian pygmy and micro stilt-form species of sundew (*Drosera* spp., Droseraceae). Aust J Bot 40:353–364

Karlsson PS, Pate JS (1992b) Contrasting effects of supplementary feeding of insects or mineral nutrients on the growth and nitrogen and phosphorus economy of pygmy species of *Drosera*. Oecologia 92:8–13

Keller H (1985) Erfahrungsheilkunde. Acta Med Emp 34:416–418

Krafft CC, Handel SN (1991) The role of carnivory in the growth and reproduction of *Drosera filiformis* and *D. rotundifolia*. Bull Torr Bot Club 118:12–19

Kreher B, Neszmélyi A, Wagner H (1990) Naphthoquinones from *Dionaea muscipula*. Phytochemistry 29:605–606

Krenn L, Länger R, Kopp B (1995) Qualitätsprüfung von Sonnentaukraut. 2. Botanische Identitätsprüfung sowie qualitative und quantitative Naphthochinonbestimmung an Handelsmustern. Dtsch Apoth Ztg 10:37–42

Krishnaswamy M, Purushothaman KK (1980) Plumbagin: a study of its anticancer, antibacterial and antifungal properties [5, hydroxy, 2-methyl, 1,4 naphthoquinone from *Plumbagin zeylanica*]. Indian J Exp Biol 18:876–877

Krishnamoorthy V, Thompson RH (1969) A new binaphthaquinone from *Drosera ramentacea*. Phytochemistry 8:1591–1594

Kubo I, Taniguchi M, Chapya A, Tsujimoto K (1980) An insect antifeedant and antimicrobial agent from *Plumbago capensis*. Planta Med 0 (Suppl):185–187

Kubo I, Uchida M, Klocke JA (1983) An insect ecdysis inhibitor from the African medicinal plant, *Plumbago capensis* (Plumbaginaceae); a naturally occurring chitin synthetase inhibitor. Agric Biol Chem 47:911–913

Kukulczanka K (1991) Micropropagation and in vitro germplasm storage of Droseraceae. Bot Gard Micropropag News 1:38–42

Kukulczanka K, Czastka B (1988) Regeneration abilities of *Drosera* sp. Acta Hortic 226:631–634

Lakshmi VV, Padma S, Polasa H (1987) Elimination of multidrug-resistant plasmid in bacteria by plumbagin, a compound derived from a plant. Curr Microbiol 16:159–161

Länger R, Pein I, Kopp B (1995) Glandular hairs in the genus *Drosera* (Droseraceae). Plant Syst Evol 194:163–172

Le Clercq J, Angenot L (1984) A propos du *Drosera peltata* et de la standardisation de la teinture de *Drosera*. J Pharm Belg 39:269–274

Lecoufle M (1990) Carnivorous plants: care and cultivation. Blandford, London

Lloyd FE (1942) The carnivorous plants. Chronica Botanica Company, Waltham, Massachusetts

Lloyd JB, Middlebrook G (1944) Bacteriostatic activity of some new derivatives of diaminophenylsulphone and naphthoquinones against tubercle bacillus. Am Rev Tuberc 49:539–542

Lowrie A (1989) Carnivorous plants of Australia II. University of Western Australia Press, Nedlands

Luckner R, Lucker M (1970) Naphthochinonderivate aus *Drosera ramentacea* Burch. ex Harv & Sond. Pharmazie 25:261–265

Macbeth AK, Winzor FL (1935) The identity of droserone. J Chem Soc 51:334–336

Macbeth AK, Price JR, Winzor FL (1935) The synthesis of droserone. J Chem Soc 51:325–327

McNally SF, Stewart A, Wilson UE (1988) The stimulation of acid-phosphatase activity in the stalked gland of *Drosera rotundifolia*. Ann Bot 61:289–292

Middleton E, Kandaswami C (1993) The impact of flavonoids on mammalian biology: implications for immunity, inflammations and cancer. In: Harborne JB (ed) The flavonoids: advances in research since 1986. Chapman & Hall, London, pp 619–652

Mitchell MJ, Smith SL (1988) Effects of the chitin synthetase inhibitor plumbagin and its 2-demethyl derivative juglone on insect ecdysone 20-monooxygenase activity. Experientia 44:990–991

Murashige T, Skoog F (1962) A revised medium for rapid growth and bio-assays with tobacco tissue cultures. Physiol Plant 15:473–497

Nahrstedt A (1980) Absence of cyanogenesis from Droseraceae. Phytochemistry 19:2757–2758

Nessler CL, Mahlberg PB (1979) Ultrastructure of laticifers redifferentiated organs on callus from *Papaver somniferum* (Papaveraceae). Can J Bot 57:675–685

Ragazzi L, De Biasi M, Pandolfo L, Chinellato A, Caparrotta L (1993) In vitro effects of naphthoquinones isolated from *Drosera* species. Pharmacol Res 27:87–88

Ramachandran Nair AG, Shanmugasundaram P, Madhusudhanam KP (1990) Naphthoquinones and flavonols from leaves of *Drosera peltata*. Fitoterapia 61:85–86

Ray PG, Majumdar SK (1976) Antimicrobial activity of some Indian plants. Econ Bot 30:317–320

Reinert RA, Mohr HC (1967) Propagation of *Cattleya* by tissue culture of lateral bud meristems. Proc Am Soc Hortic Sci 91:664–671

Repčák M, Galambosi B (1994) Ontogenetic aspects of 7-methyljuglone accumulation in some *Drosera* L. species. Biol Plant 36 (Suppl):232

Repčák M, Galambosi B (1996) Flavonoids in some *Drosera* species. Biologia 51:98–99

Reynolds T (1987) Comparative effects of alicylic compounds and quinones on inhibition of lettuce fruit germination. Ann Bot 60:215–223

Šamaj J, Bobák M, Erdelský K (1990) Vacuolar system in the cells of organogenic callus cultures of *Papaver somniferum* L. in vitro. Acta FRN Uni Comen; Physiol Plant 25:21–32

Šamaj J, Bobák M, Ovečka M, Krištín J, Blehová A (1994) Extracellular matrix in early stages of plant regeneration in vitro. Cell Biol Int 18:545

Šamaj J, Bobák M, Blehová A, Krištín J, Auxtová-Šamajová (1995) Developmental SEM observations on an extracellular matrix in embryogenic calli of *Drosera rotundifolia* and *Zea mays*. Protoplasma 186:45–49

Šamaj J, Bobák M, Ovečka M, Blehová A, Pretová A (1997) Structural features of plant morphogeneis in vitro. Veda, Bratislava

Sampara-Rumantir N (1971) Rossoliside. Pharm Weekbl 106:653–664

Schmid G (1912) Beiträge zur Ökologie der insektivoren Pflanzen. Flora 104:335–383

Schölly T, Kapetanidis I (1989a) Droseron-5-glucosid – ein neues Heterosid aus den oberirdischen Teilen von *Drosera rotundifolia* L. (Droseraceae). Pharm Acta Helv 64:66–67

Schölly T, Kapetanidis I (1989b) Flavonol and naphthoquinone glycosides of *Drosera rotundifolia*. Planta Med 55:611–612

Schölly T, Kapetanidis I (1993) Flavonoide aus *Drosera rotundifolia* L. (Droseraceae). Sci Pharm 61:277–282

Schulze W, Schulze ED (1990) Insect capture and growth of the insectivorous *Drosera rotundifolia* L. Oecologia 82:427–429

Simola LK (1978a) The effect of several amino acids and some inorganic nitrogen sources on the growth of *Drosera rotundifolia* in long and short-day conditions. Z Pflanzenphysiol 90:61–68

Simola LK (1978b) Dipeptides as nitrogen for *Drosera rotundifolia* in aseptic culture. Physiol Plant 44:315–318

Slack A (1980) Carnivorous plants. Edburg Press, London

Slack A (1985) Karnivoren: Biologie und Kultur der insektenfangenden Pflanzen. Ulmer, Stuttgart

Small JGC, Hendrikz B (1974) Evidence for nitrate reductase activity in the carnivorous plant *Drosera aliciae* Hamet. S Afr J Sci 70:156–157

Small JGC, Onraet A, Grieson DS, Reynolds G (1977) Studies of insect-free growth, development and nitrate assimilating enzymes of *Drosera aliciae* Hamet. New Phytol 79:127–133

Smith TJ, Yang CS (1994) Effect of food phytochemicals on xenobiotic metabolism and tumorigenesis. In: Huang MT, Osawa T, Ho T, Rosen RT (eds) Food phytochemicals for cancer prevention I. Fruits and vegetables. ACS Symp Ser 546, Am Chem Soc, Washington, DC, pp 17–48

Stewart CN, Nilsen ET (1992) *Drosera rotundifolia* growth and nutrition in a natural population with special reference to the significance of insectivory. Can J Bot 70:1409–1416

Svensson BM (1995) Competition between *Sphagnum fuscum* and *Drosera rotundifolia*: a case of ecosystem engineering. Oikos 74:205–212

Taiz L, Zeiger E (1991) Plant physiology. Benjamin Cummings, Redwood City

Takahashi H (1988) Ontogenetic development of pollen tetrads of *Drosera capensis* L. Bot Gaz 149:275–282

Takahashi H, Sohma (1982) Pollen morphology of Droseraceae and its related taxa. Sci Rep Tohoku Univ 4th Ser Biol 38:81–156

Thomson RH (1949) Naturally occuring quinones: a synthesis of droserone. J Chem Soc 5:1277–1278

Thomson RH (1987) Naturally occurring quinones, 3rd edn. Butterworth Scientific Publications, London

Thum M (1989a) The significance of carnivory for the fitness of *Drosera* in its natural habitat. 1. The reactions of *Drosera intermedia* and *Drosera rotundifolia* to supplementary feeding. Oecologia 75:472–480

Thum M (1989b) The significance of carnivory for the fitness of *Drosera* in its natural habitat. 2. The amount of captured prey and its effect on *Drosera intermedia* and *Drosera rotundifolia*. Oecologia 81:401–411

Van der Vijver LM, Lotter AP (1971) The constituents in the roots of *Plumbago auriculata* Lam. and *Plumbago zeylanica* L. responsible for antibacterial activity. Planta Med 20:8–13

Vinkenborg J, Sampara-Rumantir N, Ufelie OF (1969) Presence of hydroplumbagin glucoside in *Drosera rotundifolia*. Pharm Weekbl 104:45–49

Vinkenborg J, Sampara-Rumantir N, Ufelie OF (1970) Rossoliside, a D-glucoside from *Drosera rotundifolia*. Pharm Weekbl 105:414

Walt G (1972) A dictionary of economic products of India, vol 4. Periodical Experts, New Delhi

Wawrosch C, Steinberger B, Markotai J, Kopp B (1993) In vitro propagation of *Drosera* species. Planta Med (Suppl) 59:A653

Wawrosch C, Markotai J, Steinberger B, Kopp B (1996) In vitro-Vermehrung von Sonnentau-Arten. Sci Pharm 64:709–717

Weiss RF (1991) Herbal medicine. 6th edn. Beaconsfield Publishers, Beaconsfield

Wurm G, Grimm H, Geres U, Schmidt H (1984) Plumbagin: Reactivität, Toxizität und antimicrobielle Aktivität des in *Drosera* und *Plumbago* Arten vorkommenden Naturstoffe. Dtsch Apoth Ztg 124:2128–2132

Zenk MH, Fürbringer M, Steglich W (1969) Occurrence and distribution of 7-methyljuglone and plumbagin in the Droseraceae. Phytochemistry 8:2199–2200

VIII *Fagara zanthoxyloides* Lam. (Rutaceae): In Vitro Culture and the Production of Benzophenanthridine and Furoquinoline Alkaloids

E. Couillerot[1], C. Caron[2], C. Trentesaux[3], J.C. Chenieux[4], and J.C. Audran[1]

1 Introduction

Fagara zanthoxyloides (family Rutaceae) is a West African tree used in traditional medicine. Its major interest appears to be the presence in the roots of the specific alkaloid fagaronine (Messmer et al. 1972; Barret and Sauvaire 1992). Systematic confusion between the genera *Fagara* and *Zanthoxylum* is frequent due to the similar morphologic characteristics of the two. The opposite single/double perianth is used to distinguish *Zanthoxylum* / *Fagara* respectively (Engler and Prantl 1896). *Fagara zanthoxyloides* Lam. (synonyms: *Zanthoxylum zanthoxyloides* Waterm., *Fagara senegalensis* A. Chev., *Zanthoxylum senegalense* A. Chev., *Zanthoxylum polyganum* Schum.; comon names: ouo, kaingué, busan, goro-gna in African languages) is a thorny shrub or tree 3 to 8 m high (Fig. 1). Cylindric stems exhibit a very hard and pale yellow wood with a reddish-brown colored bark. Branchlets and leaf rachises are armed with sharp recurved prickles. The leaves are imparipinnate alternate (5 to 9 pairs of leaflets). The leaflets are not accuminate (a characteristic of *F. zanthoxyloides*), oblong or oblong-oblanceolate, sometimes emarginate at the apex, subacute at the base, 5–10 cm long, 2–4 cm broad and glabrous. Flowers, clustered panicles, are white and unisexual. Fruits, the size of a pea, are capsuled with two valve-like openings containing a single shiny black seed. The germination rate is low and germination relatively slow. Its area of distribution is the littoral from Senegal to Nigeria (Hutchinson et al. 1958; d'Aubreville 1959; Bouquet and Debray 1974; Ake-Assi and Adjanohoun 1979). In West Africa, roots of *F. zanthoxyloides* are commonly used for tooth cleaning as a chewing stick (Lewis 1983), and it was proved that the root extract exhibited a high antimicrobial activity against oral flora (El Said et al. 1971).

In traditional medicine, the plant has various uses (Table 1). The antisickling activity of root extracts on sickle cell anemia has been associated

[1] Laboratoire de Biologie et Physiologie Végétales, EA 2069, UFR Sciences, Université de Reims Champagne-Ardenne, BP 1039, 51687 Reims Cedex 2, France
[2] Laboratoire de Pharmacognosie, UPRES-A 6013, UFR Pharmacie, Université de Reims Champagne-Ardenne, CRPB, BP 1039, 51687 Reims Cedex 2, France
[3] Laboratoire de Biochimie et Biologie Moléculaire, EA 2063 IFR 53, UFR Pharmacie, Université de Reims Champagne-Ardenne, 51 rue Cognacq-Jay, 51096 Reims Cedex, France
[4] Laboratoire de Biologie Cellulaire et Biochimie Végétale, EA 1370, UFR Pharmacie, Université de Tours, 31 avenue Monge, 37200 Tours, France

Biotechnology in Agriculture and Forestry, Vol. 43
Medicinal and Aromatic Plants XI (ed. by Y.P.S. Bajaj)

Fig. 1. *Fagara zanthoxyloides* Lam

Table 1. Examples of traditional uses of *Fagara zanthoxyloides* Lam. (Kerharo and Adam 1973; Adjanohoun 1986; for additional information see also Irvine 1930; Dalziel 1937; Paris and Moyse-Mignon 1947; Bouquet and Debray 1974)

Part of the plant used	Treated symptoms
Roots	Worming powder
	Gingivitis
	Pharyngitis
	Dysmenorrhoea[a]
	Analgesic toothpick
	Sickle cell anemia[a]
	Snake bite[a]
Root bark	Scabies
Roots/leaves	Gastroenteritis
	Diarrhoea
	Fever
Fresh leaves	Leucorrhea
Bark	Laxative and stomachic
	Cough
Stem bark	Gonorrhea
Pulp	Rheumatisms

[a] In association with other plants.

with the presence of benzoic acid derivatives (Edwards et al. 1979). Clinical experiments are in progress in different countries of West Africa and positive antisickling effect were obtained by oral treatment of children with hard capsules of an association of *Fagara zanthoxyloides* and *Calotropis procera* (Ouattara et al. 1992).

In addition to antitumor activity towards lymphocytic leukemias in vivo and towards colon 26 (Messmer et al. 1972; Douros and Suffness 1981), fagaronine has been shown to induce the cell differentiation and growth inhibition of various hematopoietic cell lines (Comoë et al. 1987, 1988). Due to the difficulty in obtaining fagaronine using plant root extraction or chemical synthesis, plant and tissue cultures of *Fagara zanthoxyloides* have been investigated in order to study the biosynthesis and the production of the potent antileukemic molecule in vitro and, widely, the alkaloid metabolism of this plant.

2 Chemical Constituents and Their Pharmacological Properties

Among the different groups of secondary metabolites isolated from the genus *Fagara*, i.e. coumarins, amides, flavonoids, lignans, alcohols, ketons, organic acids (Comoë 1987; Molinatti 1994), the group of alkaloids appears to be the most interesting from a biological activity point of view. Major alkaloids of *F. zanthoxyloides* are derivatives of the two aromatic amino acids tyrosine and tryptophan and from anthranilate (Table 2). The two alkaloid groups found in tissue cultures of *F. zanthoxyloides*, i.e. benzophenanthridines and furoquinolines are described here.

Table 2. Major alkaloids of *Fagara zanthoxyloides* Lam. (Mester 1973; Comoë 1987)

Alkaloids derived from	Alkaloid type	Name
Tyrosine	Benzyltetrahydroisoquinoline	Tembetarine
	Quaternary aporphine	Magnoflorine
		N-methylcorydine
		N-methylisocorydine
	Protoberberine	Berberine
	Benzo[c]phenanthridine	Fagaronine
		Chelerythrine
		Fagaridine
		Angoline
	Protopine	Fagarine II
Anthranilate	Hydroquinoline	Atanine
	Furoquinoline	Skimmianine
Tryptophane	Canthinone	Canthin 6-one

2.1 Benzophenanthridines

Chelerythrine and fagaronine exhibit a high specificity and biological activity (Fig. 2). Chelerythrine, isolated in root bark of the plant by Torto et al. (1966), occurs in several species of Rutaceae and Papaveraceae (Krane et al. 1984) and has been described as a specific protein kinase C inhibitor (Herbert et al. 1990).

Fagaronine has so far only been isolated from the roots of *F. zanthoxyloides* (Messmer et al. 1972; Comoë 1987). Furthermore, this alkaloid is the only benzophenanthridine to possess a phenolic group at C-3. Fagaronine exhibits antibacterial activities (Pezzuto et al. 1983; Smekal et al. 1986) and the quaternary nitrogen is essential to the biological activity (Pezzuto et al. 1983; Dostal and Potacek 1990). Inhibition of reverse transcriptase in RNA tumor viruses was demonstrated (Sethi and Sethi 1975; Sethi 1979, 1981) and this activity has been extended to HIV-1 and HIV-2 reverse transcriptase (Tan et al. 1991, 1992).

The major interest of fagaronine appears to be its antileukemic properties, shown first by Messmer et al. (1972) with the isolation of the molecule. Studying erythroid differentiation in human leukemic K562 cells by fagaronine, Comoë et al. (1987) showed an inhibition of 80% of growth without increasing the mortality of the treated cells. This growth inhibition was correlated with an inhibition of DNA and total proteins synthesis, a decrease in the mitotic index, a slowing down – not a blockage – of the cycle at the end of the S-phase and during G2 phase, a hemoglobin accumulation (Comoë et al. 1987, 1988). These results showed that fagaronine could be a potent molecule for the new concept of differentiation therapy. This new concept is based on the possibility of suppressing the imbalance between proliferation and differentiation which characterizes tumor cells (Trentesaux et al. 1993).

The mechanism by which fagaronine induces differentiation in K562 cells is under investigation at the molecular level by our group, but it has been

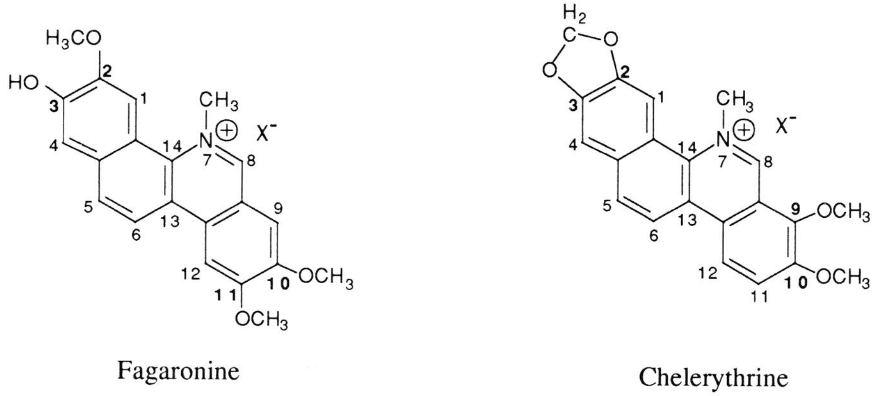

Fig. 2. Structures of fagaronine and of chelerythrine

shown that fagaronine interferes with DNA with different modalities: intercalation into initiation complexes within the DNA double strand or into double strand regions of RNA (Sethi and Sethi 1975; Sethi 1979; Pezzuto et al. 1983), inhibition of DNA, RNA and protein synthesis (Casiano Torres and Baez 1986), inhibition of DNA polymerase by intercalation, but only a weak action on RNA polymerase (Sethi and Sethi 1975), inhibition of DNA topoisomerase I and II (Larsen et al. 1993; Wang et al. 1993).

The absence of mutagenicity of fagaronine on *Salmonella typhimurium* (Cheng et al. 1977; Pezzuto et al. 1983) and *Drosophila melanogaster* (Pérez-Chiesa and Rodriguez 1993) was shown; futhermore, the last authors indicated that these results could be associated with the presence of the alkoxy groups and the quaternary nitrogen.

2.2 Furoquinolines

Skimmianine (Fig. 3) is the most common furoquinoline alkaloid of the dictamnine group found in the Rutaceae (Mester 1983) and was isolated in *F. zanthoxyloides* by Paris and Moyse-Mignon (1947). This alkaloid, which exerts sedative and hypothermic properties (Evdokimova et al. 1971), has been more recently described as a potential antileishmanial drug (Fournet et al. 1993) and as an antagonist of 5-hydroxytryptamin receptors in animals (Cheng et al. 1994). As we know, γ-fagarine (Fig. 3) has not been described in *F. zanthoxyloides*, but we have isolated and identified this alkaloid in cell suspension cultures of this plant (Couillerot et al. 1994). The presence of γ-fagarine in the roots of a native plant from the Ivory Coast by HPLC photodiode array analysis was also confirmed (personal data).

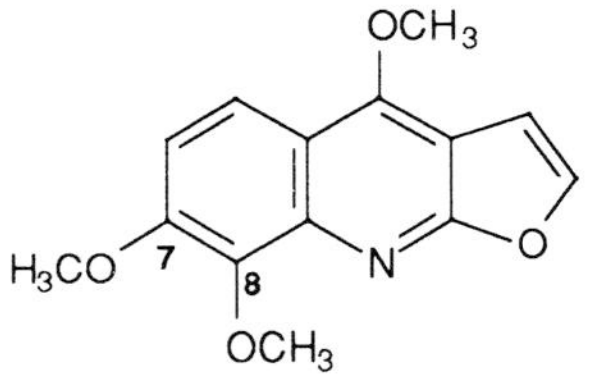

Skimmianine

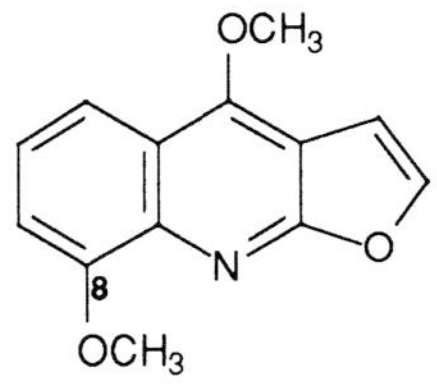

γ-fagarine

Fig. 3. Structures of skimmianine and of γ-fagarine

2.3 Biosynthesis of Benzophenanthridines and Furoquinolines

Introduction. Benzophenanthridines and furoquinolines are derived from the shikimic pathway, which branches at chorismate (Fig. 4) Benzophenanthridines are derived from phenylalanine and tyrosine: chorismate mutase

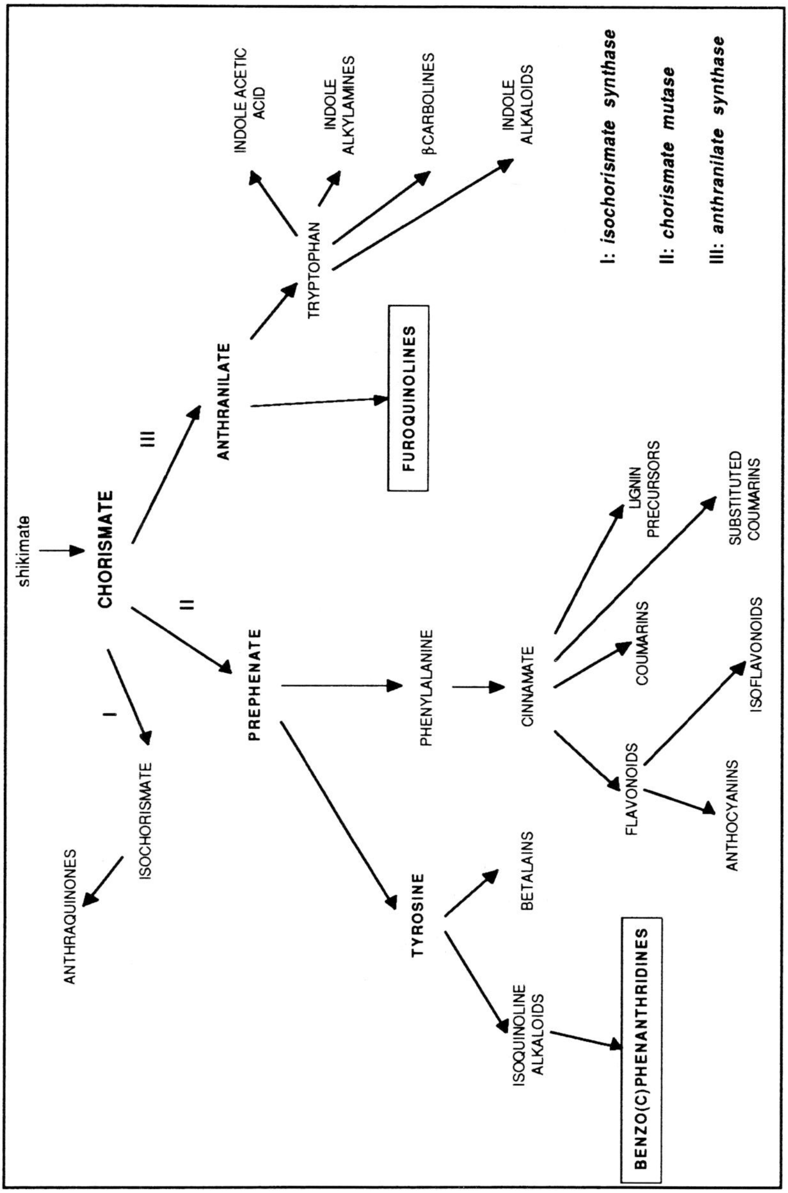

Fig. 4. Examples of some classes of secondary metabolites that arise from chorismate. (Adapted from Poulsen and Verpoorte 1991)

(CM, EC 5.4.99.5) catalyses the Claisen-like rearrangement of chorismate to prephenate which gives tyrosine and phenylalanine. In most plants studied, two isoforms of CM have been isolated, which were named CM-1 and CM-2 but the presence of a third isoenzyme (CM-3) has been detected. Isoenzyme CM-1, strongly regulated by aromatic amino acids (tryptophane, tyrosine, phenylalanine), is associated with protein synthesis in the plastidial compartment. CM-2 isoform is non-regulated and it has been postulated that it provides the precursors for the secondary metabolism in the cytosol by a single overflow mechanism (Jensen 1986; Poulsen and Verpoorte 1991, 1992; Romero et al. 1995). Identification of c-DNA, corresponding to the genes AtCM1 and AtCM2, encoding, respectively, a plastidic CM and a cytosolic CM has been reported (Eberhard et al. 1996). Nevertheless, the existence of a dual shikimate pathway for the biosynthesis of the aromatic amino acids is still disputed (for a review see Schmid and Amrhein 1995). The differential determination of the activity of the isoenzymes by the tryptophan activation rate would be an indicator of the presence of CM isoenzymes but results obtained with tryptophan activation were ambiguous (Poulsen and Verpoorte 1992; Bohlmann and Eilert 1994).

The determination of CM activity was done in both cell suspension and hairy root cultures of *F. zanthoxyloides* cultured in the dark. As expected, CM activity was present in the transformed roots and we have observed a strong correlation between root growth and the enzyme activity (Fig. 5). The specific CM activity found here was similar to the activities found in cell cultures of various other species (Poulsen and Verpoorte 1992). Regulation of CM acti-

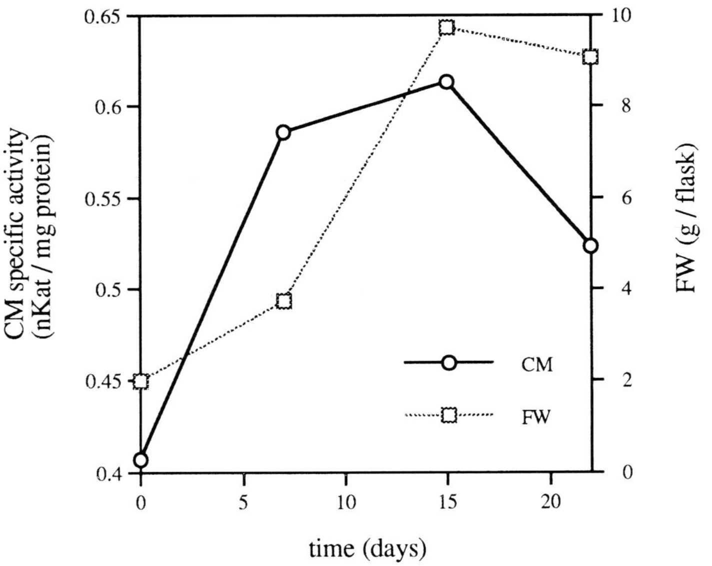

Fig. 5. Evolution of chorismate mutase (CM) specific activity and fresh weight (FW) in hairy roots of *Fagara zanthoxyloides* cultured in liquid medium. (Couillerot et al., unpubl.)

vity by tryptophan and tyrosin was effective in both cell suspension and hairy roots of *F. zanthoxyloides* (Table 3). Moreover, these results showed a regulation of CM activity in dark-cultured cells and roots, confirming a possible CM activity in non-photosynthetic plant tissues in accordance with other observations (Romero et al. 1995). But, in contrast, CM activity in *Datura stramonium* hairy roots showed that the tryptophan activation was only effective when the transformed roots were cultured under light conditions (Leuschner et al. 1995).

Benzophenanthridines. The biosynthetic pathway of benzophenanthridine is now well known since Zenk and co-workers have elucitated all the intermediates and the enzymes implicated in the biosynthesis of sanguinarine and of macarpine (review in Kutchan et al. 1991; Kutchan and Zenk 1993). Feeding experiments have clearly demonstrated that (S)-norcoclaurine, and not norlaudanosoline, was implicated in the biosynthesis leading to S-reticuline (Stadler et al. 1989; Müller and Zenk 1992). Transformation of reticuline by the berberine bridge enzyme (BBE, EC 1.5.3.9) lead to the specific benzophenanthridine pathway. The high stereoselectivity of the N-methyltransferase (methylation of the tetrahydroprotoberberine with the S configuration) is, moreover, dependent on the substitution patterns of rings A and D: the N-methylation should only occur when the methylendioxy groups have been formed (Kutchan et al. 1991). The formation of benzophenanthridines which possess none or only one methylendioxy group (as, respectively, fagaronine and chelerythrine) would involve different or modified steps of biosynthesis. The 6-hydroxylation of the protropine (Kutchan et al. 1991) or of the (S)-*cis*-N-methylstylopine (Battersby et al. 1975) appears to be a key in the benzophenanthridine pathway.

Furoquinolines. The biosynthesis of furoquinolines (Fig. 6) has been elucidated with tracer feeding experiments showing that 4-hydroxy-2-quinolone, 3-prenylquinolone and platydesmine were the precursors of dictamnine which is the specific precursor of skimmianine (Grundon 1988). From chorismate, the

Table 3. Examples of CM activity in hairy roots and suspension cells of *F. zanthoxyloides* cultured in the dark; a tryptophan activation and, especially, a tyrosin inhibition of CM activity were observed. (Couillerot et al., unpubl.)

Cultures	CM activity (nKat/mg protein)		
	Standard	+ Tryptophan	+ Tyrosine
Hairy roots			
7 days old	0.934 ± 0.137	1.194 ± 0.187	0.233 ± 0.056
14 days old	0.680 ± 0.100	0.963 ± 0.057	0.096 ± 0.017
Cell suspension			
10 days old	0.418 ± 0.090	0.519 ± 0.024	0.189 ± 0.023
21 days old	0.668 ± 0.049	0.805 ± 0.028	0.382 ± 0.039

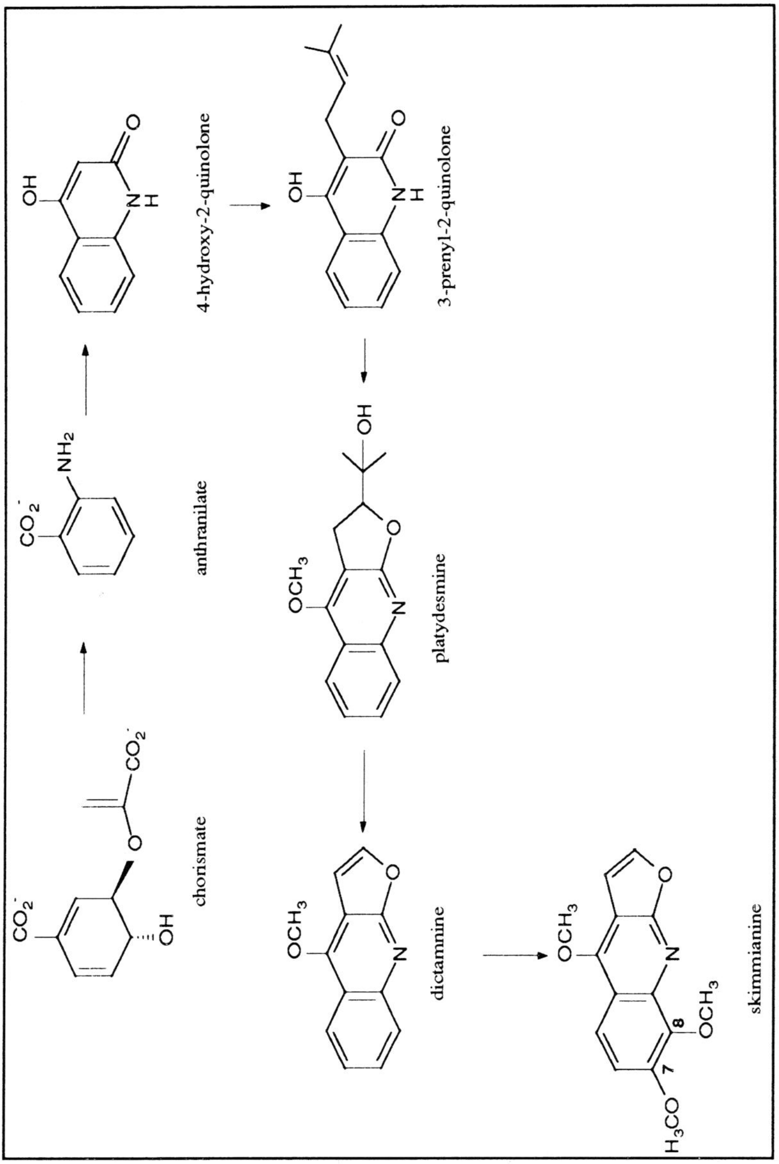

Fig. 6. Skimmianine pathway. (After Grundon 1988)

first step of the pathway is the synthesis of anthranilate by anthranilate synthase (EC 4.1.3.27).

3 In Vitro Culture Studies

3.1 Establishment of Callus, Suspension Cultures, and Hairy Roots

Callus cultures of *F. zanthoxyloides* (Fig. 7A) were obtained using an MS medium (Murashige and Skoog 1962). Leaf and stem explants from the whole plant were used, but formation of primary calli occured faster with leaves. Dark, and the presence of naphthaleneacetic acid (NAA, 26.8 μM) and benzylaminopurine (BAP, 4.4 μM) appeared to be necessary for growth and obtaining for friable calli, from which suspension cultures (Fig. 7B) were established (Couillerot et al. 1994).

Transformed roots were established by the co-culture method using leaf segments of shoot cultures (Fig. 7C) and *Agrobacterium rhizogenes* strain 9402. The other strains used (15834, A4pc and 8196) were inefficient under the conditions set. Only one root tip was obtained and then transferred on a modified MS medium (5 times reduced nitrogen source) containing cefotaxime. The transformed roots (Fig. 7D, E), named line p11Agr9402, were transferred 4 weeks later in a woody plant (WP) liquid medium (LLoyd and McCown 1980). The genetic transformation of the cultures was confirmed by the opine assay, revealing the presence of both agropine and mannopine. The same assay performed 1 year later only showed the presence of agropine; 3 years after root transformation, neither mannopine nor agropine were observed. This phenomenon could be associated with opine gene repression (Schaerer and Pilet 1991), but, as we observed with subcultures, root growth or alkaloid production were not modified.

Analysis of root growth in agar medium, expressed as length of root tips, showed an average length of 1.6 mm day^{-1} (Fig. 8A). The hairy root clone cultures in a liquid medium renewed weekly showed a biomass production of 3.2 g fresh weight $day^{-1} l^{-1}$ (Fig. 8B). The presence of numerous lateral branches was absolutely necessary in obtaining a high biomass, in contrast to the hairy roots of *Atropa belladonna*, which showed a negligible influence of branches on growth (Falk and Doran 1996).

3.2 Production of Alkaloids

3.2.1 Cell Suspension Cultures

Initial analysis of alkaloid contents in calli and cell suspensions showed the presence of chelerythrine and fagaronine, but the production of these two metabolites was lost in subcultures. Nevertheless, fagaronine could be

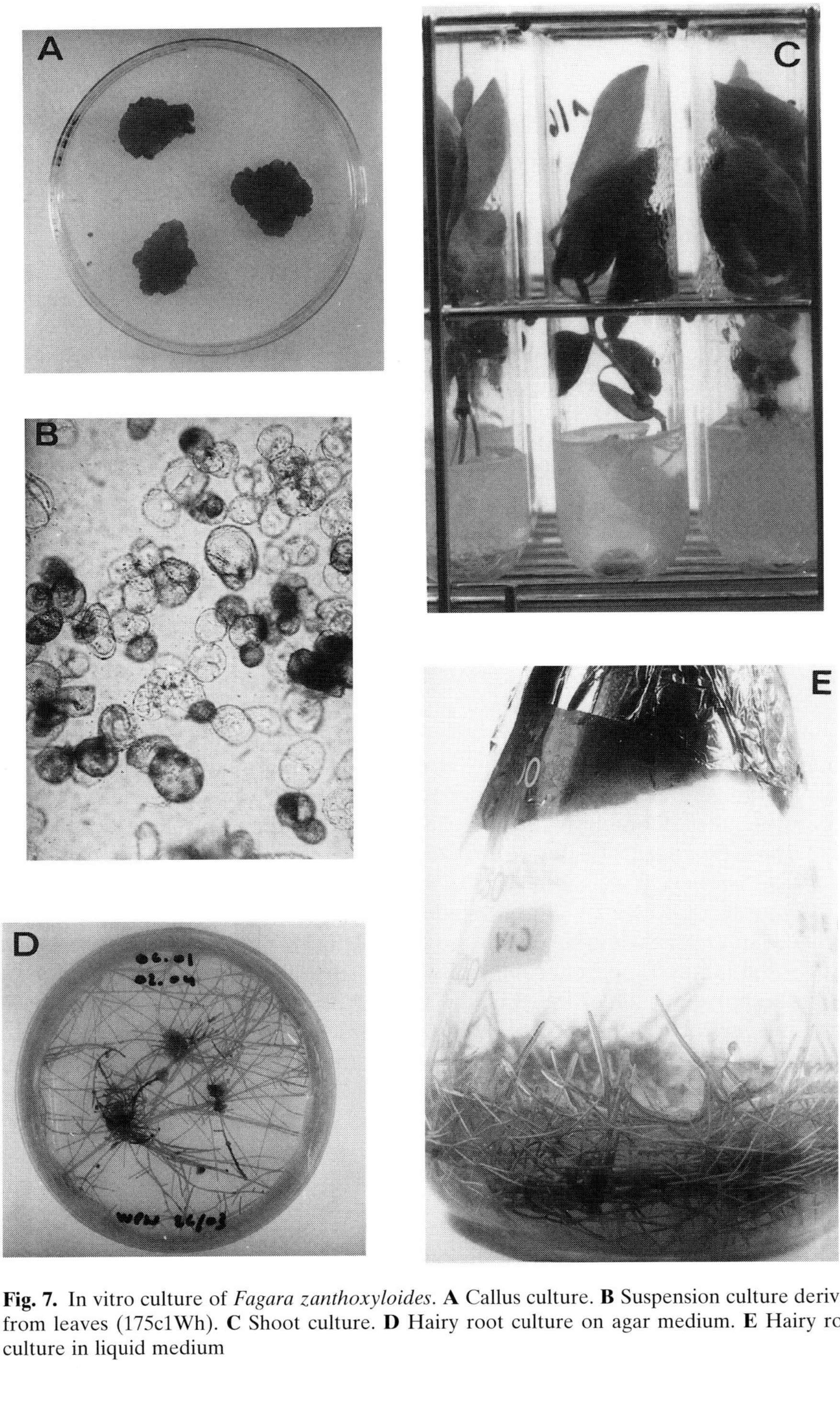

Fig. 7. In vitro culture of *Fagara zanthoxyloides*. **A** Callus culture. **B** Suspension culture derived from leaves (175c1Wh). **C** Shoot culture. **D** Hairy root culture on agar medium. **E** Hairy root culture in liquid medium

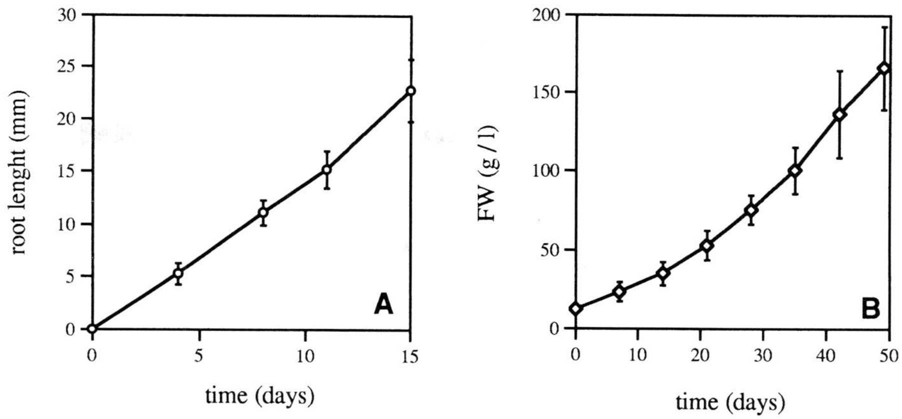

Fig. 8. A Growth curves of hairy roots of *Fagara zanthoxyloides* in solid medium and **B** in a weekly liquid renewed medium. (Couillerot 1994)

purified, and its effect on human K562 leukemic cells evaluated in comparison to the fagaronine standard (extracted from the native plant): the results clearly indicated that growth inhibition and cell differentiation were identical (Fig. 9) and confirmed the particular activity of the molecule on that cell line. Two furoquinoline alkaloids were identified as skimmianine, present in the plant, and γ-fagarine which has, as we know, never been isolated from *F. zanthoxyloides*. The production of these two alkaloids remained high and constant within the subcultures. Time courses of accumulation of skimmianine and γ-fagarine showed, respectively, a growth stage-dependent production and a growth stage-independent production (Fig. 10).

3.2.2 Hairy Roots

Extraction of alkaloids from the transformed roots gave a mixture containing four major compounds. Isolation of fagaronine and chelerythrine was performed by HPLC. They were identified by comparing their spectroscopic data (UV, MS, NMR) with authentic samples available in our laboratory. The two other compounds were isolated by means of centrifugal TLC. They were identified as furoquinoline skimmianine and γ-fagarine, by comparison of their spectral properties with reference samples. So, the alkaloid profiles were similar between transformed roots and initial cell suspension cultures, but the accumulation of the benzophenanthridine appeared to be constant in the hairy roots.

Preliminary results on the production of chelerythrine and fagaronine indicated a precocious accumulation of the alkaloids within the hairy roots with respectively 2.3 ± 0.29 and $0.95 \pm 0.12\,mg\,g^{-1}$ dry weight after 7 days of culture. At day 14, the production of the two alkaloids was equivalent with

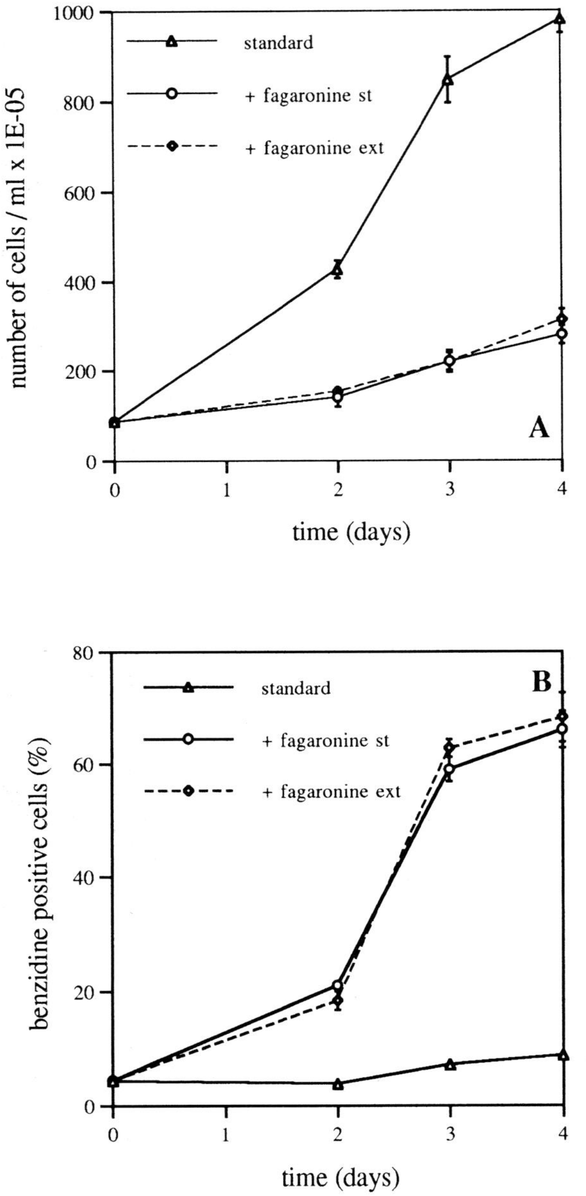

Fig. 9. Effect of fagaronine extracted from the *F. zanthoxyloides* cell suspension (fagaronine ext) on the K562 cell growth rate compared with fagaronine standard (fagaronine st) (**A**) and on the kinetics of appearance of benzidine-positive K562 cells (**B**). Cell counts from K562 cell cultures initiated at 90 × 1E-05 cells/ml and treated with 3 × 1E-06 M of fagaronine or without added fagaronine (standard). (Couillerot et al. 1994)

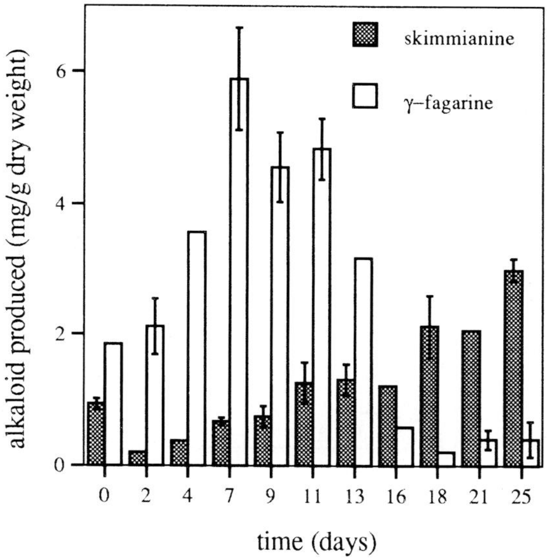

Fig. 10. Skimmianine and γ-fagarine time-course production in *F. zanthoxyloides* cells cultured for 25 days in a MS modified medium. (Couillerot 1994)

2.14 ± 0.38 and 0.77 ± 0.13 mg g^{-1} dry weight. Only traces of the two alkaloids have been found in the culture medium by HPLC analysis.

The effect of fagaronine, extracted from the Fagara hairy roots, on the human K562 cell growth rate and on the kinetics of appearance of hemoglobin synthesis-positive K562 cells (differentiated cells) showed 90% growth inhibition with a 46% differentiation for a concentration of 1×10^{-5} of fagaronine without increasing lethality (Fig. 11). These results confirm the antileukemic activity of fagaronine on the K562 human cell line.

3.2.3 Relation Between Furoquinoline Accumulation and Plant Growth Regulators

The effect of plant growth regulators on secondary metabolism, and more especially on alkaloid accumulation in cell cultures, is well established. In order to study a possible relationship between the auxin naphthaleneacetic acid (NAA), the cytokinin benzylaminopurine (BAP) and furoquinoline production, we cultured a non-habituated cell line of *F. zanthoxyloides* in either NAA-free, BAP-free or NAA/BAP-free media (Couillerot et al. 1996). Growing the cells in these three different conditions induced a slight growth reduction but did not affect the final biomass obtained (Fig. 12). Analysis of the consumption kinetics of carbon and nitrogen sources showed no major alteration in the primary metabolism of the cell line.

On the other hand, it was observed that furoquinoline production was highly dependent on BAP (Fig. 13): the total accumulation of skimmianine and

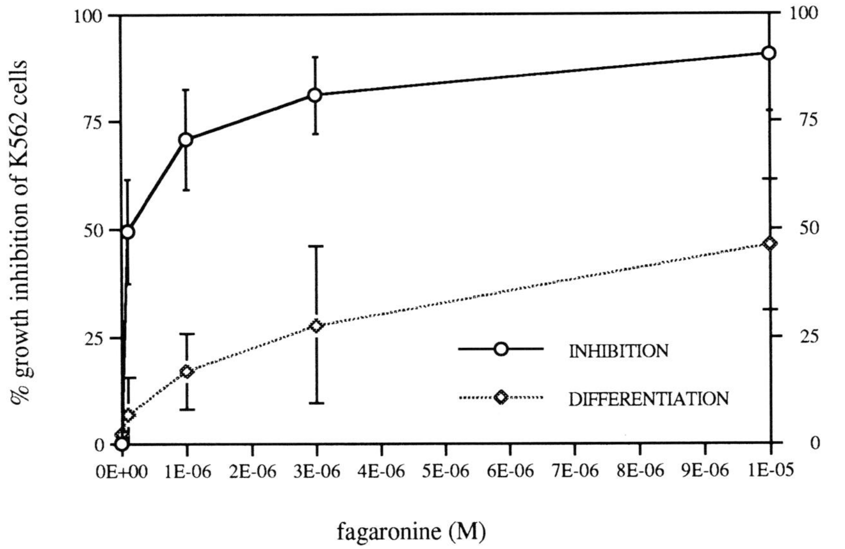

Fig. 11. Effect of fagaronine purified from hairy roots of *F. zanthoxyloides* on the growth and the differentiation of the K562 human erythroleukemic cells. (Couillerot et al., unpubl.)

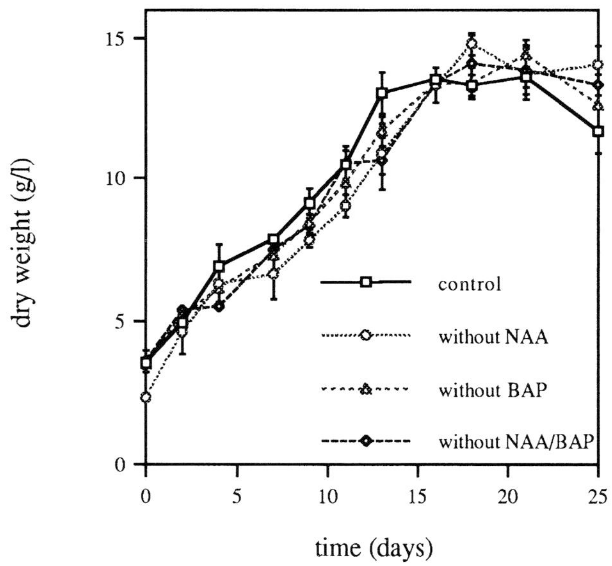

Fig. 12. Growth curves of *F. zanthoxyloides* cells cultured for 25 days in standard medium (NAA 26.8 μM; BAP 4.4 μM) and in NAA, BAP, NAA/BAP deprived media. (Couillerot et al. 1996)

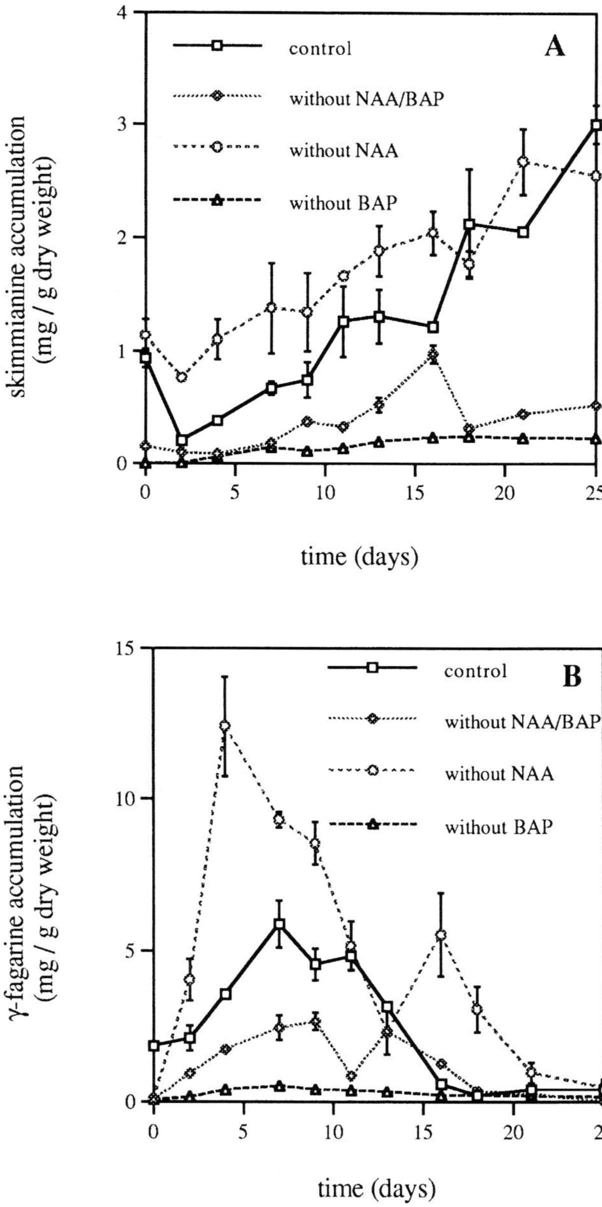

Fig. 13. A Skimmianine and **B** γ-fagarine time-course accumulations in *F. zanthoxyloides* cells cultured for 25 days in standard medium (NAA 26.8 μM; BAP 4.4 μM) and in NAA, BAP, NAA/BAP deprived media. (Couillerot et al. 1996)

γ-fagarine was 9 times lower when cells were cultured in BAP-free medium. NAA, in contrast, was shown to inhibit the production of both alkaloids. These results were confirmed by the analysis of cell production in a NAA/BAP-free medium, with levels of accumulation respectively 3.5 and 2.1 lower for skimmianine and γ-fagarine. The cytokinin BAP was shown to play an essential role in the accumulation of skimmianine and γ-fagarine within the cells of *F. zanthoxyloides*: the presence of BAP in the culture medium appeared to be necessary to obtain high levels of furoquinolines. So, exogenous cytokinin appears to be implicated in the regulation of furoquinoline biosynthesis.

4 Suggested Protocols

4.1 Induction of Primary Callus

Explants of stems and leaves were excised in March 1989 from a young shrub from the central Ivory Coast. A modified Murashige and Skoog (1962) medium, supplemented with 1.4 mM glutamine, 88 mM sucrose, 8 $g l^{-1}$ agar and either 26.8 μM NAA + 4.4 μM BAP or 5.3 μM NAA + 4.4 μM BAP was used. The cultures were placed at 28 °C under illumination (16 h day^{-1}, 20 $W m^{-2}$).

4.2 Establishment of Callus Cultures and Cell Suspensions

Once calli were obtained, they were transferred in the dark at 28 °C and subcultured over 5 weeks. Calli cultured on 5.3 μM NAA turned brown and died. So, only calli cultured on 26.8 μM NAA gave cell lines.

Cell suspension cultures were initiated by transferring callus to 80 ml of liquid medium (26.8 μM NAA, 4.4 μM BAP) in 300 ml Erlenmeyer flasks. The suspensions were subcultured weekly at a dilution of 1 : 2 and maintained in the dark at 26 °C on a gyratory shaker (100 rpm). The cell lines accumulated benzophenanthridines and furoquinolines during the first year, and then only furoquinoline accumulation was maintained.

4.3 Establishment of Transformed Root Cultures

The hairy roots were established by co-culture of leaf disks of sterile plantlets, obtained from in vitro seed germination, with *Agrobacterium rhizogenes* 9402. The bacteria were eliminated by transferring the explants on a Murashige and Skoog (1962) modified solid medium containing cefotaxime (0.2 mg/ml). Transformed roots were then cultured in WP liquid medium (LLoyd and McCown 1980) and maintained at 26 °C in the dark on a rotary shaker (100 rpm). The transformation was proved by the detection of opines as described by Petit et al. (1983) using paper electrophoresis.

4.4 Shoot Cultures

Shoots (1 cm high) were excised from 6-month-old sterile seedlings, or from young stems surface-sterilized and cultivated on Morel and Müller (1964) modified agar medium containing 30 $g l^{-1}$ sucrose. Shoot cultures are maintained at 26 °C under illumination (16 h day^{-1}, 20 $W m^{-2}$).

4.5 Quantitative Determination of Alkaloids by HPLC

Freeze-dried cultured cells or roots were ground and refluxed (cell suspension) or agitated at room temperature (roots) for 5 h with MeOH. After filtration (glass fibre filter), the extract was evaporated to dryness under reduced pressure. HPLC photodiode array analysis using a Waters high pressure liquid chromatography system equipped with a 510 pump and photodiode array detector 991 was used for alkaloid measurement. The monitor wavelength was set at 400 nm for the benzophenanthridines and at 255 nm for the furoquinolines. The alkaloids were separated using a Merck Manucart Lichrospher 60 RP Select B (5 μm) (125-4) column in MeOH-HOAc 0.05 M (44 : 56) at a flow rate of 1 ml/min.

5 Summary and Conclusions

Fagara zanthoxyloides has been described as the only plant producing the original antileukemic alkaloid fagaronine. Furthermore, the structure of this molecule is quite interesting since it is the only benzophenanthridine with a free hydroxyl group at C-3, clearly implicated in the biological activity. As a complement to the classical cytotoxic effect, our group has shown the ability of fagaronine to differentiate human leukemic cells in vitro. The action of fagaronine on the DNA topoisomerase I and II is also quite original, justifying the interest in studying and producing this molecule.

Tissue culture studies on *F. zanthoxyloides* have shown the ability of both cell suspension and hairy root cultures to accumulate the potent antileukemic alkaloid fagaronine, as well as the other benzophenanthridine chelerythrine and the furoquinoline skimmianine and γ-fagarine. The obtaining of fagaronine-producing-transformed roots using *Agrobacterium rhizogenes* confirmed the ability of hairy roots to synthesize valuable products independently from the plant and its environment. This synthesis depends on the characteristics of transformed roots, i.e. their ability to synthesize the metabolites found in the roots of the plant they are derived from, their rapid growth and their genetical stability in secondary metabolite production (Rhodes et al. 1990). Furthermore, extraction of fagaronine is much easier from root culture and does not require cutting down the tree and the consequences of this.

From a biosynthetic point of view, studies on cell suspension provide an essential complement to the understanding of alkaloid metabolism pathways in *F. zanthoxyloides*. The furoquinoline biosynthesis depends on exogenous growth regulators and the cytokinin BAP was shown to be essential for the accumulation of skimmianine and γ-fagarine alkaloids. The blocking of the benzophenanthridine pathway in the undifferentiated cells is an exciting challenge, since the feeding of (S)-reticuline restored protoberberine biosynthesis in non-producing *Thalictrum* cultures (Zenk 1991). Studying the specific chorismate mutase activity in both cell suspensions and hairy roots cultured in the dark has shown regulation of the enzyme activity by tryptophan and tyrosine.

References

Adjanohoun E (1986) Contribution aux études ethnobotaniques et floristiques au Bénin et au Togo. ACCT, Paris

Ake-Assi L, Adjanohoun E (1979) I. Contribution au recensement des plantes médicinales de Côte d'Ivoire. CRES, Université d'Abidjan

Barret Y, Sauvaire Y (1992) Fagaronine, a novel antileukaemic alkaloid. Phytother Res 6:59–63

Battersby AR, Staunton J, Wiltshire HR, Francis RJ, Southgate R (1975) Biosynthesis. XXIII. The origin of chelidonine and of other alkaloids derived from the tetra-hydroprotoberberine skeleton. J Chem Soc Perkin Trans I:1147–1156

Bohlmann J, Eilert U (1994) Elicitor induced secondary metabolism in *Ruta graveolens* L. Role of chorismate utilizing enzymes. Plant Cell Tissue Organ Cult 38:189–198

Bouquet A, Debray M (1974) Plantes médicinales de la Côte d'Ivoire. Trav Doc ORSTOM 32:156–158

Casiano Torres CA, Baez A (1986) Effects of the antitumor drugs 3-nitrobenzothialozo {3,2-a}quinolium and fagaronine on nucleic acid and protein synthesis. Biochem Pharmacol 35:679–685

Cheng CC, Engles RR, Hodgson J, Ing RB, Wood HB, Yan SJ, Cheng RKYZ (1977) Absence of mutagenecity of coralyne and related antileukemic agents: structural comparisons with the potent carcinogen 7,12-dimethylbenz[a]anthracene. J Pharm Sci 66:1781–1783

Cheng JT, Chang TK, Chen IS (1994) Skimmianine and related furoquinolines function as antagonist of 5-hydroxytryptamine receptors in animals. J Auton Pharmacol 14:365–374

Comoë L (1987) Apports de la médecine traditionnelle africaine à la chimiothérapie anticancéreuse. Activité antitumorale et différenciante de la fagaronine (*Fagara xanthoxyloïdes* Lam.) PhD Thesis, University of Reims

Comoë L, Jeannesson P, Trentesaux C, Desoize B, Jardillier JC (1987) The antileukemic alkaloid fagaronine and the human K 562 leukemic cells: effects on growth and induction of erythroid differentiation. Leuk Res 11:445–451

Comoë L, Carpentier Y, Desoize B, Jardillier JC (1988) Effect of fagaronine on cell cycle progression of human erythroleukemia K562 cells. Leuk Res 12:667–672

Couillerot E (1994) Etude de la bioproduction alcaloïdique de suspensions cellulaires et de racines transformées de *Fagara zanthoxyloides* Lam. (Rutaceae). PhD Thesis, University of Reims

Couillerot E, Caron C, Comoë L, Audran JC, Molinatti P, Zèches M, Le Men-Olivier L, Jardillier JC, Chénieux JC (1994) Benzophenanthridine and furoquinoline accumulation in cell suspension cultures of *Fagara zanthoxyloides*. Phytochemistry 37:425–428

Couillerot E, Caron C, Audran JC, Jardillier JC, Chénieux JC (1996) Furoquinoline alkaloid accumulation in *Fagara zanthoxyloides* cell cultures is highly dependent on the presence of exogenous benzylaminopurine. Plant Growth Regul 19:203–206

Couillerot E, Caron C, Trentesaux C, Chenieux JC, Audran JC (1999) Genetic transformation of *Fagara zanthoxyloides*. In: Bajaj YPS (ed) Biotechnology in agriculture and forestry, vol 44. Transgenic trees. Springer, Berlin Heidelberg New York (in press)

D'Aubreville A (1959) La flore forestière de la Côte d'Ivoire. Centre Tech For Natl 2:107–112

Dalziel JM (1937) The useful plants of West tropical Africa. Crown agents of oversea governments and administration, London, pp 308–309

Dostal J, Potacek M (1990) Quaternary benzo[c]phenanthridine alkaloids. Collect Czech Chem Commun 55:2840–2873

Douros J, Suffness M (1981) New natural products under development at the National Cancer Institute. Recent Results Cancer Res 76:153–175

Eberhard J, Ehrler TT, Epple P, Felix G, Raesecke HR, Amrhein N, Schmid J (1996) Cytosolic and plastidic chorismate mutase isozymes from *Arabidopsis thaliana*: molecular characterization and enzymatic properties. Plant J 10:815–821

Edwards AR, Heading VE, Wheeler JW (1979) Isolation and identification of new-antisickling agents from *Fagara zanthoxyloides*. Clin Res 27:239A

El Said F, Fadulu SO, Kuye JO, Sofowora EA (1971) Native cures in Nigeria. II. The antimicrobial properties of buffered extracts of chewing-sticks. Lloydia 34:172–174

Engler A, Prantl K (1896) Die natürlichen Pflanzenfamilien. Engelmann, Leipzig, pp 115–123

Evdokimova NI, Polietsev NP, Sultanov MB (1971) Pharmacological properties of the alkaloid skimmianine. Farmakol. Alkaloidov Serdechnykh Glikozidov, Sultanov MB, Tashkent, pp 167–169

Falk LR, Doran PM (1996) Influence of inoculum morphology on growth of *Atropa belladonna* hairy roots and production of tropane alkaloids. Biotechnol Lett 18:1099–1104

Fournet A, Barrios AA, Munoz V, Hocquemiller R, Cave A, Bruneton J (1993) 2-substituted quinoline alkaloids as potential antileishmanial drugs. Antimicrob Agents Chemother 37:859–863

Grundon MF (1988) Quinoline alkaloids related to anthranilic acid. In: Brossi A (ed) The alkaloids, vol 32. Academic Press, London, pp 341–439

Herbert JM, Augereau JM, Gleye J, Maffrand JP (1990) Chelerythrine is a potent and specific inhibitor of protein kinase C. Biochem Biophysic Res Commun 172:993–999

Hutchinson J, Dalziel JM (1958) Flora of West tropical Africa, 1 (2). Whitefriars Press, London, pp 683–685

Irvine FR (1930) Plants of the Gold Coast. Milford, Oxford University Press, London

Jensen RA (1986) Tyrosine and phenylalanine biosynthesis: relationship between alternative pathways, regulation and subcellular location. In: Conn EE (ed) Recent advances in phytochemistry: the shikimic acid pathway. Plenum Press, New York, pp 57–81

Kerharo J, Adam JG (1973) La pharmacopée sénégalaise traditionnelle. In: Plantes médicinales et toxiques. Vigot, Paris, pp 710–715, 802

Krane BD, Fagbule MO, Shamma M, Gozler B (1984) The benzophenanthridine alkaloids. J Nat Prod 47:1–43

Kutchan TM, Zenk MH (1993) Enzymology and molecular biology of benzophenanthridine alkaloid biosynthesis. J Plant Res 3 (Spec Issue):165–173

Kutchan TM, Dittrich H, Bracher D, Zenk MH (1991) Enzymology and molecular biology of alkaloid biosynthesis. Tetrahedron 47:5945–5954

Larsen AK, Grondard L, Couprie J, Desoize B, Comoë L, Jardillier JC, Riou JF (1993) The antileukemic alkaloid fagaronine is an inhibitor of DNA topoisomerases I and II. Biochem Pharmacol 46:1403–1412

Leuschner C, Walton JW, Herzog G, Schultz G (1995) Chorismate mutase isoenzymes in transformed roots of *Datura stramonium*. Plant Physiol Biochem 33:367–371

Lewis JR (1983) Biological activity of some Rutaceous compounds. In: Waterman PG, Grundon MF (eds) Chemistry and chemical taxonomy of the Rutales. Academic Press, London, pp 301–308

LLoyd GB, McCown BH (1980) Commercially feasible micropropagation of mountain laurel, *Kalmia latifolia* by culture. Proc Int Plant Prop Soc 30:421–427

Messmer WM, Tin-WA M, Fong HHS, Bevelle C, Farnsworth NR, Abraham DJ, Trojanek J (1972) Fagaronine, a new tumor inhibitor isolated from *Fagara zanthoxyloïdes* Lam. (Rutaceae). J Pharm Sci 61:1858–1859

Mester I (1973) The occurrence of the alkaloids in Rutaceae. Fitoterapia 44:123–152

Mester I (1983) Structural diversity and distribution of alkaloids in the Rutales. In: Waterman PG, Grundon MF (eds) Chemistry and chemical taxonomy of the Rutales. Academic Press, London, pp 31–96

Molinatti P (1994) Le genre *Fagara* (*Rutaceae*): étude bibliographique; étude chimique et structurale des métabolites secondaires de *Fagara zanthoxyloides* obtenus in vitro, et de *Fagara rhoifolia*. PhD Thesis, University of Reims

Morel G, Müller JF (1964) In vitro culture of the apical meristem of the potato. C R Acad Sci Paris 258:5250–5252

Müller MJ, Zenk MH (1992) The norcoclaurine pathway is operative in berberine biosynthesis in *Coptis japonica*. Planta Med 58:524–527

Murashige T, Skoog F (1962) A revised medium for rapid growth and bio-assays with tobacco tissue cultures. Physiol Plant 15:473–497

Ouattara A, Guissou IP, Sawadogo A, Sawadogo M (1992) Etude comparée du traitement de la crise drépanocytaire par une présentation galénique de deux plantes médicinales (*Fagara xanthoxyloides* Lam. et *Calotropis procera* Ait) et un médicament usuel de référence (dihydroergotoxine = hydergine) chez les enfants de 5 à 15 ans à l'hôpital de Ouagadougou. Pharm Afr 68:19–27

Paris MR, Moyse-Mignon H (1947) Etude préliminaire de *Fagara xanthoxyloïdes* Lam. Ann Pharm Fr 5:410–420

Pérez-Chiesa Y, Rodriguez A (1993) Absence of mutagenicity of benzo[c]phenanthridine alkaloids in somatic cells of *Drosophila melanogaster*: comparison with 7,12-dimethylbenzen[a]anthracene and chrysene. Mutat Res 298:277–283

Petit A, David C, Dahl GA, Ellis JG, Guyon P, Casse-Delbart F, Tempé J (1983) Further extension of the opine concept: plasmids in *Agrobacterium rhizogenes* co-operate for opine degradation. Mol Gen Genet 190:204–214

Pezzuto JM, Antosiak SK, Messmer WM, Slaytor MB, Honig GR (1983) Interaction of the antileukemic alkaloid, 2-hydroxy-3,8,9-trimethoxy-5-methylbenzo[c]phenanthridine (fagaronine) with nucleic acids. Chem Biol Interact 43:323–339

Poulsen C, Verpoorte R (1991) Roles of chorismate mutase, isochorismate synthase and anthranilate synthase in plants. Phytochemistry 30:377–386

Poulsen C, Verpoorte R (1992) Activities of chorismate utilizing enzymes and of enzymes involved in indole alkaloid biosynthesis in cell suspension cultures. Plant Physiol Biochem 30:105–113

Rhodes MJC, Robins RJ, Hamill JD, Parr AJ, Hilton MG, Walton NJ (1990) Properties of transformed root cultures. In: Charlwood BV, Rhodes MJC (eds) Secondary products from plant tissue culture. Clarendon Press, Oxford, pp 201–225

Romero RM, Roberts MF, Phillipson JD (1995) Chorismate mutase in microorganisms and plants. Phytochemistry 40:1015–1025

Schaerer S, Pilet PE (1991) Roots, explants and protoplasts from pea transformed with strains of *Agrobacterium tumefaciens* and *A rhizogenes*. Plant Sci 78:247–258

Schmid J, Amrhein N (1995) Molecular organization of the shikimate pathway in higher plants. Phytochemistry 39:737–749

Sethi ML (1979) Inhibition of reverse transcripatse activity by benzophenanthridine alkaloids. J Nat Prod 42:187–196

Sethi ML (1981) Screening of benzophenanthridine alkaloids for their inhibition of reverse-transcriptase activity. Can J Pharm Sci 16:29–34

Sethi VS, Sethi ML (1975) Inhibition of reverse transcriptase activity of RNA tumor viruses by fagaronine. Biochem Biophys Res Commun 63:1070–1076

Smekal E, Kubova N, Kleinwatcher V, Cushman M (1986) DNA-binding studies: selected benzophenanthridine alkaloids as binding probes. Stud Biophys 114:257–262

Stadler R, Kutchan TM, Zenk MH (1989) (S)-norcoclaurine is the central intermediate in benzylisoquinoline alkaloid biosynthesis. Phytochemistry 28:1083–1086

Tan GT, Pezzuto JM, Kinghorn AD, Hughes SH (1991) Evaluation of natural products as inhibitors of human immunodeficiency virus type 1 (HIV-1) reverse transcriptase. J Nat Prod 54:143–154

Tan GT, Miller JF, Kinghorn AD, Hughes SH, Pezzuto JM (1992) HIV-1 and HIV-2 reverse transcriptases: a comparative study of sensitivity to inhibition by selected natural products. Biochem Biophys Res Commun 185:370–378

Torto FG, Sefcovic P, Dadson BA (1966) Medicinal plants of Ghana: identity of alkaloid from *Fagara xanthoxyloides*. Tetrahedron Lett GB 2:181–183

Trentesaux C, Ngo Nyoung M, Aries A, Morceau F, Ronchi A, Ottolenghi S, Jardillier JC, Jeannesson P (1993) Increased expression of GATA-1 and NFE-2 erythroid-specific transcription factors during aclacinomycin-mediated differentiation of human erythroleukemic cells. Leukemia 7:452–457

Wang LK, Johnson RK, Hecht SM (1993) Inhibition of topoisomerase I function by nitidine and fagaronine. Chem Res Toxicol 6:813–818

Zenk MH (1991) Chasing the enzymes of secondary metabolism: plant cell cultures as a pot of gold. Phytochemistry 30:3861–3863

IX *Frangula* Species: In Vitro Culture and the Production of Anthraquinones

L. Sajc[1], N. Kovačević[2], D. Grubišić[3], and G. Vunjak-Novaković[4]

1 General Account

1.1 Distribution and Importance of the Plant

Plants belonging to the Rhamnaceae family are the main source of anthraquinone (AQ) drugs. Over 50 genera and 500 species of this family are known worldwide, but in the flora of Serbia, there are only three genera: *Paliurus* Mill., *Frangula* Mill. and *Rhamnus* L. The genus *Frangula* is presented by two species: *Frangula alnus* Mill. (syn. *Rhamnus frangula* L., Fig. 1), and *Frangula rupestris* Scop. (syn. *Rhamnus rupestris* Scop.; Tutin et al. 1968; Jovanović 1973). Only *F. alnus* is used as a source of AQ drugs.

F. alnus (alder buckthorn) is a shrub or small tree, with alternate, ovate leaves and small flowers in axillary cymes. Green fruit turn purple to black as they mature, and they contain two to three oval seeds. AQs are mainly concentrated in the bark, the surface of which is brownish to red, characteristically laminated with whitish transversal lenticels. Bark (frangulae cortex) and dry bark extract (extractum frangulae siccum) obtained from *F. alnus* are the official AQ drugs according to most European pharmacopoeias (Ph.Eur II: Europaishes Hencibuch 1979 Ph. Yug IV: Health Council of Serbia 1984), and these drugs are usually applied as monocomponent or mixed tea, pills or tablets, solutions for infusion and cold macerates (Steinegger and Hansel 1988; Bisset 1994). Bark of *Rhamnus catharticus* (common buckthorn) is widely used to fake the bark of *Frangula alnus*, but it can be identified by the method of Rauwald and Miething (1985). The fruit of *R. catharticus* is, on the other hand, widely used and approved as an official drug (Rhamnii cathartici fructus, Baccae spinae cerrinae). Frangulae cortex in different medical preparations mainly originates from the northeastern part of Europe (Poland and Russia) and from the Balkan peninsula (Yugoslavia, Bulgaria). The Yugoslav drug

[1] Department of Analytical Chemistry, Faculty of Technology, Belgrade University, Karnegijeva 4, 11000 Belgrade, Yugoslavia
[2] Department of Farmacognosy, Faculty of Pharmacy, Belgrade University, Vojvode Stepe 450, 11000 Belgrade, Yugoslavia
[3] Department of Plant Physiology, Faculty of Biology, Belgrade University, 29. novembra 142, 11000 Belgrade, Yugoslavia
[4] Division of Health Sciences and Technology, Massachusetts Institute of Technology, Cambridge, Massachusetts 02139, USA

Biotechnology in Agriculture and Forestry, Vol. 43
Medicinal and Aromatic Plants XI (ed. by Y.P.S. Bajaj)

contains up to 9% AQs (calculated as glucofrangulin A), 6% catechinic tannins and 2–3% of all flavonoids (Kovačević et al. 1991a). The fruit of *F. alnus*, not often used as an AQ drug, contains AQs mainly in the seed coats and embryos. Seeds contain 0.7% AQs (calculated as glucofrangulin A), 1.3–1.8% anthocyanidins (calculated as cyanidin), 0.5% flavonoids and 17–19% fatty oil (density 0.91 g/cm^3, refractive index 1.47, unsoapy matter 0.5, 85% unsaturated fatty acids), and the fresh fruit contain 0.10–0.16% vitamin C and 32% sugar (Kovačević et al. 1991b).

F. rupestris is an ascending or procumbent shrub with elliptical to suborbicular leaves, denticulated or sometimes entire, with 5–8 pairs of lateral veins. Flowers are in small, axillary, umbelate cymes. Red fruits become black during maturation. This shrub usually grows in the mountains of the Balkan peninsula and the northeastern part of Italy (Tutin et al. 1968; Jovanović 1973). There are no published data about the chemical constituents, application and therapeutic affects, or in vitro cultures of this plant.

Fig. 1. *Frangula alnus* Mill. (Kovačević 1995)

1.2 Medicinal Properties of Anthraquinones

The largest group of natural quinones are tricyclic anthraquinones, and a wealth of structures are known with varying numbers of substituents. Anthraquinones (AQs) are derivatives of antracen with the basic structure of 9,10-dioxoanthracene. These metabolites are synthesized as reduced compounds (anthrones, anthranols, oxoanthrones and very often as mixed dimmers of iso and heterodianthrones) and in plant tissue predominantly present as glycosides, with just a small extent of free anthraquinone aglycones.

Based on only two biosynthetic pathways, AQs can be divided into alizarin and emodin types (Fig. 2). The alizarin type has one of the rings unsubstituted and it is formed along the shikimate-o-succinylbenzoate biosynthetic pathway (Koblitz 1988; Van der Berg and Labadie 1989). These AQs, found in the family Rubiacea (*Rubia*, *Morinda*, *Galium*, *Cinchona*) and sporadically in *Digitalis* species from the Scrophulariaceae family, are used mainly as natural pigments and dyes, and also as antimalaric agents (Korunaglo et al. 1992).

AQs of emodin type have both rings substituted and a structure of 1,8-dixydroxyanthraquinone. Recent studies confirmed that these compounds occur frequently with 1,8-dihydroxynaphtalene glycosides (Evans 1996). The emodin type of AQs is synthesized via the acetate-malonate pathway (Koblitz 1988; Van der Berg and Labadie 1989) and contain emodin, aloe-emodin, chrysophanol, physcion and rhein as basic aglycones. These AQs are present in large extent in the Fabaceae (*Cassia*, *Araroba*), Rhamnaceae (*Rhamnus*, *Frangula*), Polygonaceae (*Rheum*, *Rumex*, *Fagopyrum*) and sporadically in Hypericaceae, Ericaceae, Lythraceae, Verbenaceae and Saxifragaceae.

The most frequently used AQ drugs are: Sennae folium et fructus, Rhei rhizoma, Aloe, Chrysarobinum, Frangulae cortex, Rhamni purshianae cortex, Rhamni fallacis cortex and Rhamni cathartici fructus (Deutsches Arzneibuch 1991; Bisset 1994).

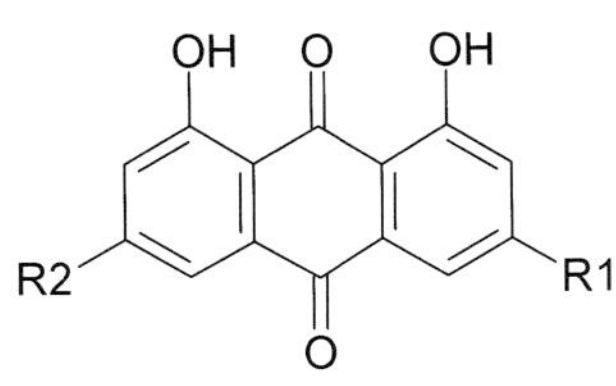

	R1	R2
emodin	CH_3	OH
aloe-emodin	CH_2OH	H
chrysophanol	CH_3	H
physcion	CH_3	OCH_3

Fig. 2. Structure of four dominant emodin type AQ aglycones. (Kovačević 1995)

Traditionally, the drugs containing, 1,8-dihydroxyanthraquinones are used as homeostatics and cathartics (Thomson 1965; Rai 1978; Suzuki et al. 1984; Honing et al. 1992). Anthraquinones have a broad spectrum of activity, including antibacterial, antifungal and antiviral (Wijnsma et al. 1985; Benzanger-Beauquesne 1989; Wood et al. 1990; Kaiil-a-Komb et al. 1992), antitumour (Kupchan and Karim 1976; Yeh et al. 1988; Westendorf et al. 1988; Andreani et al. 1992; Morier-Tessier et al. 1993; Nebois et al. 1994), hepatoprotective (Wong et al. 1988, 1989; Malterud et al. 1993) and inhibition of xantine oxidase, monoamine oxidase and catalase (Noro et al. 1987; Hiraoka et al. 1989; Moreno-Sepulveda et al. 1994).

2 Analysis of Anthraquinones

The bark, leaves and fruit of *F. alnus* and *F. rupestris* were collected at Brezovica mountain and Prizrenska Bistrica river gorge in August 1993 and 1994, and the identity of material was confirmed at the Botanical garden, Faculty of Botany, University of Belgrade.

2.1 Qualitative Analysis

The Borntrager test was used for the identification of free AQ aglycones and AQ O-glycosides in plant material, while AQ C-glycosides were analyzed by two-dimensional thin layer chromatography (TLC) using direct acid hydrolysis on the plate (Ebel and Kaal 1980).

Isolation, purification and structure determination of main aglycones, i.e. chrysophanol, physcion, emodin and aloe-emodin was performed using extraction, chromatography, UV, IR and 1H NMR spectroscopy and MS spectrometry. AQ glycosides in lyophilized and pulverized cells (85 g) were subjected to hydrolysis in boiling 2 M HCl (2 l) for 2 h. AQ aglycones were extracted from the filtrate and macerate with chloroform (5×50 ml) and the components of dry extract (3.5 g), obtained after solvent evaporation in vacuo (40 °C), separated and purified by column chromatography (Silica gel 230–400 mesh, Sigma) during elution with petrol ether, petrol ether: benzene (1:1), benzene, chloroform and methanol.

Compound 1 (10 mg, recrystallized from benzene), compound 2 (6 mg, recrystallized from benzene), and compound 3 (15 mg, recrystallized from methanol) were purified from petrol ether and petrol ether: benzene eluents by TLC (Kiselgel G60, Merck) using petrol ether: benzene (1:1) and benzene: acetic acid (8:2) solvent systems. Compound 4 (5 mg, recrystallized from toluene) was isolated from benzene and chloroform eluents by TLC, as three previous compounds, but with petrol ether: benzene: chloroform (4:5:1) and benzene: acetic acid (8:2) mixtures. Before final recrystallization, all isolated compounds were purified once more on a Sephadex LH_{20} (Sigma) column by elution with methanol. Compound 1 was recrystallized once more from n-butyl acetate to obtain orange red crystals for X-ray diffraction. According to its spectral data, it was identified as chrysophanol (Argay et al. 1996).

The basic information about the structure of each component was derived from UV/VIS (Specord M-40, Carl Zeiss, ethanol blank probe), IR (FT 1725X Perkin Elmer, solid KBr pellet), MS (Finnigan – MAT 8230, electron ionization impact 70 eV, ion source temperature 180 °C), and 1H NMR (Bruker, 600 MHz, $CDCl_3$ solution, TMS standard) spectra, as previously described (Steglich and Losel 1969; Savonius 1972; Harborne 1973; Kalidhar 1989; Danielsen et al. 1992; Danielsen and Francis 1994).

Compound 1: UV (ethanol) absorption maxima at 225, 258, 280, 290 and 432 nm, IR bands (KBr) at 3444 (aOH), 1676 (C = O free), 1627 (C = O chelating), 1559 cm^{-1} (C = C aromatic), 1H NMR spectrum ($CDCl_3$, 600 MHz) contained an H5 doublet (7.8), an H6 triplet (7.6), an H7 doublet (7.3), an H2 multiplet (7.1), an H4 multiplet (7.6) and a 3H (methyl) singlet (2.5 ppm). Its mass spectrum showed M^+ 254, and $C_{15}H_{10}O_4$ requires 254. These, and X-ray diffraction data showed that this compound was chrysophanol.

Compound 2: UV (ethanol) absorption maxima at 225, 255, 269, 290 and 440 nm, IR bands (KBr) at 3443 (aOH), 16776 (C = O free), 1622 (C = O chelating), 1577 cm^{-1} (C = C aromatic), 1H NMR spectrum ($CDCl_3$, 600 MHz) contained an H5 singlet (7.3), an H7 singlet (6.7), an H4 singlet (7.7), an H2 singlet (7.1), a 3H (metoxy) singlet (3.8), and a 3H (methyl) singlet (2.5 ppm). The basic structure of the compound was confirmed by comparison with an NMR spectra library, but the exact position of the metoxy group was not possible to locate (C1, C6 or C8). Its mass spectrum showed M^+ 284 which corresponds with physcion.

Compound 3: UV (ethanol) absorption maxima at 225, 258, 278, 288 and 430 nm, IR bands (KBr) at 3445 (aOH), 1653 (C = O free), 1627 (C = O chelating), 1559 cm^{-1} (C = C aromatic), 1H NMR spectrum ($CDCl_3$, 600 MHz) contained an H5 doublet (7.8), an H6 triplet (6.7), an H7 doublet (7.3), an H2 multiplet (7.4), an H4 multiplet (7.8) and a 2H singlet (4.8 ppm). Its mass spectrum showed M^+ 270, and the compound was identified as aloe-emodin.

NMR and IR spectra of compound 4 were not provided, due to the small amount and low solubility of the isolated and purified material. This compound showed UV (ethanol) absorption maxima at 222, 254, 267, 290 and 440 nm, and on the basis of HPLC characteristics and its mass spectrum (M^+ 270) which corresponded with that of emodin in the MS library spectra, identified as emodin. The structure of chrysophanol, physcion, emodin and aloe-emodin is shown in Fig. 2.

2.2 Quantitative Analysis

The quantity of AQ aglycones was determined by HPLC, which is widely used for the separation and identification of different drug constituents, including anthraquinone compounds (Ohshima and Takashi 1983; Rauwald and Miething 1985; Wijnsma et al. 1986; Ohshima et al. 1986; Reynolds 1994). Extraction of AQs from plant tissue was performed according to a modified procedure of Van der Berg and Labadie (1984), which permits the analysis of both free AQ aglycones and AQ glycosides. Dried and pulverized plant material (0.2–0.3 g) was extracted with ethyl ether (50 ml) under reflux for 15 min. The ether solution was dried over anhydrous sodium sulphate and ether evaporated in vacuo (30 °C) to obtain the dry residue containing free AQ aglycones (fraction A). AQ glycosides in the solid residue, remaining after the extraction of free aglycones, were subjected to hydrolysis and oxidation in a boiling solution of 1 g of $FeCl_3$ in 50 ml of 2 M HCL under reflux for 60 min. After cooling and filtration, the filtrate and macerate were extracted with ether (5 × 25 ml) and the combined ether fractions washed with distilled water (3 × 30 ml) and dried over anhydrous sodium sulphate. Ether was evaporated in vacuo (30 °C) and the dry residue containing aglycones of AQ O-glycosides obtained (fraction B).

Fractions A and B were dissolved in 1 ml of methanol and subjected to HPLC (HP 1090 Hewlett Packard, diode array detector, Hypersil RP column ODS 200 × 4.6 mm) under the following conditions: flow rate of mobile phase methanol: 1% acetic acid in water (75 : 25) 1 ml/min; temperature 40 °C; injection volume 20 ml; wavelength for spectra acquisition 280 and 430 nm. The identification of AQ aglycones was done by comparison of the main chromatographic peaks retention times and UV spectra of the test solution with those of authentic compounds. Quantification was done by an external calibration method. The calibration solution was a mixture of authentic compounds: 0.911 g aloe-emodin (Roth), 0.929 g emodin (Fluka), 1.061 g chrysophanol (Fluka) and 1.829 g physcion (Roth). Under these conditions, the four components were separated within 30–35 min and the retention times for aloe-emodin, emodin chrysophanol and physcion were 5.83, 11.82, 17.94 and 27.24 min, respectively.

The amount of anthraquinones, determined in leaves, bark and embryos of *F. alnus* and *F. rupestris* is shown in Table 1. The major aglycone in bark and leaves of *F. alnus* was emodin, whereas high extent of aloe-emodin were found in embryos. *F. rupestris* contained more free AQ aglycones than *F. alnus*, although a much lower quantity of all produced AQs. In addition, these aglycones were predominantly present in the form of aloe-emodin which is not often identified in the members of the Rhamnaceae family.

3 In Vitro Culture Studies

In vitro cultures of *F. alnus* were already established and their ability to synthesize AQs was confirmed by Van der Berg and Labadie (1984, 1988a,b, 1989), but the productivity in callus tissue (0.11% AQ glycosides) and cell suspension (0.28% AQ glycosides) was still lower than in the intact plant. Multiplication of *F. alnus* shoots was studied by Bignami (1983) and Meyer (1983). To the best of our knowledge, there are no published data on in vitro cultivation of *F. rupestris*.

Sajc et al. (1995b) have shown that the productivity of *F. alnus* plant cells can be significantly improved by immobilization and continuous liquid-liquid extraction of AQs inside the bioreactor.

3.1 Establishment of Tissue Cultures

In vitro production of anthraquinones was studied in shoot, root, callus and cell suspension cultures of *F. alnus*, and shoot, root and callus cultures of *F. rupestris*, maintained on MS (Murashige and Skoog 1962), B5 (Gamborg et al. 1968), WH (White 1963), and WPM (Lloyd and McCown 1980) basal media, supplemented with 30 and 50 g sucrose per l solid and liquid medium, respectively. The medium was solidified by the addition of 7 g/l agar. The solutions of plant growth regulators and vitamins were added to the medium prior to autoclaving (25 min at 114 °C). Plant cell cultures on solid medium were kept at 25 °C, under a long day regime (16 h) illuminated continuously by ca. 4.7 W/m^2, while cell cultures in liquid medium were placed on an orbital shaker (60 rpm) and illuminated by ca. 1.7 W/m^2.

Three different culture lines were usually maintained simultaneously, and the analyzed material was cultivated under the same conditions during at least two repeated subcultures.

The cultures of *F. alnus* and *F. rupestris* were initiated from seedlings, obtained by germination of by zygotic embryos. Surface sterilized fruit (70% v/v ethanol, 1.2% w/v NaOCl) was opened, and the seeds and embryos were excised. Isolated embryos were placed on solid MS + 2 g/l PVP (polyvinylpyrrolidone, MW 44000, BDH). Embryo germination started after approximately 5 days and seedlings were left on the same medium for 3 weeks.

3.2 *Frangula alnus*

Shoot Culture. This was established from the epicotyl of those seedlings cultivated on solid MS + IAA (indole acetic acid) 0.1 mg/l + BAP (6-benzylaminopurine) 1 mg/l. Although the influence of different plant growth regulators on shoot multiplication was studied, the best results were obtained using this IAA and BAP combination on MS (index of multiplication 7.65 ± 0.59) and WPM (index of multiplication 8.84 ± 1.11). Rooting of shoots was the most successful on MS/2 + IBA (indole-3-butyric acid) 0.1 mg/l (Fig. 3A). The highest productivity of AQs was observed in shoots grown on MS + NAA (1-naphtaleneacetic acid) 0.1 mg/l + TDZ (thidiurazon) 0.1 mg/l (1.73% of all AQ aglycones) and WPM + IAA 0.05 mg/l + BAP 0.5 mg/l (1.49% of all AQ aglycones).

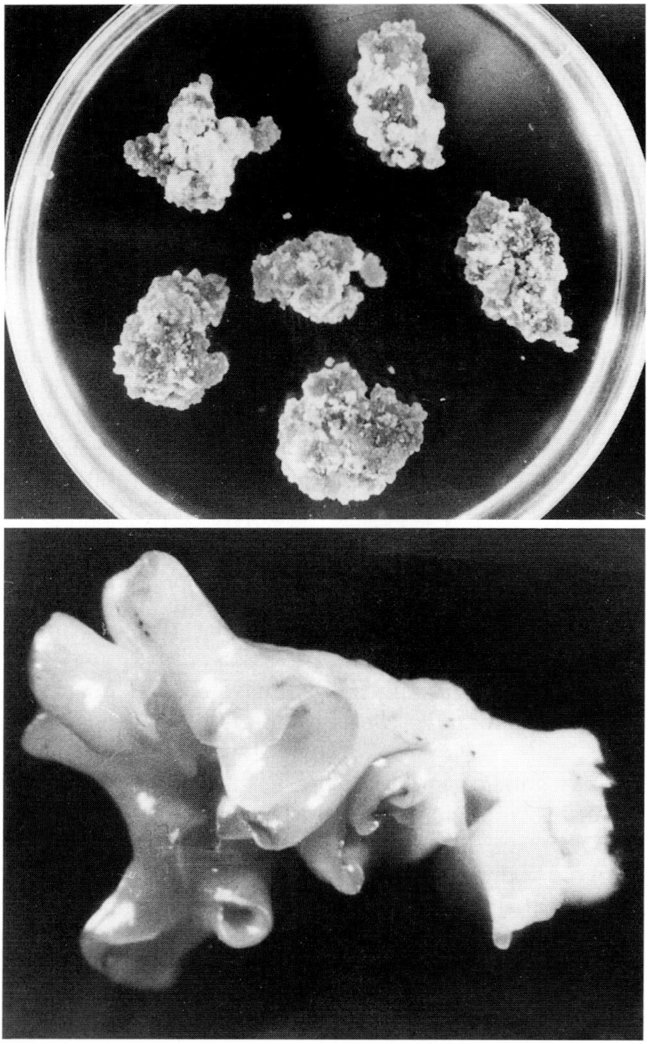

A

B

Fig. 3. **A** *Frangula alnus* Mill. callus culture (MS + IBA 10 mg/l + Kin 1 mg/l). **B** Somatic embryo. (Kovačević 1995)

Callus Culture. This was established from hypocotyl segments of seedlings (Fig. 3A). Callus formation and proliferation on these explants were the most successful on MS + IBA 10 mg/l + Kin (kinetin) 1 mg/l. This callus was green and compact, but also contained friable parts. Several different mineral and plant growth regulator combinations were used to increase callus productivity, but the AQs content was nearly constant (0.5% of all AQ aglycones), and much lower than in the intact plant. Callus

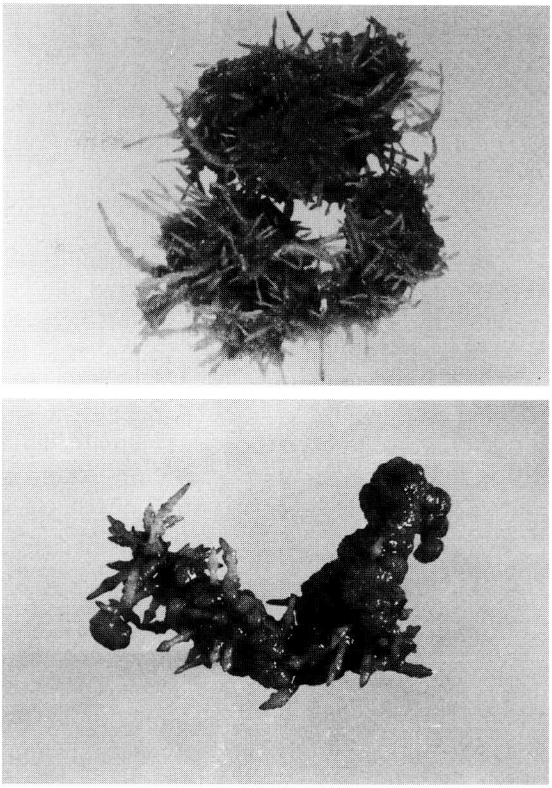

A

B

Fig. 4. *Frangula alnus* Mill. root culture (Kovačević 1995) **A** MS/2 + NAA 0.1 mg/l + GA_3 0.1 mg/l. **B** MS/2 + 2,4-D 0.1 mg/l + GA_3 0.1 mg/l

derived from zygotic embryos formed somatic embryos (Fig. 3B; Kovačević 1995).

Root Culture. This was established from the root tips of the seedlings, placed on MS/2 + IBA 0.1 mg/l + GA_3 (gibberelic acid, MW 364.4) 0.1 mg/l. Roots were cultivated in liquid MS/2, and mass yield was the best on MS/2 + IBA 1.0 mg/l + GA_3 0.1 mg/l under a long day regime or on MS/2 + NAA 1.0 mg/l + GA_3 0.1 mg/l in the dark. Although the presence of 2,4-D (2,4-dichlorophenoxyacetic acid) in the medium caused poor growth and root degeneration, the best production of AQs (1.25% of all aglycones, mainly in the form of O-glycosides) was observed in roots cultivated in the presence of 2,4-D 0.5 mg/l, under a long day regime. Low concentrations of IBA and NAA gave rise to a better morphology of the roots (Fig. 4A,B).

Cell Suspension. This was initiated from friable parts of callus tissue, placed in liquid MS + IBA 10 mg/l + Kin 1 mg/l. It was maintained later in liquid

MS + 2,4-D 1 mg/l + BAP 1.0 mg/l. Thick cell suspensions yielded 30 g/l of dry cell weight. The effects of medium composition, auxin and cytokinin type and concentration, phosphate concentration, sugar type (sucrose, glucose and maltose), fungal elicitation by *Aspergillus niger*, *Nectria ditissima* and *Nectria coccinea*, growth retardant Ancimidol 10^{-6}–10^{-7} M (Elanco) and herbicide Sandoz 9789 10^{-5}–10^{-6} M (Sandoz), and immobilisation in Ca-alginate matrix, on anthraquinone production was studied in cell suspension.

Maximum cell suspension productivity was achieved on B5 + NAA 1 mg/l + BAP 1 mg/l medium. Sugar type had no influence on cell suspension productivity. Medium composition and auxin type and concentration showed insignificant effect on cell suspension productivity, while light regime and cytokinin type and concentration mostly influenced the ratio of free AQ aglycones vs. AQ glycosides. Fungal elicitation and growth retardants did not improve cell suspension productivity. However, immobilisation of cells in Ca-alginate matrix significantly improved the production of aloe-emodin and emodin, mostly on account of chrysophanol, as well as total AQ productivity.

3.3 *Frangula rupestris*

Shoot, root and callus cultures of *F. rupestris* were obtained as described for *F. alnus* in Section 3.2. Multiplication of *F. rupestris* shoots was most successful on MS + BAP 0.5 mg/l (index of multiplication 6.20 ± 0.64), while the fastest roots and callus proliferation was observed in liquid MS + NAA 0.5 mg/l + GA_3 0.1 mg/l and MS + NAA 0.5 mg/l, respectively.

3.4 Production of Anthraquinones in Tissue Cultures

A number of studies have been conducted on the production of anthraquinones by plant cell cultures (see Suzuki and Matsumoto 1988). The ratio of free AQ aglycones vs. AQ aglycones bound in glycosides, R (%), and the individual fractions of aloe-emodin, emodin, chrysophanol, and physcion, in the sum of free AQ aglycones and AQ aglycones bound in glycosides, G ± STD (%), analyzed in samples of leaves, bark and fruit of *F. alnus* and *F. rupestris*, and in their tissue cultures maintained on different media are shown in Table 1 (Kovačević 1995).

All tissue cultures accumulated higher amounts of free AQ aglycones than corresponding plants. Also, the AQ aglycones distribution was less favourable in cell cultures when compared with distribution in plants. The quantity of AQs in *F. alnus* and *F. rupestris* cell cultures was significantly lower than in *F. alnus* bark, but all cell cultures of *F. alnus* produced more AQs than collected leaves and zygotic ambryos. High productivity of AQ glycosides was observed in *F. alnus* cell suspension.

Table 1. Anthraquinones isolated from *F. alnus* and *F. rupestris*. (Kovačević 1995)

	R[a]	G ± STD[b]	Aloe-emodin	Emodin	Chrysophanol	Physcion
		%				
F. alnus						
Leaves	25/75	0.21 ± 0.02	1.1	70.9	17.2	10.8
Bark	3/97	2.06 ± 0.26	–	78.3	12.8	8.9
Zygotic embryos	13/87	0.36 ± 0.04	49.0	34.0	16.4	0.6
Shoots	51/49	1.05 ± 0.11	3.9	17.5	21.4	57.2
Callus tissue	44/56	0.98 ± 0.10	3.1	7.3	27.3	62.3
Roots	45/55	0.54 ± 0.04	4.7	–	42.6	52.7
Cell suspension	30/70	0.23 ± 0.02	1.8	10.1	25.4	62.7
Somatic embryos	52/48	0.85 ± 0.03	6.1	7.2	73.5	13.2
F. rupestris						
Leaves	39/61	0.01 ± 0.01	81.1	4.4	14.5	–
Bark	63/37	0.03 ± 0.01	60.0	9.0	31.0	–
Fruit	69/31	0.24 ± 0.01	93.3	3.8	2.9	–
Shoots	81/19	0.29 ± 0.02	6.9	4.6	39.5	49.0
Roots	73/27	0.10 ± 0.01	6.7	–	74.3	19.0
Callus tisue	85/15	0.29 ± 0.01	3.9	1.5	39.2	55.4

[a] free AQ aglycones (%) / AQ aglycones bound in glycosides (%).
[b] percentage of all aglycones in the smaple.

3.5 Bioreactor Studies

Our studies indicate that the overall productivity of plant cells can be improved if bioconversion and downstream processing are integrated with product removal and accumulation in the extractive phase, since the medium has an inherently low capacity for extracellular products of the plant cells. This strategy is based on the addition of the extractive phase which can provide an artificial site for product accumulation and to shift equilibrium to favour product secretion by: (1) removal of products, shifting the equilibrium towards product formation; (2) removal of products from an environment which can cause their degradation, to improve product stability and; (3) removal of inhibitory products, to increase the biosynthetic rate. The related advantages include enhanced mass transfer, decreased level of product inhibition, and facilitated product recovery, and result in a reduced reaction volume for a given amount of product.

The idea for creating an artificial accumulation and conservation site for secondary plant products is derived from the actual situation in intact plants. In vivo, the synthesis and excretion of secondary products is coupled in many instances and associated with membrane transport of the product. Hydrophilic secondary products, often toxic for metabolically active proteins in the cytoplasm, are synthesized by enzymes located in membranes separating the cytoplasm, endoplasmic reticulum and diocytosomes, then secreted into the nonplasmic spaces of these organelles, and from there transported to specialized storage compartments by vesicles which fuse their membranes with those of the storage compartments. Lipophilic substances may be found, on the other hand, in membranes or cytoplasm, and they may penetrate through membranes into nonplasmic compartments by simple diffusion. An artificial accumulation site can therefore mimic similar transport phenomena in vitro cultures, not able to differentiate specialized storage compartments (Lukner 1980; Beiderbeck and Knoop 1988; Glad and Larsson 1991; Daugulis 1994; Chang and Sim 1995).

The production of anthraguinoues by immobilized plant cells was used as a model system to evaluate the performance of an extractive bioconversion in a hybrid airlift bioreactor that integrated the biosynthesis and recovery process within a single unit. The continuous aqueous liguid phase containing nutrients and plant cells – calcium alginate scatfolds was used in a batch and periodically replaced during cultivation, while the gas and the dispersed nonagueous solvent were continuously perfused through the reactor (Sajc et al. 1995a).

3.5.1 Production of Anthraquinones in Shake Flasks

Kinetics of cell growth, substrate consumption and product formation were studied in shake flask cultures of suspended and immobilized cells in order to determine the effect of silicone oil used for liquid-liquid extraction of anthraquinones from MS medium on cell viability and productivity. These experiments were carried out simultaneously, using the same amount of inoculated cell material in each culture, and four sets of 12 125-ml Erlemayer flasks, containing 30 ml of MS + 30 g/l sucrose + 2,4-D 1 mg/l + BAP 1 mg/l. The first and second sets contained cell suspension and immobilized cells without silicon oil, while the third and fourth sets contained cultures of cell suspension and immobilized cells with the addition of 3 ml of silicon oil in each flask, respectively. At 4-day intervals, starting 2 days after inoculation, three flasks in each set were harvested and dry cell weight (DCW), sugar (S) and product (P) concentration, and cell viability by staining with TTC (2,3,5-triphenyltetrazolium chloride) were determined in three samples taken from each flask (Sajc et al. 1995b).

Figure 5a shows the changes in dry cell weight, substrate and product concentration in cell suspension culture without silicon oil. Each point is an average of nine measurements on days 2, 6 10 and 14. Figures 5b and 6a,b show the respective changes in dry cell weight, substrate and product concentration in cell suspension culture with silicon oil, and immobilized cell cultures with and without silicon oil.

In suspension cell cultures without silicone oil, dry cell weight sharply increased after a short lag phase, and the stationary phase was reached

Fig. 5. Cell growth, *X*(●), substrate consumption, *S*(○), and product formation, *P*(⊙) in: **a** cell suspension, and **b** immobilised cell culture in flasks without silicone oil. (Sajc et al. 1995b)

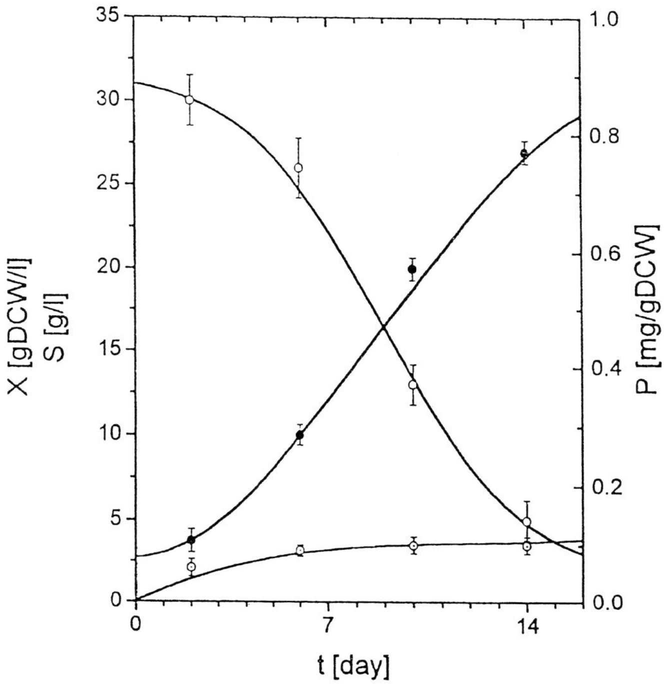

(a)

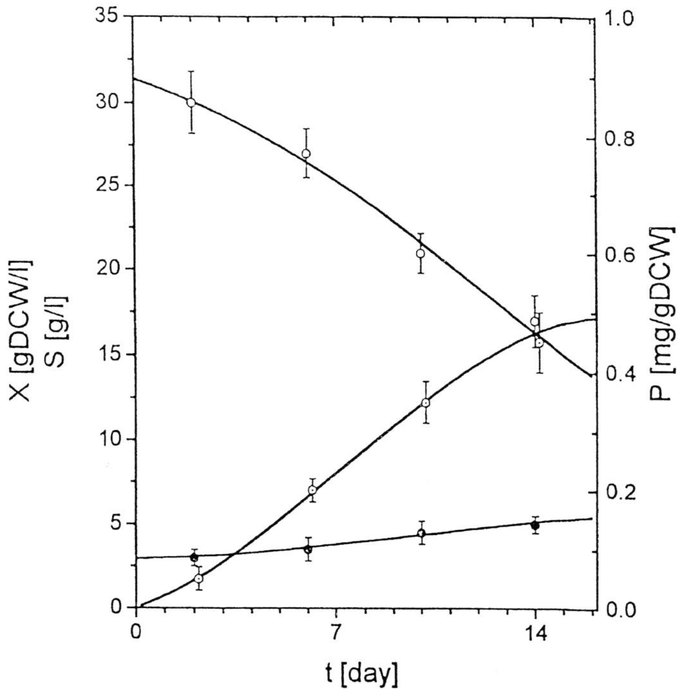

(b)

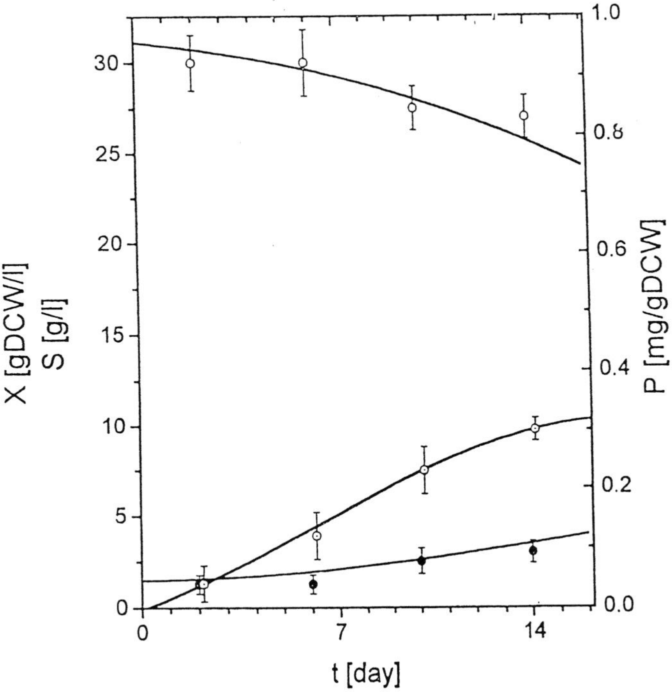

(a)

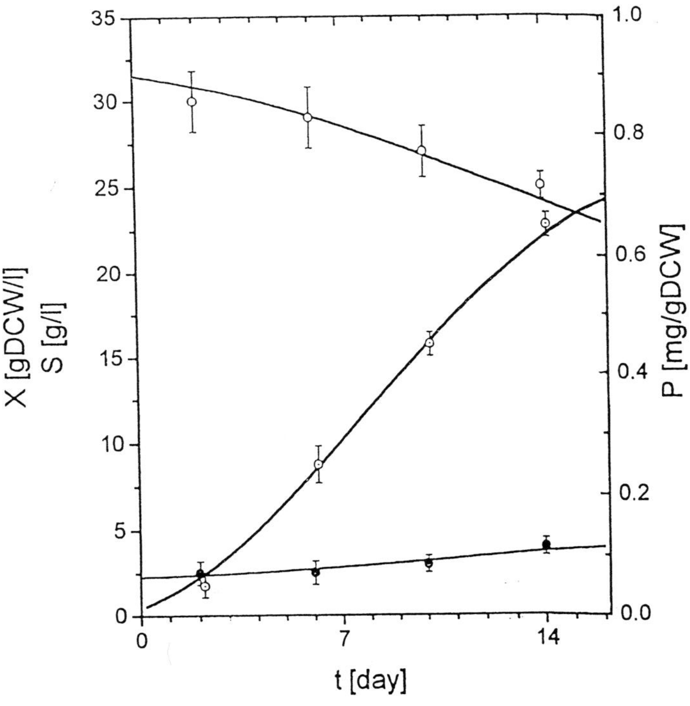

(b)

when sugar was depleted (Fig. 5a). Immobilized cells in the culture without silicon oil grew slower (Fig. 5b), but produced approximately five times more anthraquinones (0.45 mg/g DCW) when compared with the suspended cells (0.10 mg/g DCW).

In cell cultures with silicon oil, cell growth and sugar consumption were slower both in suspended (Fig. 6a) and immobilized cell cultures (Fig. 6b). Immobilized cells in the culture with silicon oil produced more anthraquinones (0.65 mg/g DCW) than immobilized cells in the culture without silicon oil and suspended cells in the culture with silicon oil. Product recovery by silicone oil improved the productivity of both suspended and immobilized cells, without affecting their viability (within ± 15% of the initial, as assessed by cell transfer on fresh medium).

3.5.2 Production of Anthraquinones in the Bioreactor

The effect of the continuous anthraquinone removal from MS medium on the kinetics of product formation by immobilized cells was determined in two bioreactor production cycles, 10 days each. The first was carried out with 40 g DCW/l. Every 48 h, the entire silicon oil volume of 50 ml was replaced by fresh silicone oil to avoid solvent saturation, since its capacity was approximately 500 mg danthron/l silicone oil at 25 °C. The second run was performed with fresh cells, 50 gDCW/l, and the entire solvent volume was changed every 24 h. Three samples were drawn from the bioreactor every 2 days, and assessed for DCW, sugar, product concentration, and cell viability (Sajc et al. 1995b).

In the first production cycle, 40 g DCW/l immobilized cells produced 6.5 mg/g DCW of anthraquinones over 10 days (Fig. 7a). This value was approximately ten times higher than that observed in the corresponding shake flask culture (i.e. immobilized cells with silicon oil). During this period, almost all sugar was consumed. In the second production cycle, performed with 50 g DCW/l and silicon oil, completely changed every 24 h, the yield of anthraquinones was 17.5 mg/g DCW (Fig. 7b).

4 Conclusion

In vitro cultures of *F. alnus* and *F. rupestris* were established and their ability to synthesise anthraquinones was confirmed by quantification of the four dominant AQ aglycones (i.e. chrysophanol, physcion, aloe-emodin and

Fig. 6. Cell growth, *X*(●), substrate consumption, *S*(○), and product formation, *P*(⊙) in: **a** cell suspension and **b** immobilised cell culture in flasks with silicone oil. (Sajc et al. 1995b)

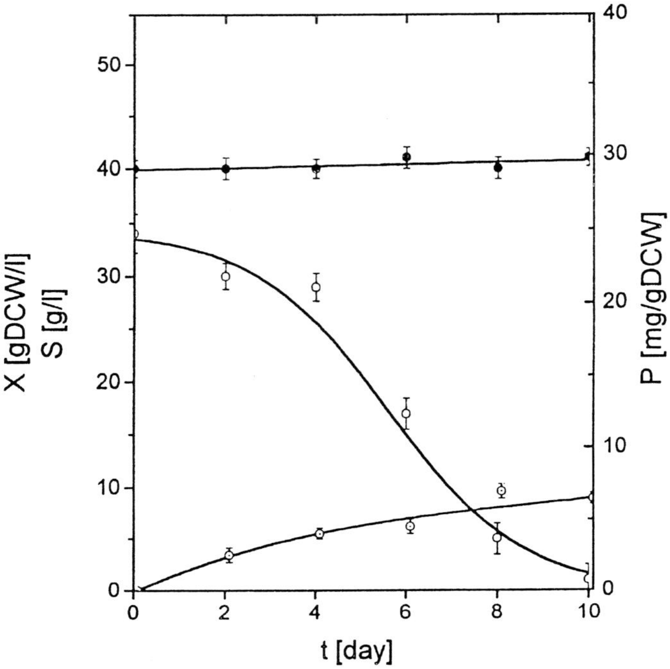

(a)

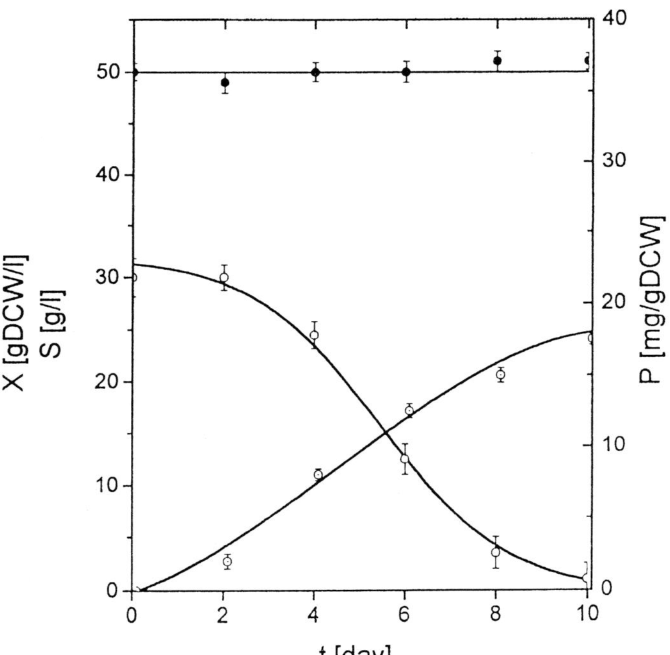

(b)

Fig. 7. Cell growth, X(●), substrate consumption, S(○), and product formation, P(⊙) in the bioreactor: **a** 40 g DCW/l; **b** 50 g DCW/l. (Sajc et al. 1995b)

emodin). The average production of AQs was generally lower in tissue cultures than in nature, and was not significantly affected by nutritional factors, light regime and elicitation. The composition of AQs synthesized in vitro differed from those in the intact plant, but all tissue cultures accumulated higher amounts of free AQ aglycones than intact plants and produced less emodin AQs than chrysophanol, physcion and aloe-emodin AQs.

However, the use of a four-phase external loop air-lift bioreactor significantly increased the productivity of immobilized *F. alnus* cells, when compared with either cells in suspension or shake flask cultures. In particular, the continuous liquid-liquid extraction of AQs from the medium inside the bioreactor with immobilized *F. alnus* cells resulted in 10–30 times higher cell productivity, compared with that of immobilized cells in flasks.

Acknowledgments. The authors wish to express their grateful thanks to Prof. Mirjana Nešković for reading an early draft of this manuscript, and for making a number of helpful suggestions. This research was supported by the Ministry of Science of Serbia (grants G1403, 03E21 and 02E36) and the Serbian Academy of Sciences and Arts.

References

Andreani A, Rambaldi M, Locatelli A, Fraccari A, Bossa R, Galatulas I (1992) Potential antitumor agents. XIX(1). Synthesis and antitumor activity of tricyclic compounds related to latifazol. Pharm Acta Helv 67:234–236

Argay G, Kalman A, Kovacevic N, Grubisic D, Ribar B (1996) Crystal structure of 1,8-dihydroxy-3-methyl-9,10-anthracendion, $C_{15}H_{10}O_4$. Z Kristallogr 211:723–724

Beiderbeck R, Knoop B (1988) Enhanced production of secondary substances: addition of artificial accumulation sites to cultures. In: Bajaj YPS (ed) Biotechnology in agriculture and forestry, vol 4. Medicinal and aromatic plants I. Springer, Berlin Heidelberg New York, pp 123–135

Benzanger-Beauquesne L (1989) Nouveautés en phytotherapie. Plant Med Phytother 23:46–76

Bignamic C (1983) In vitro propagation of *Rhamnus frangula* L. Gartenbauwissenschaften 48:272–274

Bisset NG (1994) Herbal drugs and phytopharmaceuticals. Medpharm Stuttgart; CRC Press, Boca Raton, pp 208–211, 409–411

Chang HN, Sim SJ (1995) Extractive plant cell culture. Curr Opin Biotechnol 6:209–212

Danielsen K, Francis GW (1994) An alternative solvent system for the separation of anthraquinone aglycones from *Rhubarb* on silica thin layers. Chromatographia 38:520–522

Danielsen K, Aksnes DW, Francis GW (1992) NMR study of some anthraquinones from *Rhubarb*. Magn Reson Chem 30:359–363

Daugulis A (1994) Integrated fermentation and recovery process. Curr Opin Biotechnol 5:192–195

Deutsches Arzneibuch (1991) 10th edn. Deutscher Apotheker Verlag, Stuttgart; Govi, Frankfurt

Ebel S, Kaal M (1980) Zur Analytik von Anthrachinondrogen. I. Hydrolyse der Glycoside direkt auf der DC-Platte. Planta Med 40:271–277

Europaicshes Hneibuch (1979) Ph Eur II. Deutscher Apotheker Verlag, Stuttgart; Govi Frankfurt.

Evans WC (1996) Treatise and Evan' pharmacognosy. Saunders, London, pp 232–248

Gamborg OL, Miller RA, Ojima K (1968) Nutrient requirements of suspension cultures of soybean root cells. Exp Cell Res 50:151–158

Glad M, Larsson PO (1991) New methods for separation and recovery of biomolecules. Curr Opin Biotechnol 2:413–418

Harborne JB (1973) Phytochemical methods. Chapman and Hall, London

Health Council of Serbia (1984) Ph Yug IV. Health Council of Serbia, Belgrade (in Serbocroatian)

Hiraoka A, Koike S, Sakaguchi M, Masuda Y, Terai F, Miura I, Kawasaki T, Kawai H, Sakakibara M (1989) The sennoside constituents of Rhei rhizoma and Sennae folium as inhibitors of serum monoamine oxidase. Chem Pharm Bull (Tokyo) 37:2744–2746

Honing J, Geck P, Rauwald HW (1992) Inhibition of Cl-channels as a possible base of laxative action of certain anthraquinones and anthrones. Planta Med 58:A586–A587

Jefferson RA, Karanagh TA, Temple J (1987) GUS fusion: b-glucuronidase as a sensitive and versatile gene fusion marker in higher plants. EMBO J 6:3901–3907

Jovanović B (1973) Rhamnaceae. In: Josifovic M (ed) Flora of SR Serbia, vol 5. Serbian Academy of Sciences and Arts, Belgrade, pp 358–382 (in Serbocroatian)

Kaiil-a-Komb M, Amoros M, Girre L (1992) Search for new antiviral agents of plant origin. Pharm Acta Helv 67:130–147

Kalidhar SB (1989) Structural elucidation of anthraquinones using 1H NMR glycosilation and alkylation shifts. Phytochemistry 28:3459–3463

Koblitz Z (1988) Anthraquinones. In: IK vasil, Constabel F (eds) Cell culture and somatic cell genetics of plants, vol 5. Phytochemicals in plant cell cultures. Academic Orlando, pp 113–139

Korunaglo K, Gbeassor M, Nikabu O, De Souza C, Werner W (1992) Effects of three compounds extracted from *Morinda lucida* on *Plasmodium falciparum*. Planta Med 58:533–534

Kovačević N (1995) Production of physiologically active metabolites by in vitro cultures of *Frangula alnus* Mill. and *Rhamnus catharticus* L. PhD Thesis, University of Belgrade (in Serbocroatian)

Kovačević N, Savin K, Ivanić R (1991a) Analysis of polyphenolics in *Rhamnus catharticus*, *Frangula alnus* and *F. rupestris*. Arch Pharm 41:99–102 (in Serbocroatian)

Kovačević N, Mirić M, Savin K (1991b) Fatty oil, vitamin C and sugar in *Rhamnus catharticus* and *Frangula alnus* fruit. Arch Pharm 41:181–184 (in Serbocroatian)

Kupchan SM, Karim A (1976) Tumor inhibitors. 114. aloe emodin: antileukemic principle isolated from *Rhamnus frangula* L. Lloydia 39:223–224

Lloyd G, McCown B (1980) Commercially-feasible micropropagation of mountain laurel, *Kalmia latifolia*, by use of shoot-tip culture. Proc Int Plant Prop Sci 30:421–427

Lukner M (1980) Expression and control of secondary metabolism. In: Bell EA, Charlwood BV (eds) Encyclopedia of plant physiology, vol 8. Secondary plant products. Springer, Berlin Heidelberg New York, pp 23–63

Malterud KE, Fabrot TL, Huse AE, Sund RB (1993) Antioxidant and radical scavenging effects of anthraquinones and anthrones. Pharmacology 47(1):77–85

Meyer MM (1983) A new method for propagation of woody plants from tissue culture. Am Nurseyman 6:65–70

Moreno-Sepulveda M, Vargas-Zapata R, Esquivel-Escobedo D, Waksaman de Torres N, Pineyro-Lopez A (1994) Effect of peroxisomicine and related anthraquinones on catalase activity. Planta Med 61:337–340

Morier-Tessier E, Boitte N, Helbecque N, Bernier J-L, Pommery N, Duvalet J-L, Fournier C, Hecquet B, Catteau J-P, Henichart J-P (1993) Synthesis and antitumor properties of an anthraquinone bisubstituted by the copper chelating peptide gly-gly-l-his. J Med Chem 36:2084–2090

Murashige T, Skoog F (1962) A revised medium for rapid growth and bio-assays with tobacco tissue cultures. Physiol Plant 15:473–497

Nebois P, Carneiro do Nascimento S, Boitard M, Bartoli M-H, Fillion H (1994) Synthesis and in vitro cytotoxic activity of aza-, and diazaanthraquinone derivatives. Pharmazie 49:819–821

Noro T, Noro K, Miyasa T, Kuroyanagi M, Umehara K, Ueno A, Fukushima S (1987) Inhibition of xantine oxidase by anthraquinones. Chem Pharm Bull (Tokyo) 35:4314–4316

Ohshima Y, Takashi K (1983) Separation mathod for sennosides. J Chromatogr 258:292–296

Ohshima Y, Ohno Y, Kajiyama K, Takashi K (1986) High-performance liquid chromatographic separation of *Rhubarb* constituents. J Chromatogr 360:303–306

Rai PP (1978) The production of anthraquinones in callus cultures of *Rheum palmatum*. Lloydia 41:114–116

Rauwald HW, Miething H (1985) Zur qualitativen Analytik der Rinde von Rhamnus fallax. Dtsch Apoth Ztg 125:101–105

Reynolds T (1994) A chromatographic examination of some old samples of drug aloes. Pharmazie 49:524–529

Sajc L, Obradović B, Vuković D, Bugarski B, Grubišić D, Vunjak-Novakovićc G (1995a) Hydrodynamics and mass transfer in a four-phase external loop air lift bioreactor. Biotechnol Prog 11:420–428

Sajc L, Grubišić D, Kovačević N, Bugarski B, Vuković D, Vunjak-Novaković G (1995b) Production of anthraquinones by immobilised *Frangula alnus* Mill. plant cells in a four-phase air lift bioreactor. Appl Microbiol Biotechnol 43:416–423

Savonius K (1972) The isolation and identification of some oxidized aglycones from *Frangula alnus* Mill. Farmaseuttinen Aikakauslehti 81:85–90

Steglich W, Losel W (1969) Bestimmung der Stellung von O-Substituenten bei 1,8-Dihydroxy-Anthrachinon-Derivaten mit Hilfe der NMR-Spektroskopie. Tetrahedron 25:4391–4399

Steinegger E, Hansel R (1988) Lehrbuch der Pharmakognosie und Phytopharmazie, Springer, Berlin Heidelberg New York, pp 414–421

Suzuki H, Matsumoto T (1988) Anthraquinone: production by plant cell culture. In: Bajaj YPS (ed) Biotechnology in agriculture and forestry, vol 4. Medicinal and aromatic plants I. Springer Berlin Heidelberg New York, pp 237–250

Suzuki H, Matsumoto T, Mikami Y (1984) Effects of nutritional factors on the formation of anthraquinones by *Rubia cordifolia* cells in suspension culture. Agric Biol Chem 48:603–610

Thomson RH (1965) Quinones: nature, distribution and biogenesis. In: Goodwin FW (ed) Chemistry and biochemistry of plant pigment. Academic Press, London, pp 309–332

Tutin TG, Heywood VH, Burges NA, Moore DM, Valentine DH, Walters SM, Webb DA (1968) Flora Europeae, vol 2. Cambridge University Press, London, 245 pp

Van der Berg AJJ, Labadie RP (1984) Anthraquinones, anthrones and dianthrones in callus cultures of *Rhamnus frangula* and *Rhamnus purshiana*. Planta Med 50:449–453

Van der Berg AJJ, Labadie RP (1988a) *Rhamnus* spp.: in vitro production of anthraquinones, anthrones and dianthrones. In: Bajaj YPS (ed) Biotechnology in agriculture and forestry, vol 4. Medicinal and aromatic plants I. Springer, Berlin Heidelberg New York, pp 513–528

Van der Berg AJJ, Radema MH, Labadie RP (1988b) Effects of light on anthraquinone production in *Rhamnus purshiana* suspension cultures. Phytochemistry 27:415–417

Van der Berg AJJ, Labadie RP (1989) Quinones. In: Dey M, Harborne JB (eds) Methods in plant biochemistry, vol 1. Plant phenolics. Academic Press, London, pp 413–491

Westendorf J, Poginsky B, Marquardt H, Kraus LJ (1988) Possible carcinogenicity of anthraquinone containing medicinal plants. Planta Med 54:562–563

White PR (1963) The cultivation of animal and plant cell, 2nd edn. Ronald Press, New York

Wijnsma R, Go JTKA, Van Weerden IN, Harkes PAA, Verpoorte R, Svendsen AB (1985) Anthraquinones as phytoalexins in cell and tissue cultures of *Cinchona* sp. Plant Cell Rep 4:241–244

Wijnsma R, Van Vliet TB, Harkes PAA, Van Groningen HJ, Van der Heijden R, Verpoorte R, Svendsen AB (1986) A method for the quantitative determination of anthraquinones and alkaloids in cell and tissue cultures of *Cinchona* sp. Planta Med 52:80–84

Wong SM, Wagner H, Benze S, Antus S (1988) Hepato-protective activities of coumarins, anthraquinones, naphtopyrone glycosides and iridoid glycosides. Planta Med 54:566–567

Wong SM, Seligmann O, Wagner H (1989) Isolation and structural elucidation of new antihepatotoxic naphto-g-pyrone glycosides, naphto-a-pyrone glycosides and anthraquinone glucosides from seeds of *Cassia tora*. Planta Med 55:112–113

Wood S, Hoffman J, Weber N, Anderson D, North J, Murray B, Sidewell R, Huges B (1990) Antiviral activity of naturally occuring anthraquinones and anthraquinone derivatives. Planta Med 56:651–652

Yeh SF, Chon T-C, Liu T-S (1988) Effects of anthraquinones of *Polygonatum cruspidatum* on HL-60-cells. Planta Med 54:413–414

X *Hyacinthus orientalis* L.: In Vitro Culture and the Production of Anthocyanin and Other Secondary Metabolites

K. Hosokawa

1 Introduction

Hyacinth (*Hyacinthus orientalis* L., Fig. 1A–B) family Liliaceae, is classified as an ornamental, and is native to the Mediterranean region, chiefly to its eastern region. Since it was introduced to eastern Europe from Turkey (Darlington et al. 1951), improvement in many cultivars has been made. Although more than 2000 cultivars existed in Victorian times, only 187 are currently listed in the International Register (Stebbings 1996). Out of these, about 70 are mainly cultivated as pot and garden flowers, with various flower colors i.e. blue, violet, red, pink, yellow and white. Double-flowered cultivars are not as common as single-flowered types. Anthocyanins are one of the important factors for expression of flower colors, and 20 anthocyanins were isolated from hyacinth flowers (Fig. 2, Harborne 1964; Hosokawa et al. 1995a–c).

The essential oils found only in the flowers have been used in cosmetics, toiletries and soap fragrances. Hyacinths have long been cultivated on a commercial scale for the extraction of their flower oil, and the hyacinth absolute, a very expensive flower oil, was used in high class fragrances and other floral perfumes. Recently, this raw material has lost its importance because it is not readily available on the international market and has now been replaced by synthetic hyacinth compounds (Lawrence 1978; Anonis 1985). Fifty-six compounds have been identified in the essential oils from the flowers (Table 1; Kaiser and Lamparsky 1977).

About 95% of the world's hyacinth production is in the Netherlands and 167 million bulbs were exported in 1989/1990 (Nowak and Rudnicki 1993), 32.3 million bulbs to England, 31.9 million bulbs to the USA and 31.6 million bulbs to France between May 1992 and June 1993 (Shimizu 1996).

For commercial propagation, seeds are not used because it takes 5 years to reach the flowering stage. So two methods, cross-cutting and scooping, have been used mainly for enhancing the rate of propagation (Hartmann et al. 1990). The cross-cutting method in which the basal plate is cut to half the

Iwate Biotechnology Research Center, 22-174-4 Narita, Kitakami, Iwate, 024, Japan. Present address: Tsukuba Medicinal Plant Research Station, National Institute of Health Sciences, 1 Hachimandai, Tsukuba, Ibaraki 305-0843, Japan

Biotechnology in Agriculture and Forestry, Vol. 43
Medicinal and Aromatic Plants XI (ed. by Y.P.S. Bajaj)

1

4

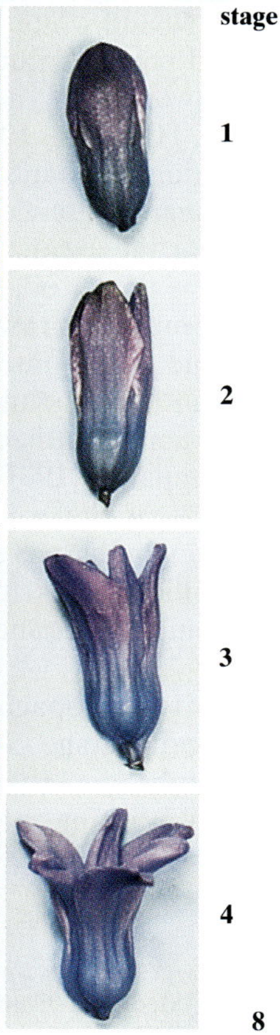

8

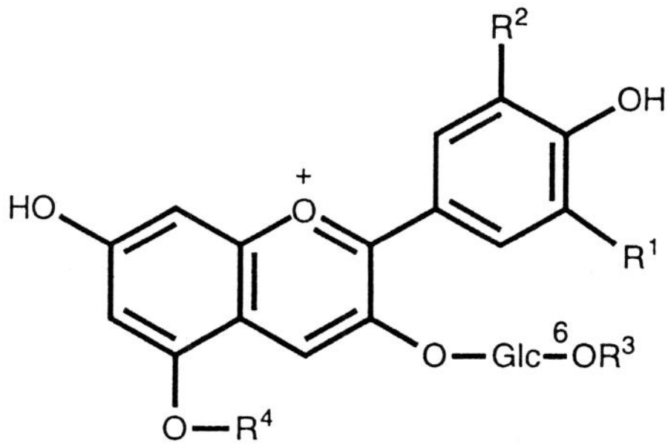

Glc; β-D-glucopyranosyl
cCou; *cis-p*-coumaroyl
tCou; *trans-p*-coumaroyl
Caf; caffeoyl
Fer; feruloyl
Mal; malonyl
Ace; acetyl

	R^1	R^2	R^3	R^4
1	OMe	OH	tCou	Glc$\overset{6}{—}$Mal
2	OH	OH	tCou	Glc
3	OH	OH	Caf	Glc$\overset{6}{—}$Mal
4	OH	OH	cCou	Glc$\overset{6}{—}$Mal
5	OH	OH	tCou	Glc$\overset{6}{—}$Mal
6	OH	H	tCou	Glc
7	OH	H	tCou	Glc$\overset{6}{—}$Mal
8	H	H	H	H
9	H	H	H	Glc
10	H	H	cCou	Glc
11	H	H	tCou	Glc
12	H	H	H	Glc$\overset{6}{—}$Mal
13	H	H	cCou	Glc$\overset{6}{—}$Mal
14	H	H	tCou	Glc$\overset{6}{—}$Mal
15	H	H	Caf	Glc
16	H	H	Fer	Glc
17	H	H	Fer	Glc$\overset{6}{—}$Mal
18	H	H	Caf	Glc$\overset{6}{—}$Mal
19	H	H	tCou	Glc$\overset{6}{—}$Ace
20	H	H	tCou	Glc$\overset{4}{—}$Mal

Fig. 2. Anthocyanins isolated from blue and red flowers of hyacinth. (Harborne 1964; Hosokawa et al. 1995a–c)

◀————————————————————

Fig. 1. A *Hyacinthus orientalis* L. cv. Delft Blue and **B** cv. Hollyhock used in the experiments
Fig. 4. Regeneration of flower buds from immature tepal explants of *H. orientalis* L. cv. Delft Blue. **A** Flower buds regenerated from tepal explants isolated from a bulb that had been stored at 3–5 °C. Explants were cultured for 8 weeks on medium that contained 2 mg/l BA and 0.1 mg/l 2,4-D. **B** A double flower, 1 month of subculture on medium that contained 2 mg/l BA. (Hosokawa and Fukunaga 1995)
Fig. 8. The four stages of development of hyacinth flowers. (Hosokawa and Fukunaga 1995)

Table 1. Constituents and compositions (%) in the absolute of hyacinth flowers. (Kaiser and Lamparsky 1977)

Substance	%	Substance	%
Alkanes		Esters	
2-Undecanone	0.04	*cis*-3-Hexenyl acetate	0.01
		n-Heptanal diethyl acetate	0.01
Ketones		Methyl benzoate	0.03
p-Methoxy acetophenone	0.10	Benzyl formate	0.05
6,10,14-trimethyl pentadecan-2-one	0.10	Ethyl benzoate	0.02
		Benzyl acetate	8.1
Aldehydes		Methyl salicylate	0.10
n-Heptanal	0.02	Phenylethyl salicylate	0.04
Benzaldehyde	0.10	Eethyl salicylate	0.02
Cinnamaldehyde	0.30	Phenylpropyl acetate	0.05
		2-Methoxy methyl benzoate	0.10
Alcohols		2-Methoxy ethyl benzoate	0.02
n-Hexanol	0.02	2-Methoxy benzyl acetate	0.02
cis-3-Hexenol	0.02	Benzyl tiglate	0.03
1-Octen-3-ol	0.70	Cinnamyl acetate	0.70
Heptanol	0.01	*p*-Methoxy phenyl ethyl acetate	0.30
Linalool	0.02	Methyl veratrate	0.01
Benzyl alcohol	40.0	Veratryl acetate	0.03
Phenylethyl alcohol	3.7	3,4,5-Trimethoxymethyl benzoate	0.02
Phenyl propyl alcohol	0.80	Benzyl benzoate	6.0
cis-Cinnamyl alcohol	0.20	Phenylethyl benzoate	1.20
trans-Cinnamyl alcohol	11.0	Benzyl salicylate	0.20
p-Methoxy phenyl ethyl alcohol	1.20		
p-Methoxyphenylpropyl alcohol	0.02		
Aromatics		Ethers	
1,2-dimethoxybenzene	0.02	Methylbenzyl ether	0.01
1,4-dimethoxybenzene	0.20		
1,2,4-Trimethoxybenzene	3.0	Others	
Estragole	0.02	Methyl-*N*-methylanthranilate	0.30
		Elemicin	0.60
Terpenes		Indole	0.15
Myrcene	0.01		
cis-β-Ocimene	0.01		
trans-β-Ocimene	0.02		
α-Farnesene	0.1		
Geraniol	0.03		
trans-α-Farnesene	0.60		
Methyl eugenol	1.5		
Eugenol	0.20		

depth of the scales produces larger bulbs, but fewer numbers of bulbils are formed. On the other hand, the scooping method in which the basal plate of the bulb is cut out to expose the scales, produces smaller, but more bulbils. By these methods, 20 to 50 small bulbils per bulb are derived.

2 In Vitro Approaches

2.1 Review of Tissue Culture

2.1.1 In Vitro Propagation

In vitro propagation techniques have greatly helped clonal propagation and the production of virus-free plants (Takayama 1991). The in vitro studies conducted so far on hyacinth are summarized in Table 2. Propagation by regeneration of bulblets or adventitious shoots has been done using various explants such as leaves, scales, inflorescence stalks, ovaries and flower buds. The attempts to produce virus-free plants have also been reported (Asjes et al. 1974; Matsumoto 1983; Blom-Barnhoorn et al. 1986). Further, rapid clonal propagation for commercial applications has been attempted (Takayama 1991; Takayama et al. 1991). In spite of many reports on propagation, the techniques developed are not applicable to commercial production. Some problems for utilizing these techniques for industrial processes are pointed out by Takayama (1991).

2.1.2 Regeneration of Floral Organs

Lu et al. (1986) showed that floral buds could be induced to form from perianth explants of cv. Delft Blue. Further, they (Lu et al. 1988) studied cultural conditions for regeneration of floral organs, such as tepals, stamens and ovules, and demonstrated the importance of both the developmental stage of perianth explants used (Tables 3, 4) and of the plant growth regulator composition of the medium for the regeneration of different floral organs by perianths of hyacinth (Table 5). Further, regimes of temperature are important factors for regeneration of stamens and for their maturity (Lu et al. 1990, 1991).

2.2 Establishment of Cultures and Regeneration of Flowers in Vitro

Bulbs (ca. 6 cm in diameter) of *H. orientalis* cv. Delft Blue and cv. Hollyhock obtained in September, were kept at 5 to 10 °C in darkness for 2 months and then stored at 3 to 5 °C. Bulbs treated in this way were used for in vitro culture experiments. Tepal explants were dissected aseptically, as described by Lu et al. (1988), from bulbs that had been treated as described above (Fig. 3A–E). For regeneration of flowers, the explants were cultured for 2 months on basal medium that contained 2 mg/l BA and 0.1 mg/l 2,4-D. The basal medium, MS (Murashige and Skoog 1962), was supplemented with 30 g/l sucrose and 0.5 g/l casamino acid. The regenerated flower buds (Fig. 4A) were subcultured for 1 month on the basal medium plus 2 mg/l BA for the further development of the regenerated flower buds (Fig. 4B). The regeneration of flower buds

Table 2. Summary of in vitro culture studies on hyacinth (see also Paek and Thorpe 1989)

Cultivar	Explant	Basal medium	Organic (mg/l)	Sucrose (g/l)	Plant growth regulators (mg/l) or [M]	Response	Reference
Pink Pearl	Scale	Heller (1×, 1/2×) Knop (1×, 1/2×)		0–40	KIN (0–10) IAA (0–10) GA_3 (0–10)	Bulblet regeneration	Pierik and Woets (1971) Pierik and Ruibing (1973)
King of the Blues Nobel Queen of the Pinks	Meristem tips	Karstens		20 (Glucose)	NAA (0.1)	Virus-free plantlet	Asjes et al. (1974)
Lady Derby	Bulblet	MS		30	BA (10) and NAA (1)		Saniewski et al. (1974)
Ostara Princes Irene	Scale Basal tissue Leaf Stem Ovary	MS salt	Myoinositol (100) Thiamine-HCL (1)	20	IAA (0, 0.5, 2, 8) NAA (0, 0.03, 0.12)	Plantlet regeneration	Hussey (1975)
Pink Pearl	Scale	MS		30	BA (10) and NAA (1)	Bulblet regeneration	Pierik and Post (1975)
Pink Pearl	Scale	Knop (1/2macro) or Heller (micro)		30	IAA (0–10) IBA (0–10) NAA (0–10) GA_3 (0–10) GA_{4+7} (0–10) ABA (0–10) Ethephon (0–120)	Bulblet regeneration	Pierik and Steegmans (1975)
Pink Pearl	Flower bud	MS		30	BA (10) and NAA (1)	Bulblet regeneration	Saniewski (1975)
Lady Derby	Ovary	MS		30	BA (10) and NAA (1)	Bulblet-like regeneration	Saniewski et al. (1974)
La Victoire Delft Blue Lady Derby Jan Bos	Scale	MS salt	Thiamine (0.5) Inositol (100)	30	BA (0.5) 2,4-D (0.001–0.005) IAA (0.5)	Adventitious shoot Bulblet regeneration	Tamura (1978)
City of Haarlem	Flower stem	MS		30	IBA (0.01–0.1) NAA (0.1)	Bulblet differentiation	Chung et al. (1981)
Anna Marie	Immature embryo	MS salt	Inositol (100)	30	BA (0.1–3) and	Bulblet regeneration	Kim et al. (1981)

Delft Blue	flower bud Scape		Glycine (2) Nicotinic acid (0.5) Pyridoxine-HCl (0.5) Thiamine-HCl (0.4)		NAA (0.1–3)		
Jan Bos	Flower stalk Leaf Scale	MS		30	KIN/BA [$0–10^{-5}$] and NAA/2,4-D [$0–10^{-6}$]	Bulblet regeneration	Lee et al. (1982a)
Jan Bos	Anther Flower stalk Leaf Ovary Scale	MS		30	2,4-D [10^{-8}]	Bulblet regeneration	Lee et al. (1982b)
Jan Bos	Flower bud Scale Stem	MS		30	KIN (0–2) NAA (0–2)	Bulblet regeneration	Paek and Choi (1982)
Jan Bos	Leaf Scale Stalk	MS		30	BA [10^{-5}] and 2,4-D [10^{-8}] BA [10^{-5}] and NAA [10^{-7}]	Bulblet regeneration	Bae et al. (1983)
Jan Bos	Scale	MS (1/2×) MS		30		Bulblet regeneration	Paek et al. (1983)
Carnegie Anna Marie Lady Derby Ostara	Scale Stalk	MS		30	IAA (0.1–10) IBA (0.1–10) NAA (0.1–5)	Bulblet regeneration	Chung et al. (1983a)
City of Haarlem Blue Jacket	Scale Leaf	MS		30	IAA (1) IBA (1)	Bulblet regeneration	Chung et al. (1983b)
Schotel	Stem tip Anther Flower bud	MS		30	KIN (0.1) and IAA (1) KIN (1) and IAA (1)	Virus-free plantlet	Matsumoto (1983)
Carnegie City of Haarlem Jan Bos Marie Lady Derby	Scale	MS		30	KIN (1) IAA (10) NAA (1)	Bulblet regeneration	Amaki et al. (1984)

Table 2. *Continued*

Cultivar	Explant	Basal medium	Organic (mg/l)	Sucrose (g/l)	Plant growth regulators (mg/l) or [M]	Response	Reference
Lady Derby	Scale Leaf	MS		30 50	IAA (1) IBA (1)	Bulblet regeneration	Chung et al. (1984a)
Jan Bos	Scale Leaf Stalk	MS		30	IBA (0 or 1)	Bulblet regeneration	Chung et al. (1984b)
Rosalie	Scale	MS-salt	Inositol (100) Thiamine-HCl (0.4)	30	IBA [$0.05–5 \times 10^{-6}$] with Virazole [200]	Virus-free plantlet	Blom-Barnhoorn et al. (1986)
Delft Blue	Scale Petal Leaf	MS	Casamino acid (500)	30	BA (0.1–4), BA (1) and NAA (0.1–1)	Regeneration of vegetative Flower bud	Lu et al. (1986)
Carnegie City of Haarlem Jan Bos Lady Derby Ostara	Scale Leaf Stalk	MS		30	IAA (1) IBA (1)	Bulblet regeneration	Paek et al. (1987)
Carnegie	Leaf	MS MS (1/2×)		30 30	BA (5) NAA (0.5) BA (0.1–1) KIN (0.1–1)	Bulblet regeneration	Bach and Cecot (1988)
Delft Blue	Perianth	MS	Casamino acid (500)	30	BA (2) and 2,4-D (0.1)	Regeneration of stamen	Lu et al. (1990)
Jan Bos	Scale	MS-salt	Inositol (100) Glycine (2) Nicotinic acid (0.5) Pyridoxine-HCl (0.5) Thiamine-HCl (0.4)	30	BA (4.4–44 μM) and NAA (0.54–5.4 μM)	Bulblet regeneration	Takayama et al. (1991)
Delft Blue	Perianth	MS	Casamino acid (500)	30–40	GA^3 (1–2)	Production of essential oils	Hosokawa and Fukunaga (1995)
Delft Blue	Regenerated flower	MS	Casamino acid (500)	30	GA^3 (1)	Production of anthocyanins	Hosokawa et al. (1996a)
Hollyhock	Regenerated flower	MS	Casamino acid (500)	30	GA^3 (1)	Production of anthocyanins	Hosokawa et al. (1996b)

Table 3. Developmental stages of *Hyacinthus* perianth explants. (Lu et al. 1988)

Stage	Length of terminal bud (mm)	Pollen	
		Diameter (μm)	Stage
I	39–47	44–48	Early binucleate
II	52–55	48–41	Mid-binucleate
III	80–92	50–52	Late binucleate

Table 4. Effects of exogenous hormones on differentiation of floral organs from stage III perianth explants of *Hyacinthus*. Data were scored at 74 days post-inoculation. (Lu et al. 1988)

Hormone (mg/l)			No. of explants	Average No. per explant		
BA	Zeatin	2,4-D		Tepals	Stamens	Ovules
2	–	–	24	1.9	1.7	17.2
2	–	0.1	24	8.4	2.7	1.9
2	–	0.5	36	11.0[a]	0	0
2	–	1.0	36	3.3[a]	0	0
–	2	–	34	3.0	3.0	0.2
–	2	0.1	26	4.6	1.2	0.1
–	2	0.5	21	3.2[a]	0	0
–	2	1.0	29	0.7[a]	0	0

[a] Aberrant, thickened, transparent and slow growing.

Table 5. Effects of the developmental stage of perianth explants of *Hyacinthus* on differentiation of floral organs. Cultures were grown on medium with 2 mg/l zeatin and 0.1 mg/l 2,4-D. Data scored 3 months post-inoculation. (Lu et al. 1988)

Developmental stage[a]	No. of explants	Average No. per explant			
		Floral buds	Tepals	Stamens	Ovules
I	35	5.9	17.4	0	0
II	111	4.9	14.8	0.14	0.01
III	36	4.4	13.3	0.04	0.47

[a] The stages correspond to the stages in Table 4.

(during cultivation for 2 months on the basal medium plus 2 mg/l BA and 0.1 mg/l 2,4-D) and the development (during cultivation for 1 month on the basal medium plus 2 mg/l BA) of regenerated flowers were repeated in successive subcultures (Fig. 4A,B). Consequently, many regenerated flowers that were obtained serially were used in experiments for the production of essential oils and anthocyanins.

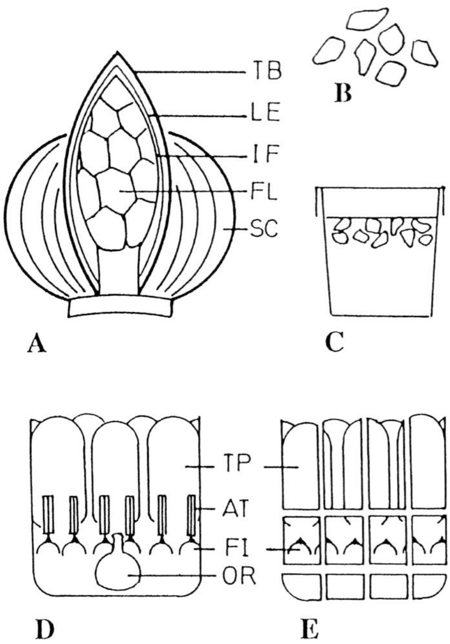

Fig. 3A–E. Diagram for excision and sterilization of perianth explants from a hyacinth bulb. **A** Longitudinal section of bulb. *TB* Terminal bud of bulb; *LE* leaf; *IF* inflorescence; *FL* flower; *SC* scale of bulb. **B** Flowers excised from the inflorescence. **C** The flowers are sterilized in 1% sodium hypochlorite for 10 min, then rinsed three times with sterilized distilled water. **D** The flower is cut open. *TP* tepal; *AT* anther; *FI* filament; *OR* ovary. **E** All anthers and the ovary were removed from the flower, the perianth was cut into 12 pieces, and the middle 4 pieces containing filaments were used for culture. (Lu et al. 1988)

2.3 Production of Secondary Metabolites by Flowers Regenerated in Vitro

2.3.1 Essential Oils (Hosokawa and Fukunaga 1995)

Essential oils were produced by flowers grown in the field during their normal development, whereas no such oils were detected in the regenerated flowers of cv. Delft Blue (Fig. 4B). Effect of eight plant growth regulators (kinetin, 2,4-D, IAA, NAA, GA_3, ABA, ACC and ethephon) were investigated at 0–10 mg/l for the production of essential oils. Among them, only GA_3 had a marked stimulatory effect on the production of essential oils in the regenerated flowers. The effects of GA_3 at various concentrations (0–5 mg/l), the concentrations of sucrose (0–40 g/l) and the regimes of temperature (15–35 °C) on the production of essential oils were examined. The results are shown in Figs. 5–7. The maximum concentration and content of essential oils (185 μg/fresh weight flowers and 392.4 μg/culture) was obtained at 20 °C after 3 weeks of culture on basal medium that contained the optimal concentration of GA_3 (1 mg/l) and sucrose (30 g/l).

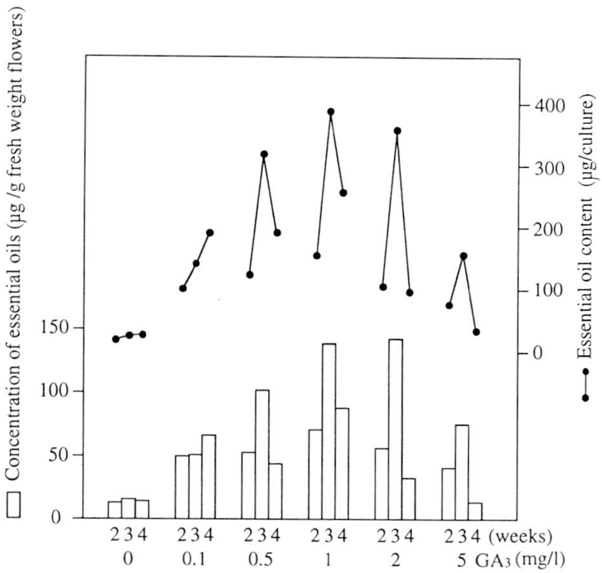

Fig. 5. Effects of the concentration of GA_3 on the concentration of essential oils and total essential oil content in regenerated flowers. The regenerated flowers were cultured on MS medium that contained 30 g/l sucrose and various concentrations of GA_3 at 25 °C for 2 to 4 weeks under 16-h day length. (Hosokawa and Fukunaga 1995)

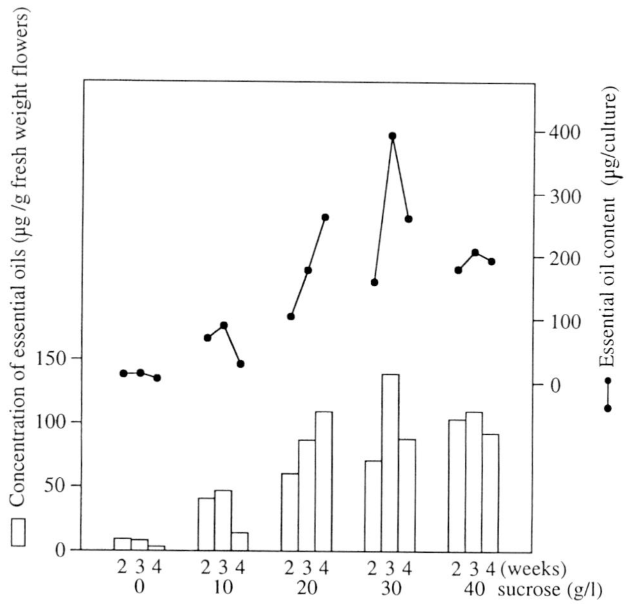

Fig. 6. Effects of the concentration of sucrose on the concentration of essential oils and total essential oil content in regenerated flowers. The regenerated flowers were cultured on MS medium that contained 1 mg/l GA_3 and various concentrations of sucrose at 25 °C for 2 to 4 weeks under 16-h day length. (Hosokawa and Fukunaga 1995)

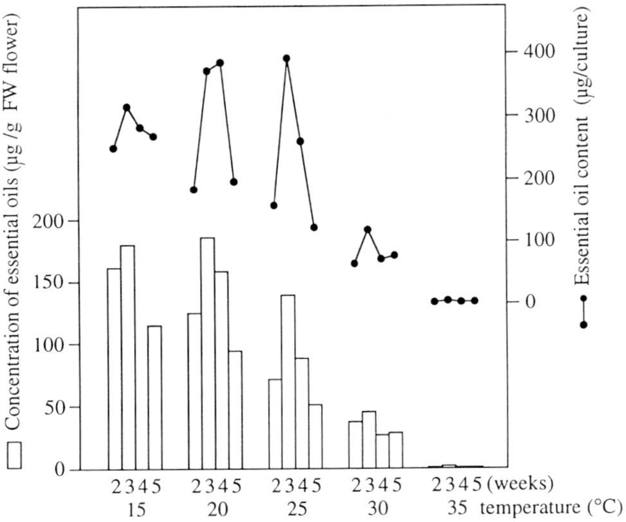

Fig. 7. Effects of temperature on the concentration of essential oils and total essential oil content in regenerated flowers. The regenerated flowers were cultured on MS medium that contained 1 mg/l GA_3 and 30 g/l sucrose at various temperatures for 2 to 5 weeks under 16-h day length. (Hosokawa and Fukunaga 1995)

Prior to comparison of the types and levels of each constituent of the essential oils from the regenerated flowers with those from flowers that had been collected from field-grown plants, essential oils from flowers at four different stages (Fig. 8) of field-grown plants were analyzed. The essential oils were produced in the flowers after anthesis (stages 3 and 4) and they were not detected in the younger buds (stages 1 and 2). The constituents and the levels of each constituent of the essential oils from the regenerated flowers were compared with those from flowers at stages 3 and 4, and each constituent of the essential oils was identified and quantified by analysis of mass spectral data and by comparison of retention times with those of authentic compounds (Table 6). The results revealed that nine compounds could be extracted from the regenerated flowers and six (stage 3) and ten (stage 4) compounds could be extracted from field-grown flowers. Four compounds, namely, 1-hepten-3-ol, benzyl alcohol, phenethyl alcohol and cinnamyl alcohol, were common to both types of flowers. In the essential oils from regenerated flowers, phenethyl alcohol was the major constituent (75%). In contrast, two compounds, phenethyl alcohol (stage 3, 55%; stage 4, 48%) and cinnamyl alcohol (stage 3, 23%; stage 4, 29%) were the major constituents of the essential oils from flowers grown in the field. Regenerated flowers yielded cinnamyl alcohol at a level of only 0.6% (1.1 μg/g fresh weight flowers), which was quite different from the level in the field-grown flowers (stage 3, 22% and 48.7 μg/g fresh weight flowers; stage 4, 29% and 82.5 μg/g fresh weight flowers). Four (1-pentanol, 4-hexen-1-ol, cinnamyl acetate and unknown-2) and five

Table 6. Constituents and the level of each constituent in essential oils from hyacinth flowers, as determined by GLC and GC-MS. (Hosokawa and Fukunaga 1995)

Compound	Retention time (min)	Mass spectral data: M^+ and fragment ions	Essential oils [µg/g fresh weight (%)]: Regenerated[a]	Field-grown: Stage 3	Field-grown: Stage 4
1. 1-Pentanol	20.51	70(M^+), 55, 42, 31	1.3 (0.70)		
2. 4-Hexen-1-ol	24.80	100(M^+), 82, 67, 55, 41	1.1 (0.60)		
3. 1-Hepten-3-ol	26.78	85, 72, 57	3.1 (1.68)[b]	26.8 (12.56)[b]	6.0 (2.11)[b]
4. Benzaldehyde	30.31	106(M^+), 105, 77, 51			2.1 (0.74)
5. Phenylacetaldehyde	33.89	120(M^+), 91, 65		7.0 (3.30)	5.8 (2.04)
6. Benzyl acetate	36.08	150(M^+), 108, 91, 43		7.2 (3.37)	6.6 (2.32)
7. Benzyl alcohol	39.95	108(M^+), 107, 91, 79, 77, 51	9.0 (4.87)	5.7 (2.67)	4.5 (1.58)
8. Phenethyl alcohol	40.94	122(M^+), 91, 65	139.4 (75.43)	117.9 (55.27)	136.9 (48.10)
9. 3-Phenyl propanol	44.26	136, 118, 117, 91			3.6 (1.26)
10. Unknown-1	44.62	168, 153, 125, 110, 95, 93			4.7 (1.65)[c]
11. Cinnamyl acetate	47.24	132(M^+), 104, 103, 78, 77, 51, 50	5.0 (2.71)		
12. Cinnamyl alcohol	49.84	134(M^+), 115, 105, 92, 91, 78, 77	1.1 (0.60)	48.7 (22.83)	82.5 (28.99)
13. Unknown-2	55.83	136, 91, 65	18.2 (9.84)[c]		
14. Benzyl benzoate	57.47	212(M^+), 105, 91, 77	6.6 (3.57)		31.9 (11.21)
Total content			184.8 (100)	213.3 (100)	284.6 (100)

[a] Flowers regenerated in vitro were cultured at 20 °C for 3 weeks on MS medium that contained 30 g/l sucrose, 0.5 g/l casamino acids and 1 g/l GA_3.
[b] The level was calculated from a standard linear curve for 4-hexen-1-ol.
[c] The level was calculated from a standard linear curve for phenethyl alcohol.

(benzaldehyde, phenylacetaldehyde, benzyl acetate, 3-phenylpropanol and unknown-1) compounds were minor constituents that were specific to regenerated and field-grown flowers, respectively. The contents of esters produced by field-grown flowers were 7.2 μg/g fresh weight flowers (3.4%) at stage 3 and 38.5 μg/g fresh weight flowers (13.5%) at stage 4, whereas the corresponding value for regenerated flowers was 11.6 μg/g fresh weight flowers (6.3%). These results suggest that esterase activities in the regenerated flowers were weak and resembled those in the flowers at stage 3 in the field. The reproducibility of the constituents and contents of the essential oils in the field-grown flowers at stages 3 and 4, as well as that of the essential oils in the regenerated flowers was confirmed in experiments conducted over the course of 2 years.

As mentioned above, Hosokawa and Fukunaga (1995) reported the first establishment of a stable system for production of essential oils, which consist mainly of aromatic alcohols and their esters, in flowers regenerated in vitro directly from tepal explants of *H. orientalis* L. Although essential oils in flowers are important natural products and many attempts have been made to produce them in tissue culture (*Jasminum officinale*, Banthorpe et al. 1986a; *Rosa* sp., Banthorpe et al. 1986b; *J. sambac*, Tomoda et al. 1976; *Lavandula officinalis*, Banthorpe et al. 1986a; Webb et al. 1984; and *Matricaria chamomilla*, Collin 1988), the productivity was extraordinarily low in each case. Up to now, the essential oils generated in tissue culture were mainly terpenoids (Mulder-Krieger et al. 1988). However, aromatic alcohols and their esters were mainly produced in the regenerated flowers of *H. orientalis*.

2.3.2 Anthocyanins (Hosokawa et al. 1996a,b)

Anthocyanins were synthesized only at the distal ends of the regenerated flowers of cv. Delft Blue that were grown on basal medium containing 2 mg/l BA. The effects of eight plant growth regulators were also investigated for the production of anthocyanins as in the case of the production of essential oils in regenerated flowers. Among them only GA_3 also had a potent stimulatory effect on the production of anthocyanins in the regenerated flowers. GA_3 (0–2 mg/l), sucrose (0–40 g/l) and temperature (10–35 °C) were investigated for the regenerated flowers (Hosokawa et al. 1996a). The results are shown in Figs. 9–11. The maximum concentration of anthocyanins (410 μg/fresh weight flowers) was obtained at 15 °C after 3 weeks of culture on basal medium that contained the optimal concentration of GA_3 (1 mg/l) and sucrose (30 g/l).

One major and six minor anthocyanins as pigments of the regenerated flowers were detected by HPLC analysis. The anthocyanins (**1–5**, **7**, **14**) were identical with those of field-grown flowers (Hosokawa et al. 1996a). The concentration of anthocyanins that were produced in the regenerated flowers was 1.2 times higher than that obtained from field-grown flowers (stage 4) and there were minor differences in the relative amounts (Table 7).

The regenerated flowers in cv. Delft Blue which mainly contained delphinidin derivatives produced the same anthocyanins as the field-grown flowers mentioned above. However, in *Impatiens balsamina*, anthocyanidins

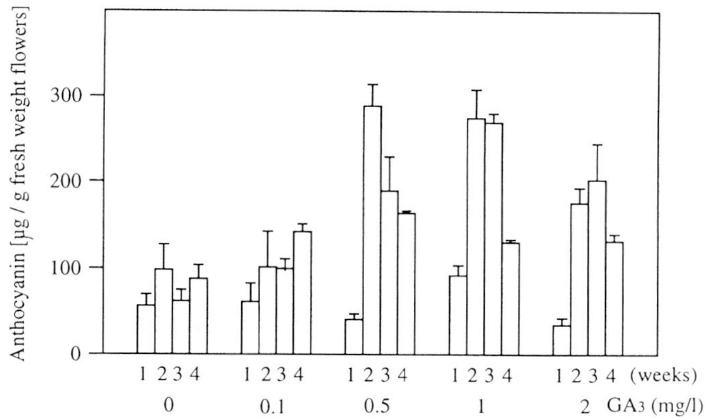

Fig. 9. Effects of the concentration of GA_3 on production of anthocyanins in regenerated flowers. The data are given as means of five samples from different flasks. *Bars* indicate standard deviations. The regenerated flowers were cultured on MS medium that contained 30 g/l sucrose and various concentrations of GA_3 at 25 °C for 1 to 4 weeks under 16-h day length

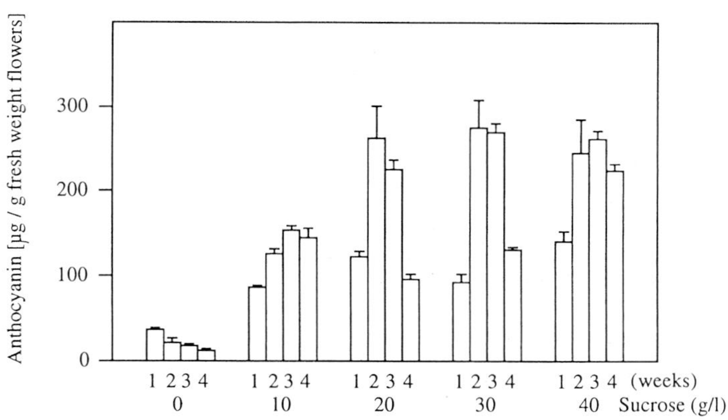

Fig. 10. Effects of the concentration of sucrose on production of anthocyanins in regenerated flowers. The data are given as means of five samples from different flasks. *Bars* indicate standard deviations. The regenerated flowers were cultured on MS medium that contained 1 mg/l GA_3 and various concentrations of sucrose at 25 °C for 1 to 4 weeks under 16-h day length.

that were not found in intact flowers were produced in in vitro culture and the difference occurred in two genotypes which mainly contained pelargonidin derivatives as floral pigments (Klein and Hagen 1961). It therefore seemed necessary to examine the anthocyanin profile of regenerated flowers of hyacinth, whose field-grown flowers contained mainly pelargonidin derivatives, in order to see whether it was anomalous or not. The anthocyanin profile of the regenerated flowers of cv. Hollyhock was compared with that of field-grown

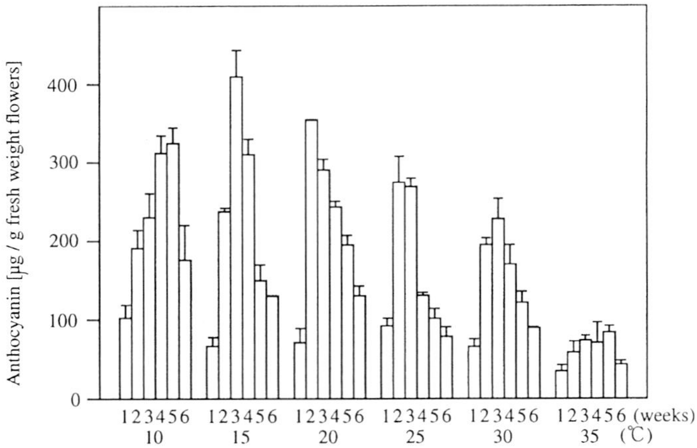

Fig. 11. Effects of temperature on production of anthocyanins in regenerated flowers. The data are given as means of five samples from different flasks. *Bars* indicate standard deviations. The regenerated flowers were cultured on MS medium that contained 1 mg/l GA_3 and 30 g/l sucrose at various temperatures for 1 to 6 weeks under 16-h day length

Table 7. Analytical HPLC data for the anthocyanins in the flowers regenerated in vitro and field-grown flowers of *Hyacinthus orientalis* L. cv. Delft Blue. (Hosokawa et al. 1996a)

Anthocyanin[a]	Regenerated flowers		Field-grown flowers	
	Rt (min)	Composition (Area % at 533 nm)	Rt (min)	Composition (Area % at 533 nm)
Del 3-cCouG 5-MG (**4**)	5.54	3.3	5.71	5.4
Del 3-CafG 5-MG (**3**)	7.07	0.2	6.81	0.2
Del 3-tCouG 5-G (**2**)	12.08	3.6	12.06	2.1
Del 3-tCouG 5-MG (**5**)	14.56	78.9	14.61	82.9
Pet 3-tCouG 5-MG (**1**)	19.93	4.5	19.81	5.6
Cya 3-tCouG 5-MG (**7**)	23.77	8.2	23.96	2.6
Pel 3-tCouG 5-MG (**14**)	33.61	1.3	33.26	1.2
Total[b]		410		340

[a] Del, delphinidin; Cya, cyanidin; Pel, pelargonidin; G, β-D-glucoside; cCou, *cis-p*-coumaroyl; tCou, *trans-p*-coumaroyl; Caf, caffeoyl; M, malonyl. The compound numbers correspond to the numbers in Fig. 2.

[b] Total anthocyanin concentration is in μg per g fresh weight flowers.

flowers which mainly contained pelargonidin derivatives (Hosokawa et al. 1996b).

Nine anthocyanins were detected in red regenerated flowers of cv. Hollyhock by HPLC. All anthocyanins (**6**, **7**, **11–14**, **17**, **18**, **20**) were identical with those of field-grown flowers (Table 8). The concentration of anthocyanins produced in the regenerated flowers was 6.2 times less than that obtained from field-grown flowers. The composition, on the other hand, was significantly

Table 8. Analytical HPLC data for the anthocyanins in the flowers regenerated in vitro and field-grown flowers of *Hyacinthus orientalis* L. cv. Hollyhock. (Hosokawa et al. 1996b)

Anthocyanin[a]	Regenerated flowers		Field-grown flowers	
	Rt (min)	Composition (Area % at 510 nm)	Rt (min)	Composition (Area % at 510 nm)
Pel 3-G 5-MG (**12**)	4.11	5.1	4.12	3.0
Pel 3-cCouG 5-MG (**13**)	8.76	0.3	8.61	0.6
Pel 3-CafG 5-MG (**18**)	11.76	0.3	11.61	1.4
Cya 3-tCouG 5-G (**6**)	13.81	1.8	13.36	6.6
Cya 3-tCouG 5-MG (**7**)	20.86	30.4	20.81	2.3
Pel 3-tCouG 5-G (**11**)	23.86	5.5	23.66	6.5
Pel 3-tCouG 5-MG (**20**)	34.66	1.2	26.01	1.0
Pel 3-tCouG 5-MG (**14**)	34.34	53.2	33.61	73.8
Pel 3-FerG 5-MG (**17**)	41.60	2.2	41.16	4.8
Total[b]		350		2170

[a] Cya, cyanidin; Pel, pelargonidin; G, β-D-glucoside; cCou, *cis-p*-coumaroyl; tCou, *trans-p*-coumaroyl; Caf, caffeoyl; Fel, feruloyl; M, malonyl. The compound numbers correspond to the numbers in Fig. 2.

[b] Total anthocyanin concentration is in μg per g fresh weight flowers.

different and the difference was found in an increase in the relative level of **7**, a cyanidin derivative, in the regenerated flowers.

Production of anthocyanins in flowers cultured in vitro have been limited in *Impatiens balsamina* (Klein and Hagen 1961), "Baccara" rose (Zieslin et al. 1974) and *Petunia hybrida* (Weiss and Halevy 1989), and the flowers used in in vitro culture were intact flowers or flower buds in each case. However, as mentioned above, Hosokawa et al. (1996a,b) reported a stable system for production of anthocyanins in the regenerated flowers of hyacinth, as for the production of essential oils.

2.4 Extraction and Structure of Secondary Metabolites

2.4.1 Essential Oils (Hosokawa et al. 1995, Table 6)

Extraction. About 1 g each of fresh whole flowers that had been regenerated in vitro, and field-grown flowers were extracted twice with 50 ml of diethyl ether. Each pooled extract was dehydrated overnight with anhydrous Na_2SO_4 and then concentrated under a stream of dry N_2 gas. The concentrate was dissolved in *n*-pentane and allowed to adsorb on to an activated Sep-Pak Silica cartridge. The cartridge was washed with 5 ml of *n*-pentane and then essential oils were eluted with 5 ml of diethyl ether. The eluate was concentrated under a stream of N_2 gas and the concentrate was dissolved in 0.5 ml of dichloromethane. 2-Hydroxybutanol was used as the internal standard for subsequent analysis.

Determination of Concentration and Content. Essential oils were separated on a gas chromatograph equipped with a Supelcowax 10 fused-silica column (0.75 mm i.d. × 60 m). The concentration of each compound in the essential oils was determined from standard linear calibration curves obtained with authentic samples. Essential oil content (EOC) was defined and calculated by application of the following formula: EOC = [concentration of essential oil] × [fresh weight of culture sample (g)]/[fresh weight of inoculum (g)].

Identification. Analysis was performed on a mass spectrometer combined with a gas chromatograph. The chromatograph was equipped with a Supelcowax 10 fused-silica capillary column (0.25 mm i.d. × 30 m). For identification of the various compounds, retention times and spectral data were compared with those for authentic compounds that had been purchased from commercial sources.

2.4.2 Anthocyanins

Determination of Concentration. (Hosokawa et al. 1996a) Anthocyanins were extracted from 1 g of fresh flowers with 1% HCl in methanol (5 ml, ×3) at 4 °C. Since the absorption of chlorophyll in acidic methanol at λ_{max} (533 nm) is about 25% of that at 657 nm (Mancinelli et al. 1991), the concentration of anthocyanin (µg/fresh weight of flowers (g)] in flowers was calculated as follows: [A533–0.25 × A657], based on delphinidin 3-*O*-(6-*O*-*trans*-*p*-coumaroyl-β-D-glucoside)-5-*O*-(6-*O*-malonyl-β-D-glucoside) as standard.

Isolation and Identification of Anthocyanins. (Hosokawa et al. 1995a–c, Fig. 2) Freeze-dried flowers were extracted with a mixture of ethanol, acetic acid and water (10:1:9, v/v) at 4 °C. The concentrated extract was loaded onto a column of Amberlite XAD-7 and washed with 5% acetic acid. The anthocyanins were eluted by 50% methanol that contained 5% acetic acid. For further purification, the crude anthocyanins were purified by preparative high performance liquid chromatography (HPLC) on an ODS column. Anthocyanins were separated with two solvent systems: 25–60% acetonitrile-acetic acid-water (5:4:11, v/v) containing 0.5% trifluoroacetic acid and then 10–25% methanol containing 15% acetic acid and 0.5% trifluoroacetic acid. The structure of each anthocyanin purified was determined by one- and two-dimensional NMR techniques and UV-visible spectral methods.

3 Summary and Conclusions

The propagation of hyacinth with artificial methods, such as cross-cutting and scooping, is limited to the propagation rate of bulbs. Many in vitro propagation techniques have been attempted for rapid clonal propagation (Table 2), which have demonstrated that plantlets are able to be regenerated from various explants and the important factors for regeneration are the type and concentration of plant growth regulators. However, these techniques are not applicable to commercial production in practice.

Since Lu et al. (1986) showed that floral buds could be induced from perianth explants of *Hyacinthus*, the conditions for regeneration of floral organs have been investigated (Lu et al. 1988, 1990, 1991). For regeneration of each floral organ, both the developmental stage of perianth explants used and the plant growth regulator combination were important factors. Further, the regimes of temperature were an important factor for regeneration of stamens and for their maturity.

By using the in vitro flowering system of hyacinth, the production of secondary metabolites, anthocyanins and essential oils was investigated. The types and levels of each constituent of the secondary metabolites from the regenerated flowers were compared with those from field-grown flowers.

The constituents of essential oils produced in the regenerated flowers of cv. Delft Blue were mainly aromatic alcohols and their esters. Although not all constituents detected in them were the same as those of field-grown flowers, the main constituent, phenethyl alcohol, was the same as that of field-grown flowers in quantity.

Anthocyanins in the regenerated flowers of cv. Delft Blue were the same as those of field-grown flowers in constituents and in their ratios. In the case of cv. Hollyhock, the constituents of anthocyanins produced in both types of flowers were also the same. However, the ratio of their anthocyanins differed significantly between the regenerated and the field-grown flowers.

In vitro flowering systems are useful for the production of secondary metabolites which are exclusively produced in flowers, although the secondary metabolites are not always the same between regenerated and field-grown flowers.

References

Amaki W, Shinohara Y, Hayata Y, Sano H, Suzuki Y (1984) Effects of bulb desiccation and storage on the in vitro propagation of hyacinth. Sci Hortic 23:353–360

Anonis DP (1985) Hyacinth in perfumery. Perfumer Flavorist 10:17–20

Asjes CJ, Bunt MH, Van Slogteren DHM (1974) Production of hyacinth mosaic virus-free hyacinths and lily symptomless virus-free lilies by meristem tip culture. Acta Hortic 36:223–228

Bach A, Cecot A (1988) Micropropagation of *Hyacinthus orientalis* cv. Carnegie in a double-phase culture medium. I. Effect of cytokinins on growth and development. Acta Hortic 226:607–610

Bae SK, Lee BK, Eun JS (1983) Clonal propagation of hyacinths by in vitro culture. III. Bulblet culture, callus subculture, and organ formation from excised tissues. J Korean Soc Hortic Sci 24:249–257

Banthorpe DV, Branch SA, Njar VCO, Osborne MG, Watson DG (1986a) Ability of plant callus cultures to synthesize and accumulate lower terpenoids. Phytochemistry 25:629–636

Banthorpe DV, Grey TJ, Poots I, Fordham WD (1986b) Monoterpene metabolism in cultures of *Rosa* species. Phytochemistry 25:2321–2326

Blom-Barnhoorn GJ, Van Aartrijk J, Van der Linde PCG (1986) Effect of virazole on the production of hyacinth plants free from hyacinth mosaic virus (HMV) by meristem culture. Acta Hortic 177:571–574

Chung JD, Chun CK, Suh YK, Rhee EM (1981) In vitro propagation of *Hyacinthus orientalis* L. I. Effect of auxins on callus formation and bulblet differentiation from flower organ and flower stem tissue. J Korean Soc Hortic Sci 22:146–151

Chung JD, Chun CK, Suh YK, Rhee EM (1983a) In vitro propagation of *Hyacinthus orientalis* L. III. Effect of auxin concentrations and light or dark incubation on organogenesis from flower stalk tissue and comparison of totipotency between bulb scale and flower stalk tissue. J Korean Soc Hortic Sci 24:76–85

Chung JD, Chun CK, Suh YK, Rhee EM (1983b) In vitro propagation of *Hyacinthus orientalis* L. IV. Organ redifferentiation through the subculture of bulblet scale and leaf tissue segments originated from flower stalk tissue culture. J Korean Soc Hortic Sci 24:162–168

Chung JD, Chun CK, Phee EM (1984a) In vitro propagation of *Hyacinthus orientalis* L. VI. Effects of sucrose concentration, pH value, bulblet age and chilling treatment of bulblet on organogenesis from the bulblet scale and leaf tissue. J Korean Soc Hortic Sci 25:72–79

Chung JD, Chun CK, Soo O (1984b) Variation of totipotency from individual organ parts of *Hyacinthus orientalis* cultured in vitro at different growth stages. J Korean Soc Hortic Sci 25:297–304

Collin HA (1988) Flavors. In: Constabel F, Vasil IK (eds) Cell culture and somatic cell genetics of plants, vol 5. Academic Press, New York, pp 569–585

Darlington CD, Hair JB, Hurcombe R (1951) The history of the garden hyacinths. Heredity 5:233–252

Harborne JB (1964) Plant polyphenols. XI. The structure of acylated anthocyanins. Phytochemistry 3:151–160

Hartmann HT, Kester DE, Davies FT Jr (1990) Plant propagation principles and practices, 5th edn. Prentice-Hall, Englewood, NJ

Hosokawa K, Fukunaga (1995) Production of essential oils by flowers of *Hyacinthus orientalis* L. regenerated in vitro. Plant Cell Reports 14:575–579

Hosokawa K, Fukunaga Y, Fukushi E, Kawabata J (1995a) Seven acylated anthocyains in the blue flowers of *Hyacinthus orientalis*. Phytochemistry 38:1293–1298

Hosokawa K, Fukunaga Y, Fukushi E, Kawabata J (1995b) Acylated anthocyanins from red *Hyacinthus orientalis*. Phytochemistry 39:1437–1441

Hosokawa K, Fukunaga Y, Fukushi E, Kawabata J (1995c) Five acylated pelargonidin glucosides in the red flowers of *Hyacinthus orientalis*. Phytochemistry 40:567–571

Hosokawa K, Fukunaga Y, Fukushi E, Kawabata J (1996a) Production of acylated anthocyanins by blue flowers of *Hyacinthus orientalis* regenerated in vitro. Phytochemistry 41:1531–1533

Hosokawa K, Fukunaga Y, Fukushi E, Kawabata J (1996b) Acylated anthocynanins in red flowers of *Hyacinthus orientalis* regenerated in vitro. Phytochemistry 42:671–672

Hussey G (1975) Propagation of hyacinths by tissue culture. Sci Hortic 3:21–28

Kaiser R, Lamparsky D (1977) Nouveaux constituants de l'absolue de jacinthe et leur comportemment olfactif. Perfumes Cosm Aromes 17:71–79

Kim YJ, Hasegawa PM, Bressan RA (1981) In vitro propagation of hyacinth. Hortscience 16:645–647

Klein AO, Hagen CW (1961) Anthocyanin production in detached petals of *Impatiens balsamina* L. Plant Physiol 36:1–9

Lawrence BM (1978) Progress in essential oils. Perfumer Flavorist 3:45–50

Lee BK, Bae SK, Eun JS (1982a) Studies on the propagation of hyacinth in vitro culture. I. Effects of auxin and cytokinin on the bulblet formation in bulb scale, leaf and flower stock culture. Korean J Plant Tissue Cult 9:13–20

Lee BK, Bae SK, Eun JS (1982b) Studies on the propagation of hyacinth in vitro culture. II. Regenerative capacity of bulblets from tissues, cultural conditions and parts of bulb scales. Korean J Plant Tissue Cult 9:21–26

Lu W, Guo Z, Wang X, Cui C (1986) Studies on directly redifferentiating floral buds by explants of *Hyacinthus orientalis* L. – control of morphogenesis of floral and vegetative buds. Sci Sin [B] 29:935–946

Lu W, Enomoto K, Fukunaga Y, Kuo C (1988) Regeneration of tepals, stamens and ovules in explants from perianth of *Hyacinthus orientalis* L. Importance of explant age and exogenous hormones. Planta 175:478–484

Lu W, Zhu Y, Enomoto K, Fukunaga Y (1990) Effects of temperature on induction of regenerative stamens in in vitro culture, microsporogenesis and pollen development in *Hyacinthus orientalis* L. Acta Bot Sin 32:832–840

Lu W, Zhu Y, Enomoto K, Fukunaga Y (1991) Cytological observations on microsporogenesis and pollen development of regenerated stamen in hyacinth. Acta Bot Sin 33:343–349

Mancinelli AL, Rossi F, Moroni A (1991) Cryptochrome, phytochrome, and anthocyanin production. Plant Physiol 96:1079–1085

Matsumoto M (1983) Propagation of virus-free hyacinths by tissue culture. Bull Toyama Veg Ornamental Crops Res Stn 1:9–22

Mulder-Krieger T, Verpoorte R, Svendsen AB, Scheffe JJC (1988) Production of essential oils and flavours in plant cell and tissue cultures. A review. Plant Cell Tissue Organ Cult 13:85–154

Murashige T, Skoog F (1962) A revised medium for rapid growth and bio-assays with tobacco tissue cultures. Physiol Plant 15:473–497

Novak J, Rudnicki RM (1993) Hyacinthus. In: De Hertogh A, Le Nard M (eds) The physiology of flower bulbs. Elsevier, Amsterdam, pp 335–347

Paek KY, Choi SL (1982) Studies on the clonal propagation by the culture of bulb scale segments, inflorescence stem and flower-bud in hyacinth. Korean J Plant Tissue Cult 9:47–55

Paek KY, Thorpe TA (1989) Hyacinth. In: Ammirato PV, Evans DA, Sharp WR, Bajaj YPS (eds) Handbook of plant cell culture, vol 5. Ornamental species. McGraw-Hill, New York, pp 479–508

Paek KY, Chun CK, Chung JD (1983) Factors influencing regeneration of bulblets and effect of chilling treatment of parent bulblets on scale segments of hyacinth in vitro. J Korean Soc Hortic Sci 24:68–75

Paek KY, Lee CW, Choi JK, Hong YP (1987) Effect of explant source, breaking of apical dominance and dormancy on shoot proliferation and bulbing of hyacinth, narcissus and lily in vitro. J Korean Soc Hortic Sci 28:88–98

Pierik RLM, Post AJM (1975) Papid vegetative propagation of *Hyacinthus orientalis* L. in vitro. Sci Hortic 3:293–297

Pierik RLM, Ruibing MA (1973) Regeneration of bulblets on bulb scale segments of hyacinth in vitro. Neth J Agric Sci 21:129–138

Pierik RLM, Steegmans HHM (1975) Effect of auxins, cytokinins, gibberellins, abscisic acid and ethephon on regeneration and growth of bulblets on excised bulb scale segments of hyacinth. Physiol Plant 34:14–17

Pierik RLM, Woets J (1971) Regeneration of isolated bulb scale segments of hyacinth. Acta Hortic 23:423–428

Saniewski M (1974) Root and bulblet regeneration by flower buds of *Hyacinthus orientalis* L. in tissue culture. Bull Acad Pol Sci Ser Sci Biol 23:121–123

Saniewski M (1975) Studies on the physiology of hyacinth bulbs. VII. Root and bulblet-like regenerations from the ovary wall of *Hyacinthus orientalis* L. Experientia 31:302–303

Saniewski M, Novak J, Rudnicki R (1974) Studies on the physiology of hyacinth bulbs (*Hyacinthus orientalis* L.). IV. Hormonal regeneration of induction of roots and bulblets in *Hyacinthus orientalis* L. grown in culture. Plant Sci Lett 2:373–376

Shimizu M (ed) (1996) *Hyacinthus* and *Crocus*. Shukan Hanahyakka Fleur, vol 47. Kodansha, Tokyo, p 22 (in Japanese)

Stebbings G (1996) Heaven scent. Garden 121:68–72

Takayama S (1991) Mass propagation of plants through shake- and bioreactor-culture techniques. In: Bajaj YPS (ed) Biotechnology in agriculture and forestry, vol 17. High-tech and micropropagation I. Springer, Berlin Heiderberg, New York, pp 495–515

Takayama S, Amo T, Fukano M (1991) Rapid clonal propagation of *Hyacinthus orientalis* bulbs by shake culture. Sci Hortic 45:315–321

Tamura S (1978) Adventitious bud formation from excised hyacinth bulb scale in vitro. J Jpn Soc Hortic Sci 46:501–508

Tomoda G, Matsuyama J, Likubo H (1976) Tissue culture of fragrant plants. Bull Fac Agric Tamagawa Univ 16:16–22

Webb JK, Banthorpe DV, Waton DG (1984) Monoterpene synthesis in shoots regenerated from callus cultures. Phytochemistry 23:903–904

Weiss D, Halevy AH (1989) Stamens and gibberellin in the regulation of corolla pigmentation and growth in *Petunia hybrida*. Planta 179:89–96

Zieslin N, Biran I, Halevy AH (1974) The effect of growth regulators on the growth and pigmentation of "Baccara" rose flowers. Plant Cell Physiol 15:341–349

XI *Hypericum patulum*: In Vitro Culture and Production of Xanthones and Other Secondary Metabolites

K. Ishiguro, H. Oku, and K. Isoi

1 Introduction

Plants of the genus *Hypericum* (Guttiferae) have been used as traditional medicines in various parts of the world, i.e. South Africa (Madeira, Rocha et al. 1995a), Russia (Taylor et al. 1996a). Europe (Dingermann 1995; Ernst 1995; Cott 1995; Anonymous 1996), Nepal (Taylor et al. 1996b), China (Wu et al. 1996), and Japan (Minami et al. 1996). Recently, there have been reports of the following actions in this species: antifungal (Rocha et al. 1994), antibiotic (Rocha et al. 1995b), antiviral (Diwu 1995), antidepressive (Hansgen et al. 1994; Payk 1994; Ernst 1995; Kerb et al. 1996), antiinflammatory (Recio et al. 1995), anticancer (Diwu 1995) and monoamine oxidase inhibitor (Bladt and Wagner 1994; Thiede and Walper 1994). What has been noted is the significant variation in the secondary metabolites (hypericine, flavonoids, procyanidins and tannins) in *Hypericum* species depending on the plant origin.

H. patulum occurs naturally in southwest and central China and Taiwan. The showy yellow flowers, arranged in terminal cymes, blossom in early summer. The flowers (Fig. 1) have been used in Chinese herbal medicine for the treatment of hepatitis, bacterial diseases and nasal hemorrhage (Chang Su New Medical College 1977). The five constituents, betulinic acid (triterpenoid), toxyloxanthone B (a pyranoxanthone), quercetin, quercitrin and astiblin (flavonoids) from the aerial parts of intact plants have also been reported (Chen et al. 1993). The importance of this plant is due to the xanthones, which are expected to possess pharmacological properties such as monoamine oxidase inhibition, cytotoxicity and antitumor activity (Peres and Nagem 1997).

School of Pharmaceutical Sciences, Mukogawa Women's University, Koshienkyuban-cho, Nishinomiya, 663-8179, Japan

Biotechnology in Agriculture and Forestry, Vol. 43
Medicinal and Aromatic Plants XI (ed. by Y.P.S. Bajaj)

Fig. 1. Intact plant of *H. patulum*

2 In Vitro Approaches

2.1 Tissue Culture Work on *Hypericum* Species

Significant work has been done on tissue culture of *Hypericum erectum* (Yazaki and Okuda 1994), *H. perforatum* (Cellarova et al. 1995), *H. canariensis* (Mederos et al. 1996) and other species of *Hypericum* (Kartnig et al. 1996). A recent review on *H. perforatum* L., which occurs naturally in Europe and has been used in herbal medicines against sores and burns or as an antibiotic, has been published by Cellarova et al. (1995). An effective regeneration pattern has been reported from seedlings of *H. perforatum* and studies on the variability of several morphological characters and histological structures in regenerants have been carried out. Hypericin, pseudohypericin and their derivatives, procyanidins and flavonoids were isolated from regenerated plants (Cellarova et al. 1992). The hypericin is particularly prevalent in intact plants of *H. perforatum* and its content in somaclones derived from a particular genotype varies considerably and is significantly higher in the second year of cultivation (Cellarova et al. 1995). Brutovska et al. (1994) also studied the effects of the non-ionic surfactant Pluronic F-68 on the growth of shoots regenerated from seedlings of *H. perforatum*.

Work on *H. erectum* T., the species most widely distributed in Japan, has been published by Yazaki and Okuda (1990, 1994). They established multiple

Table 1. Summary of various studies conducted on in vitro culture and secondary metabolites in *Hypericum patulum*

Reference	In vitro system used	Observations
1. Ishiguro et al. (1993)	Suspension cultures	γ-Mangostin (**2**), toxyloxanthone B (**4**) and paxanthone (**5**)[a] from EtOAc fraction of MeOH extract
2. Ishiguro et al. (1995a)	Suspension cultures	1,3,6,7-Tetrahydroxy-8-(3-methyl-2-butenyl)-9H-xanthene-9-one (**1**), paxanthone B (**6**)[a] and garcinone B (**7**) from EtOAc fraction of MeOH extract
3. Ishiguro et al. (1995b)	Suspension cultures	Morusignin D (**9**) and paxanthonin (**11**)[a] from $CHCl_3$ extract
4. Ishiguro et al. (1996)	Suspension cultures	Tripteroside (**3**), demethylpaxanthonin (**10**)[a] and padiaxanthone (**12**) from EtOAc fraction of MeOH extract
5. Ishiguro et al. (1997)	Suspension cultures	Patulone (**13**)[a] from $CHCl_3$ extract, 1,3,5,6-tetrahydroxyxanthone (**8**) and (−)-epicatechin (**15**) from EtOAc fraction of MeOH extract
6. Ishiguro et al. (1998)	Suspension cultures	Paglucinol (**14**)[a], oleanolic acid and β-sitsterol from $CHCl_3$ extract
7. Takamiya et al. (1998)	Suspension culture, root culture	Formation of anthocyanins and procyanidius

[a] New xanthone.

shoot cultures and discussed culture conditions for the production of procyanidins. Polyphenolic compounds, (−)-epicatecin and its condensates, procyanidin B2, procyanidin C1, cinnamtannin A2, hyperin and quercitrin were also isolated from these cultures. The hypericin was detected in regenerated plants, but not in the tissue cultures. Kartnig and Brantner (1990; Kartnig et al. 1996) reported on the production of hypericin, pseudohypericin and flavonoids in cell cultures of various *Hypericum* species and their chemotypes. Tissue and cell suspension cultures from *H. patulum* have been established by our group and several xanthones have been isolated from the cultures (Ishiguro et al. 1993, 1995a,b, 1996, 1997, 1998; see Table 1).

2.2 Establishment of Tissue Cultures and Suspension Cultures

H. patulum Thunb. was grown in Mukogawa Women's University medicinal plant garden and identified by Dr. G. Yoneda (Faculty of Pharmaceutical Sciences, Osaka University). The stamen, corolla, ovary, pistil, calyx and leaves were surface-sterilized by treatment with 70% ethanol for 30 s, further sterilized with 1% sodium hypochlorite solution containing a few drops of Tween 20 for 2 min, and then washed with sterile distilled water. These parts were cultured for callus induction on LS solid medium (Linsmaier and Skoog

1965) containing 3% sucrose, 1.4% agar and supplemented with various combinations of growth regulators: 10^{-6} or 10^{-5} M 2,4-dichlorophenoxyacetic acid (2,4-D) alone, 10^{-6} or 10^{-5} M 2,4-D combined with 10^{-8}, 10^{-7} or 10^{-6} M kinetin or α-naphthaleneacetic acid (NAA). All the media were adjusted to pH 5.5 with 0.5 M NaOH before autoclaving for 20 min at 121 °C. The cultures were maintained in the dark at 27 °C.

Callus formation was obtained in media containing 10^{-5} M 2,4-D and up to 10^{-7} M NAA or kinetin. The addition of NAA was suitable for callus induction but not for callus increase, which could be induced better by kinetin. The stamen was the most suitable plant part for callus formation, followed by the corolla (Fig. 2). Scarcely any callus formation occurred with the leaf under these conditions.

The methanol extract of induced callus was tested for antimicrobial activities against *Staphylococcus aureus* and *Escherichia coli* by the agar-well method (Hartly and Adamson 1973, see Sect. 3). About 80% of the induced callus showed significant antimicrobial activities against *S. aureus*. The methanol extracts (400 mg/ml) of the induced callus showed half the activity of the corresponding part of the intact plant. The callus of Hp-126 (Fig. 3), which was induced from the corolla on medium containing 2,4-D (10^{-5} M) and NAA (10^{-7} M), showed a good growth rate and the most potent activity against *S. aureus.* Callus samples were transferred to fresh regeneration media at 3-month intervals.

The callus tissue of Hp-126 was transferred into liquid LS medium containing 3% glucose, 10^{-5} M 2,4-D and 10^{-7} M kinetin (pH 5.5) to initiate suspension cultures, and subcultured about every 2 weeks for 1 year (Fig. 4). The suspension cultures were maintained at 27 °C in the dark on a rotary shaker (100 rpm). These cultures also showed significant activities against *S. aureus.*

Figure 5 shows a correlation between the time course of cell growth and xanthone production in the suspension-cultured cells. Cell growth was estimated at different glucose concentrations from the optical rotation value for the medium. The amounts of xanthones in the cells were determined by HPLC analysis (see Sect. 2.3.3).

2.3 Extraction and Structure of Xanthones

2.3.1 Extraction and Fractionation

The cells were collected by filtering the suspension cultures at stationary phase. The cells were washed with water, dried under a hot air flow at 60 °C, and then extracted with methanol. The methanol extracts were dissolved in a minimal amount of water and partitioned between ethyl acetate and water. Compounds **1–8**, **10**, **12** and **15** (Fig. 6) were isolated from the EtOAc fraction by repeated flash column chromatography on silica gel using a $CHCl_3$-MeOH gradient system followed by chromatography with LH-20 or recrystallization. (Ishiguro et al. 1993, 1995a,b, 1996, 1997). A further investigation of the

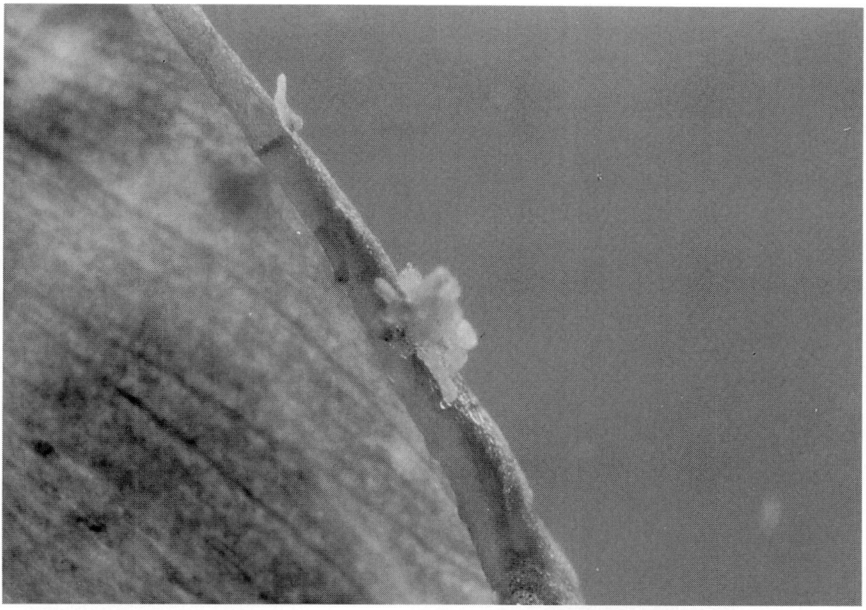

2

3

Fig. 2. Callus formation from corolla
Fig. 3. Callus tissue of Hp-126 from corolla

Fig. 4. The suspension cultures of Hp-126

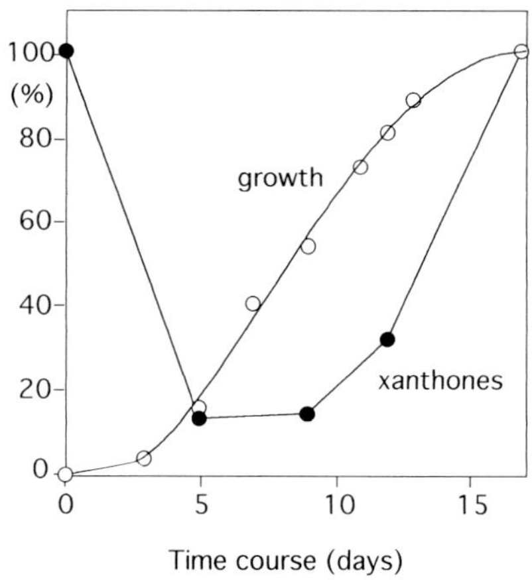

Fig. 5. Time course of cell growth and xanthone production in suspension cultured cells of *H. patulum*

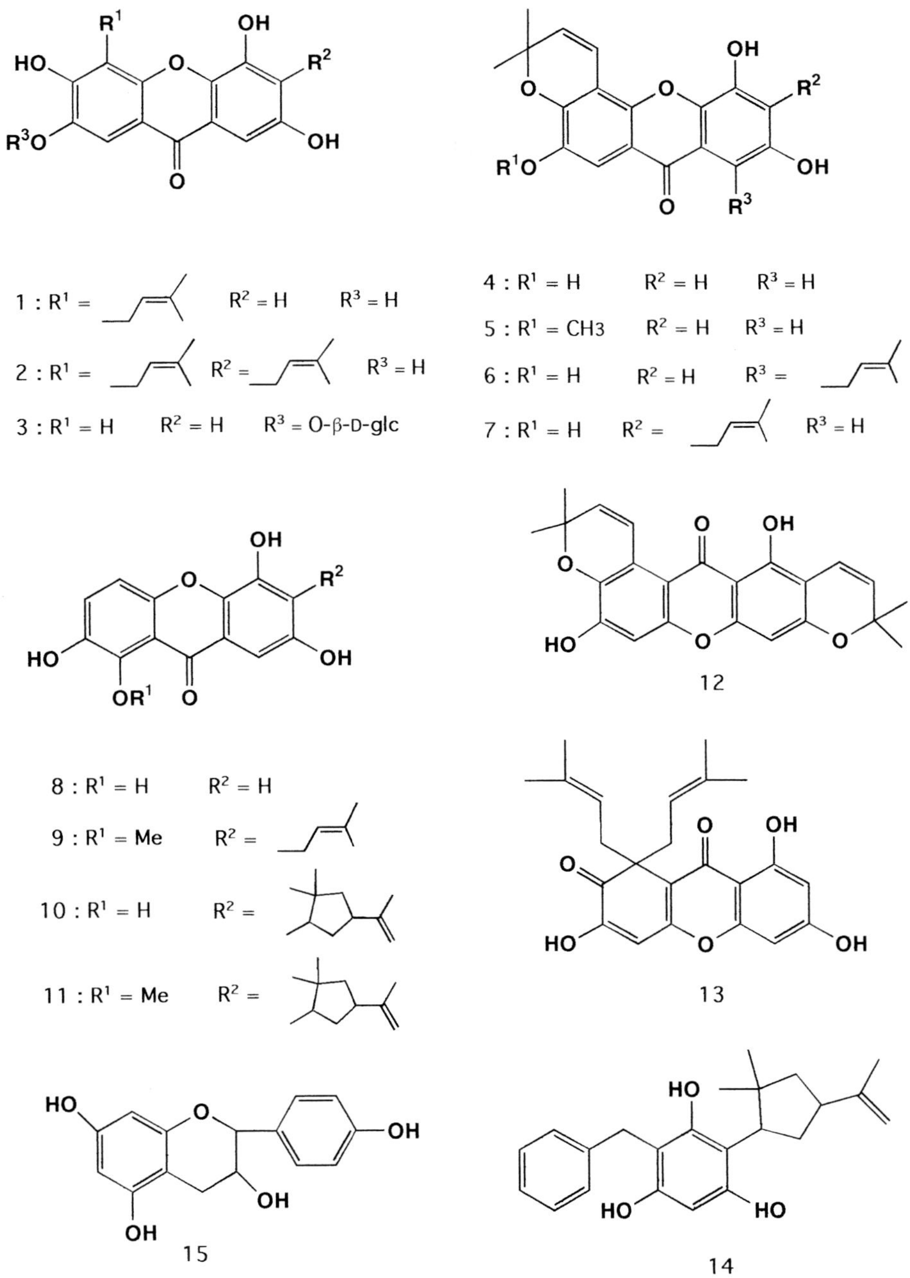

Fig. 6. Structures of xanthone derivatives from cell suspension cultures of *H. patulum*

chloroform extracts revealed the presence of compounds **9**, **11**, **13** and **14** (Ishiguro et al. 1995b, 1997, 1998).

2.3.2 Structure of Compounds ***1–15***

Spectroscopic methods were used to identify the structures of six new xanthones: **5**, paxanthone; **6**, paxanthone B; **10**, demethylpaxanthonin; **11**, paxanthonin; **13**, patulone; and **14**, paglucinol. Other compounds were identified by comparison with authentic specimens or spectroscopic data: 1,3,6,7-tetrahydroxy-8-(3-methly-2-butenyl)-9H-xanthen-9-one (**1**) from *Hypericum androsaemum* Linn. (Guttiferae), γ-mangostin (**2**) (Jefferson et al. 1970) isolated from *Garcinia mangostana* Linn. (Guttiferae), tripteroside (**3**) (Lin et al. 1982), toxyloxanthone B (**4**) from *Toxylon pomiferum* Rafin. (Moraceae, Kirtany and Paknikar 1975), garcinone B (**7**) from *Garcinia mangostana* Linn. (Guttiferae, Sen et al. 1980), 1,3,5,6-tetrahydroxanthone (**8**) from *Tripterospermum taiwanense* (Guttiferae, Sia et al. 1995), morusignin D (**9**) from *Morus insignis* Bur. (Moraceae, Hano et al. 1990), 5, 13-dihydroxy-3,3,10,10-tetramethyl-3H,10H,14H-dipyrano[3,2-a:2′,3′-i] xanthen-14-one (**12**) synthesized by the oxidation cyclization of garcinone B (**7**) (Sen et al. 1982) and (−)-epicatechin (**15**). Betulinic acid, toxyloxanthone B (**4**), quecetin, quercitrin and astilbin have been isolated from intact plants of *H. patulum* (Chen et al. 1993).

2.3.3 HPLC Analysis

The cells were harvested 19 days after addition of yeast elicitor (YE). Dried callus tissues were extracted with chloroform, and the extract was evaporated to dryness under reduced pressure and then dissolved in methanol. The methanol solution was filtered and the filtrate was used directly for HPLC analysis as injections of 5 μl onto a reverse-phase column (μ-Bondasphere RP-18, 3.9 mm i.d. × 150 mm, 5-μm particle size, Waters). Chromatography was performed using methanol –5% aqueous acetic acid as a gradient from 68% to 75% methanol in 10 min. A flow rate of 0.9 ml/min was used. Spectrophotometric detection was at 262 nm. Under these conditions, base line resolution was achieved for all metabolites of interest.

The HPLC protocol gave high resolution chromatograms to allow simultaneous analysis of the six main xanthones (**1**, **2**, **4**, **5**, **6** and **7**) in unfractionated extracts. Mono-isoprenyl xanthones, 1,3,6,7-tetrahydroxy-8-prenyl xanthone (**1**), toxyloxanthone B (**4**) and paxanthone (**5**), have retention times between 9.6 and 15.9 min. Di-isoprenyl derivatives, γ-mangostin (**2**), paxanthone B (**6**) and garcinone B (**7**), have retention times between 29 and 37.6 min. Of the two groups of xanthones, the dimethyl chromene derivatives, **4** and **7**, have longer retention times than their corresponding prenylated xanthones. This method is simple, sensitive and highly reproducible. The examination of such profiles over time should provide a dynamic view of the many concurrent molecular events related to biological phenomena.

2.3.4 Metabolic Changes by Treatment with Yeast Elicitor (YE)

We now have the experimental tools to determine the levels of xanthones and their metabolic shifts induced by YE (0.1 g/300 ml medium) for 19 days at growth phase. Measurements of the ratio of the integration units for the peak area at 262 nm revealed that the ratio of compounds **1** and **4** (4:1) changed to 0.8:1 and the ratio of **2** and **7** (8.5:1) changed to 2.5:1. This indicates that treatment of the cultures with YE resulted in reduced production of **1** and **2**, with a concomitant increase in **4** and **7**.

These results suggest that the cyclization of the isoprenyl side chain or the activation of cyclase is stimulated in response to environmental stress.

The elicitor-mediated induction or activation of the cyclase which catalyzes the transformation of prenyl to chromen, has been described for soybean tissue cultures (Welle and Grisebach 1988; Kochs and Grisebach 1989). By analogy to glyceollin formation from its prenylated precursor, it seems likely that the conversion of the prenylated xanthones, **1** and **2** to **4** and **7**, respectively, involves cytochrome P-450 dependent monooxygenase.

3 Antibacterial Activity of Xanthones

Antimicrobial activity was tested against both *Staphylococcus aureus* 209P IF0-3061 and *Escherichia coli* NIHJ. The in vitro antimicrobial activity was determined by the agar-well method. Plates were prepared with 4-mm-deep sensitivity test agar (STAb: Eiken) medium which was previously inoculated with 1% (v/v) of a test organism grown overnight at 37 °C in heart infusion broth (HIB: Nissui). Holes (8 mm diameter) were bored in the medium with a sterile borer, and filled with 40 μl solution (1000 μg/ml) of test samples. The plates were sealed and then incubated for 18 h at 37 °C. The activity was expressed by the diameter of the inhibition: diameter 9–14 mm (+), 15–20 mm (++; Table 2). Compounds **2** and **13** which possess two prenyl groups exhibited (++), compounds **1**, **6**, **7** and **9** which possess a prenyl group and compounds **10**, **11** and **14** which possess a cyclopenetane ring exhibited (+) whereas the inactive compounds were **4**, **5**, **8**, **12** and **15** which have no prenyl

Table 2. The antimicrobial activities of xanthone derivatives from cell suspension cultures of *H. patulum* against *Staphylococcus aureus*

Test compounds	Activity	Test compounds	Activity	Test compounds	Activity
(**1**)	+	Paxanthone B (**6**)	+	Paxanthonin (**11**)	+
γ-Mangostin (**2**)	++	Garcinone B (**7**)	+	(**12**)	−
Tripteroside (**3**)	++	(**8**)	−	Patulone (**13**)	++
Toxyloxanthone B (**4**)	−	Morusignin D (**9**)	+	Paglucinol (**14**)	+
Paxanthone (**5**)	−	Demethylpaxanthonin (**10**)	+	Epicatechin (**15**)	−

+ Diameter of inhibition 9–14 mm; ++ diameter of inhibition 15–20 mm.

or cyclopentane moiety. Thus the prenyl group or cyclopentane ring is an important factor for the antimicrobial activity of xanthone derivatives.

4 Discussion and Conclusions

Co-occurence of 1,3,5,6- and 1,3,6,7-tetraoxygenated xanthones in Guttiferae (Sultanbawa 1980) and Moraceae (Nomura and Hano 1994) has been reported, and their co-existence in cell suspension cultures of *H. patulum* has been shown for the first time by Ishiguro et al. (1995b). On the basis of the isolation of maclurin, 2,4,6,3′,4′-pentahydroxy benzophenone, together with 1,3,5,6- and 1,3,6,7-tetrahydroxyxanthone from the heart-wood of *Symphonia globulifera* (Guttiferae), the biogenetic role of maclurin as a precursor for xanthones has been discussed (Locksley et al. 1967). The first tracer studies on the biosynthesis of xanthones in the Guttiferae were reported by Bennett and Lee in 1988. Cinnamic acid, benzoic acid, *m*-hydroxybenzoic acid and 4′-deoxymaclurin as well as malonic acid were efficient precursors to mangostin in *Gardenia mangostana*, suggesting that the 1,3,5,6- and 1,3,6,7-tetraoxygenated xanthones may be formed by the 6-oxygenation of either the 1,3,5- or the 1,3,7-trioxygenated xanthone derived from 4′-deoxy maclurin (Bennett and Lee 1989). The xanthones from *H. patulum* cultures (Ishiguro et al. 1993, 1995a,b) could be biosynthesized from 4′-deoxymaclurin and/or maclurin, although it remains to be seen whether prenylation or cyclopentane ring addition occurs at the benzophenone or xanthone stage. Ollis et al. (1965) suggested that prenylation in xanthones usually occurs at the xanthone stage but not at the benzophenone stage. However, the isolation of compound **14** suggests that the irregular 5-membered ring monoterpene moiety (cyclopentane ring) of compounds **10** and **11** is introduced at the benzophenone stage. Thus, prenylation could occur at the benzophenone

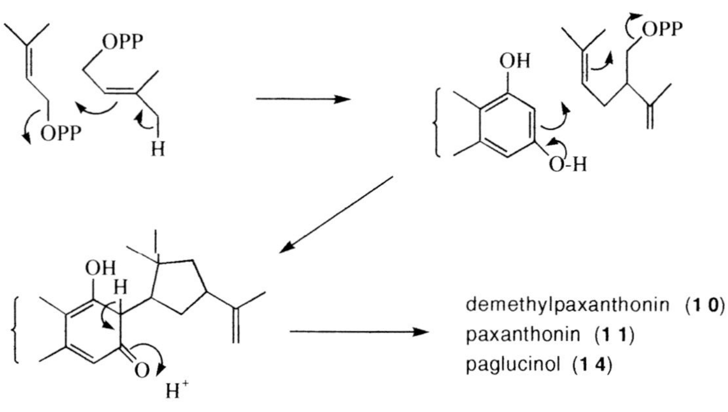

Fig. 7. Probable biogenetic route for cyclopentane-ring formation

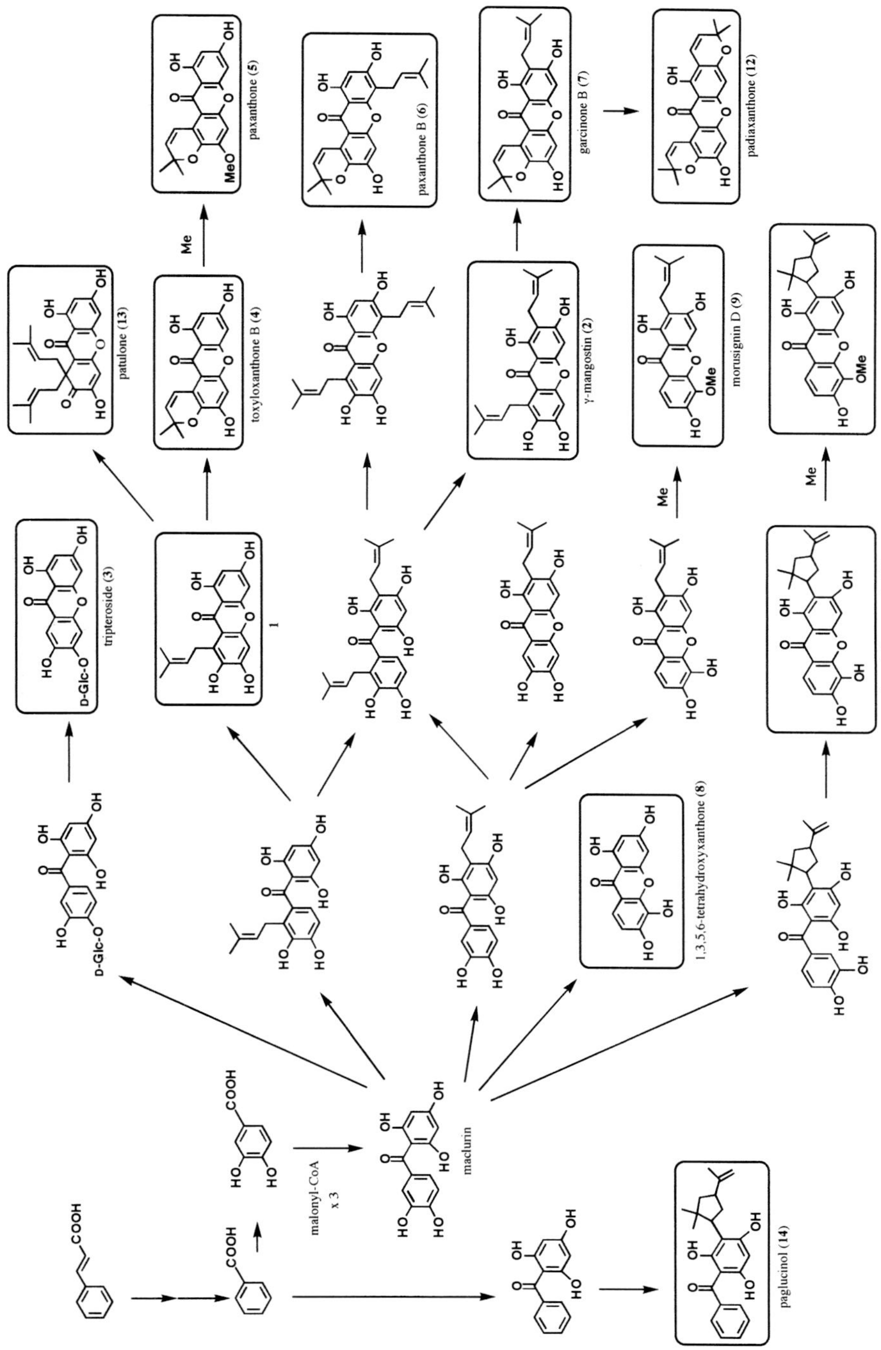

Fig. 8. Proposed biosynthetic pathway. Isolated compounds in *box*; others: possible intermediate compounds

stage. We postulate a possible biosynthetic route containing lavandulol-PP for the cyclopentane-ring formation as shown in Fig. 7.

Bennett et al. (1989) have reported that 2-prenylation is most common but is only found in the presence of 1,3-dioxygenation, suggesting that the prenyl group inhibits reduction of the 3-hydroxyl group, which is very often absent in simple xanthones. The irregular 5-membered ring monoterpene moiety of compounds **10** and **11** in *H. patulum* is also present at the 2-position in the presence of 1,3-dioxygenation, suggesting that the monoterpene group as well as the prenyl group inhibits reduction of the 3-hydroxyl group. A plausible biosynthetic pathway of the xanthones in *H. putulum* is summarized in Fig. 8.

Futher investigation of xanthone-related enzymes should yield better understanding of the field. Recently, Beerhues (1996) isolated the enzyme catalyzing the formation of the C_{13} skeleton, named benzophenone synthase, from cultured cells of *Centaurium erythraea.*

There are many reviews of biological studies on xanthones (Bennett et al. 1989; Hostettmann and Hostettmann 1989; Peres and Nagem 1997). Biological activities such as the antiplatelet effect of tripteroside (**3**) (Teng et al. 1989) and 1,3,5,6-tetrahydroxyxanthone (**8**) (Lin et al. 1992), and the antiinflammatory (Gopalakrishnan et al. 1980) and antifungal (Gopalakrishnan et al. 1997) effects of γ-mangostin (**2**) have been reported. Suspension cultures should reveal a variety of biological effects of xanthones, especially when studied using the newly developed antianphylactic (Ishiguro and Fukumoto 1997) or antipruritic (Ishiguro and Oku 1997) assays.

References

Anonymous (1996) *Hypericum perforatum* as source of drugs used in treating depression. Herba Pol 42:192–198

Beerhues L (1996) Benzophenone synthase from cultured cells of *Centaurium erthraea.* FEBS Lett 383:264–266

Bennett GJ, Lee HH (1988) The biosynthesis of mangostin: the origin of the xanthone skeleton. J Chem Soc Chem Commun: 619–620

Bennett GJ, Lee HH (1989) Xanthones from Guttiferae. Phytochemistry 28:967

Bladt S, Wagner H (1994) Inhibition of MAO by fractions and constituents of *Hypericum* extract (Review). J Geriatr Psychiatry Neurol 7 Suppl 1:S57–S59

Brutovska R, Cellarova E, Davey MR, Power MR (1994) Stimulation of multiple shoot regenerated from seedlings leaves of *Hypericum perforatum* by Pluronic F-68. Acta Biotechnol 14:347–353

Cellarova E, Kimakova K, Brutovska R (1992) Multiple shoot formation and phenotypic changes of Ro regenerants in *Hypericum perforatum* L. Acta Biotechnol 12:445–452

Cellarova E, Kimakova K, Daxnerova Z, Martonfi P (1995) *Hypericum perforatum* (St. John's wort): in vitro culture and the production of hypericin and other secondary metabolites. In: Bajaj YPS (ed) Biotechnology in agriculture and forestry vol 33. Medicinal and aromatic plants VIII. Springer, Berlin Heidelberg New York, pp 261–275

Chang Su New Medical College (1977) Dictionary of Chinese crude drugs. Shanghai Scientfic Technological Publishers, Shanghai

Chen ZT, Chen MT, Kuoh CS, Chen CM (1993) Constituents of *Hypericum patulum.* Zhoonghua Yaoxue Zazhi 45:337–342

Cott J (1995) Natural product formulations available in Europe for psychotropic indications. Psychopharmacol Bull 31(4):745–751

Dingermann T (1995) Phytopharmaceuticals in old age. Crataegus, ginkgo, hypericum, and kava-kava preparations. Pharm Z 140(23):9–14, 16

Diwu Z (1995) Novel therapeutic and diagnostic applications of hypocrellins and hypericins (Review). Photochem Photobiol 61(6):529–539

Ernst E (1995) St. John's wort as antidepressive therapy (Review). Fortschr Med 113:354–355

Gopalakrishnan C, Shankaranarayanan D, Kameswaran L, Nazimudeen SK (1980) Effect of mangostin, a xanthone from *Gardenia mangostana* Linn. in immunopathological and inflammatory ractions. Indian J Exp Biol 18(8):843–846

Gopalakrishnan G, Banumathi B, Suresh G (1997) Evaluation of the antifungal activity of natural xanthones from *Garcinia mangostana* and their synthetic derivatives. J Nat Prod 60:519–524

Hano Y, Okamoto T, Nomura T, Momose Y (1990) Cudraflavones C and D, two new prenylflavones from the root bark of *Cudrania tricuspidata.* Heterocycles 31(7):1339–1344

Hansgen KD, Vesper J, Ploch M (1994) Multicenter double-blind study examining the antidepressant effectiveness of the *Hypericum* extract LI 160. J Geriatr Psychiatry Neurol 7 Suppl 1:15–18

Hartly E, Adamson DMC (ed) (1973) British Pharmacopoeia. Her Majesty's Stationery Office, London

Hostettmann K, Hostettmann M (1989) Xanthones. Methods Plant Biochem 1:493–508

Ishiguro K, Fukumoto H (1997) A practical and speedy screening method for murine anaphylaxis: on the antianaphylactic effect of *Impatiens balsamina L.* Phytother Res 11:48–50

Ishiguro K, Oku H (1997) Antipruritic effect of flavonol and 1,4-naphthoquinone derivatives from *Impatiens balsamina L.* Phytother Res 11:343–347

Ishiguro K, Fukumoto H, Nakajima M, Isoi K (1993) Xanthones in cell suspension cultures of *Hypericum patulum.* Phytochemistry 33(4):839–840

Ishiguro K, Nakajima M, Fukumoto H, Isoi K (1995a) Co-occurrence of prenylated xanthones and their cyclization products in cell suspension cultures of *Hypericum patulum.* Phytochemistry 38(4):867–869

Ishiguro K, Nakajima M, Fukumoto H, Isoi K (1995b) A xanthone substituted with an irregular monoterpene in cell suspension cultures of *Hypericum patulum.* Phytochemistry 39(4):903–905

Ishiguro K, Fukumoto H, Suitani A, Nakajima M, Isoi K (1996) Prenylated xanthones from cell suspension cultures of *Hypericum patulum.* Phytochemistry 42(2):435–437

Ishiguro K, Nagareya N, Suitani A, Fukumoto H (1997) A prenylated xanthone from cell suspension cultures of *Hypericum patulum.* Phytochemistry 44(6):1065–1066

Ishiguro K, Nagareya N, Fukumoto H (1998) A phloroglucinol derivative from cell suspension cultures of *Hypericum patulum.* Phytochemistry 47(6):1041–1043

Jefferson A, Quillinan AJ, Scheinmann F, Sim KY (1970) Studies in the xanthone series. Aust J Chem 23:2539

Kartnig T, Brantner A (1990) Secondary constituents in cell cultures of *Hypericum perforatum* and *Hypericum maculatum.* Planta Med 56:634

Kartnig T, Gruber A, Sauer H (1989) Comparative phytochemical investigation of *Hypericum* species. Planta Med 55:215

Kartnig T, Gobel I, Heydel B (1996) Production of hypericin, pseudohypericin and flavonoids in cell cultures of various *Hypericum* species and their chemotypes. Planta Med 62(1):51–53

Kerb R, Brockmoller J, Staffeldt B, Ploch M, Roots I (1996) Natural product formulations available in Europe for psychotropic indications. Antimicrob Agents Chemother 40(9):2087–2093

Kirtany JK, Paknikar SK (1975) Structure of toxyloxanthone-B: a reinterpretation of spectral data. Indian J Chem 13:104

Kochs G, Grisebach H (1989) Phytoalexin synthesis in soybean: purification and reconstitution of cytochrome P450 3,9-dihydroxypterocarpan 6a-hydroxylase and separation from cytochrome P450 cinnamate 4-hydroxylase. Arch Biochem Biophys 273(2):543–553

Lin C, Chang M, Arisawa M, Shimizu M, Morita N (1982) A xanthone glycoside from *Tripterospermum taiwanense* and rutin from *Gentiana flavo-maculata*. Phytochemistry 21:948–949

Lin C, Liou S, Ko F, Teng C (1992) γ-Pyrone compounds. II. Synthesis and antiplatelet effects of tetraoxygenated xanthones. J Pharm Sci 81:1109–1112

Locksley HD, Moor I, Scheinmann F (1967) Extractives from guttiferae. IV. The significance of maclurin in xanthone biosynthesis. Tetrahedron 23:2229–2234

Linsmaier EM, Skoog F (1965) Organic growth factor requirements of tobacco tissue cultures. Physiol Plant 18:100–127

Mederos S, Lopez-Bazzocchi I, Ravelo A, Gonzalez A (1996) Hypericin from clonal propagation of *Hypericum canariensis* L. Plant Tissue Cult 6:7–13

Minami T, Suzuki A, Nagashima Y, Fukuda Y (1996) Cosmetics for massage to promote circulation. Jpn Kokai Tokkyo Koho JP 08291038 A2 26

Nomura T, Hano Y (1994) Isoprenoid-substituted phenolic compounds of moraceous plants Nat Prod Rep 11:205–218

Ollis WD, Ramsay MVJ, Sutherland IO, Mongolsuk S (1965) The constitution of gambogic acid. Tetrahedron 21:1453–1470

Payk TR (1994) Treatment of depression. J Geriatr Psychiatry Neurol 7 Suppl 1:S3–S5

Peres V, Nagem TJ (1997) Trioxgenated naturally occurring xanthones. Phytochemistry 44:191–214

Recio M, Giner RM, Manez S, Rios JL, Marston A, Hostettmann K (1995) Screening of tropical medicinal plants for antiinflammatory activity. Phytother Res 9:571–574

Rocha L, Marston A, Kaplan MA, Stoeckli-Evans H, Thull U, Testa B, Hostettmann K (1994) An antifungal gamma-pyrone and xanthones with monoamine oxidase inhibitory activity from *Hypericum brasiliense.* Phytochemistry 36(6):1381–1385

Rocha L, Marston A, Kaplan MA, Stoeckli-Evans H, Thull U, Testa B, Hostettmann K (1995a) The ethnopharmacology of Madeira and Porto Santo Islands, a review. J Ethnopharmacol 46:73–93

Rocha L, Marston A, Potterat O, Kaplan MA, Stoeckli-Evans H, Hostettmann K (1995b) Antibacterial phloroglucinols and flavonoids from *Hypericum brasiliense.* Phytochemistry 40:1447–1452

Sen AK, Sarkar KK, Majumder PC, Banerji N (1980) Isolation of three new minor xanthones from *Garcinia mangostana* Linn. Indian J Chem 19B:1008

Sen AK, Sarkar KK, Mazumder PC, Banerji N, Uusvuori R, Hase TA (1982) The structures of garcinones A, B and C: three new xanthones from *Gardenia mangostana.* Phytochemistry 21:1747

Sia G-L, Bennett GJ, Harrison LJ (1995) Minor xanthones from the bark of *Cratoxylum cochinchinense.* Phytochemistry 38:1521–1528

Sultanbawa MUS (1980) Xanthonoids of tropical plants. Tetrahedron 36:1465–1506

Takamiya, Tanaka N, Touno K, Terahara N, Shimomura K, Ishimaru K (1998) Anthocyanin and procyanidin in *Hypericum patulum* tissue cultures. Jpn J Food Chemistry 5(1):3–8

Taylor RS, Manandhar NP, Hudson JB, Towers GH (1996a) The immunotropic properties of biologically active products obtained from Klamath weed. Eksp Klin Farmakol 59:51–54

Taylor RS, Manandhar NP, Hudson JB, Towers GH (1996b) Antiviral activities of Nepalese medicinal plants (Review). J Ethnopharmacol 52:157–163

Teng CM, Lin CN, Ko FN, Cheng KL, Huang TF (1989) Novel inhibitory actions on platelet thromboxane and inositolphosphate formation by xanthones and their glycosides. Biochem Pharmacol 38(21):3791–3795

Thiede HM, Walper A (1994) Inhibition of MAO, COMT by hypericum extracts and hypericin. J Geriatr Psychiatry Neurol 7 Suppl 1:S54–S56

Welle R, Grisebach H (1988) Induction of phytoalexin synthesis in soybean: enymatic cyclization of prenylated pterocarpans to glyceollin isomers. Biochem Biophys 263:191–198

Wu QL, Liao YH, Wang SP, Wang LW, Feng YX, Yang JS, Xiao PG (1996) New pyrones from *Hypericum japonicum.* Chin Chem Lett 7(12):1095–1096

Yazaki K, Okuda T (1990) Procyanidins in callus and multiple shoot cultures of *Hypericum erectum.* Planta Med 56:490–491

Yazaki K, Okuda T (1994) *Hypericum erectum* Thunb. (St. John's wort): in vitro culture and the production of procyanidins. In: Bajaj YPS (ed) Biotechnology in agriculture and forestry, vol 26, Medicinal and aromatic plants VI. Springer, Berlin Heidelberg New York, pp 167–178

XII *Jamesoniella autumnalis* (Liverwort): Culture and Production of Metabolites

K.P. Adam

1 General Account

1.1 Introduction

Liverworts are known to be a rich source of natural products (Zinsmeister et al. 1991; Asakawa 1995). A remarkable number of bioactive compounds have been described so far from this plant category (Asakawa 1990). Since less than 5% of known species have been phytochemically investigated in detail, one may expect that quite a few interesting substances remain to be isolated. However, the analysis of a large number of species is hindered by the difficulty in obtaining sufficient amounts of plant material for isolating and testing components for potential biological and pharmacological activity. The small size and the fact that liverworts grow mixed with other bryophytes make their purification a difficult and time-consuming task. Furthermore, numerous species can usually be obtained only in small amounts from their natural habitats. In addition to these diffculties, there is the problem that many of them are spread over wide areas and occur only in small populations, making it quite impossible to obtain sufficient plant material from field collections for a detailed chemical analysis. Furthermore, the bryophyte locations, especially in the tropics, are often accessible only with difficulty. Field cultivation of bryophytes similar to that of higher plants is not practicable; even if the required ecological conditions were created, mixed populations would result.

A possibility for overcoming this shortage of plant material is growing liverworts in aseptic cultures. They can be grown in vitro on suitable nutrient media under aseptic conditions, as surface or submerged cultures. Bryophyte cultures offer effectively the same advantages as in vitro cultures of higher plants. Plants from all climatic zones can be grown in cultures. A comparatively constant production of homogenous material can be achieved through controlled culture conditions, giving sufficient amounts for subsequent analysis (Becker 1994; Adam 1996).

In the course of our studies on liverwort constituents, axenic cultures of the liverwort *Jamesoniella autumnalis* (DC.) Steph. (Jungermanniaceae) have

FR 12.3, Pharmakognosie und Analytische Phytochemie Universität des Saarlandes, 66041 Saarbrücken, Germany

Biotechnology in Agriculture and Forestry, Vol. 43
Medicinal and Aromatic Plants XI (ed. by Y.P.S. Bajaj)

been induced from field collected plants. In vitro propagation of the liverwort has made available the large amounts of plant material necessary for a thorough phytochemical analysis (Fig. 1).

1.2 The Plant

J. autumnalis is a common liverwort found throughout most temperate forest regions of the Northern Hemisphere. Plants usually form green and somewhat opaque patches, becoming reddish brown in sunlight. The prostrate shoots (ca. 1–4 cm long) with ascending apices have many rhizoids all along the stem and are simple or sparingly branched. The leaves of rotund-quadrate to shortly oblong shape consist of a single cell layer. They are quite obliquely inserted at the stem. Underleaves are reduced exept below the gynoecia and at the shoot apices. Cells contain 7–15 colorless, finely granulose oil bodies of spherical to ellipsoidal shape. The plant is dioecious with a cylindrical perianth, plicate in the upper quarter with an irregularly toothed mouth. The capsule is of ovoid shape.

The species frequently occurs over acidic rocks, such as noncalcareous sandstones, quartz conglomerate or granite, and can be found on decaying logs and on humus-rich soils. *J. autumnalis* grows mixed with other bryophyte species or forms homogeneous patches up to hand size (Schuster 1969).

Fig. 1. Aseptic culture of *J. autumnalis*

2 Chemical Constituents

The secondary metabolite pattern of *J. autumnalis* has been well investigated. Due to limited natural abundance, most of the known secondary metabolites have been isolated from aseptic cultures of the liverwort (Fig. 2).

The majority of the reported constituents are terpenoids. Among the sesquiterpenes, the common compounds α-phellandrene, p-cymene,

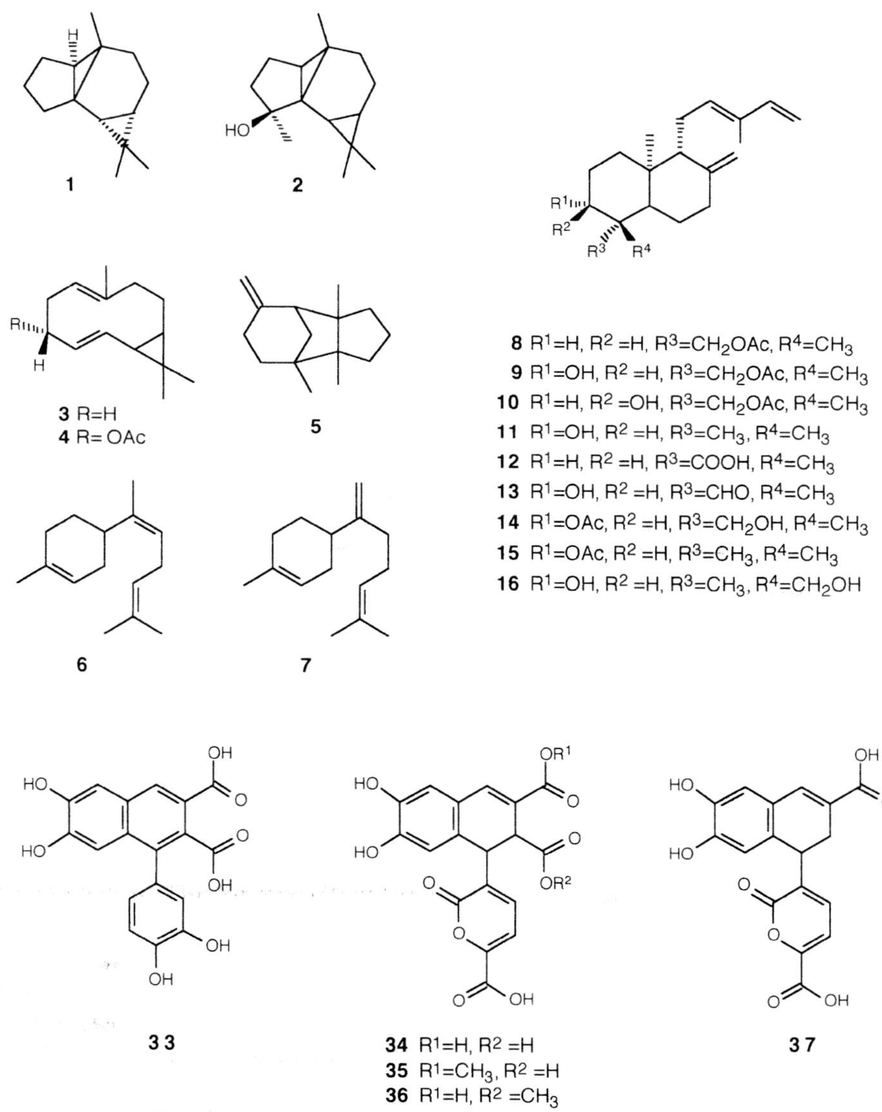

Fig. 2. Chemical constituents of in vitro cultures of *J. autumnalis*

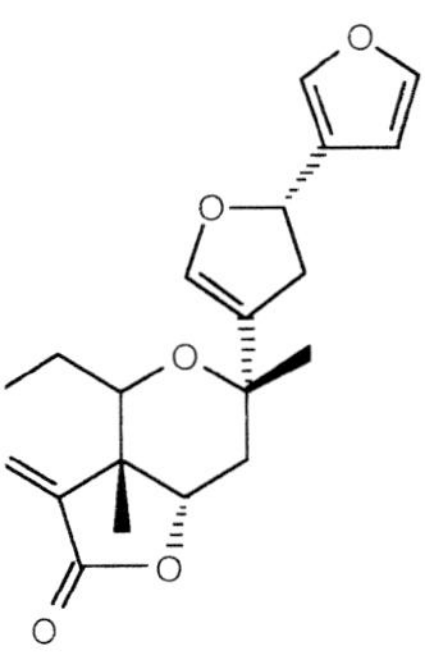

17

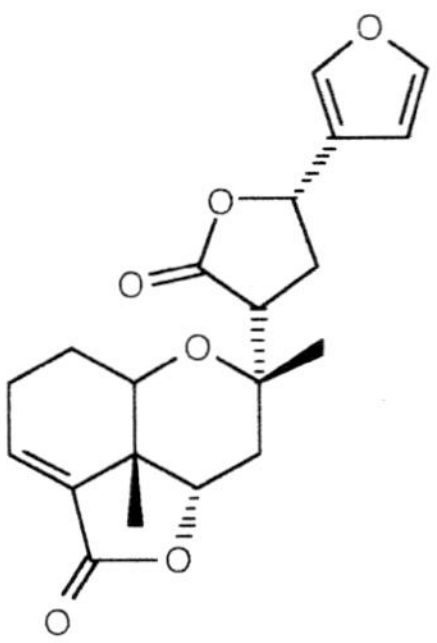

18

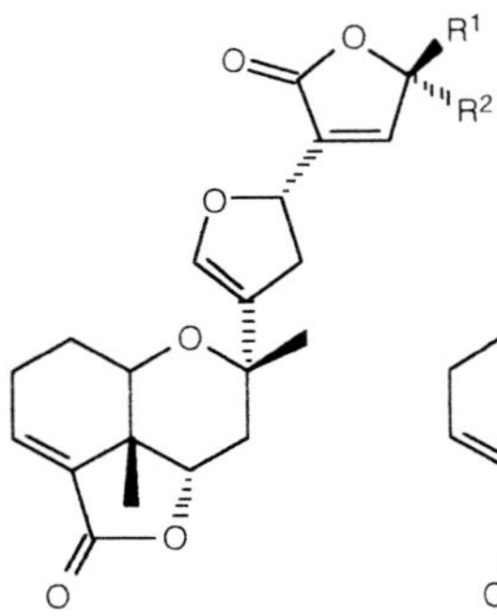

19 R^1=OH, R^2=H
20 R^1=H, R^2=OH

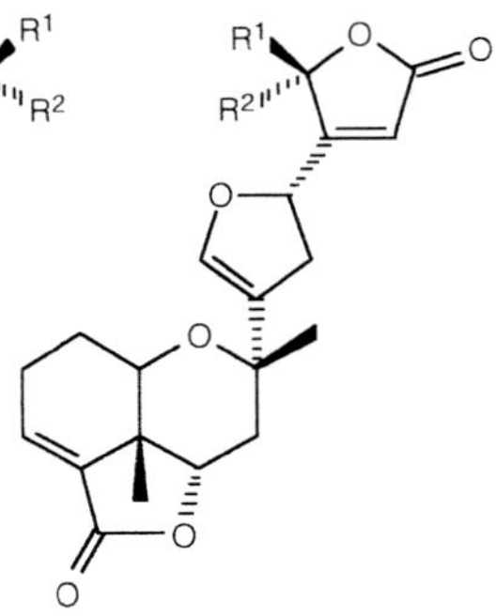

21 R^1=OH, R^2=H
22 R^1=H, R^2=OH

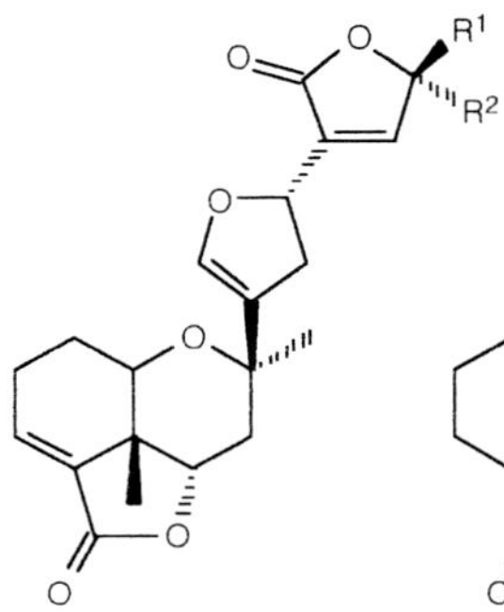

23 R^1=OH, R^2=H
24 R^1=H, R^2=OH

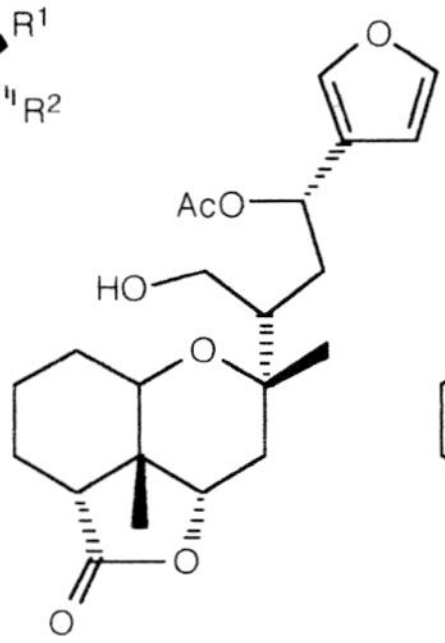

25

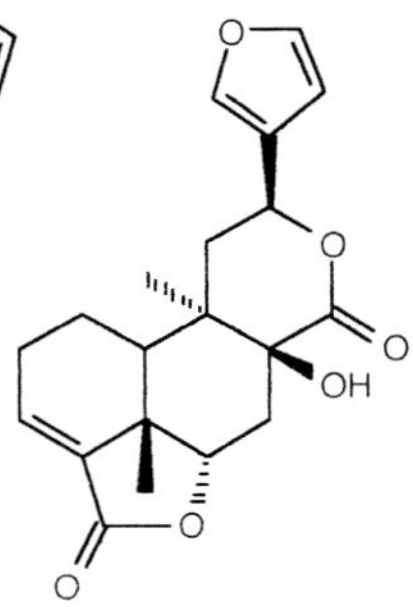

26

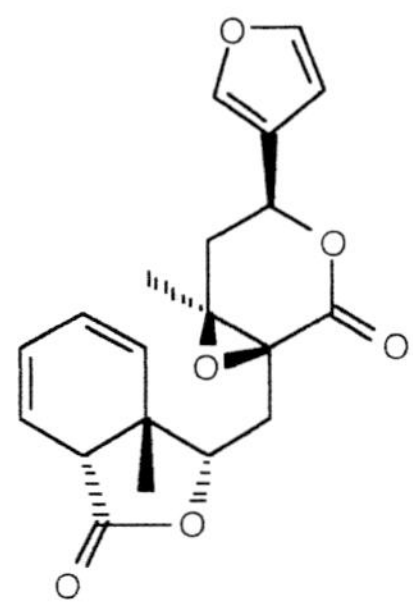

27

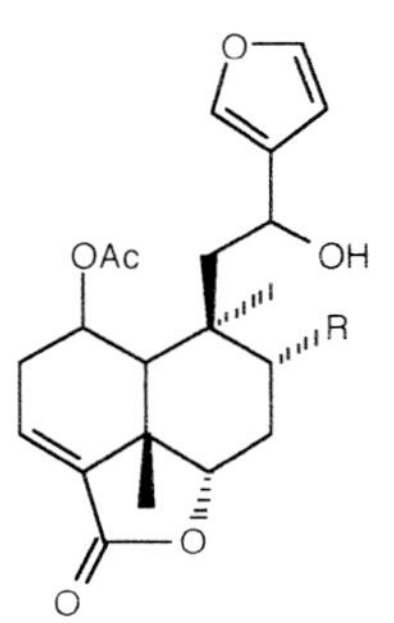

28 R=CH_3
29 R=CH_2OH

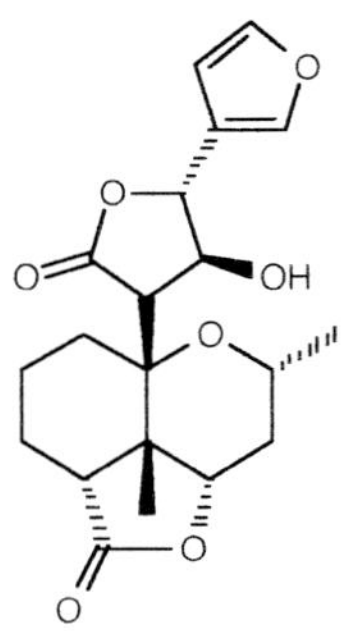

30

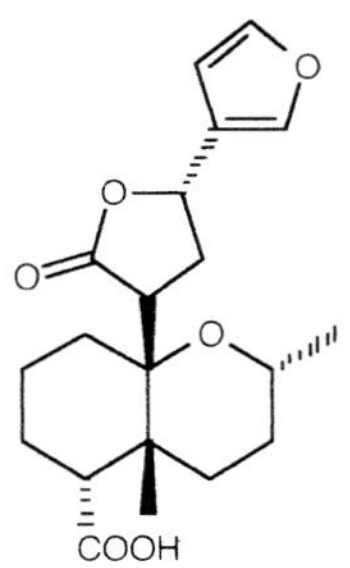

31

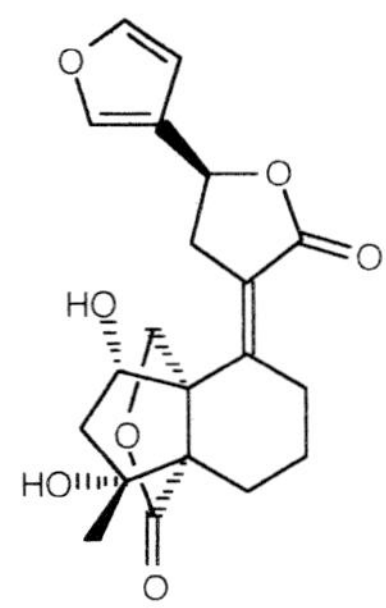

32

Fig. 2. *Continued*

β-chamigrene, anastreptene (**1**), spathulenol (**2**), bicyclogermacrene (**3**), 3α-acetoxybicyclogermacrene (**4**), β-barbatene (**5**), α-bisabolene (**6**) and β-bisabolene (**7**), have been described from collected material and cultures. (Asakawa et al. 1980; Blechschmidt and Becker 1995).

Diterpenes are the most important group of constituents of *J. autumnalis*. Several labdanes (**8–16**) and a large series of clerodanes (**26**, **28**, **29**) and secoclerodanes (jamesionellides A–J, **17–25**, **27**, **30–32**) with unusual structures have been found in field material and cultures of the liverwort (Blechschmidt and Becker 1992; Tazaki et al. 1994, 1995a, 1998). Collected *J. autumnalis* from Japan contained common diterpenes of the kaurane type (Nagashima et al. 1990). In the liverwort cultures from Europe, kaurane derivatives could not be detected (Tazaki et al. 1995a). This fact reflects the chemical variability of the species. As a member of the triterpene group, cycloartenol linolic acid ester has been isolated from cultured *J. autumnalis* (Blechschmidt 1991). Furthermore, a cyclolignan (**33**) and lignan derivatives with a modified phenyl ring (**34–37**) are remarkable constituents of *J. autumnalis* cultures (Tazaki et al. 1995b). A similar compound, scapaniapyrone A, has only been reported from the liverwort *Scapania undulata* (Mues et al. 1988).

3 In Vitro Culture Studies (Table 1)

3.1 Growth Characteristics and Medium Improvement of Aseptic Cultures

Several cultures of different origin of *J. autumnalis* were generated from field-collected gametophytes (for protocol, see below) and routinely subcultured on B5 medium (Gamborg et al. 1968) whose microelements had been replaced by those of the MSK-2 medium (Katoh et al. 1980). The medium was solidified with 0.9% agar and contained 2% sucrose. The cultures were kept in 200-ml Erlenmeyer flasks at 20 ± 1.5 °C under constant light (2000 ux). *J. autumnalis* grew as a completely differentiated surface culture. The growth rate varied quite significantly among the cultures of different origin. The growth curve of

Table 1. Summary of the in vitro studies conducted on *Jamesionella autumnalis*

References	Observations/Remarks
1. Blechschmidt (1991)	Induction of aseptic cultures; medium improvement; growth characteristics
2. Blechschmidt and Becker (1992)	New labdane and secoclerodane diterpenes from cultures
3. Tazaki et al. (1994)	New secoclerodane diterpene from cultures
4. Blechschmidt and Becker (1995)	Sesquiterpene formation in aseptic cultures
5. Tazaki et al. (1995a)	New secoclerodane diterpenes from cultures
6. Tazaki et al. (1995b)	New lignan derivatives from cultures

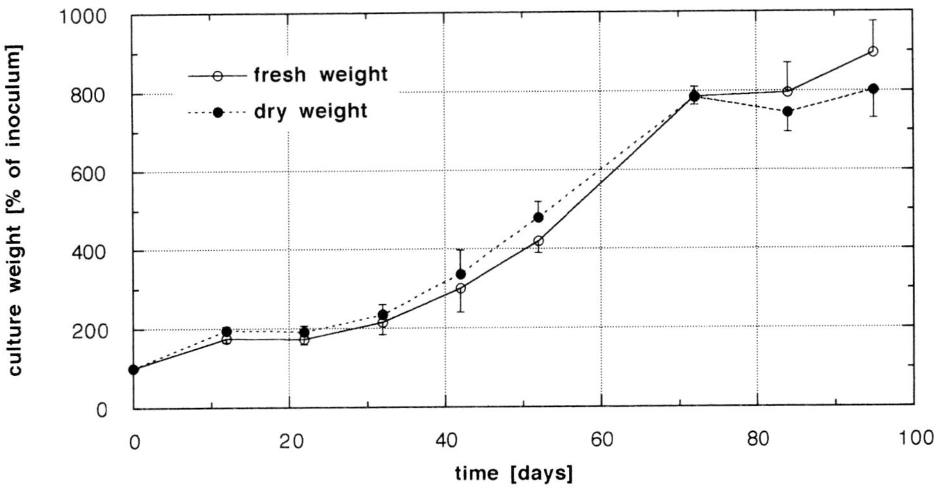

Fig. 3. Growth curves of aseptic cultures of *J. autumnalis*, culture strain Kirkel. (Blechschmidt and Becker 1995)

the fastest growing strain (Kirkel) showed a sigmoidal shape with a rather long lag phase and a stationary phase after 70 days (Fig. 3). The dry weight increased eightfold within 80 days, whereas the fresh weight only increased sevenfold. Using the same medium with only half of the mineral salt concentration resulted in nearly identical dry weight and fresh weight curves over the whole growth period. Growth during a day-night cycle (16 h light, 8 h dark) reduced the final dry weight to 80% of the control culture kept under continuous light (Blechschmidt 1991). Compared with in vitro cultures of other liverworts, the growth rate of *J. autumnalis* is quite low (Sauerwein and Becker 1990; Wurzel and Becker 1990; Morais and Becker 1991; Adam and Becker 1993; Adam 1996). Nevertheless, the respectable amount of 815 g of dry weight of the plant that was used in the phytochemical analysis has been generated by in vitro culture.

3.2 Sesquiterpene Formation in Aseptic Cultures

Sesquiterpenes are the major essential oil components of *J. autumnalis*. The formation of the six sesquiterpenes **1–3** and **5–7** has been studied in aseptic cultures of different origin and the respective field material. Figure 4 shows a comparison of the sesquiterpene pattern of two culture strains (Kirkel and Limbach) with the corresponding naturally grown plants. Field material from different habitats contained the same compounds but displayed qualitative variations. The patterns and total amounts of sesquiterpenes in cultures and the corresponding field material were quite similar. The latter observation can be explained by the fact that the plants remain fully differentiated in the cultures. The sesquiterpene composition during the culture period has also

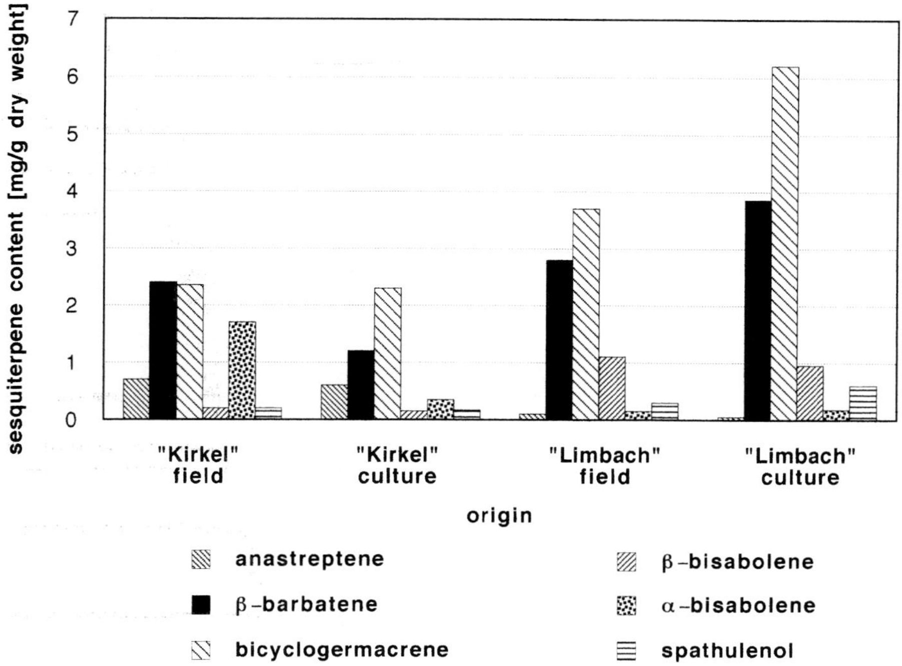

Fig. 4. Sesquiterpene content of field-collected material and corresponding cultures of *J. autumnalis*. (Blechschmidt and Becker 1995)

been analyzed (Fig. 5). It could be demonstrated that the sesquiterpene pattern was quite stable over the whole growth period. Furthermore, the absolute concentration of the compounds also remained at the same level at all measuring points (start, 13, 52, 84 days of growth) (Blechschmidt and Becker 1995).

4 Summary

J. autumanalis is a good example for demonstrating the value of aseptic cultures for the phytochemical analysis of bryophytes. Starting from surface disinfected gametophytes, well growing sterile cultures could be obtained. Propagation of the cultures led to larger amounts of plant material, which facilitated a thorough phytochemical analysis of the plant. A large series of new clerodane and secoclerodane diterpenoids (jamesionellides A–J) with unusual structures could be isolated from these cultures. The lignan derivatives with a pyrone ring system are also remarkable constituents of *J. autumnalis*. All these compounds have only been accessible through working on large amounts of plant material that would not have been obtainable from field collections.

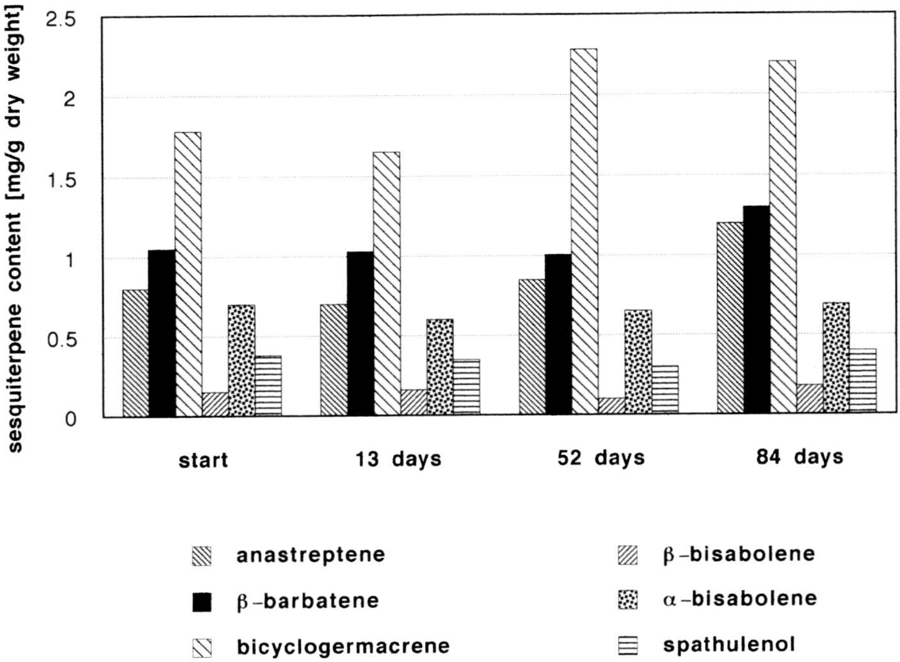

Fig. 5. Sesquiterpene content of *J. autumnalis*, cultures during growth, culture strain Kirkel. (Blechschmidt and Becker 1995)

The comparison between sesquiterpene composition in cultures and in collected plant material revealed that *J. autumnalis* synthesizes the same compounds in vitro as in the field. Due to differentiated growth, the sesquiterpene pattern in cultures is similar to that of field material. Furthermore, it was shown that the sesquiterpene content in the cultures remains at the same level over the whole growth period.

5 Protocol for the Establishment of Aseptic Cultures of Folious Liverworts

Two methods are generally used for the establishment of aseptic cultures from bryophytes. The most successful technique involves starting up cultures from spore capsules which contain sterile spores. Capsules are soaked for 2 min in 2% NaOCl solution containing 1% Tween 80. The capsules are then washed with sterile water. The washed capsules are opened with sterile forceps and the spores are spread out on MSK2 (Katoh et al. 1980) or B5 medium (Gamborg et al. 1968) of which the micro elements have been replaced by those of MSK2 medium (Katoh et al. 1980) containing 2% glucose or sucrose and 0.9% agar.

In the event of not having spore capsules available, sterile cultures can be started from the gametophyte by using surface sterilization procedures (Basile 1972). For the latter procedure, an optimized protocol has been developed for *J. autumnalis*. Short apical tips of approximately 5 mm length were cut from the plants and washed several times with sterile water. Then, the plant parts were immersed in a 1% NaOCl-solution containing 1% Tween 80 and exposed to the disinfectant for 90 to 120s. After the NaOCl-treatment, the tips were washed 3 times with sterile water. Finally, the plant tissue was transferred to modified B5 medium (see above) with half of the mineral salt concentration, solidified with 0.9% agar, containing 1–2% sucrose. All procedures can be carried out in large reagent tubes. Following this protocol, up to 40% of the treated tips resulted in sterile cultures that were further propagated on modified B5 solid medium. This protocol has been successfully used for the generation of axenic cultures of several folious liverworts. A similar protocol can be applied for starting axenic cultures of thallose liverworts, e.g. *Marchantia polymorpha* (Adam 1996).

References

Adam KP (1996) *Marchantia polymorpha* (liverwort) culture and production of metabolites. In: Bajaj YPS (ed) Biotechnology in agriculture and forestry, vol 37. Medicinal and aromatic plants IX. Springer, Berlin Heidelberg New York, pp 186–201

Adam KP, Becker H (1993) Bisbibenzyl formation in aseptic cultures of *Marchantia polymorpha* L. Z Naturforsch [c] 48:838–842

Asakawa Y (1990) Terpenoids and aromatic compounds with pharmacological activities from bryophytes. In: Zinsmeister HD, Mues R (eds) Bryophytes, their chemistry and chemical taxonomy. Procedings of the Phytochemical Society of Europe. Clarendon Press, Oxford, pp 369–410

Asakawa Y (1995) Chemical constituents of the bryophytes. In: Herz W, Kirby GW, Moore RE, Steglich W, Tamm C (eds) Progress in the chemistry of organic natural products, vol 65. Springer, New York, pp 1–618

Asakawa Y, Suire C, Toyota M, Tokunaga M, Takemoto T, Hattori S, Mitutani M (1980) Chemosystematics of bryophytes. V. The distribution of terpenoids and aromatic compounds in European and Japanese Hepaticae. J Hattori Bot Lab 48:285–303

Basile DV (1972) A method for surface sterilizing small plant parts. Bull Torrey Bot Club 99:313–316

Becker H (1994) Secondary metabolites from bryophytes in vitro cultures. J Hattori Bot Lab 76:283–291

Blechschmidt M (1991) Sterilkultur und sekundäre Inhaltsstoffe des Lebermooses *Jamesoniella autumnalis* (DC) Steph. PhD Thesis, University of Saarbrücken

Blechschmidt M, Becker H (1992) *Ent*-labdanes and furanoditerpenes from the liverwort *Jamesoniella autumnalis*. J Nat Prod 55:111–121

Blechschmidt M, Becker H (1995) Comparison of sesquiterpenes from field collected material and in vitro cultures of *Jamesoniella autumnalis* (DC.) Steph. Flavour Fragrance J 10:187–191

Gamborg OL, Miller RA, Ojima K (1968) Plant cell cultures. I. Nutrient requirements of suspension cultures of soybean root cells. Exp Cell Res 50:151–158

Katoh K (1988) Isolation and maintainance of callus and cell suspension cultures of bryophytes. In: Glime JM (ed) Methods in bryology. Hattori Botanical Laboratory, Nichinan, pp 99–105

Katoh K, Ishikawa M, Miyake K, Ohta Y, Hirose Y, Iwamura T (1980) Nutrient utilization and requirement under photoheterotrophic growth of *Marchantia polymorpha*: improvement of the culture medium. Physiol Plant 49:241–247

Morais RMSC, Becker H (1991) Growth and secondary product formation of in vitro cultures from the liverwort *Reboulia hemispherica*. Z Naturforsch [c] 46:28–32

Mues R, Huneck S, Conolly JD, Rycroft DS (1988) Scapaniapyrone A, a novel aromatic constituent of the liverwort *Scapania undulata*. Tetrahedron Lett 29:6793–6796

Murashige T, Skoog F (1962) A revised medium for rapid growth and bio-assays with tobacco tissue cultures. Physiol Plant 15:473–497

Nagashima F, Toyota M, Asakawa Y (1990) Terpenoids from some Japanese liverworts. Phytochemistry 29:2169–2174

Tazaki H, Blechschmidt M, Huch V, Veith M, Becker H (1994) A furanoditerpenoid from the liverwort *Jamesoniella autumnalis*. Phytochemistry 37:491–494

Tazaki H, Zapp J, Becker H (1995a) Diterpenes from in vitro cultures of the liverwort *Jamesionella autumnalis*. Phytochemistry 39:859–868

Tazaki H, Adam KP, Becker H (1995b) Five lignan derivatives from in vitro cultures of the liverwort *Jamesoniella autumnalis*. Phytochemistry 40:1671–1675

Tazaki H, Nabeta K, Becker H (1998) Clerodane-type diterpenoids from axenic cultures of the liverwort *Jamesionella autumnalis*. Phytochemistry 48:681–685

Sauerwein M, Becker H (1990) Growth, terpenoid production and antibacterial activity of an in vitro culture of the liverwort *Fossombronia pusilla*. Planta Med 56:364–367

Schuster RM (1969) The Hepaticae and Anthocerotae of North America, vol 2. Columbia University Press, New York, pp 818–827

Wurzel G, Becker H (1990) Growth and terpenoid production of an axenic culture from the liverwort *Ricciocarpos natans*. Z Naturforsch [c] 45:13–18

Zinsmeister HD, Becker H, Eicher T (1991) Bryophytes, a source of biologically active, naturally occuring material? Angew Chem Int Ed Engl 30:130–147

XIII *Karwinskia* Species (Buckthorn): In Vitro Culture, Micropropagation, and the Production of Toxin Anthracenone

D. LIŠKOVÁ[1], A. LUX[2], A. PIÑEYRO-LÓPEZ[3], and R. LUJÁN-RANGEL[3]

1 Introduction

1.1 Distribution and Importance of *Karwinskia* Species

The genus *Karwinskia* (buckthorn family, Rhamnaceae) was named after the Bavarian botanist Wilhelm Frederick von Karwinski, who collected plant material for herbaria in Mexico in the first half of the last century. Medicinal use of some species of this family, mainly *Rhamnus* spp. and *Frangula* spp., are well known in Europe, some of them being used since Ancient Greek times, or even earlier.

Species of the genus *Karwinskia* grow in the southern USA, Mexico, Central America, north of Columbia in South America, and in Cuba, Haiti and the Dominican Republic. They are, according to Standley (1923), trees or shrubs (Fig. 1), often spined, the leaves are simple, usually stipulate, entire or toothed. The flowers are perfect or polygamo-dioecious, small, inconspicuous, greenish or yellowish, usually in axillary cymes. The calyx is four or five lobate, the petals, four or five (or none) are cucculate or involute. The stamens are four or five, localized opposite the petals. The fruit is one- to four-celled, capsular or drupaceous. The wood is hard and strong.

There are 15 described species, ten of which grow in Mexico: *K. humboldtiana* (Schult) Zuccarini, *K. mollis* Schlechtendal, *K. umbellata* (Cav) Schlechtendal, *K. subcordata* Schlechtendal, *K. rzedowskii* Fernández, *K. affin humboldtiana* (Roem and Schult) Zuccarini, *K. venturensii* Fernández, *K. calderoni* Standley, *K. parvifolia* Rose, *K. latifolia* Standley. A survey of all Mexican species, their geographical distribution, local names and uses in traditional medicine was published by Fernández-Nava (1992). *Karwinskia* species are known in Mexico mostly for their toxins. Intoxications with these compounds, mostly from fruits, are clinically manifested with paralytic syn-

[1] Institute of Chemistry, Slovak Academy of Sciences, Dúbravská cesta 9, 842 38 Bratislava, Slovakia
[2] Department of Plant Physiology, Faculty of Natural Sciences, Comenius University, Mlynská dolina, 842 15 Bratislava, Slovakia. Present address: Arid Land Research Center, Tottori University, 1390 Hamasaka, Tottori 680, Japan
[3] Departamento de Farmacología y Toxicología, Facultad de Médicina, U.A.N.L. Apdo. Postal 146, Col. del Valle, Monterrey, N.L. 66220, México

Biotechnology in Agriculture and Forestry, Vol. 43
Medicinal and Aromatic Plants XI (ed. by Y.P.S. Bajaj)

Fig. 1. Fruits of *Karwinskia humboldtiana* ripening in the greenhouse at the Department of Plant Physiology in Bratislava, Slovakia

dromes (Escobar and Nieto 1965; Puertolas et al. 1984). However, effects on lung (pulmonary edema), liver (fatty liver degradation), colon, kidney, and hepatic, cardiac and skeletal muscle necrosis are known (Dewan et al. 1964; Charlton and Pierce 1970; Weller et al. 1980; Bermudez et al. 1986; Garza-Ocañas et al. 1992). Malformations on mouse embryos evoked by one of these toxins, T-514, were observed (Martínez de Villareal et al. 1990). There is evidence of their effects on celullar and subcellular level – as cytostatic agents acting in the proliferation of cells (Velazco et al. 1996), and in mitochondrial and peroxisomal degradation (Sepulveda-Saavedra et al. 1992; Acosta-Garcia et al. 1993). The toxin T-514 has an inhibitory effect on lipid peroxidation stimulated by Fe^{2+}-ADP complexes. From these finding it is possible to con-

Table 1. Content of toxins in the genus *Karwinskia*. (Waksman et al. 1989)

Species	% T-544	% T-496	% T-514
K. humboldtiana[a]	1.60	0.80	0.70
K. humboldtiana[b]	0.70	0.58	0.36
K. mollis	1.33	1.08	0.95
K. umbellata	0.75	0.50	0.05
K. subcordata	1.51	0.98	1.07
K. latifolia	–	1.10	0.50
K. parvifolia	–	1.58	2.04
K. calderonii	0.75	2.00	0.70
K. rzedowskii	–	1.60	–
K. venturensii	–	0.11	–
K. affin humboldtiana	0.83	0.40	0.33

[a] *K. humboldtiana* – Villa de García, Nuevo León, Mexico.
[b] *K. humboldtiana* – Linares, Nuevo León, Mexico.

clude that it might have antioxidant activity in the in vitro systems used (Garza-Ocañas et al. 1990). Some toxins isolated from the roots of *K. humboldtiana* showed antimicrobial activities (Mitscher et al. 1985).

Analysis of *Karwinskia* toxins showed the presence of anthracenones. Dreyer et al. (1975) isolated four of these from the fruits of *K. humboldtiana* (T-496, T-514, T-516, T-544). Dominguez et al. (1976) found the toxins T-544 and T-516 in roots. The survey of *Karwinskia* toxins and their effects studied at the Faculty of Medicine, U.A.N.L. Monterrey, Mexico demonstrated their potential use, T-514 named peroxisomicine (toxin A_1) being the most important of all. This toxin is included in a screening programme of the National Cancer Institute in Bethesda, Maryland, USA, and there is a patent granted by the EC (European Patent Office no. 0467407) focused on the investigation of T-514.

The difficulties in T-514 isolation and purification resulted in screening, with the aim of finding out the most convenient species for its production. It was shown by Waksman et al. (1989) that alongside *K. humboldtiana*, *K. parvifolia* is traditionally used (Table 1). These two species are thus of special interest, especially *K. parvifolia*, since it has the highest content of T-514, and also has the most convenient composition of toxins, with no T-544 which is the principal substance with a neurotoxic effect.

K. humboldtiana (the most common names are coyotillo and tullidora, Standley 1923; Bermudez et al. 1986) is an evergreen shrub or a small tree distributed in Mexico, in Baja California, Tamaulipas, Veracruz, Yucatan, Oaxaca, and in the USA, in western Texas in semi-desert conditions (Rzedowski 1978). Its leaves are oblong to oval 1–6.5 cm long. The umbels are short pedunculate, or sessile. The fruit is 6 to 9 mm long, almost spherical, blackish when ripe. The fruit mesocarp is edible (sweet, because of the glucose content), but the seeds are highly toxic with paralytic effects. The use of seeds in traditional medicine is mainly as an anticonvulsive.

K. parvifolia is an evergreen shrub growing along the Pacific coast in states Sinaloa and Sonora in the type of vegetation called deciduous seasonal forest (Rzedowski 1978). According to Standley (1923), *K. parvifolia* is only a form of *K. humboldtiana* with very small leaves, but the leaves vary too much in size through the range of the plant to permit the use of this as a specific character. Other characteristics are similar to *K. humboldtiana*. The common name of this plant is cacachila, and according to Fernandez-Nava (1992) it has no local use, and the plants are considered to be dangerous.

Karwinskia species do not represent typical xerophytic shrubs and their method of drought tolerance remains unclear. However, these plants are successfully competitive even in the areas heavily damaged by overstocking of domestic livestock. Anthracenone toxins are their most important defense against animal ingestion. These toxins are likely to be related to the secretory pigments present in secretory structures in all parts of these plants (Lux and Earl 1989). Brilliant red and yellow lipoid pigments of canals and cavities in different organs are shown in Fig. 2. Secretory canals and cavities represent special structures in *Karwinskia*.

To understand the biology and toxin production of *Karwinskia* species we examined its anatomy and some physiological characteristics. It was shown by Lux and Earl (1989) that the leaves of *K. humboldtiana* are dorsiventral, hypostomatal with slightly sunken stomata and a crystalline wax coating on the epidermis. The mesophyll is organized into a palisade and spongy parenchyma. Small lateral veins are vertically translucent. In the leaves, as well as in other organs there are abundant internal secretory structures: idioblasts with druse crystals, enlarged palisade cells with solitary crystals, mucilage cells and receptacles, secretory canals beneath the veins and extensive secretory cavities in the mesophyll (Fig. 2a,b). Some of these features are charasteristic for various species of the family Rhamnaceae (Metcalfe and Chalk 1950). The same leaf structure as that of *K. humboldtiana* is also found in *K. parvifolia*. The root systems of both *K. humboldtiana* and *K. parvifolia* are well developed and branched (Sadloňová 1996). Higher order lateral roots develop faster in *K. humboldtiana*. The difference in the development of root systems between *K. humboldtiana* and *K. parvifolia* existing in natural conditions resulted in the differences of root formation of in vitro cultures. The anatomy of roots of both species is similar (Fig. 2d,e). Secretory canals and cavities are present in stems, too (Fig. 2c). All types of secretory structures observed in the vegetative organs are present also in generative organs (Hanáčková, unpubl.; Ruíz-Ordóñez et al. 1992, 1993; Hanáčková et al. 1994a,b).

Secondary metabolism is related to the primary metabolism in plants (Bell and Charlwood 1980; Ford et al. 1994). In this connection some physiological analyses were performed, and the results indicate (Masarovičová and Lux 1995) that the daily course of CO_2 exchange, as well as absolute values of the maximal daily net photosynthetic rate and dark respiration rate, are in good agreement with values found in common Middle-European trees *Fagus sylvatica* (Masarovičová and Štefančík 1990), or *Quercus* spp. (Masarovičová 1992).

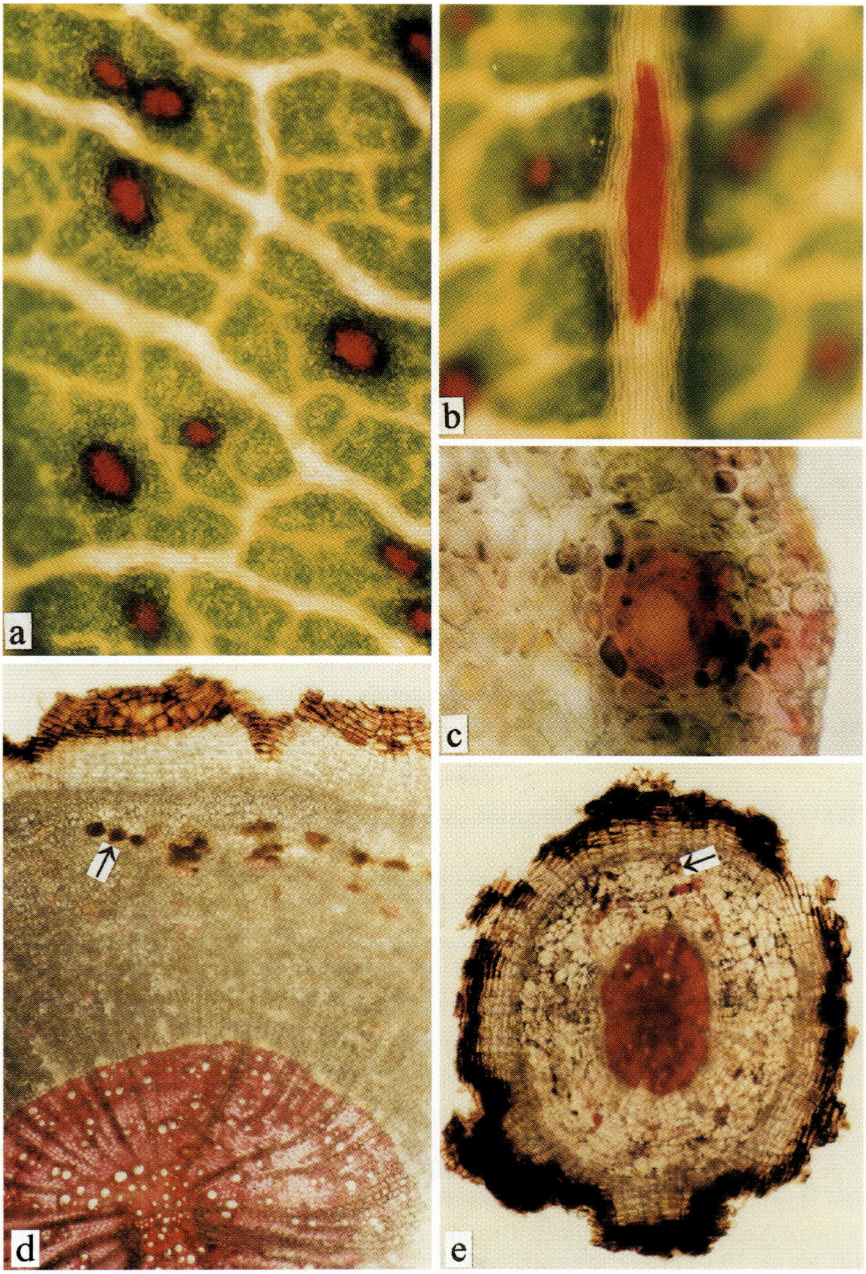

Fig. 2. Secretory structures in various vegetative organs of *Karwinskia parvifolia*. **a** Cavities in leaf mesophyll, **b** Canals beneath the vein with brilliant red pigments seen in light microscopy (unstained, both ×70). **c** Canal within the primary cortex of young stem, note yellow lipoid content inside (hand section, unstained, ×270). **d** Secondary thickening root of 3-year-old plant with secretory cells in the cortex (*arrow*), hand section, lignified cell walls stained red with floroglucine + HCl (×70). **e** Lateral root of the same plant as in **d**, secretory cells in cortex are sparse (*arrow*), preparation as in **d** (×70)

1.2 Conventional Practices for *Karwinskia* Propagation and Production of Their Medicinal Compounds

Initial studies demonstrating the relative importance of *K. humboldtiana* for the production of anthracenone substance T-514 with cancerostatic effects (Piñeyro-López 1990), has stimulated extensive investigation of this species at the Department of Toxicology and Pharmacology, Faculty of Medicine, U.A.N.L, Monterrey, Mexico since 1981. For the first experiments, ripe and unripe fruits were collected from nature, mainly in the locality Villa de García in the vicinity of Monterrey, state Nuevo León, Mexico. Subsequent screening of the T-514 content has shown great variability of this toxin content, depending on the time of collection, locality and also differing from one year to another (Waksman et al. 1989; Santoyo et al. 1991). This stimulated further investigation and experiments with cultivation of these plants. Luján-Rangel et al. (1993a–d) have shown the possibility of growing this plant in an experimental field, and its growth dependence on water, nutrition and low temperatures. The same author also tried vegetative propagation and grafting (Luján-Rangel, unpubl.). Later, when it was demonstrated that *K. parvifolia* is more suitable for the production of the T-514 because of its high content and convenient toxins composition, subsequent studies were concentrated on this species, including cultivation experiments (Luján-Rangel et al. 1993b,d). Cultivation of *Karwinskia* species in European greenhouse conditions has also been successful (Fig. 1). Growth from seeds and development of the seedlings in the first 2 years was documented by Sadloňová (1996).

Despite these observations, up until now, collection of unripe fruits from nature remains the most important, if not the only source of toxins, especially T-514. The cost of fruit collection is $6 (US)/1 kg of dry fruits containing 60 g of T-514. However, due to difficulties of separation, only 0.1% of the pure toxin is obtained. The purification costs are $1000 (US)/1 g of pure T-514, not including equipment costs.

2 In Vitro Approaches

2.1 Review of Tissue Cultures/Biotechnological Studies

The first experiments with *Karwinskia* cultures in vitro were realized in the Laboratory of Plant Tissue Cultures at the Institute of Chemistry, Slovak Academy of Sciences, Bratislava in collaboration with the Department of Plant Physiology, Faculty of Natural Sciences, Comenius University, Bratislava, in 1990. For initiation of these cultures, various source materials have been used, explants from leaves and stems were tried for these experiments (see Table 2). Apical meristems of the stems also gave good results, but the most suitable sources in this case were isolated zygotic embryos (Fig. 3a). In 1996, in vitro cultures were established at the Department of Histology, Faculty of Medicine, U.A.N.L. Monterrey, Mexico.

Table 2. Summary of in vitro studies conducted on *Karwinskia* species

Reference	Species	Observation
Lišková et al. (1990–1992, unpubl.)	*K. humboldtiana*	Screening of culture conditions and primary explants (leaves, stems, embryos) for cultures in vitro
Lux et al. (1992)	*K. humboldtiana*	Microscopical studies of secretory structures in plants in vivo and cultures in vitro
Lišková et al. (1993)	*K. humboldtiana* *K. parvifolia*	Screening of regeneration potential of zygotic embryos
Lišková et al. (1994)	*K. humboldtiana*	Isolation of embryos, induction of shoots, shoot multiplication, rooting, induction of root cultures, callus induction
Lux et al. (1994) Hanáčková et al. (1994a,b)	*Karwinskia spp.*	Structural analysis of in vitro and in vivo plants
Lux et al. (1994, unpubl.)	*K. parvifolia* *K. humboldtiana*	Transfer of in vitro induced plants to greenhouse conditions
Nahálka, Kákoniová, Lišková and Lux (1995, unpubl.)	*K. humboldtiana*	Experiments on the induction of suspensions and shoot cultures cultivated in liquid medium
Lux et al. (1995)	*Karwinskia spp.*	Structural studies on in vitro cultures
Lux and Lišková (1996, unpubl.)	*K. parvifolia* *K. humboldtiana*	Meristem cultures
Lux and Lišková (1996)	*K. parvifolia*	Micropropagation
Mikuš et al. (1996) Lux et al. (1996a) Lux et al. (1996b)	*K. humboldtiana* *K. parvifolia*	Toxin production in shoots and roots in plants in vivo and in vitro – comparison
Lux et al. (1998)	*K. parvifolia*	Isolation of embryos, induction of shoots, shoot multiplication, rooting

2.2 Establishment of Tissue Cultures/Cell Suspensions

In vitro studies were aimed at the two most interesting species – *K. humboldtiana* and *K. parvifolia* (Lišková et al. 1994; Lux et al. 1997/98). In vitro cultures were induced from isolated zygotic embryos in different stages of maturation depending on the grade of ripeness, from unripe (green) to ripe (dark) dry fruits from the field. In all cases, the embryos were already completely developed. *K. humboldtiana* fruits were collected mainly in the locality of Monterrey-Saltillo; *K. parvifolia* fruits in different regions of the state Sinaloa, Mexico – Sauz, Batosori, Melchor Ocampo. Before use, all seeds were stored at 4 °C for at least 1 month. For the preparation of sterile material, surface sterilization of such fruits is of great importance because of high possibility of contamination through the soft mesocarp. The fruits of *K. humboldtiana* and *K. parvifolia* were sterilized twice; before and after imbibition in sterile distilled water, with a sterilization and wetting agent, and

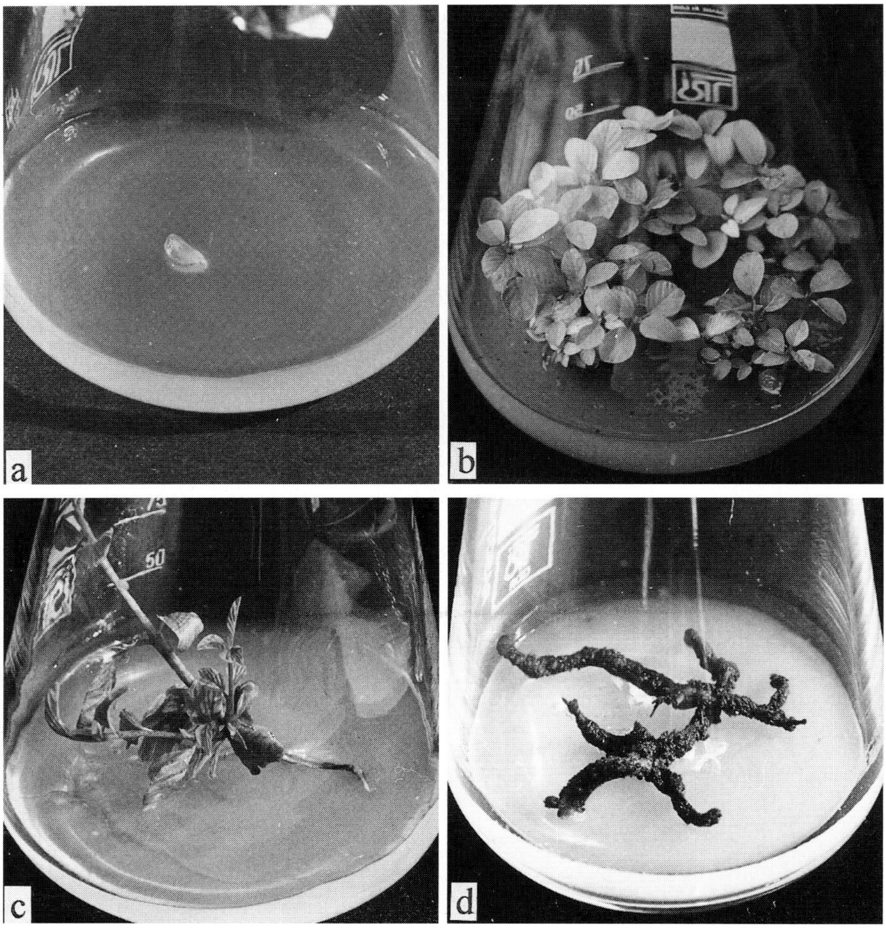

Fig. 3. Cultures of *Karwinskia* species. **a** Isolated zygotic embryo (3–4 mm). **b** Multiplication of *Karwinskia humboldtiana* shoots. **c** *Karwinskia parvifolia*, proliferation of shoots from the base of the stem. **d** Initiation of root cultures from *Karwinskia humboldtiana* plantlets grown in vitro

a commercial bleach, 20% v/v (Lišková et al. 1994). The imbibition time varied between 18–24 h depending on seed dryness. After the isolation of embryos from seeds, they can be post-sterilized in a commercial bleach if necessary, but the solution in this case must be a 10 times lower concentration (Lux et al. 1998). A screening of media for the cultivation of isolated embryos was made for both species. The best results were obtained on the woody plant medium, WPM (Lloyd and McCown 1981), supplemented with 0.06 μM IAA, 0.03 μM GA_3, and 2 μM BA. The development and growth of embryos was positively influenced if they were preincubated for a short period (48–72 h) in dim light (4 $\mu mol\, m^{-2} s^{-1}$) at a higher temperature of 28 °C (Fig. 4). These conditions, also used for isolated spruce embryos (Lišková, unpubl.), seem to decrease the stress evoked with the isolation procedure. After the preincubation period the

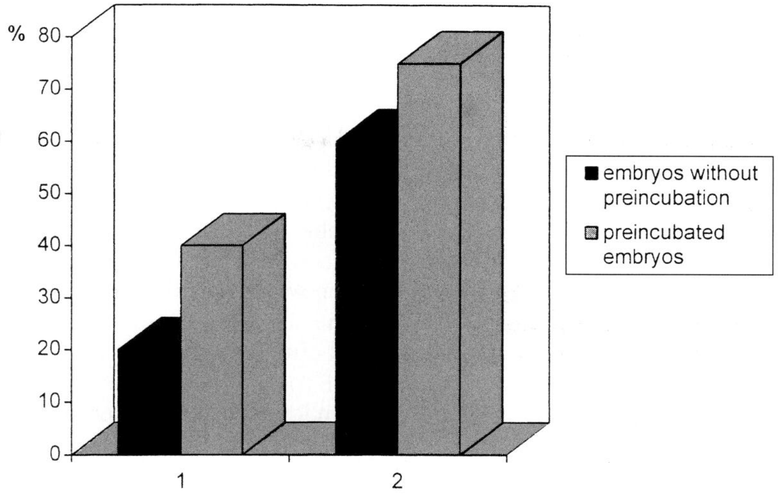

Fig. 4. Effect of preincubation on isolated *K. humboldtiana* and *K. parvifolia* zygotic embryos *1 K. parvifolia*; *2 K. humboldtiana*

Table 3. Shoot multiplication in *Karwinskia* cultures in vitro **A** and their significance **B**

A		**B**	
Species	Shoot mutiplication/ explant means ± SE	Comparison of plant material	Significance
K. humboldtiana	16.10 ± 2.26	*K.h.*:*K.p.* (Sauz)	a
K. parvifolia (Sauz)	2.12 ± 0.24	*K.h.*:*K.p.* (Bat)	a
K. parvifolia (Bat)	5.28 ± 0.50	*K.h.*:*K.p.* (M.O.)	a
K. parvifolia (M.O.)	0.80 ± 0.04	*K.p.* (Sauz):*K.p.* (Bat)	c
		K.p. (Sauz):*K.p.* (M.O.)	a
		K.p. (Bat):*K.p.* (M.O.)	a

a = highly significant ($P < 0.01$); c = non-significant ($P > 0.05$).
K.h. = *K. humboldtiana.*
K.p. = *K. parvifolia.*

embryos were transferred to conditions of 16-h photoperiod, 23 °C and 60% humidity.

The basal medium also served for the multiplication of shoots which were initiated from nodal areas of shoot explants excised from primary shoots developed from embryos cultured in vitro. The multiplication values for *K. humboldtiana* (Fig. 3b) and *K. parvifolia* (Fig. 3c) are shown in Table 3. The multiplication potential, expressed in percentages, is about 95% for *K. humboldtiana* cultures, and 58% or less for *K. parvifolia* cultures, depending on the collection area. The multiplication of shoots in *K. humboldtiana* cultures did not decrease in long-term experiments, which can be seen from the mean number of shoots in the first year of culture, 16.1 ± 2.26, compared with the mean number in the second year 22.6 ± 1.58 per explant. There was a seasonal effect in the regeneration potential in both species, with a preference for spring. Microscopical analysis of the stems and leaves showed similar anatomy and the presence of the same internal secretory structures as was observed in plants in vivo (Fig. 2; Lux and Earl 1989; Lux and Hanáčková, unpubl.). After the separation of single shoots of *K. humboldtiana* from primary explants, rooting on WPM medium was initiated, supplemented with 10 μM NAA, to produce a mean number of 7.7 ± 2.10 roots on each shoot. Rooting was also possible on the basal medium with a lengthening of the subculture period to 8–9 weeks, or on WPM medium without growth hormones. However, in the first case the appearance of roots is more rapid, and they also differ morphologically, being swollen and very fragile (Lišková et al. 1994, Fig. 3d). The initiation of roots seems to be easier using shoot explants consisting of 3–4 shoots 2–3 cm long. A quite different root formation was observed in the cultures of *K. parvifolia*. In this species, the rooting was very low compared with *K. humboldtiana*. Differences were also found due to seed origin. The only success appeared in cultures initiated from seeds from the Sauz locality (89% of embryos formed roots) on WPM medium supplemented with NAA. However, in this case the shoots produced long firm roots. Results shown here indicate a specificity in responses to culture conditions and in multiplication potential of two different species of the genus *Karwinskia* and specificity even of the same species originating from different localities.

Long-term cultures of *K. humboldtiana* and *K. parvifolia* (Sauz) as well as their transfer to semisterile conditions in perlite (Fig. 5a), and subsequently to a greenhouse, are without substantial problems (Fig. 5b,c,d). The majority of plants (about 90%) are able to adapt to these new conditions when tempered by a gradual increasing of the photo period, with a lowering of humidity (during about 1 month). This demonstrates the high adaptability of these plants, probably related to the severe semiarid conditions of their areas of origin.

Root cultures (Fig. 6a–c) were grown on MS medium (Murashige and Skoog 1962) supplemented with 10 μM NAA, 14 μM GA_3, 5 μM kinetin, 30 g l^{-1} sucrose, and 100 mg l^{-1} casein hydrolysate. Callus from these root cultures was initiated on MS medium supplemented with 10 μM 2,4-D and 10 μM BA. Recently, better results with the callus growth were obtained on WPM medium supplemented with 2,4-D and NAA in the ratio 1:1 (10 μM; Fig. 6d).

Light microscopy demonstrated that some callus cells were filled with dark pigmented homogeneous or granular material (Fig. 6e). However, the majority of cells did not possess this dark material. The TEM showed that dark granular material or big globular structures were present inside the vacuoles (Fig. 6f). Because of the presence of unidentified dark substances in vacuoles, the accumulation of some secondary metabolites in callus tissue could be assumed.

From these callus cultures it is possible to derive cell suspensions. In the liquid medium shoot cultures were also cultivated, and their production of toxins was examined (Nahálka, unpubl.).

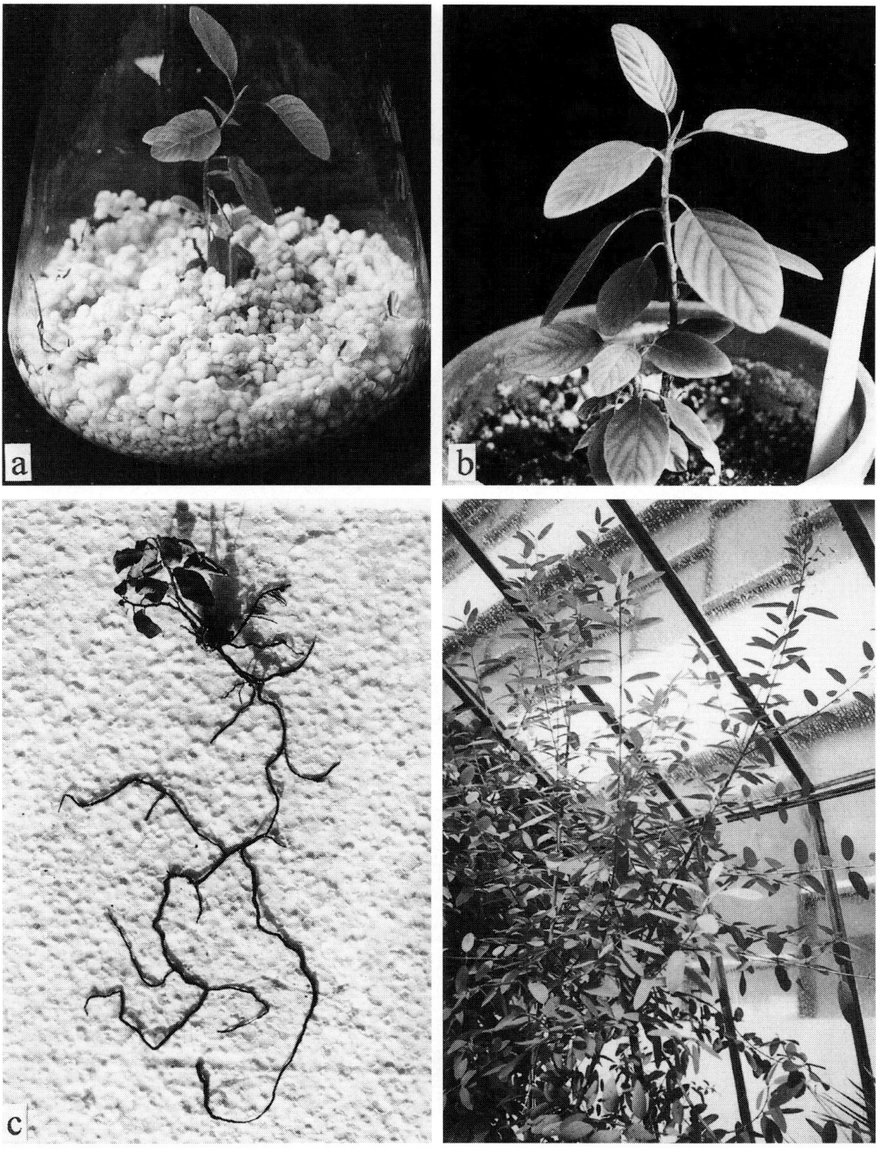

Fig. 5. *Karwinskia humboldtiana* and *Karwinskia parvifolia* – in vitro and ex vitro cultivation. **a** *Karwinskia parvifolia* under semisterile conditions cultivated in perlite. **b** *Karwinskia parvifolia* in a pot at the time of adaptation to greenhouse conditions. **c** *Karwinskia humboldtiana* plant developed in vitro and grown for 4 months in greenhouse, note well-developed, long roots. **d** *Karwinskia humboldtiana* from in vitro conditions growing for the 3rd year in the greenhouse at the Department of Plant Physiology, Comenius University in Bratislava, Slovakia

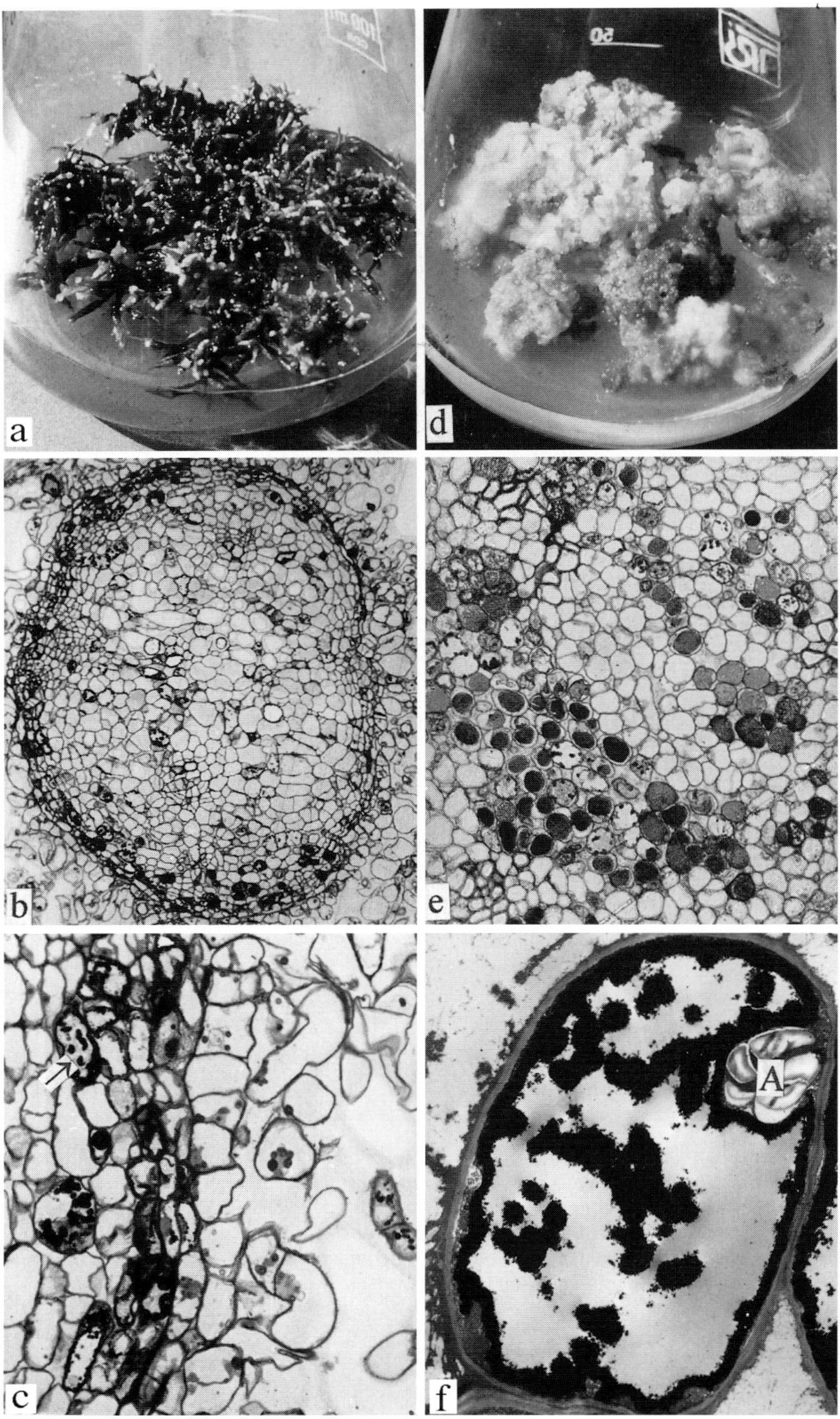
a
d
b
e
c
A
f

2.3 Extraction and Structure of *Karwinskia* Secondary Metabolites

Karwinskia species, as well as some species of the genus *Cassia* (Kinjo et al. 1994), are the only plants reported in the literature to be a source of dimeric tetrahydro-anthracenones. The extraction of these toxins from *Karwinskia* species, predominately from *K. humboldtiana* and *K. parvifolia*, is in principle similar from all parts of the plant. Most studies are focused on the extraction, isolation, and identification of toxins from fruits (Waksman et al. 1989; Rivas et al. 1990; Waksman and Ramírez 1992; Piñeyro-López et al. 1994), however, attention is also paid to other parts of the plant – roots, leaves, stems (Dreyer et al. 1975; Dominguez et al. 1976; Waksman et al. 1993; Yussim et al. 1995; Argalášová 1996). In these parts, new anthracenonic compounds – dimeric 9,10-dihydroxy anthracenones – with interesting biological activities, have been discovered.

The improved method for quantification of T-514 (peroxisomicine A_1) and other dimeric anthracenonic compounds was described for the first time by Salazar et al. (1996). A summary of these procedures is introduced here.

The extracts were prepared from air dried ground fruits of *K. parvifolia* with petroleum ether and ethyl acetate at room temperature. This extract containing dimeric anthracenones was evaporated to dryness under N_2 and redissolved in methanol. An HPLC analytical system for the separation, identification, and quantification of anthracenonic compounds was used. The system employed uses reverse phase (C-8 and C-18) and diode array detection, and is suitable for the analysis of plant extracts as well as biological fluids. Only this reverse system exhibited satisfactory properties. It was necessary to use a gradient for the analysis of plant extracts because of the long retention time of some of the compounds (e.g. T-496, which can be eluted only in 100% methanol). The use of TLC (silica gel) and quantification of this type of compounds by means of densitometry has been reported previously (Guerrero et al. 1987).

The chemical structure, UV and IR spectrum of the most investigated toxin (T-514) of *Karwinskia* species is shown in Fig. 7 (Piñeyro-López et al. 1994).

Structures of other important dimeric anthracenones (T-510, iso T-514, T-516, T-544, T-496) isolated from plants of the genus *Karwinskia* are shown in Fig. 8.

Fig. 6. Root and callus cultures of *Karwinskia humboldtiana*. **a** Root culture established on WPM medium containing NAA, GA_3 and kinetin. **b** Transverse section of root from the root culture with abnormal histological organization, the stele is absent and the surface is covered with callus tissue (semithin section stained according to Lux 1981; ×70). **c** Detail from **b**, note secretory idioblasts with dark material in vacuoles – *arrow* (semithin section stained according to Lux 1981; ×270). **d** Callus culture growing on WPM medium supplemented with 2,4-D and NAA, 1:1. **e** Semithin section from the callus, note numerous secretory cells with variable content (stained according to Lux 1981; ×130). **f** Detail of the secretory cell with electrondense material inside vacuole (*A* amyloplast, from Lišková et al. 1994), transmission electron microscopy (×2700)

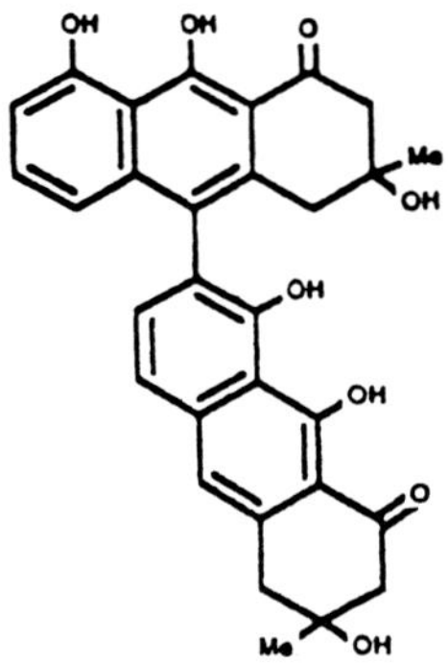

Chemical structure of toxin T-514.

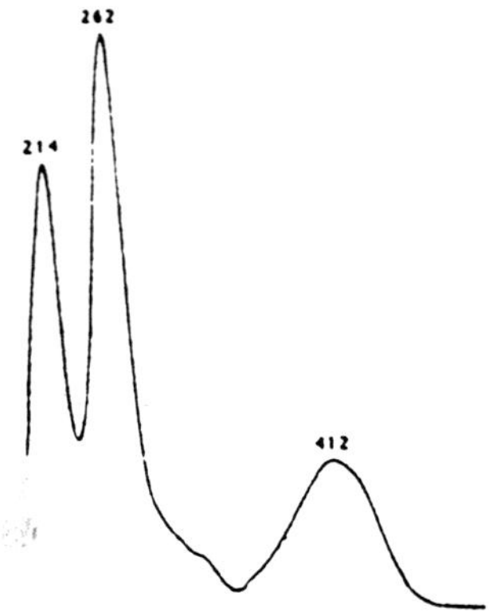

UV-Vis spectrum of toxin T-514.

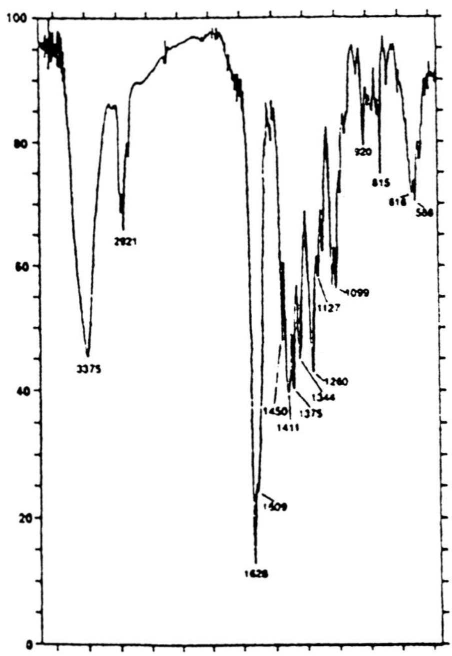

IR spectrum of toxin T-514.

Fig. 7. The chemical structure, UV and IR spectrum of the most investigated toxin (T-514) of *Karwinskia* species is shown in Fig. 7 (Piñeyro-López et al. 1994)

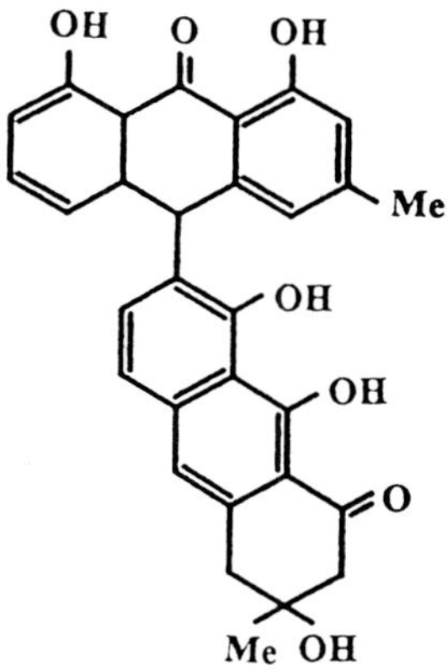

T - 496

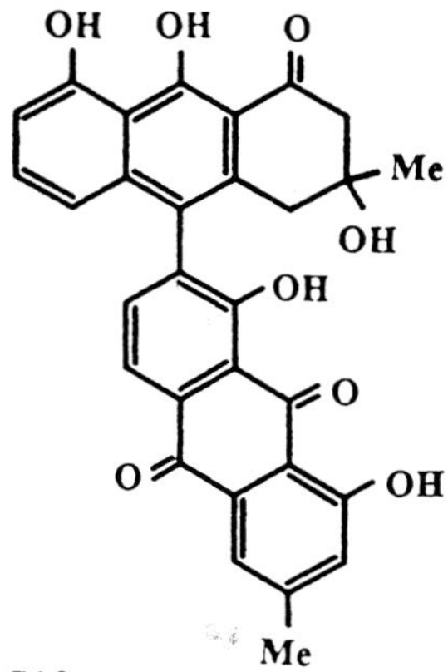

T - 510

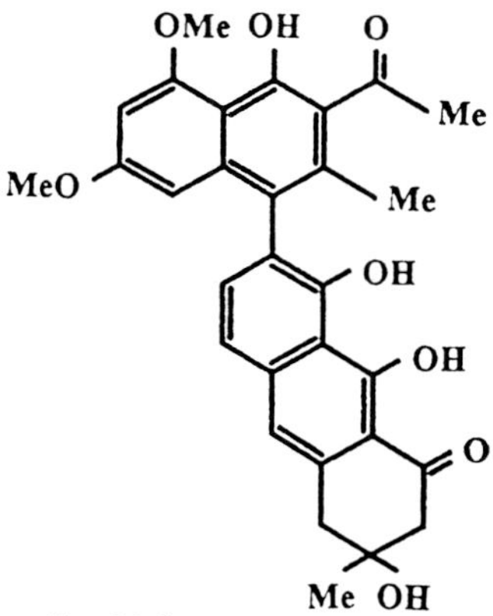

T - 516

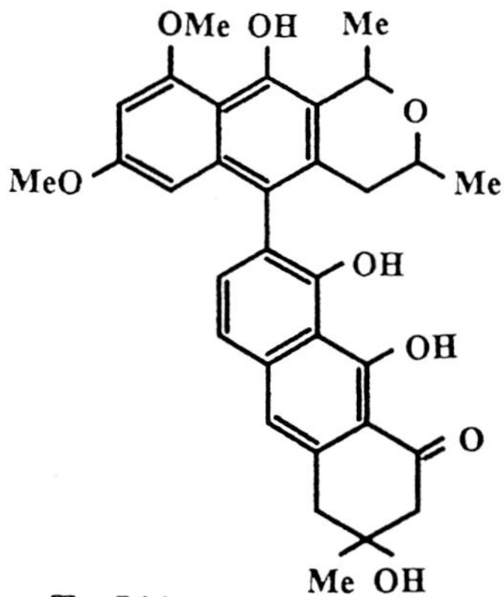

T - 544

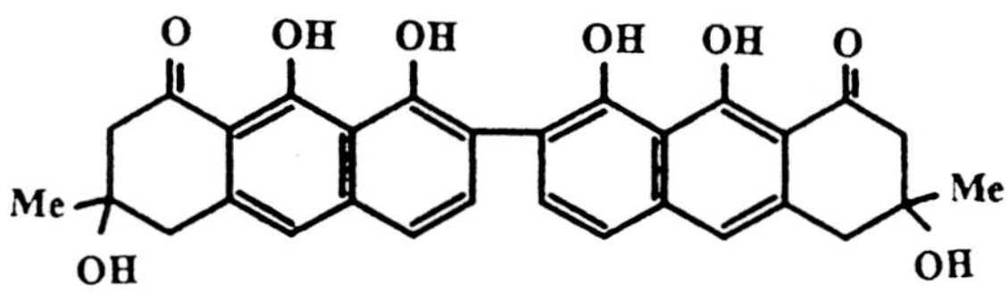

T - iso514

Fig. 8. Structures of other important dimeric anthracenones (T-510, iso T-514, T-516, T-544, T-496) isolated from plants of the genus *Karwinskia* are shown in Fig. 8

Preliminary studies on the mechanism of action of the toxins indicate that the peroxisome is the target organelle in yeast and liver cells of rats and monkeys (Sepulveda-Saavedra et al. 1992, 1995). Further investigations have shown that this substance does not competitively inhibit peroxisomal catalase (Moreno-Sepulveda et al. 1995). Some malignant cells contain fewer peroxisomes and have low catalase activity compared with normal cells. This could explain the selectivity observed in in vitro studies (Piñeyro-López et al. 1994). A relationship between structure and degree of inhibition of certain anthracenonic compounds is assumed.

3 Summary

The genus *Karwinskia* (Rhamnaceae) with its 15 species described, as well as some species of the genus *Cassia*, are the only plants reported to produce dimeric anthracenones – toxins manifesting paralytic effects in mammalian cells. At present, these toxins which are characterized by anti-neoplastic activity are predominantly isolated from unripe fruits of *K. humboldtiana* and *K. parvifolia* collected in the field. For the effective isolation and purification, especially of the toxin T-514 (peroxisomicine A_1), the improved method of Salazar et al. (1996) is used, which is an HPLC analytical system employing reverse phase (C-8 and C-18) and diode array detection. Besides T-514, there are other compounds with similar effects, or they show other possible uses with their antioxidant, or antimicrobial characteristics.

The possibility of using plants grown in experimental fields, as well as in vitro cultures is examined for the production of suitable plant material with satisfactory quality and amounts of toxins demanded. In in vitro conditions, tissue (*K. humboldtiana* callus), as well as organ cultures (*K. humboldtiana* and *K. parvifolia* shoot and root cultures) were established, and after multiplication of shoots and their rooting, a successful transfer to a greenhouse was achieved.

Working from the knowledge that chemotherapy in general and cancer chemotherapy in particular, is designed to be selectively toxic, studies on plants of the genus *Karwinskia*, as well as on the characteristics and effects of their toxins in mammalian cells have brought some new information to this field of research. Preliminary investigations showed selective toxicity of T-514 for certain types of malignant cells in vitro and in vivo (Velazco-Campos et al. 1993; Piñeyro-López et al. 1994). T-514 is more toxic to all neoplastic cells than to the most sensitive normal cells. These findings encourage effort to explore further the potential of the anti-neoplastic effects of these compounds on other types of tumoral cells of human origin. Peroxisomicine A_1 (T-514) is currently under preclinical screening for its potential anti-neoplastic effect (Salazar et al. 1996) in Mexico.

Acknowledgements. This research was supported by the Slovak Grant Agency for Science, VEGA, no. 1/4102/97, and CONACYT-Consejo National de Ciencia y Technología, Mexico, no. 3095 and 4705.

References

Acosta-García RL, Sepulveda-Saavendra J, Piñeyro-López A (1993) Caracterización ultraestructural y morfometría del daño inducido en las mitocondrias del epitelio del tubo contorneado proximal por efecto de la T-514, extraidos de *Karwinskia parvifolia.* XI. Encuentro de Investigación Biomédica, Monterrey, Universidad Autonoma de Nuevo León, Mexico, p 110

Argalášová K (1996) Cutures in vitro and secondary metabolites analysis of *Karwinskia* species. Diploma Thesis, Comenius University, Bratislava (in Slovak)

Bell EA, Charlwood BV (1980) Secondary plant products. Springer, Berlin, Heidelberg, New York

Bermudez MV, Gonzalez-Spencer D, Guerrero M, Waksman N, Piñeyro-López A (1986) Experimental intoxication with fruit and purified toxins of buckthorn (*Karwinskia humboldtiana*). Toxicology 24:1091–1097

Charlton K, Pierce K (1970) Neuropathy in goats caused by experimental coyotillo (*Karwinskia humboldtiana*). I. Poison. Pathol Vet 7:385–407

Dewan NL, Henson JB, Dollahit JW, Bridges C (1964) Toxic myo-degeneration in goats produced by feeding mature fruits from the coyotillo plant (*Karwinskia humboldtiana*). Pathol Vet 46:215–226

Dominguez XA, Temblador S, Cedillo ME (1976) Estúdio quimico de la raíz de tullidora (*Karwinskia humboldtiana* /Zucc./). Rev Latinoam Quim 7:46–48

Dreyer D, Arai I, Bachman C, Anderson W, Smith R, Daves D (1975) Toxins causing non inflammatory paralytic neuropathy. Isolation and structure elucidation. J Am Chem Soc 97:4986

Escobar A, Nieto D (1965) Aspectos neuropatológicos de la intoxicación con *Karwinskia humboldtiana.* Estudio experimental. Gac Med Mex 95:163–177

Fernandez-Nava R (1992) Nombres comunes, usos y distribución geográfica del género *Karwinskia* (Rhamnaceae) en México. An Inst Biol Univ Nac Auton Mex Ser Bot 63:1–23

Ford YY, Fox GG, Ratcliffe RG, Robins RJ (1994) In vivo ^{15}N NMR studies of secondary metabolism in transformed root cultures of *Datura stramonium* and *Nicotiana tabacum.* Phytochemistry 36:333–339

Garza-Ocañas L, Torres-Alanis O, Piñeyro-López A (1990) Effect of toxin T-514 of *K. humboldtiana* on liver microsomal lipid peroxidation stimulated by Fe^{2+} ADP complex and NADPH in vitro. Eur J Pharmacol 183:76–77

Garza-Ocañas L, Hsieh GC, Acosta D, Torres-Alanis O, Piñeyro-López A (1992) Toxicity assessment of toxins T-514 and T-544 of buckthorn (*Karwinskia humboldtiana*) in primary skin and liver cell cultures. Toxicology 73:191–201

Guerrero M, Piñeyro A, Waksman N (1987) Extraction and quantification of toxins from *Karwinskia humboldtiana* (tullidora). Toxicon 25:565

Hanáčková Z, Lux A, Torres-Cepeda T, Piñeyro-López A (1994a) Formación de las estructuras de secreción interna en *Karwinskia parvifolia.* XII. Encuentro de Investigación Biomédica, Monterrey, Universidad Autonoma de Nuevo León, Mexico, p 203

Hanáčková Z, Torres-Cepeda T, Lux A, Lišková D (1994b) Comparación morfológica y ultraestructural de las hojas de *Karwinskia parvifolia* cultivadas in vivo e in vitro. XII. Encuentro de Investigación Biomédica, Monterrey, Universidad Autonoma de Nuevo León, Mexico, p 28

Kinjo J, Ikeda T, Watanabe K, Nohara T (1994) An anthraquinone glycoside from *Cassia angustifolia* leaves. Phytochemistry 37:1685–1687

Lišková D, Ruiz-Ordóñez J, Lux A, Piñeyro-López A (1992) Cultivo de *Karwinskia humboldtiana* in vitro. X. Encuentro de Investigación Biomédica, Universidad Autonoma de Nuevo León, Monterrey, Mexico, p 185

Lišková D, Lux A, Ruiz-Ordóñez J, Piñeyro-López A (1993) Regeneration potential of zygotic embryos of *Karwinskia humboldtiana* and *Karwinskia parvifolia* in vitro. 6th Conference of Plant Embryologists of Slovakia, Czech Republic and Poland, Nitra – Slovakia, p 42 (in Slovak)

Lišková D, Ruiz-Ordóñez J, Lux A, Piñeyro-López A (1994) Tissue culture of *Karwinskia humboldtiana* – a plant producing toxins with antitumoural effects. Plant Cell, Tissue Organ Cult 36:339–343

Lloyd GB, McCown BH (1980) Commercially feasible micropropagation of mountain laurel, *Kalmia latifolia*, by use of shoot-tip culture. Int Plant Propagators Soc Comb Proc 30:421–427

Lujan-Rangel R, Gonzalez-Alanis RJ, Piñeyro-López A (1993a) *Karwinskia humboldtiana* y su fructificación. XI. Encuentro de Investigación Biomédica, Universidad Autonoma de Nuevo León, Monterrey, Mexico, p 96

Lujan-Rangel R, Gonzalez-Alanis RJ, Piñeyro-López A (1993b) Efecto del frio sobre *Karwinskia parvifolia*. XI. Encuentro de Investigación Biomédica, Universidad Autonoma de Nuevo León, Monterrey, Mexico, p 97

Lujan-Rangel R, Waksman de Torres N, Gonzalez-Alanis RJ, Piñeyro-López A (1993c) La humedad y su importancia en la producción de la T-514. XI. Encuentro de Investigación Biomédica, Universidad Autonoma de Nuevo León, Monterrey, Mexico, p 111

Lujan-Rangel R, Gonzalez-Alanis RJ, Piñeyro-López A (1993d) Enfermedades del genero *Karwinskia*. XI. Encuentro de Investigación Biomédica, Universidad Autonoma de Nuevo León, Monterrey, Mexico, p 112

Lux A (1981) Rapid method for staining of semithin sections from plant material. Biologia 36:753–757 (in Slovak)

Lux A, Earl E (1989) The leaf structure of coyotillo, *Karwinskia humboldtiana* (Rhamnaceae). Publ Biol FCB Univ Auton Nuevo Leon 3:83–90

Lux A, Lišková D (1996) Micropropagation of *Karwinskia parvifolia* (Rhamnaceae) and its transfer to greenhouse conditions. Biologia 51(1):85

Lux A, Ruiz-Ordóñez J, Lišková D, Piñeyro-López A (1992) Estructuras de secreción de *Karwinskia humboldtiana* cultivada in vivo e in vitro. X. Encuentro de Investigación Biomédica, Monterrey, Universidad Autonoma de Nuevo León, Mexico, p 157

Lux A, Liš ková D, Ruiz-Ordóñez J, Piñeyro-López A (1994) Culture and structural analysis of *Karwinskia* species in vitro. Biol Plant 36:87

Lux A, Lišková D, Hanáčková Z (1995) Cultures in vitro of *Karwinskia* species producing cancerostatic substances and their structural analysis. Recent Advances in Plant Biotechnology, Nitra, Slovakia, p 89

Lux A, Sadloňová K, Lišková D, Masarovičová E, Mikuš M, Hanáčková Z, Ruiz-Ordóñez J, Piñeyro-López A (1996a) Study of *Karwinskia* roots. Conference on Progress in Plant Science from Plant Breeding to Growth Regulation, Mosonmagyaróvár, Hungary, pp 53–57

Lux A, Lišková D, Ruiz-Ordóñez J, Sadloňová K, Mikuš M, Hanáčková Z, Piñeyro-López A (1996b) Morfología, anatomía, respiración y metabolitos de la raiz de *Karwinskia*. XIV. Congresso National de Investigación Biomédica, Monterrey, Universidad Autonoma de Nuevo León, Mexico, p 76

Lux A, Lišková D, Piñeyro-López A, Ruiz-Ordóñez J, Kákoniová D (1997/98) Micropropagation of *Karwinskia parvifolia* and the transfer of plants to glasshouse conditions. Biol Plant 40:143–147

Martinez de Villareal L, Velazco-Campos R, Piñ eyro-López A, Gonzalez-Alanis R (1990) Effects of toxin T-544 from the *Karwinskia humboldtiana* (buckthorn) plant upon mouse embryos explanted at 11 days. Toxicon 28:449–452

Masarovičová E (1992) Morphological, physiological, biochemical and productional characteristics of three oak species. Acta Physiol Plant 14:99–106

Masarovičová E, Lux A (1995) Photosynthetic and respiratory activity of *Karwinskia humboldtiana*, species producing substances with antitumoural effects. In: Mathis P (ed) Photosynthesis: from light to biosphere. Kluwer, Dordrecht, pp 785–788

Masarovičová E, Štefančík L (1990) Some ecophysiological features in sun and shade leaves of tall beech trees. Biol Plant 32:374–387

Metcalfe CR, Chalk L (eds) (1950) Anatomy of the dicotyledons, 2 vols. Clarendon Press, Oxford

Mikuš M, Lišková D, Lux A, Hanáčková Z, Masarovičová E, Piñeyro López A (1996) *Karwinskia humboldtiana* – toxin production, physiology and structure of shoots in vitro and in vivo. 10th FESPP Congr – From molecular mechanisms to the plant: an integrated approach, Florence, Italy, p 308

Mitscher LA, Gollapudi SR, Oburn DS, Drake S (1985) Antimicrobial agents from higher plants: two dimethyl-benzisochromans from *Karwinskia humboldtiana*. Phytochemistry 24:1681–1683

Moreno-Sepulveda M, Vargas-Zapata R, Esquiveles-Cobedo D, De Torres NW, Piñeyro-López A (1995) Effect of peroxisomicine and related anthracenones on catalase activity. Planta Med 61:337–340

Murashige T, Skoog F (1962) A revised medium for rapid growth and bio-assays with tobacco tissue cultures. Physiol Plant 15:473–497

Piñeyro-López A (1990) Toxicidad selectiva in vitro de la toxina T-514 de la *Karwinskia humboldtiana*. PhD Thesis, University of Monterrey, Mexico

Piñeyro-López A, Martinez de Villareal L, Gonzalez-Alanis R (1994) In vitro selective toxicity of toxin T-514 from *Karwinskia humboldtiana* (buckthorn) plant on various human tumor cell lines. Toxicology 92:217–227

Puertolas M, Nava O, Medina H, Lopez F, Oyervides J (1984) Poliradiculoneuritis por *Karwinskia humboldtiana*. Rev Med (México) 22:25–31

Rivas VG, Torres S, Waksman NT (1990) Dimeric anthracenones from *Karwinskia* species. Planta Med 6:562–563

Ruiz-Ordóñez J, Torres-Cepeda TE, Piñeyro-López A (1992) The leaf mesophyll of coyotillo (*Karwinskia humboldtiana* Zucc.) morphological characterization. 50th Annu Meet of the Electron Microscopy Society of America, Boston, pp 1–3

Ruiz-Ordóñez J, Lux A, Torres-Cepeda TE, Piñeyro-López A (1993) Estructuras de secreción interna durante el proceso de prefloración-fructificación en *Karwinskia parvifolia*. Un estudio morfologico. XI. Encuentro de Investigación Biomédica, Universidad Autonoma de Nuevo León, Monterrey, Mexico, p 109

Rzedowski J (eds) (1978) Vegetación de México. Limusa, México

Sadloñová K (1996) Development of *Karwinskia humboldtiana* and *Karwinskia parvifolia* roots. Diploma Thesis, Comenius University, Bratislava (in Slovak)

Salazar ML, Piñeyro A, Waksman N (1996) A reverse-phase HPLC method for quantification of peroxisomicine and other anthracenonic compounds. J Liquid Chromatogr Related Technol 19:1391–1403

Santoyo A, Waksman N, Fernández R, Piñeyro A (1991) Peroxisomicina un producto de interes farmacologico busqueda de la fuente mas adecuada de extraccion. IX. Encuentro de Inverstigación Biomédica, Monterrey, Universidad Autonoma de Nuevo León, Mexico, pp 1–8

Sepulveda-Saavedra J, Van der Klei IJ, Keizer I, Piñeyro-López A, Harder W, Veehuis M (1992) Studies on the effect of toxin T-514 on the integrity of peroxisomes in methylotropic yeasts. FEMS Microbiol Lett 91:207–212

Sepulveda J, Vargas R, Ballesteros R, Piñeyro A, Moreno M (1995) Effect of peroxisomicine on liver catalase activity in the albino mouse. Mol Biol Cell 6:621–627

Solereder H (1908) Systematic anatomy of the dicotyledons, 2 vols. Clarendon Press, Oxford

Standley PC (1923) Trees and shrubs of Mexico. Contrib Nat Herb 23:710–717

Velazco-Campos M, Montero-Montoja R, Rojas del Castillo E, Espinoza-Aguirre J, Torres-Alanis O, Rodrigez-Carranza R, Piñeyro-López A, Ostrosky-Wegman P (1993) Efecto de la toxina T-514 sobre parametros de proliferación celular y genotoxicidad y su acción antitumoral en el sistema P-388. XI. Encuentro de Investigación Biomédica, Universidad Autonoma de Nuevo León, Monterrey, Mexico, p 113

Velazco MD, Montero R, Rojas E, Gonsebatt ME, Sordo M, Piñeyro A, Ostrosky-Wegman P (1996) Genotoxic effects of *Kawinskia humboldtiana* toxin T-514 in peripheral-blood lymphocytes. Anticancer Drugs 7:710–715

Waksman NT, Ramirez RD (1992) Isolation of a new dymeric anthracenone from *Karwinskia parvifolia*. Rev Latinoam Quím 23:25–27

Waksman N, Martinez L, Fernandez R (1989) Chemical and toxicological screening in genus *Karwinskia*. Rev Latinoam Quím 20:27–29

Waksman NT, Salazar ML, Cisneros L, Garcia VM, Lujan RR, Piñeyro A (1993) Estudio quimico de hojas, raíz y tallo de *Karwinskia parvifolia* y *K. humboldtiana*. XI. Encuentro de Investigación Biomédica, Universidad Autonoma de Nuevo León, Monterrey, Mexico, p 204

Weller RO, Mitchell J, Daves D (1980) Experimental and clinical neurotoxicology. Williams and Wilkins, Baltimore

Yussim LF, Lara OR, Benavides A, Hernandez B, Fernandez R, Tamariz J, Zepeda G (1995) Anthracenones and anthraquinones from roots of five species of the genus *Karwinskia*. Phytochemistry 40:1429–1431

XIV *Lactarius* Species (Mushrooms): In Vitro Culture and the Production of Sesquiterpenes and Other Secondary Metabloites

L.S. Trifonov[1] and P. Chakravarty[2]

1 Introduction

The genus *Lactarius* (family Russulaceae, class Basidiomycotina) is one of the largest and best known in North America and Europe (Burlingham 1908, 1910, 1913; Hesler and Smith 1979). Tropical species of *Lactarius* have been studied by several authors and are now known to occur in Asia, Africa, and central and South America (Singer 1975; Singer et al. 1983; Montoya et al. 1996). There are more than 250 species distributed in both temperate and tropical regions of the world. Several species of *Lactarius* are edible, known for their delicacy, and are popular all over the world.

Lactarius spp. are known as milk cap fungi because they produce a milky or juicy fluid or latex when broken. However, some older fruiting bodies or a few species that grow in dry or very wet weather may lack latex. Some species may have variable colors of latex, whereas in other species the color of latex will change when exposed to air. The genus *Lactarius* has both edible and poisonous species. In general, the species which have a peppery taste and have white latex that turns yellow or purple when exposed to air are poisonous. However, the bright color of the juice is considered to be an indicator that the species is edible. For example, *L. deliciosus*, a much coveted mushroom in Europe, has orange latex that turns green in contact with air. A typical *Lactarius* sp. consists of a cap, gills, and a stalk or stem (Fig. 1). The gills of the mushroom bear spores.

Species of *Lactarius* are known to occur under conifer and broadleaf trees and form mycorrhizal associations with a wide range of temperate and tropical trees. Most of the land plants form mycorrhizal associations with a specialized group of soil fungi. In this association, both partners (plant and fungus) benefit from one another, i.e. the fungus derives nutrients and carbohydrates from the host plant, and in turn, the fungus helps the plant to obtain nutrients and water more efficiently from the soil. These fungi are considered to be far-reaching extensions of the root system. The fungal mycelia explore and extract nutrients from a volume of soil far beyond the capabilities of the roots alone. Some nutrients (mainly phosphate) are translocated through the mycelia to the mycorrhizae, where they are released to the roots for host utilization. Several different types of mycorrhizae are known, but ectomycorrhizae (colonize out-

[1] Department of Chemistry, University of Alberta, Edmonton, Alberta T6G 2G2, Canada
[2] Canadian Forest Service, Northern Forestry Centre, 5320-122 Street, Edmonton, Alberta T6H 3S5 Canada

Biotechnology in Agriculture and Forestry, Vol. 43
Medicinal and Aromatic Plants XI (ed. by Y.P.S. Bajaj)

Fig. 1. Fruiting bodies of *Lactarius deliciosus*. (Courtesy of Dr. L.J. Hutchison)

side the roots) and endomycorrhizae (colonize inside the roots) are the most common. Several species of *Lactarius* form ectomycorrhizae with pines, spruces, firs, birches, alders, oaks, and several tropical tree species. Besides enhancing nutrient uptake, mycorrhizal fungi also help the plant to take up water and increase the drought resistance of young seedlings (Duddridge et al. 1980; Parke et al. 1983), detoxify certain soil toxins (Zak 1971), help seedlings to withstand high soil temperature (Marx and Bryan 1971) or extreme acidity (Marx 1976), and reduce the incidence of root pathogenic fungi (Zak 1964; Marx 1972, 1973; Chakravarty and Unestam 1987a,b; Chakravarty and Hwang 1991; Duchesne 1994). It is now a well-accepted fact that tree seedlings need mycorrhizal associations for survival in the field when planted out. Nonmycorrhizal seedlings do not survive or grow well until indigenous fungi form mycorrhizae on their roots.

In addition to being used as food, various species of *Lactarius* are reported to have antifeedant activity against pests (Daniewski et al. 1995), antitumor activities (Vidari et al. 1995), and antibacterial and anti-fungal activities (Anke and Sterner 1991); see also Table 1.

2 In Vitro Cultures of *Lactarius* spp.

Cultures of *Lactarius* species can be grown in a variety of ways: from mushroom tissue, ectomycorrhizae, or from spores. Mushrooms collected from the field should be placed into waxed paper bags to prevent drying. Isolation

Table 1. List of some biologically active secondary metabolites reported in *Lactarius*

Secondary metabolite	Biological activity	References
Necatorin (**1**)	Mutagenic	Klamann et al. (1989) Sterner et al. (1982)
2-Isoprenylhydroquinone (**4a**)	Antifungal	De Bernardi et al. (1992)
2-Geranylgeranylhydroquinone (**5a**)	Cytotoxic	Vidari et al. (1995b)
Flavidulol A, B, C (**6a, 7, 6b**)	Immunosuppresive	Fujimoto et al. (1993)
Lactarorufin A (**13**)	Antifeedant	Daniewski et al. (1993, 1995)
3-O-ethylfurandiol (**14**)	Antifeedant	Daniewski et al. (1993, 1995)
Uvidin A (**15**)	Cytotoxic, antifeedant,	Garlaschelli et al. (1994)
Isovelleral (**16**)	Cytotoxic, antifeedant, antibacterial, antifungal	Vidari et al. (1995a)[a]; Sterner et al. (1985a)
9-Hydroxyisovelleral (**17**)	Antifeedant, mutagenic	Sterner et al. (1987)
Velleral (**18**)	Antifeedant, antibacterial, antifungal	Vidari et al. (1995a)[a]
Piperdial (**19**)	Antibacterial, antifungal	Sterner et al. (1985b)
Lactaroscrobiculide A (**20**)	Cytotoxic	De Bernardi et al. (1993)
Chrysorrhedial (**21**)	Cytotoxic, antifeedant	De Bernardi et al. (1993)
Chrysorrheal (**22**)	Cytotoxic	De Bernardi et al. (1993)

[a] See also references therein.

should be done as soon as possible; otherwise mushrooms should be kept in a refrigerator. According to Trappe (1977), best results are often obtained when isolations are done immediately in the field. For ectomycorrhizae, roots should be collected from the field and placed together with a little moist soil, humus, or moss in a tightly closed container. Isolation should be done as soon as possible, since mycorrhizal roots can dehydrate rapidly.

2.1 Isolation of *Lactarius* spp.

2.1.1 Isolation from Mushrooms

Young clean mushrooms are cut with a shallow slit across the middle of the cap surface along the length of one side of the stipe. With a fine tipped sterilized scalpel, small pieces of healthy tissue are cut and placed directly onto the nutrient agar in plates or tubes. Isolations should be tried from more than one mushroom for a better chance of success. Cultures of *Lactarius* can be grown both on synthetic and semisynthetic nutrient media. Some of the common media used are PDA (potato dextrose agar, 39 g Difco or BBL potato dextrose agar and 1 l distilled water), MEA (malt extract agar, 20 g malt extract, 1 g yeast extract, 15 g agar, 1 l distilled water), carrot agar (400 g carrot extract, 15 g agar, 1 l distilled water), Hagem agar medium (malt extract 5 g, glucose 5 g, NH_4Cl 0.5 g, KH_2PO_4 0.5 g, $MgSO_4.7H_2O$ 0.5 g, $FeCl_3$ 1 ml of 1%, distilled water 1 l, pH 4.5–4.8, Modess 1941), Melin-Norkrans' nutrient medium (0.5 g $CaCl_2$, 0.025 g NaCl, 0.5 g KH_2PO_4, 0.025 g $(NH_4)_2HPO_4$, 0.15 g $MgSO_4.7H_2O$, 1.2 ml $FeCl_3$ (1%), 25 μg thiamine HCl, 2.5 g glucose, 15 g agar, 1 l distilled water, pH 5.0–5.5; Norkrans 1949), and modified Melin-Norkrans nutrient (MMN) medium (0.05 g $CaCl_2$, 0.025 g NaCl, 0.5 g KH_2PO_4, 0.25 g

Fig. 2. Submerged culture of *Lactarius deliciosus* in liquid MMN medium

$(NH_4)_2HPO_4$, 0.15 g $MgSO_4.7H_2O$, 1.2 ml $FeCl_3$ (1%), 100 μg thiamine HCl, 10 g glucose, 3 g malt extract, 1 l distilled water, pH 5.0–5.5; Marx 1969). The authors used MMN medium to isolate several species of *Lactarius* and obtained excellent growth of the fungi. Once tissues are plated onto the nutrient medium the Petri plates or tubes are incubated at 20 to 25 °C. After 3–5 days, culture plates or tubes are observed under a stereomicroscope for initial fungal growth and contamination. Several uniform cultures of *Lactarius* emerging in different plates or tubes prove initial evidence of successful isolation. If there is contamination, attempts should be made to transfer pure mycelia of *Lactarius* onto new plates or tubes. Heavily contaminated plates should be discarded. Cultures that grow well in the plates or tubes will be ready for transfer after 3 to 5 weeks. In order to identify a species of *Lactarius*, microscopical, physiological, and biochemical tests are necessary (Hutchison 1991). *Lactarius* spp. can also be grown in liquid media in a flask (Fig. 2) or fermentor for mass production of inoculum for isolation of metabolite produced by the fungi.

2.1.2 Isolation from Ectomycorrhizae

Ectomycorrhizal roots are rinsed under tap water to remove adhering debris and are then placed in a plastic vial containing a mild detergent and shaken vigorously for 4 min. The mycorrhizal roots are rinsed off with tap water, treated with 30% H_2O_2 for 15–20 s, and then washed several times with sterile distilled water. Roots are aseptically transferred to MMN nutrient media in plates or tubes and incubated as described above. Species of *Lactarius* should be properly identified once grown in the nutrient media.

2.1.3 Isolation from Spores

Spores are collected by keeping a fruit body cap or pieces of a cap under the lid of a sterilized glass Petri plate, so that the spores are cast onto the bottom of the plate. The plate can be replaced a few times to obtain several spore prints (Fries 1978). To each Petri plate containing spore prints, 0.5 ml of sterile distilled water is added and mixed thoroughly. From this spore suspension, 50 μl is added to each Petri plate containing sterilized MMN nutrient agar medium and incubated at 20–25 °C. The growing mycelia of *Lactarius* sp. should be properly identified.

2.1.4 In Vitro Synthesis

Ectomycorrhizae are synthesized under controlled conditions in order to study the effect of ectomycorrhizal fungi on host plants, their ability to form ectomycorrhizae, biochemical responses of fungal symbionts, and genetic control of ectomycorrhiza formation. Various techniques have been used by several authors to synthesize ectomycorrhizae in vitro (Peterson and Chakravarty 1991). Using Erlenmeyer flasks with sand as a root substrate, and a nutrient solution, Melin (1921, 1922) was the first to synthesize an ectomycorrhiza in vitro. Later, other workers (Hacskaylo 1953; Marx and Zak 1965; Chakravarty and Unestam 1987a,b; Chakravarty et al. 1990) modified Melin's technique by replacing sand with vermiculite, fine peat moss, brick pellets, or a mixture of vermiculite-peat moss-perlite as root substrates. Flasks containing root substrate moistened with nutrient solution are autoclaved. The mouth of the flasks are stoppered with absorbent cotton plugs or with aluminium foil. When the substrates have cooled, one or more aseptic seeds or seedlings are introduced into each flask under sterile conditions. Fungal inoculum can be added as a mycelial slurry or as agar plugs of actively growing fungal mycelium at the same time as introducing seeds/seedlings of after seedling establishment. Similarly, Mason jars, test tubes, Petri dishes, etc. can be used. In this system of in vitro synthesis, seedlings can be kept sterile. Recently, mycorrhizae have also synthesized under semi-sterile or non-sterile conditions using growth pouches, plastic chambers, plastic pots, root-trainers, or in a hydroponic system (Peterson and Chakravarty 1991). After the desired growth, seedlings are harvested and mycorrhizal colonization is estimated.

2.2 Production of Fungal Inoculum and Inoculation of Seedlings

Species of *Lactarius* can be produced on a large scale for inoculation of seedlings. Marx and Kenney (1982) described detailed procedures for production of ectomycorrhizal fungi. Alexander (1981) reported significantly increased growth of Sitka spruce seedlings when inoculated with *L. rufus* (Fr.) Fr. Ectomycorrhizal fungi on certain sites are more beneficial to trees than other fungal species that occur naturally on such sites. Large quantities of inoculum of *Lactarius* spp. can be produced either as spores or vegetative inoculum. For spore inoculum, whole or chopped mushrooms are dried before use. The vegetative matrix of the mushroom is killed by desiccation and contains only spore inoculum. This dried inoculum can be applied to the seedlings for ectomycorrhiza formation. The disadvantages of spore inoculum are the lack of standard laboratory tests to determine spore viability, and a large quantity of mushrooms is required to inoculate nurseries. For large scale inoculation, vegetative inoculum of fungi is used. Several methods are used to produce vegetative inocula: in liquid media, in cereal grains, or in peat moss. The authors obtained excellent vegetative growth of *Lactarius* spp. on a peat:vermiculite (1:5) mixture. Fine particles of vermiculite are removed by sieving it through a fine mesh. The mixture is moistened with liquid MMN medium and then autoclaved. The authors have successfully used 2-l Erlenmeyer flasks as well as autoclavable plastic containers. When cooled, slightly blended mycelium from a liquid culture is added to the peat-vermiculite mixture. In addition, several agar-mycelia discs are cut into small pieces and evenly spaced through the mixture. The peat-vermiculite mixture is now thoroughly mixed and incubated at 20–25 °C. Many species of *Lactarius* have slow growth (e.g. *L. deliciosus* Fr. [Fr.] S.F. Gray) and take several months to colonize the substrate. When the fungus has colonized the entire substrate and is free of microbial contamination, the inoculum is ready to be used. The presence of fungus should be confirmed by isolation from the substrate.

The nutrients from the peat-vermiculite inoculum are to be removed before inoculation of the seedlings. The authors found rapid colonization of inoculum by saprophytic fungi if nutrients were not removed. Five liters of inoculum is wrapped in several layers of cheesecloth and kept under running cool tapwater for 2 to 3 min. Excess water is removed by squeezing the inoculum in the cheesecloth by hand. The inoculum can be air-dried after washing. The inoculum is applied before seeds are sown, at the same time as seeds are sown, when cuttings are propagated, or after seedlings emerge. The inoculum is either surface broadcast in the seedbeds or applied as layers in the root zone. Surface broadcasting requires large quantities of inoculum for uniform distribution on the surface of seedbeds, whereas layering of inoculum requires small quantities of inoculum since it stays below the root systems of seedlings.

3 Secondary Metabolites Isolated from *Lactarius* Species

3.1 Chemical Groups of Metabolites

A wide range of secondary metabolites belonging to different chemical groups has been reported, and several reviews covering the literature up until 1993 has appeared (Ayer and Browne 1981; Turner and Aldridge 1983; Vidari and Vita-Finzi 1995). Some of the secondary metabolites reported from *Lactarius* species are listed in Table 1.

3.1.1 Sesquiterpenes

More than 100 sesquiterpenes belonging to nine different types of carbon skeletons have been isolated from *Lactarius* species. Studies on their biosynthesis, conducted by several groups over the years, have led to a good understanding of the nature of cyclization and rearrangement reactions interconnecting most of the sesquiterpenes which have a common precursor, farnesyl pyrophosphate (Fig. 3). Several related metabolites derived from the basic skeletons after bond cleavage, loss of one carbon atom or cyclization are presented in Fig. 4.

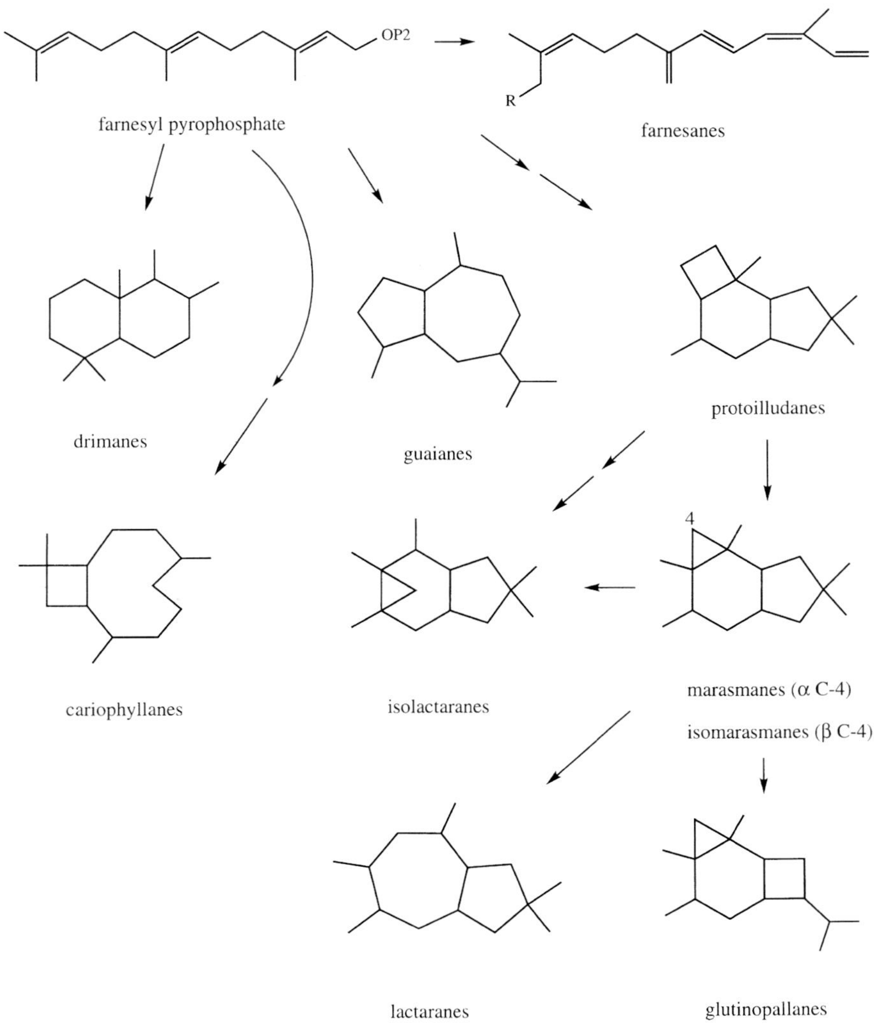

Fig. 3. Basic skeletons and biosynthetic relationships of sesquiterpene metabolites from *Lactarius* species

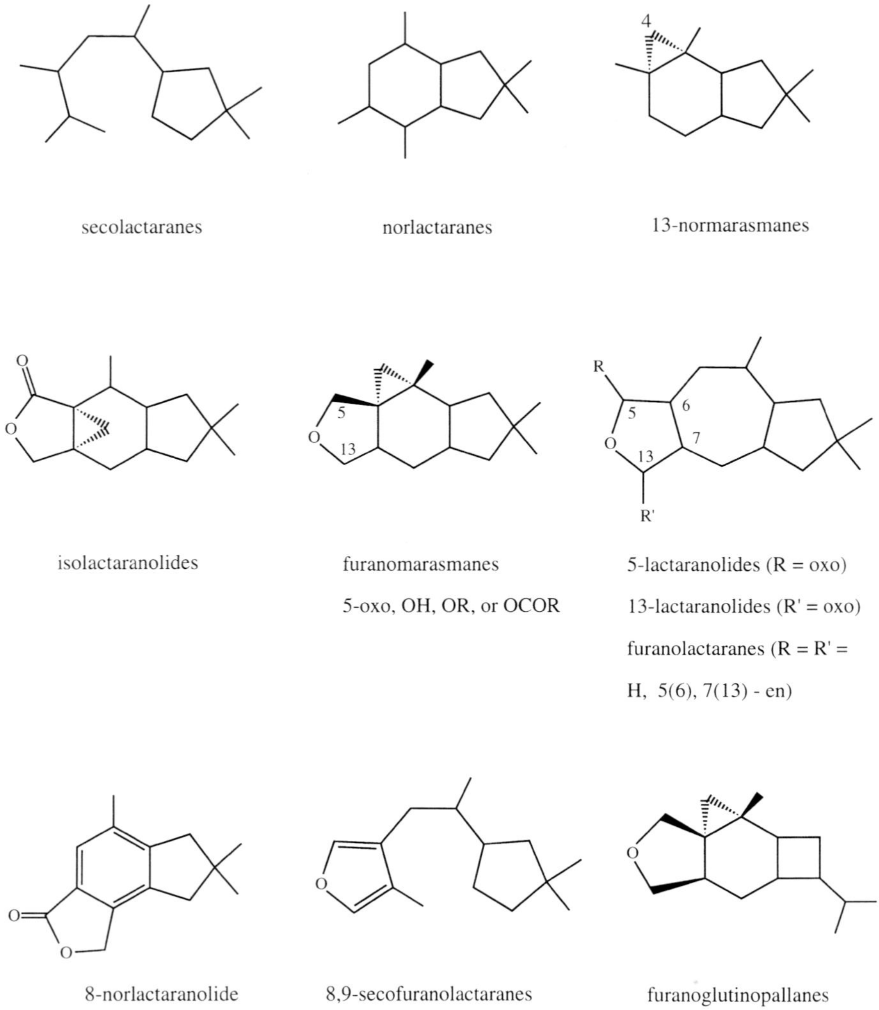

Fig. 4. Seco-, nor-, and heterocyclic sesquiterpenes from *Lactarius*

Marasmanes, lactaranes, and secolactarane sesquiterpenes are ubiquitous among *Lactarius* species, while drimanes, farnesanes, glutinopallanes, isolactaranes, and especially guaianes, have been isolated from certain *Lactarius* species and are considered good chemotaxonomical markers. Interestingly, farnesane, drimane, cariophyllane, and guaiane sesquiterpenes are also common plant metabolites. The metabolites derived from the protoilludane skeleton (Fig. 3) have been isolated only from Basidiomycotina.

A small group of farnesane metabolites was reported recently (Gamba-Invernizzi et al. 1993). Their handling requires special conditions and modification of classical isolation techniques, since they are thermally and photochemically unstable and polymerize readily. Only one cariophyllane sesquiterpene (4,5-epoxy, 12-hydroxy, 8(15)-ene) has been reported so far (Daniewski et al. 1981). In contrast, more than 36 drimanes, mostly fatty acid esters of drimane alcohols, have been reported from *L. uvidus* (Garlaschelli et al. 1994) and *L. flavidus* (Vidari and Vita-Finzi 1995).

Guaianes are responsible for the orange to blue color of the milky juice of many *Lactarius* species of Section Dapetes (*L. deliciosus*, *L. indigo*, etc.). Due to their polyunsaturated nature, most of the guaianes polymerize readily when exposed to air and light. The natural form of some of these sesquiterpenes is esters with fatty acid, which hydrolyze quickly. Interestingly, some of the guaiane sesquiterpenes have been reported from European, but not from Indian or Californian specimens. An adequate explanation of this has not yet been found. Only two protoilludane sesquiterpenes have been isolated, both from *L. violascens* (Vidari and Vita-Finzi 1995).

The marasmanes and 13-normarasmanes are represented each by more than a dozen compounds, while only two isomarasmanes have been reported so far. Recently, four new marasmane hydroxy lactones were reported (Daniewski et al. 1996). It has been demonstrated that several metabolites from the marasmane series are labile compounds and decompose or rearrange readily to form furanolactaranes, especially when extracted with alcohols (List and Hackenberg 1973) or chromatographed on aluminium oxide (Sterner et al. 1985a,b). All furanomarasmane sesquiterpenes (Fig. 4) bear an additional oxygen atom at C-5. In some representatives, the lactol hydroxyl group at C-5 is esterified with fatty acids (Sterner et al. 1983; Hansson and Sterner 1991).

Glutinopallanes are represented by only two fatty acid-coupled furane lactols (Favre-Bonvin and Gluchoff-Fiasson 1988), closely related to the corresponding furanomarasmanes (Fig. 4). Only one isolactarane lactone is known, while the lactarane and the 5-lactaranolides are represented by nine and 23 metabolites, respectively. The only known 8-norlactarane could be an artifact, derived from the corresponding 5-lactaranolide. Several naturally occuring and synthetic 13-lactaranolides, as well as three 8,9-secofuranolactaranes are also known (Vidari and Vita-Finzi 1995).

3.1.2 Alkaloids, Prenylated Phenols, Chromenes, Benzofuranes, and Indoles

The search for metabolites responsible for the strong mutagenic effect of *L. necator* led to the isolation of one monomeric (necatorin **1**) and several 4,4′-O-dimeric dibenzonaphthiridinone alkaloids. These unique alkaloids are responsible for the color and the biological activity of *L. necator* (Klamann et al. 1989, Fig. 5).

A number of prenylated phenols with general formulae **4–7** have been isolated from the edible *L. flavidulos*, *L. fuliginosus*, and *L. picinus* (Fig. 6). A trimer of compound **4a**, as well as a benzofurane (**9**) derived from **4a** have been found in *L. fuliginosus*. Flavidulol B (**7**), as well as the 8,8′-O-dimers of

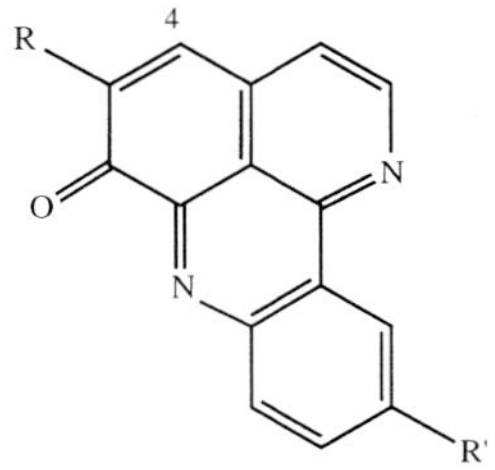

1 R = R' = OH (necatorin)

2 R = OH, R' = H, 4,4' - dimer

3 R = R' = OH, 4,4' - dimer

Fig. 5. Alkaloids from *Lactarius necator*

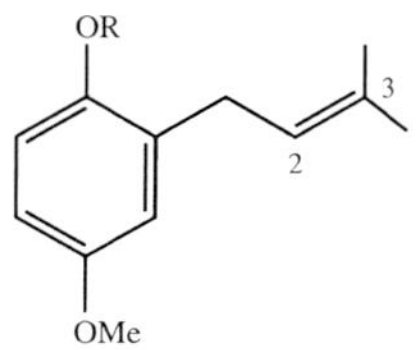

4a R = H

4b R = stearoyl

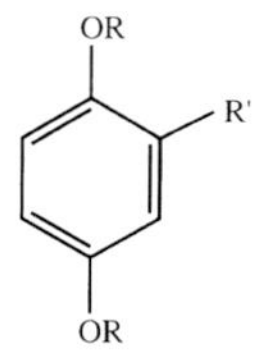

5a R = H, R' = geranylgeranyl

5b R = $C_{13\text{-}17}H_{27\text{-}35}CO$,

R' = geranylgeranyl

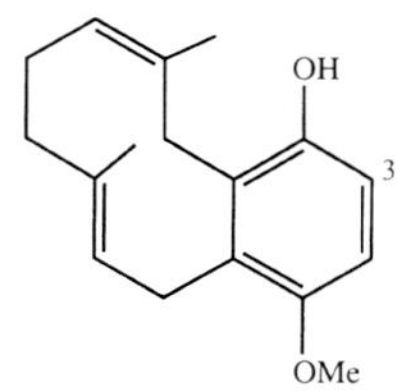

(**6a**) flavidulol A

(**6b**) flavidulol C,

3,3' - dimer

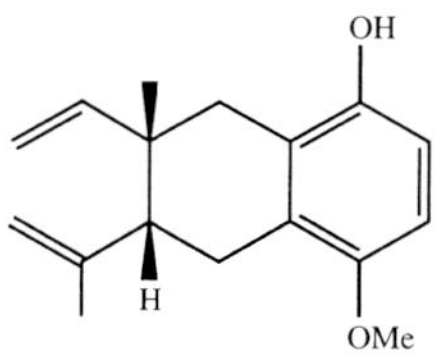

(**7**) flavidulol B

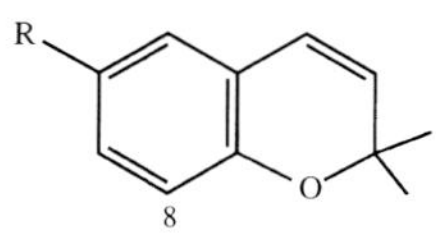

8a R = H

8b R = OMe

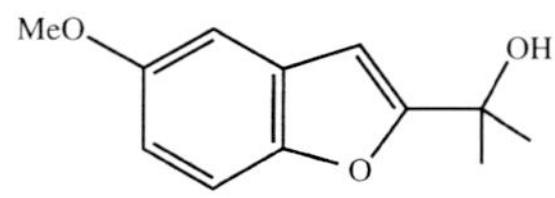

9

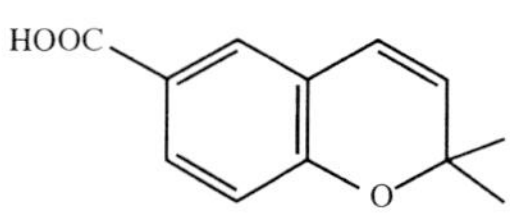

10

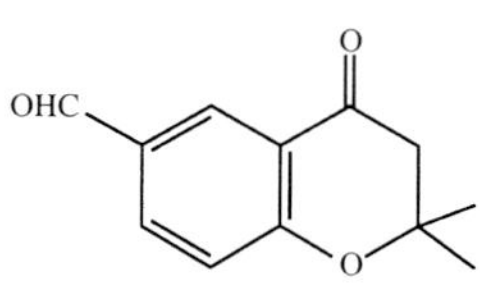

11

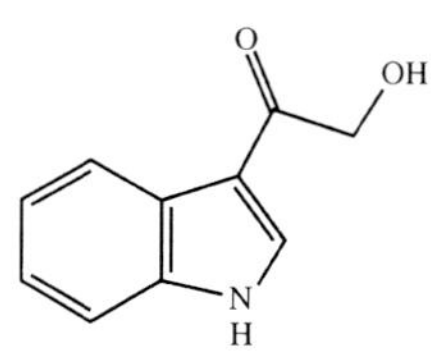

12

Fig. 6. Prenylated phenols and indoles from *Lactarius* species

8 may be artifacts, products of rearrangement of **6a** or dimerization of **8**, respectively.

Recently, Ayer and Trifonov (1994) isolated aromatic compounds from liquid malt extract cultures of *L. deliciosus*. Among the metabolites produced under these conditions were the known plant metabolite anofinic acid (**10**), the closely related new 2,2-dimethylchroman-4-one-6-carboxaldehyde (**11**), and a small amount of the simple new indole derivative, 3-hydroxyacetylindole (**12**) (see Fig. 6).

3.1.3 Miscellaneous Metabolites

Palmitic, stearic, oleic, linoleic, and 6-oxo-stearic acid (lactarinic acid) and their esters have been frequently isolated from *Lactarius* species. Terpenes, steroids, volatiles, and non-protein amino acids have also been reported (Turner and Aldridge 1983). A new heptanorergostane metabolite was recently added to the group of steroids (Kobata et al. 1994). Simple amides such as N-phenyl-2-naphthylamide and heptanamide have been isolated from *L. hysginus* (Dong et al. 1992).

3.2 Occurrence of Metabolites of *Lactarius* Species and Biosynthesis

Many irregularities in the occurrence of metabolites in *Lactarius* species have been observed. In several instances, metabolites have not been detected in species from which they have been reported earlier. Whether this is due to dealing with easily confused subspecies, "habitat-induced differences", or due to following different isolation procedures, is still not completely understood. Typical for *Lactarius* are fast enzymatic processes leading to hydrolysis of fatty acid esters, cyclization reactions, and skeletal rearrangements, occurring especially after the fruiting bodies have been injured.

A complete list of secondary metabolites isolated from *Lactarius* species have been compiled (as of the end of 1993) by Vidari and Vita-Finzi (1995). Recent biosynthetic studies conducted by Sterner in Sweden with ^{18}O and ^{2}H precursors shed light on the intimate interconversions and biosynthesis of sesquiterpene metabolites of *L. vellereus* (Sterner et al. 1985a; Hansson and Sterner 1991; Hansson et al. 1993).

3.3 Other Studies

Extensive studies on protein, lipid, carbohydrate, fiber, and vitamin D content of edible *Lactarius* species have been published (Kasuga et al. 1993; Mattila et al. 1994). Insight into the structure and biosynthesis of liquid rubber (*cis*-polyisoprene) was obtained. It has been found that trans, trans-farnesyl pyrophosphate is the biosynthetic precursor of cis-polyisoprene and that the termination of polymerization is caused by esterification of polyisoprenyl pyrophosphate (Tanaka and Kawahara 1994). The metals content, which is a growing concern in connection with commercial utilization of *Lactarius* species, has been studied. Pronounced accumulation of manganese but not of copper or zinc was reported for *L. nebularis* (Vetter 1994). The contents of selenium (Vetter 1993), aluminium, cadmium, lead (Mandic et al. 1992), arsenic (Stijve and Bourqui 1991), and radioactive cesium (Vobecky and Tethal 1993) have also been studied but no alarmingly high levels of these metals have been detected. Recently, the hydrolysis of stearoylvelutinal to hot-tasting velutinal catalyzed by human saliva was studied (Hashimoto et al. 1994).

3.4 Biological Activities of Metabolites Produced by *Lactarius* Species

Although the biological activities of most of the metabolites isolated from *Lactarius* species have not been tested, numerous metabolites have shown interesting biological activities. Most of the work in this respect has been conducted by Daniewski in Poland, and Sterner in Sweden. Some of the biologically active metabolites from *Lactarius* species are listed in Table 1.

3.4.1 Antifeedant Activity

Significant but not universal deterrent activity has been found for lactarorufin A (**13**) and 3-O-ethylfurandiol (**14**, obvious artifact, Fig. 7) against some storage pests (e.g. *Sitophilus granarius*, *Tribolium confusum*, etc.). Detailed structure-activity studies have shown that inversion of the configuration at C-8, hydrogenation, acetylation, and shifting the carbonyl group from C-13 to C-5 decreases the activity (Daniewski et al. 1993). A high degree of selectivity of deterrent activity against individual species has been documented in many instances.

Sesquiterpene dialdehydes with a marasmane (**16**, **17**) and lactarane (**18**, **19**) skeleton (Fig. 7), usually having a very strong pungent taste, are among the strongest antifeedants, not only against insects and fishes (Sterner 1985), but also against mammals such as opossum (Camazine et al. 1983). A recent structure-antifeedant activity study included 53 sesquiterpenes from *Lactarius* species (Daniewski et al. 1995).

3.4.2 Mutagenic Activity

Testing for mutagenic effect is imperative in the face of ever increasing importance of wild mushrooms as food. Of all the species tested in Ames *Salmonella typhimurium* tester, strain *L. necator* had the highest mutagenic activity. The alkaloid necatorin (**1**) isolated from this species has mutagenic activity comparable to that of aflatoxins (Sterner et al. 1982; Suortti and Wright 1983).

Isovelleral (**16**) and 9-hydroxyisovelleral (**17**) (Fig. 7) have also shown significant mutagenic activity. Interestingly, the analog of isovelleral with a

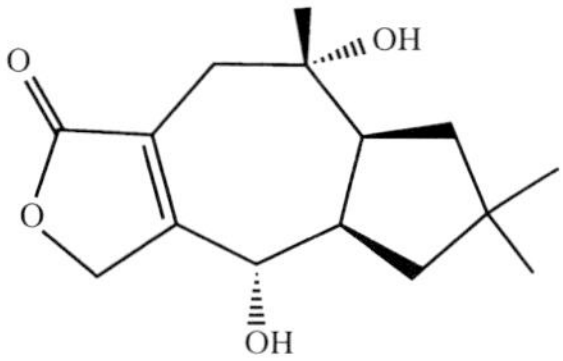

(**13**) lactarorufin A

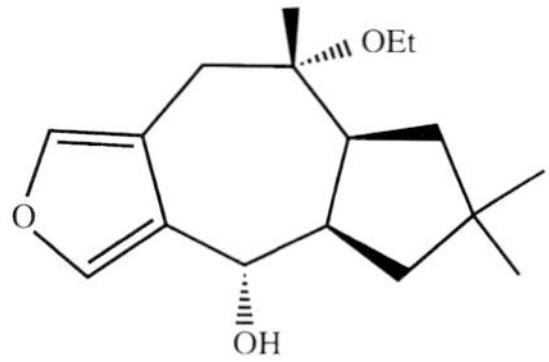

(**14**) 3-O-ethylfurandiol

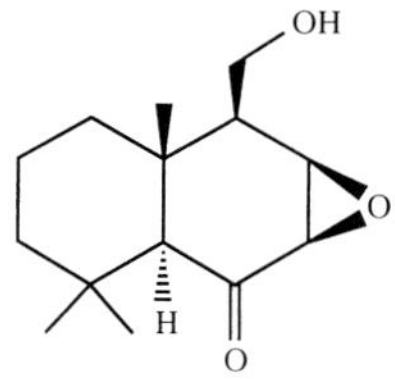

(**15**) uvidin A

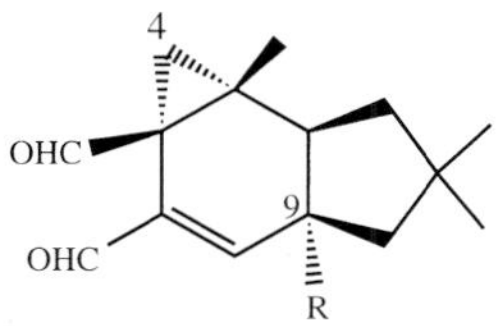

(**16**) isovelleral (R = H)

(**17**) 9-hydroxyisovelleral (R = OH)

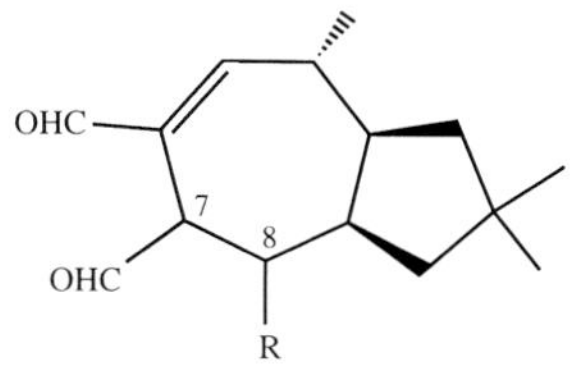

(**18**) velleral (R = H, 7(8)-en)

(**19**) piperdial (R = α-OH, 7α-H, 8β-H)

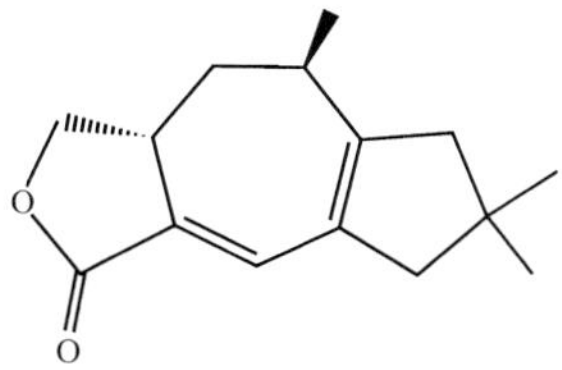

(**20**) lactaroscrobiculide A

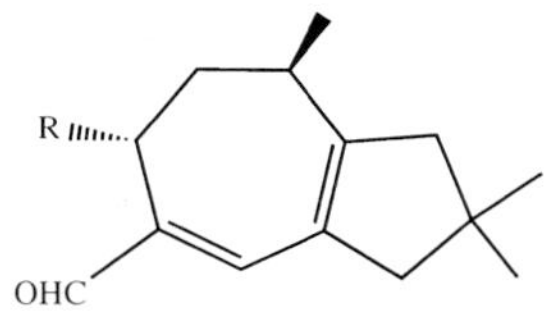

(**21**) chrysorrhedial (R = CHO)

(**22**) chrysorrheal (R = CH_2OH)

Fig. 7. Biologically active sesquiterpenes from *Lactarius* species

β-cyclopropane ring has a much lower activity (Sterner et al. 1987; Anke and Sterner 1991).

3.4.3 Cytotoxic Activity

Several guaiane sesquiterpenes isolated from edible, commercially important species like *L. deliciosus*, *L. sanguifluus*, and *L. deterrimus* have shown weak cytotoxic activity. Since these compounds are thermally labile and do not survive food processing, they do not pose a risk to humans (Anke et al. 1989b). Immunosuppressive activity has been reported for phenols **6a**, **6b**, and **7** (Fig. 6; Fujimoto et al. 1993). Recently, the phenol **5a** was found to be a highly active antitumor antibiotic (Vidari et al. 1995). Cytotoxic activity has been observed for uvidin A (**15**) (Garlaschelli et al. 1994), lactaroscrobiculide A (**20**), chrysorrhedial (**21**), and chrysorrheal (**22**) (De Bernardi et al. 1993).

3.4.4 *Antibacterial and Antifungal Activity*

The most active antibacterial and antifungal metabolite reported from *Lactarius* species so far is isovelleral (**16**) (Anke and Sterner 1991), with MIC 0.5–5 μg/ml. The same compound has cytotoxic activity and arrests germination of seeds at a concentration of 100 μg/ml (Anke et al. 1989a). Reduction of one or both aldehydic groups significantly reduces the antibacterial and antifungal activity (Sterner et al. 1985a,b). Several guaianes, heterocyclic marasmanes, and lactaranes, as well as the phenols **4a** and **6a** (Fig. 6) have shown moderate antibacterial activity.

3.4.5 *Other Biological Activities*

The dialdehydic sesquiterpenes responsible for the hot-tasting and pungent taste of several species of *Lactarius* have been studied in detail (Sterner et al. 1985a; De Bernardi et al. 1993). These dialdehydes are the chemical basis for the deterrent activity against predators. The farnesane sesquiterpenes having a disagreeable taste, as well as those responsible for gastrointestinal upset to humans have been identified.

4 Summary and Conclusions

Species of *Lactarius* are found both in tropical and temperate parts of the world and are used as food, being considered a delicacy. Several species form ectomycorrhizal associations with conifer seedlings and have been used by forest nurseries to inoculate seedlings for better growth. The secondary metabolites isolated so far from more than 50 species of *Lactarius* belong to the terpenes (mostly sesquiterpenes with nine different carbon skeletons), alkaloids, prenylated phenols, fatty acids, indoles, amides, and amino acids. Although a small number of these metabolites have been tested for biological activity, immunosuppressive (phenols), cytotoxic, antibacterial, antifungal, and antifeedant activities (sesquiterpenes with lactarane, marasmane or drimane skeletons) have been documented. Mutagenic activity has been attributed to some alkaloids and sesquiterpene dialdehydes. Significant insight has been obtained over the years into the chemical interconversions of many sesquiterpenes as well as their biological activities.

Since *Lactarius* species are ectomycorrhizal fungi, they cannot be grown for commercial mushroom production in artificial media. However, vegetative mycelia of these species can be easily grown in nutrient media and metabolites can be isolated. The biological activities of some of the metabolites make them of potential use as templates for new drugs and pesticides. Screening for mutagenic activity of edible and especially commercially important mushrooms has to be extended to all other species. So far, less than 25% of all known *Lactarius* species have been studied for secondary metabolites. Rigorous screening of the metabolites of *Lactarius* species is needed in order to exploit their potential use for medicinal, pharmaceutical, agricultural, and forestry purposes.

References

Alexander IJ (1981) The *Picea sitchensis* + *Lactarius rufus* mycorrhizal association and its effect on seedling growth and development. Trans Br Mycol Soc 76:417–423

Anke H, Sterner O (1991) Comparison of the antimicrobial and cytotoxic activities of twenty unsaturated sesquiterpene dialdehydes from plants and mushrooms. Planta Med 57:344–346

Anke H, Hillen-Maske E, Steglich W (1989a) Antibiotics from Basidiomycetes. XXX. 9β-hydroxymarasmic acid and other sesquiterpenoids from submerged cultures of a Basidiomycete. Z Naturforsch [c] 44:1–6

Anke H, Bergendorff O, Sterner O (1989b) Assays of the biological activities of guaiane sesquiterpenoids isolated from the fruit bodies of edible *Lactarius* species. Food Chem Toxic 27:393–397

Ayer WA, Browne LM (1981) Terpenoid metabolites of mushrooms and related Basidiomycetes. Tetrahedron 37:2199–2248

Ayer WA, Trifonov LS (1994) Aromatic compounds from liquid cultures of *Lactarius deliciosus*. J Nat Prod 6:839–841

Burlingham GS (1908) A study of the *Lactariae* of the United States. Mem Torrey Bot Club 14:1–109

Burlingham GS (1910) The *Lactariae* of North America – fascicles I and II. Mycologia 2:27–36

Burlingham GS (1913) The *Lactariae* of the Pacific coast. Mycologia 5:305–311

Camazine SM, Resch JR, Eisner T, Meinwald J (1983) Mushroom chemical defense: pungent sesquiterpenoid dialdehyde antifeedant to opossum. J Chem Ecol 9:1439–1447

Chakravarty P, Hwang SF (1991) Effect of an ectomycorrhizal fungus, *Laccaria laccata* on *Fusarium* damping-off in *Pinus banksiana* seedlings. Eur J For Pathol 21:97–106

Chakravarty P, Unestam T (1987a) Mycorrhizal fungi prevent disease in stressed pine seedlings. Phytopathol Z 118:335–340

Chakravarty P, Unestam T (1987b) Differential influence of ectomycorrhizae on plant growth and disease resistance in *Pinus sylvestris* seedlings. Phytopathol Z 120:104–120

Chakravarty P, Peterson RL, Ellis BE (1990) Integrated control of *Fusarium* damping-off in red pine seedlings with *Paxillus involutus* and fungicides. Can J For Res 20:1283–1288

Daniewski WM, Grieco PA, Huffman JC, Rymkiewicz A, Wawrzun A (1981) Isolation of 12-hydroxycaryophillene-4,5-oxide, a sesquiterpene from *Lactarius camphoratus*. Phytochemistry 20:2733–2734

Daniewski WM, Gumulka M, Ptaszynska K, Skibicki P, Bloszyk E, Drozdz B, Stromberg S, Norin T, Holub M (1993) Antifeedant activity of some sesquiterpenoids of the genus *Lactarius*. Eur J Entomol 90:65–70

Daniewski WM, Gumulka M, Przesmycka D, Ptaszynska K, Bloszyk E, Drozdz B (1995) Sesquiterpenes of *Lactarius* origin, antifeedant structure – activity relationships. Phytochemistry 38:1161–1168

Daniewski WM, Gumulka M, Truszewska D, Jacobsson U, Norin T (1996) Four new monohydroxylactones from *L. vellereus*. Phytochemistry 41:1093–1096

De Bernardi M, Vidari G, Vita-Finzi P, Fronza G (1992) The chemistry of *Lactarius fuliginosus* and *Lactarius picinus*. Tetrahedron 48:7331–7344

De Bernardi M, Garlaschelli L, Toma L, Vidari G, Vita-Finzi P (1993) The chemical basis of hot-tasting and yellowing of the mushrooms *Lactarius chrysorrheus* and *Lactarius scrobiculatus*. Tetrahedron 49:1489–1504

Dong D, Wang H, Li G (1992) Chemical research on the metabolites of *Lactarius hysginus*. Tianran Chanwu Yanjiu Yu Kaifa 4:44–47 (Chem Abstr 118: 18924u)

Duchesne LC (1994) Role of ectomycorrhizal fungi in biocontrol. In: Pfleger FL, Linderman RG (eds) Mycorrhizae and plant health. American Phytopathological Society, St Paul, Minnesota, pp 27–45

Duddridge JA, Malibari A, Read DJ (1980) Structure and mycorrhizal rhizomorphs with special reference to their role in water transport. Nature 287:834–836

Favre-Bonvin J, Gluchoff-Fiasson K (1988) Structures of two glutinopallal esters, new natural sesquiterpenoids from *Lactarius glutinopallens*. Phytochemistry 27:286–287

Fries N (1978) Basidiospore germination in some mycorrhiza-forming hymenomycetes. Trans Br Mycol Soc 70:319–324

Fujimoto H, Nakayama Y, Yamazaki M (1993) Identification of immunosuppressive compounds of a mushroom, *Lactarius flavidulus*. Chem Pharm Bull (Tokyo) 41:654–658

Gamba-Invernizzi A, Garlaschelli L, Rossi A, Vidari G, Vita-Finzi P (1993) New farnesane sesquiterpenes from *Lactarius porninsis*. J Nat Prod 56:1948–1953

Garlaschelli L, Mellerio G, Vidari G, Vita-Finzi P (1994) Fungal metabolites. XXXII. New fatty acid esters of drimane sesquiterpenes from *Lactarius uvidus*. J Nat Prod 57:905–910

Hacskaylo E (1953) Pure culture synthesis of pine mycorrhizae in terra-lite. Mycologia 45:971–975

Hansson T, Sterner O (1991) Studies of the conversion of sesquiterpenes in injured fruit bodies of *Lactarius vellereus*. A biomimetic transformation of stearoylvelutinal to isovelleral. Tetrahedron Lett 32:2541–2544

Hansson T, Pang Z, Sterner O (1993) The conversion of [12-$^{2}H_{3}$]-labeled velutinal in injured fruit bodies of *Lactarius vellereus*. Further insight into the biosynthesis of the Russulaceae sesquiterpenes. Acta Chem Scand 47:403–405

Hashimoto T, Tanaka H, Asakawa Y (1994) Stereostructure of plagiochiline A and conversion of plagiochiline A and stearoylvelutinal into hot-tasting compounds by human saliva. Chem Pharm Bull (Tokyo) 42:1542–1544

Hesler LR, Smith AH (1979) North American species of *Lactarius* (Pers.:Fr.) S.F. Gray. Doc Mycol 10:1–85

Hutchison LJ (1991) Description and identification of cultures of ectomycorrhizal fungi found in North America. Mycotaxon 42:387–504

Kasuga A, Fujihara S, Aoyagi Y, Kawai H, Sasaki H, Sugahara T (1993) Mushroom composition. Joshi Eiyo Daigaku Kiyo 24:97–105 (Chem Abstr 121: 229327b)

Klamann JD, Fugmann B, Steglich W (1989) Alkaloidal pigments from *Lactarius necator* and *Lactarius atroviridis*. Phytochemistry 28:3519–3522

Kobata K, Wada T, Hayashi Y, Shibata H (1994) Studies on chemical components of mushrooms. Part III. Velomolide, a novel norsterol from the fungus *Lactarius volemus*. Biosci Biotechnol Biochem 58:1542–1544

List P, Hackenberg H (1973) Sharp tasting substance from *Lactarius vellereus*. Z Pilzkd 39:97–102

Mandic ML, Grgic J, Grgic Z, Seruga M (1992) The natural levels of aluminium, cadmium, and lead in wild mushrooms in eastern Croatia. Dtsch Lebensm Rundsch 88:76–77

Marx DH (1969) The influence of ectotrophic mycorrhizal fungi on the resistance of pine roots to pathogenic infections. I. Antagonism of mycorrhizal fungi to root pathogenic fungi and soil bacteria. Phytopathology 59:153–163

Marx DH (1972) Ectomycorrhizae as biological deterrents to pathogenic root infection. Annu Rev Phytopathol 10:429–454

Marx DH (1973) Mycorrhizae and feeder root diseases. In: Marks GC, Kozlowski TT (eds) Ectomycorrhizae – their ecology and physiology. Academic Press, New York, pp 351–382

Marx DH (1976) Use of specific mycorrhizal fungi on tree roots for forestation on disturbed surface areas. In Utz KA (ed) Conference on forestation of disturbed areas. Burmingham, Alabama. USDA Forestry Service, Altanta, Georgia, pp 109–116

Marx DH, Bryan WC (1971) Influence of ectomycorrhizae on survival and growth of aseptic seedlings of loblolly pine at high temperatures. For Sci 17:37–41

Marx DH, Kenney DS (1982) Production of ectomycorrhizal fungus inoculum. In: Schenck NC (ed) Methods and principles of mycorrhizal research. American Phytopathological Society, St Paul, Minnesota, pp 131–146

Marx DH, Zak B (1965) Effect of pH on mycorrhizal formation of slash pine in aseptic culture. For Sci 11:66–75

Mattila PH, Piironen VI, Uusi-Rauva EJ, Koivistoinen PE (1994) Vitamin D contents in edible mushrooms. J Agric Food Chem 42:2449–2453

Melin E (1921) Über die Mykorrhizenpilze von *Pinus silvestris* L. und *Picea abies* (L.) Karst. Svensk Bot Tidskr 15:192–203

Melin E (1922) Untersuchungen über die *Larix* Mycorrhiza. I. Synthese der Mykorrhiza in Reinkulture. Svensk Bot Tidskr 16:161–196

Modess O (1941) Zur Kenntnis der Mykorrhizabildner von Kiefer und Fichte. Symb Bot Ups 5:1–147

Montoya BL, Bandala VM, Guzman G (1996) New and interesting species of *Lactarius* from Mexico including scanning electron microscope observation. Mycotaxon 57:411–424

Norkrans B (1949) Some mycorrhiza-forming *Trichoderma* species. Svensk Bot Tidskr 43:485–490

Parke JL, Linderman RG, Black CH (1983) The role of ectomycorrhiza in drought tolerance of Douglas-fir seedlings. New Phytol 95:83–95

Peterson RL, Chakravarty P (1991) Techniques in synthesizing ectomycorrhiza. In: Norris JR, Read DJ, Varma AK (eds) Methods in microbiology, vol 23. Techniques for the study of mycorrhiza. Academic Press, San Diego, pp 75–106

Singer R (1975) Tropical Russulaceae: *Lactarius* sect. Polysphaerophorini in the Gulf area. Nova Hedwigia 26:897–901

Singer R, Araujo I, Ivory MH (1983) The ectotrophically mycorrhizal fungi of the neotropical lowlands, especially central Amazonia. Cramer, Vaduz

Sterner O (1985) The Russulaceae sesquiterpenes. PhD Thesis, University of Lund

Sterner O, Bergman R, Franzen C, Kesler E, Nilsson L (1982) Mutagens in larger fungi. II. The mutagenicity of commercial pickled *Lactarius necator* in the *Salmonella* assay. Mutat Res 104:233–237

Sterner O, Bergman R, Kesler E, Nilsson L, Oluwadiya J, Wickberg B (1983) Velutinal esters of *Lactarius vellereus* and *Lactarius necator*. The preparation of free velutinal. Tetrahedron Lett 24:1415–1418

Sterner O, Bergman R, Kihlberg J, Wickberg B (1985a) The sesquiterpenes of *Lactarius vellereus* and their role in a proposed chemical defense system. J Nat Prod 48:279–288

Sterner O, Bergman R, Franzen C, Wickberg B (1985b) New sesquiterpenes in a proposed Russulaceae chemical defense system. Tetrahedron Lett 26:3163–3166

Sterner O, Carter RE, Nilsson L (1987) Structure-activity relationships for unsaturated dialdehydes. I. The mutagenic activity of 18 compounds in the *Salmonella*/microsome assay. Mutat Res 188:169–174

Stijve T, Bourqui B (1991) Arsenic in edible mushrooms. Dtsch Lebensm Rundsch 87:307–310

Suortti T, Wright A (1983) Isolation and mutagenic fraction from aqueous extract of the wild mushroom *Lactarius necator*. J Chromatogr 255:529–532

Tanaka Y, Kawahara S (1994) Structure of cis-polyisoprene from *Lactarius* mushrooms. Acta Biochim Pol 41:303–309

Trappe JM (1977) Selection of fungi for ectomycorrhizal inoculation in nurseries. Annu Rev Phytopathol 15:203–222

Turner WB, Aldridge DC (1983) Fungal metabolites II. Academic Press, London

Vetter J (1993) Selenium content of some higher fungi. Acta Aliment 22:383–387

Vetter J (1994) Copper, manganese and zinc contents of some edible mushrooms. Z Lebensm Unters Forsch 198:469–472

Vidari G, Vita-Finzi P (1995) Sesquiterpenes and other secondary metabolites of genus *Lactarius* (Basidiomycetes): chemistry and biological activity. In: Atta-ur-Rahman (ed) Studies in natural products chemistry, vol 17. Elsevier, Amsterdam pp 153–206

Vidari G, Vita-Finzi P, Zanochi AM (1995) Fungal metabolites. XXXIV. A bioactive tetraprenylphenol from *Lactarius ligniotus*. J Nat Prod 58:893–896

Vobecky M, Tethal T (1993) Contamination of mushrooms by radiocesium in Bohemia during 1986–1990. Radiochim Acta 60:165–168

Zak B (1964) Role of mycorrhizae in root disease. Annu Rev Phytopathol 2:888–901

Zak B (1971) Detoxification of autoclaved soil by a mycorrhizal fungus. Research note PNW-159. USDA Forestry Service, Portland, Oregon

XV *Lactuca virosa* L. (Bitter Lettuce): In Vitro Culture and Production of Sesquiterpene Lactones

A. STOJAKOWSKA, J. MALARZ, and W. KISIEL

1 General Account

1.1 Distribution and Importance

Lactuca virosa L. (family Asteraceae), also known as *Lactuca silvestris* Lam. or *Wiestia virosa* C.H. Schultz, is indigenous to south west Europe, North Africa and West Asia, and is also distributed in other regions of the Northern Hemisphere, including America. It grows singly or in clusters, on rocky soil and as a ruderal plant, on plains and uplands up to 1000m above sea level (Hallier 1887; Hegi 1929; Bremer 1994).

L. virosa is a biennial herb with a fusiform, thick, branched root. In the first year of vegetation it produces a rosette of basal leaves (Fig. 1) and in the second year an erect stem (60–150cm high) branched in the upper part (Fig. 2). Its flowers are yellow, ligulate, collected in detached, elongated, pyramidal panicles. The flowering period falls in June-August. Leaves are bluish-green, with prickles on the bottom side along the veins, basal – egg-shaped or oblongly oval (15–20cm long), upper – small sagittate. The roots, leaves and stem produce white latex which darkens in the open air.

The latex from aerial parts known as *Lactucarium germanicum*, *Opium frigidum* or lettuce opium was mentioned in Dutch, Spanish and German (until 1890) pharmacopoeias. It was known as an antitussive and sedative remedy in ancient times (Dioscorides, Hippocrates, Galen) and was used to produce analgesia before the introduction of chloroform. In 1799 Cox from Philadelphia noticed similarity of pharmacological effects caused by *lactucarium* to those produced by poppy opium. In the 19th century *lactucarium*, as well as tinctures, extracts and syrups prepared from the plant were widely used in official medicine. The interest in the plant at that time led to the first isolation of two major constituents of *lactucarium* – lactucin and lactucopicrin, by Ludwig and Kromayer in 1847 (cited by Späth et al. 1951). Later, initial pharmacological and clinical studies were carried out and it was generally accepted that the compounds were responsible for the opiate resembling activity of the plant (Hegi 1929; Forst 1937). In the 1930s, Knoll A.G. pharmaceutical company introduced Latucyl – an antitussic drug containing 0.2g of the two sesquiterpene lactones per dose. The drug was successfully

Institute of Pharmacology, Polish Academy of Science, 12 Smetna Street, 31-343 Krakow, Poland

Biotechnology in Agriculture and Forestry, Vol. 43
Medicinal and Aromatic Plants XI (ed. by Y.P.S. Bajaj)

Fig. 1. *Lactuca virosa* L., first year of cultivation

used in the therapy of coughs in the course of long-term diseases such as emphysema and tuberculosis (Hessen 1937; Behrens 1938).

1.2 Present Status and Cultivaton

Nowadays, galenic preparations of *L. virosa* are used in cough mixtures and as mild sedatives, mainly in homeopathy (Da Legnano 1968; List and Hörhammer 1973; Weiss 1974; Levis and Elvin-Levis 1977). However, until recently the plant lacked detailed analysis of its constituents and more modern pharmacological investigations. Another study (Gromek et al. 1992) suggested that lactucin and lactucopicrin are not the only pharmacologically active agents in the plant preparations. For example, a freeze-dried centrifuged latex from roots of 1-year-old plants, lacking the two compounds, reduced the spontaneous locomotor activity in mice at a dose of 2 mg/kg and showed an analgesic effect at a dose of 15 mg/kg, and lactucin itself was much less active in the applied tests.

L. virosa was formerly cultivated in Western Europe and Russia. At present it is grown in Germany, France and in the Netherlands. The seeds germinate easily. Plants cultivated in the Garden of Medicinal Plants of the Institute of Pharmacology, Polish Academy of Sciences, Cracow, yielded 1.9 kg of leaves and 2.2 kg of roots (fresh weight) from 1 m^2 (17 plants) at the end of the first year of vegetation. In the second year, the fresh weight of shoots doubled whilst the weight of roots remained unchanged (Binek 1995).

Fig. 2. *Lactuca virosa* L., 2-year-old plants

L. virosa, due to its multiple virus resistance has been successfully used in breeding programs for the introduction of resistance genes into commercial varieties of the garden lettuce *Lactuca sativa* L. (Tamaki et al. 1995).

1.3 Secondary Metabolites

L. virosa has not been phytochemically studied in detail. Earlier studies were focused on the constituents of *lactucarium*. According to the literature (Hegnauer 1964, 1989; List and Hörhammer 1973) it contains sesquiterpene lactones (lactucin and lactucopicrin), triterpenes (taraxasterol and its acetate, β-amyrin, germanicol and isolupeol), organic acids, carbohydrates and proteins. Oil from the seeds yielded fatty acids (oleic, linolic, palmitic and stearic), α-tocopherol and squalene. The presence of a mydriatic substance of unknown structure was also reported.

Table 1. Yields of sesquiterpene lactones (mg % fresh weight) isolated from 1-year-old roots of *Lactuca virosa* L. intact plant and transformed root culture of the plant

One-year-old roots		Transformed roots	
8-Desoxylactucin + jacquinelin	1.64	8-Desoxylactucin + jacquinelin	1.58
Lactuside A	4.90	Lactuside A	0.90
Macrocliniside A	0.64	Macrocliniside A	0.90
Crepidiaside B	0.62	Crepidiaside B	2.71
Ixerin F	0.62	Ixerin F	3.39
11β,13-Dihydrolactucin	0.48	Scorzoside	1.13
8 Minor lactones	ca. 0.6	Picriside A	1.36

Sesquiterpene lactones were the only extensively investigated secondary metabolites of *L. virosa*. Studies on lactucin structure were completed in the 1950s after more than 100 years since the isolation of this compound (Späth et al. 1951; Wessely et al. 1951; Holzer and Zinke 1953; Barton and Narayanan 1958; Dolejš et al. 1958). The stereochemistry of lactucin was finally revised in 1973 (Bachelor and Ito 1973). A revision of lactucopicrin structure was published in 1977 by Pyrek. In the last decade, Gromek (1989a,b, 1991) isolated and identified, in addition to lactucin and lactucopicrin, three further lactucin derivatives (11β,13-dihydrolactucin, 8-desoxylactucin and its 11,13-dihydro derivative, jacquinelin) and a melampolide glycoside, lactuside A. The composition and total sesquiterpene lactone content depended on the organ analyzed and the age of the plant (lactucin and lactucopicrin were not found in 1-year-old plants).

A detailed phytochemical study of the 1-year-old roots of *L. virosa*, which were proved to be active in the pharmacological tests mentioned above, has been completed very recently (Kisiel and Barszcz 1997). From the roots, 15 guaiane- and germacrane-type sesquiterpene lactones were isolated. Of the guaianolides, lactucin-like and zaluzanin C-like compounds were identified. The former show a characteristic UV maximum at 254 nm and are easier to detect than the latter which possess only end UV absorption. Of the germacranolides, germacrolide- and melampolide-like derivatives were characterized (Table 1; Fig. 3).

Studies on the in vitro cultures and secondary metabolites of the plant are summarized in Table 2.

2 In Vitro Culture Studies

Although numerous in vitro culture studies (Alconero 1988) have been conducted on *Lactuca sativa* (common lettuce), i.e. protoplast culture (Enomoto and Ohyama 1989), micropropagation (Pink 1992), genetic transformation (Enomoto and Ohashi 1993); other species of *Lactuca* have not received much interest. The sesquiterpene lactones are undesired constituents of the garden lettuce due to their bitter taste (van Beek et al. 1990), and are sometimes

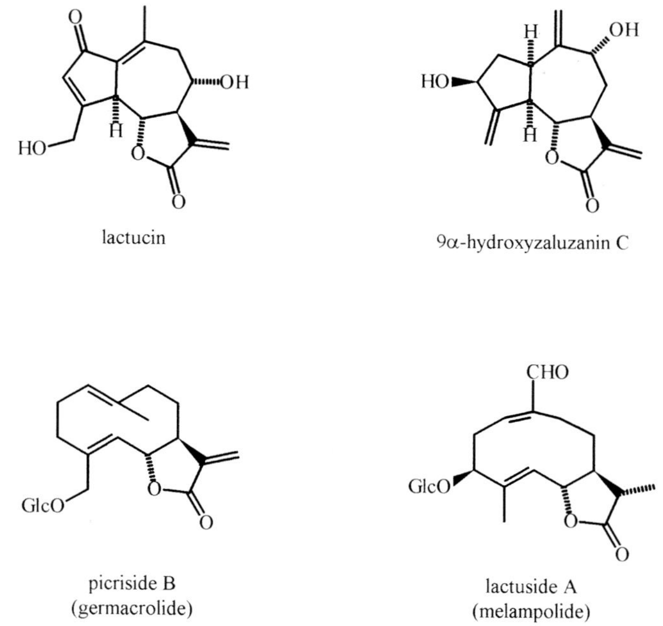

Fig. 3. Skeletal types of sesquiterpene lactones from *Lactuca virosa*

monitored in commercial varieties of the plant. However, the analysis is usually limited to quantification of lactucin-like guaianolides. Biosynthesis of the compounds in hairy root culture of *Lactuca floridana* (L.) Gaertn. was studied recently (Song et al. 1995). Our work on *Lactuca virosa* (Stojakowska et al. 1994, 1997; Kisiel et al. 1995; Kisiel and Barszcz 1997) is discussed here.

2.1 Callus Culture and Cell Suspension

Seeds of *L. virosa* were surface sterilized with 70% ethanol followed by 4% solution of Dimanin C (Bayer), rinsed with sterile distilled water and placed onto a hormone-free agar medium containing 1/4 of MS (Murashige and Skoog 1962) macro- and micronutrients for germinating. To obtain callus, hypocotyls of aseptic seedlings were inoculated on the MS medium containing sucrose (30 g/l), 2,4-D (5 mg/l) and kinetin (2 mg/l), solidified with agar (7 g/l). The calli were transferred to the medium with lowered concentrations of the growth regulators (1 mg/l 2,4-D and 0.3 mg/l kinetin) and maintained either as a stationary culture on the agar medium, or transferred to the liquid medium to obtain a cell suspension. The callus was cultivated under continuous illumination (40 μmol/m^{2}s, cool white fluorescent tubes) at 28 ± 2 °C and subcultured every 4 weeks. The cell suspension was kept under permanent light, initially of intensity 22 μmol/m^{2}s, subsequently 40 μmol/m^{2}s, on a gyratory shaker (120 rpm) and was subcultured every 2 weeks. The culture was also

Table 2. Secondary metabolite and in vitro culture studies on *Lactuca virosa* L.

Reference	Studies conducted/Remarks
Ludwig and Kromayer (1847, 1862, cited by Späth et al. 1951), Kromayer (1861, cited by Späth et al. 1951)	First isolation of two bitter compounds from lactucarium named lactucin and lactucopicrin
Forst (1937), Hessen (1937), Behrens (1938)	Pharmacological and clinical research on lactucarium and latucyl (Knoll)
Barton and Narayanan (1958), Dolejš et al. (1958) Bachelor and Ito (1973) Pyrek (1977)	Structure determination of lactucin and lactucopicrin
Gromek (1989a,b, 1991)	Isolation of three further lactucin-like guaianolides and one germacranolide from 1- and 2-year-old plants of *L. virosa*
Gromek et al. (1992)	Estimation of sedative and analgesic activity of lactucin, jacquinelin and some preparations from roots and aerial parts of *L. virosa*
Stojakowska et al. (1994)	Establishment of *L. virosa* tissue culture. Isolation of three zaluzanin C-like guaianolides from callus and cell suspension cultures of the plant
Kisiel et al. (1995)	Isolation of 8 guaiane- and germacrane-type sesquiterpene lactones from *Agrobacterium rhizogenes* transformed root culture of *Lactuca virosa*
Stojakowska et al. (1997)	Monitoring of ixerin F (guaianolide) accumulation in relation to the growth cycle of *L. virosa* cell suspension culture
Kisiel and Barszcz (1997)	Isolation of 15 guaiane- and germacrane-type sesquiterpene lactones from 1-year-old roots of *L. virosa*
Malarz (1997)	Attempts to increase sesquiterpene lactone accumulation in a transformed root culture of *L. virosa*

successfully cultivated in a medium containing 1 mg/l 2,4-D and 0.2 mg/l kinetin.

The dissimilation curve from the cell suspension (Fig. 4) was determined according to Schripsema et al. (1990). The curve depicts the weight loss of the suspension culture due to dissimilation and is correlated with the growth curve of the culture determined by measurements of dry weight increments (Schripsema and Verpoorte 1995).

Growth index (GI), calculated as a ratio of the final dry weight of the culture to the dry weight of inoculum, reached 7.2 (initially 2.1) and 3.4 for cell suspension and callus culture, respectively, after 3 weeks from subculturing.

2.2 Establishment of Transformed Root Culture

The transformed roots of *L. virosa* were induced in leaf explants from sterile grown seedlings. The leaf fragments were placed, the upper surface down, onto the solidified, hormone-free MS medium. The veins of the explants were

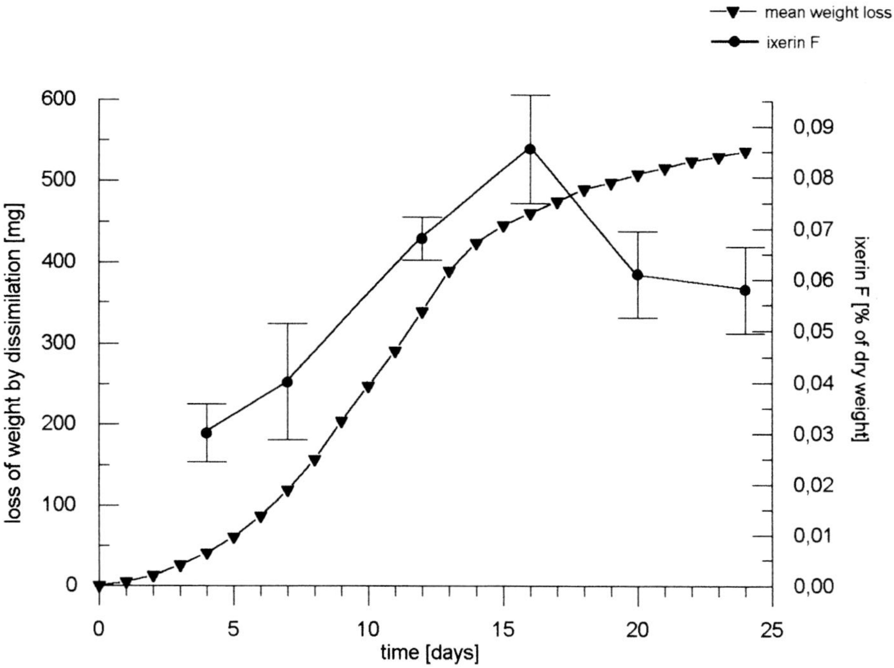

Fig. 4. Time course of ixerin F accumulation during the growth cycle of *Lactuca virosa* cell suspension

wounded with a scalpel immersed previously in the bacterial suspension, prepared by inoculation and 24h culture of *Agrobacterium rhizogenes* LBA 9402 in the liquid YMB medium (Hooykaas et al. 1977). After 1 week, roots appeared at the sites of infection. On the contrary, control explants (non-infected wounded explants) produced no tissue and wounded sites became necrotic. Apical parts of the individual roots were transferred to the hormone free liquid MS medium, containing macronutrients reduced to half concentration, 30mg/l sucrose and 0.5g/l cefotaxime. From the tenth passage on, the roots were cultivated in the medium without antibiotic. The cultures were grown in the dark, on a gyratory shaker (110rpm), at 25 ± 3°C. Attempts to cultivate the roots in a photoperiod or under permanent light, even of low intensity (22μmol/m^{2}s), remained unsuccessful because of rapid browning and necrosis of the cultured roots. Opine detection, performed according to Morgan et al. (1987), confirmed the presence of agropine and mannopine in extracts from the obtained root clones.

2.3 Growth Studies of Transformed Roots

Different compositions of basal salts in the culture media were tested, of which the most favourable for biomass increment were the MS medium and its modification with macronutrients reduced to half concentration, as well as NH_4^+-free MS formulation (Sigma) (GI 3.5–5.0). Further reduction of macro- and micronutrient content in the culture medium (1/2 MS, 1/4 MS) led to a decrease in GI (1.8–2.7). Results obtained with B5 medium (Gamborg et al. 1968), preferably used for root cultivation, were unsatisfactory (GI 2.0). Supplementation of the nutrient medium with GA_3 (10 mg/l) or its combination with BA (2.25 and 10 μg/l) did not result in increasing biomass production.

The final dry weight of the roots was strongly influenced by sucrose concentration in the culture medium. Four concentrations of sucrose were tested (20, 30, 40 and 50 g/l), giving averaged growth indices: 2.5, 3.5, 4.0 and 4.7, respectively.

The biomass accumulation in the root culture could also be improved by very careful preparation of inoculum and removing pieces of the browning tissue. Under the optimized conditions, GI of the culture reached 6.

2.4 Isolation and Identification of Sesquiterpene Lactones

The material for phytochemical analysis was harvested 3 weeks after the last subculturing. The lyophilized tissue was exhaustively extracted with ethanol at room temperature and chromatographed on a silica gel column using benzene-ethyl acetate (up to 50%) followed by chloroform-methanol (up to 20%) mixtures of increasing polarity. Hexane-ethyl acetate (up to 50%) and ethyl acetate-methanol (up to 10%) gradient solvent systems can also be used for elution of the column. The relevant fractions were further separated either by preparative TLC on silica gel, or by semipreparative RP-HPLC on a Delta-Pak C_{18} cartridge column (particle size 15 μm, 25 × 100 mm) coupled to a UV photodiode array detector. Separation of sesquiterpene lactone glycoside mixtures by RP-HPLC was achieved using methanol-water (7:13) at a flow rate 3 ml/min. All the isolated compounds were identified by comparison of their spectra (300 MHz ^{1}H NMR and EI or FAB mass spectra) with those of authentic samples or with literature data.

2.5 Determination of Sesquiterpene Lactone Content

The lyophilized, pulvered plant material (0.5 g) was extracted twice with 25 ml of ethanol at room temperature. The extract was concentrated in vacuo to the volume of ca. 10 ml and left to stand overnight at 4 °C. The solution was filtered and the filtrate was evaporated to dryness. The residue was dissolved in 1 ml of methanol and filtered through a membrane filter (0.22 μm) or centrifuged (20 000 rpm, 5 min, 4 °C). The filtrate or supernatant (5 μl) was subjected to RP-HPLC on the μ-Bondapack C_{18} column (particle size 10 μm, 21 × 300 mm) coupled to the photodiode array detector. The methanol-water (7:13) solvent system at a flow rate 0.5 ml/min was employed. Absorbance was monitored at 210, 235 or 254 nm, depending on the analyzed compound.

2.6 Sesquiterpene Lactones in Callus and Cell Suspension

Both callus and cell suspension synthesized only zaluzanin C-like guaianolides (Stojakowska et al. 1994). From the plant material, 9α-hydroxy-11β,13-dihydrozaluzanin C, its 3β-D-glucopyranoside (ixerin F), and zaluzanin C-3β-D-glucopyranoside (vernoflexuoside) were isolated and identified (Fig. 5). In the spent liquid medium only the first of the above-mentioned compounds was detected (ca. 4 mg/l). Cells from the suspension culture, kept under optimal conditions, accumulated neolignans, derivatives of coniferyl alcohol, up to 0.2% dry weight (Kisiel and Stojakowska, unpubl.).

Studies on accumulation of ixerin F (the main sesquiterpene lactone) in the cells during the growth cycle of the culture revealed that the content of the compound increased from 0.03% dry weight (4th day) to 0.09% in the early stationary phase (14th day) and then started to decrease to reach 0.05% at the 25th day of culture (Stojakowska et al. 1997; Fig. 4).

2.7 Sesquiterpene Lactones in Transformed Roots

The transformed roots were capable of synthesizing a wide spectrum of sesquiterpene lactones. In addition to zaluzanin C derivatives, formerly found in the callus and cell suspension, lactucin derivatives and lactuside A were also isolated (Kisiel et al. 1995). The yields of the sesquiterpene lactones and their structures are shown in Table 1 and Fig. 6, respectively. As in the undifferentiated cultures, ixerin F was the main sesquiterpene lactone found in the transformed roots. Lactuside A, making up ca. 50% of sesquiterpene lactones in the roots of 1-year-old intact plants, was present in the transformed roots in smaller amounts. Although the major sesquiterpene lactones (yields higher than 0.5 mg % fresh weight) isolated from the normal and transformed roots are represented by the same structures, the quantities of the individual compounds found in the transformed roots differed substantially from those of 1-year-old roots (see Table 1). The sesquiterpene lactones were not ex-

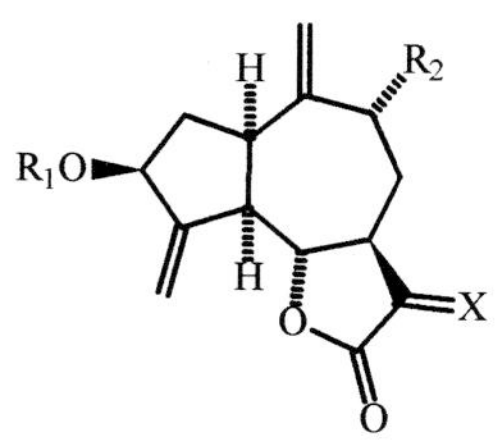

R_1	R_2	X	
H	OH	H,αMe	9α-hydroxy-11β,13-dihydrozaluzanin C
Glc	OH	H,αMe	ixerin F
Glc	H	CH_2	vernoflexuoside

Fig. 5. Chemical structures of sesquiterpene lactones isolated from the callus and cell suspension cultures of *Lactuca virosa*

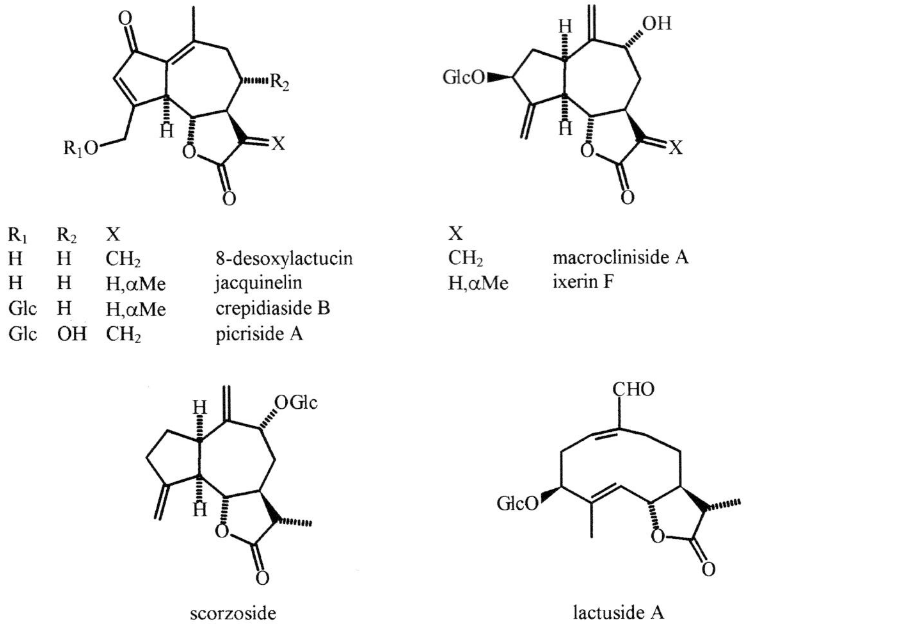

Fig. 6. Chemical structures of sesquiterpene lactones isolated from the transformed roots of *Lactuca virosa*

creted into the culture medium. Along with the sesquiterpene lactones, other secondary metabolites were isolated from the transformed roots. They were identified as α- and β-amyrin, lupeol, taraxasterol and stigmasterol.

2.7.1 Optimization

Factors influencing sesquiterpene lactone accumulation in the transformed root culture of *L. virosa* were studied by Malarz (1997). The contents of lactuside A, crepidiaside B and 8-desoxylactucin/jacquinelin mixture in the culture were monitored by analytical HPLC.

The roots grown in the medium containing 1/2 of MS micro- and macronutrients accumulated 50% more of the above mentioned sesquiterpene lactones than those grown in the MS medium and its modification with only the macronutrients reduced to half concentration. This effect, however, was accompained by a diminished growth rate. None of the growth regulators tested (GA_3, BA) stimulated biosynthesis of the investigated compounds. Although 5% sucrose in the nutrient medium, in the range of concentration tested (see Sect. 2.3), was optimum for biomass accumulation, the highest sesquiterpene lactone content was detected in the roots cultivated in the medium supplemented with 40 g/l of sucrose.

Addition of pectinase (12 units/flask, i.e., 20 ml of the medium), at the 6th day after inoculation of the roots into the fresh medium, caused transient increase of lactuside A (80% after 24 h) and crepidiaside B (40% after 48 h) contents, followed by prolonged reduction of the sesquiterpene lactone accumulation. The sesquiterpene lactones were not found in the culture medium. The dose of the enzyme employed did not affect the growth rate of the culture.

Methyl jasmonate at a concentration of 100 μM caused transient accumulation of lactuside A (20% and 30% more than in the control after 24 h and 48 h, respectively) and crepidiaside B (60% more than in the control after 24 h). However, 72 hours after addition of jasmonate no difference in the sesquiterpene lactone content was observed between the treated culture and the control.

3 Summary and Concluding Remarks

Pharmacological investigations carried out in the 1930s and more recently by Gromek et al. (1992), confirmed the opiate resembling, though weaker, activity of lactucin in some *L. virosa* preparations. However, the compounds responsible for the pharmacological activity of lactucin-free preparations are still unknown.

Research on in vitro production of sesquiterpene lactones, revealed that the transformed root culture of *L. virosa*, unlike the undifferentiated cultures, accumulated the compounds characteristic for roots of the intact plant. In the course of 4 year studies, yields of lactuside A and crepidiaside B in the transformed roots cultivated in an MS medium with a reduced concentration of macronutrients, with 3% sucrose, remained relatively stable (71 ± 16.2 and 285 ± 45.2 mg % of dry weight, respectively).

The major sesquiterpene lactones isolated from roots of *L. virosa* need further biological and pharmacological studies. The studies may be focused not only on their activity in the CNS; zaluzanin C was shown to be an antitumor agent (Jolard et al. 1974), and dehydrozaluzanin C demonstrated antiprotozoal activity in vitro (Fournet et al. 1993). Derivatives of zaluzanin C were isolated from both undifferentiated and organized cultures of the plant. Vernoflexuoside (glucozaluzanin C), detected in the callus and cell suspension, exhibited antimitotic activity in vitro (Kohlmünzer et al. 1976). Aglycones of macrocliniside A and vernoflexuoside showed cytotoxic activity (Seto et al. 1988).

L. virosa transformed root culture is a biosynthetically fully competent system for production of sesquiterpene lactones. It gives an opportunity to carry out biosynthetic studies and to improve the yields of the desired compounds. Attempts to stimulate sesquiterpene lactone production revealed that the biosynthetic potential of the culture has not been fully exploited (Malarz 1997).

References

Alconero R (1988) Lettuce (*Lactuca sativa* L.). In: Bajaj YPS (ed) Biotechnology in agriculture and forestry, vol 6. Crops II. Springer, Berlin Heidelberg New York, pp 351–369

Bachelor FW, Ito S (1973) A revision of the stereochemistry of lactucin. Can J Chem 51:3626–3630

Barton DHR, Narayanan CR (1958) Sesquiterpenoids. X. The constituents of lactucin. J Chem Soc 963–971

Behrens I (1938) Über unsere Erfahrungen mit Latucyl. MMW 85:139

Binek E (1995) Próba uprawy salaty jadowitej w warunkach krajowych. Wiad Ziel 5:9

Bremer K (1994) Asteraceae, cladistics and classification. Timber Press, Portland, Oregon, pp 157–201

Da Legnano LP (1968) Le piante medicinali nella cura delle malattie umane. Edizioni Mediterranee, Rome, p 826

Dolejš L, Souček M, Horak M, Herout V, Šorm F (1958) On terpenes. XCIV. The structure of lactucin. Collect Czech Chem Commun 23:2195–2199

Enomoto S, Ohashi Y (1993) Transformation of lettuce (*Lactuca sativa* L.) In: Bajaj YPS (ed) Biotechnology in agriculture and forestry, vol 33. Plant protoplasts and genetic engineering IV. Springer, Berlin Heidelberg New York, pp 237–247

Enomoto S, Ohyama K (1989) Regeneration of plant from protoplasts of lettuce and its wild species. In: Bajaj YPS (ed) Biotechnology in agriculture and forestry, vol 8. Plant protoplasts and genetic enginerring I. Springer, Berlin Heidelberg New York, pp 217–226

Forst AW (1937) Der Giftlattich, eine vergessene, alte Heilpflanze. MMW 84:1251–1254

Fournet A, Munoz V, Roblot F, Hoequemiller R, Cave A, Gantier J-C (1993) Antiprotozoal activity of dehydrozaluzanin C, a sesquiterpene lactone isolated from *Munnozia maronii* (Asteraceae). Phytother Res 7:111–115

Gamborg OL, Miller RA, Ojima K (1968) Nutrient requirements of suspension cultures of soybean root cells. Exp Cell Res 50:151–158

Gromek D (1989a) Sesquiterpene lactones of *Lactuca virosa* L. PhD Thesis, Pharmaceutical Faculty, Medical Academy of Cracow (in Polish)

Gromek D (1989b) Sesquiterpene lactones from *Lactuca virosa* L. Pol J Chem 63:297–301

Gromek D (1991) Lactuside A from *Lactuca virosa* L. Pol J Chem 65:11–12

Gromek D, Kisiel W, Klodzińska A, Chojnacka-Wójcik E (1992) Biologically active preparations from *Lactuca virosa* L. Phytother Res 6:285–287

Hallier E (1887) Flora von Deutschland, vol 30. Köbler, Gera-Untermhaus, pp 266–268

Hegi G (1929) Illustrierte Flora von Mittel-Europa, vol 6. Lehmanns Munich, pp 122–125

Hegnauer R (1964) Chemotaxonomie der Pflanzen, vol 3. Birkhäuser, Basel

Hegnauer R (1989) Chemotaxonomie der Pflanzen, vol 8. Birkhäuser, Basel

Hessen W (1937) Latucyl, ein neues Hustenbekämpfungsmittel. Dtsch Med Wochenschr 62:1731–1732

Holzer K, Zinke A (1953) Über die Bitterstoffe der Zichorie (*Cichorium intybus* L.). II. Mitteilung: Lactucin und Lactucopikrin. Monatsh Chem 84:901–909

Hooykaas PJJ, Klapwijk PM, Nuti MP, Schilperoort RA, Rorsch A (1977) Transfer of the *Agrobacterium tumefaciens* Ti plasmid to virulent *Agrobacteria* and to *Rhizobium explanta*. J Gen Microbiol 98:477–484

Jolad SD, Wiedhopf RM, Cole JR (1974) Tumor-inhibitory agent from *Zaluzania robinsonii* (Compositae). J Pharmacol Sci 63:1321–1322

Kisiel W, Barszcz B (1997) Minor sesquiterpene lactones from *Lactuca virosa*. Phytochemistry 46:1241–1243

Kisiel W, Stojakowska A, Malarz J, Kohlmünzer S (1995) Sesquiterpene lactones in *Agrobacterium rhizogenes*-transformed hairy root culture of *Lactuca virosa*. Phytochemistry 40:1139–1140

Kohlmünzer S, Grzybek J, Kisiel W (1976) Biological activity of vernoflexuoside on the basis of *Allium* test. Experientia 32:172–173

Levis WH, Elvin-Levis MPF (1977) Medical botany, plants affecting man's health. Wiley, London, p 446

List PH, Hörhammer L (1973) Hagers Handbuch der pharmazeutischen Praxis, vol 5. Springer, Berlin Heidelberg New York, pp 432–434

Malarz J (1997) Optimization of sesquiterpene lactone biosynthesis in hairy root culture of *Lactuca virosa* L. PhD Thesis, Pharmaceutical Faculty, Collegium Medicum, Jagiellonian University, Cracow (in Polish)

Morgan AJ, Cox PN, Turner DA, Peel E, Davey MR, Gartland KMA, Mulligan BJ (1987) Transformation of tomato using an Ri plasmid vector. Plant Sci 49:37–49

Murashige T, Skoog F (1962) A revised medium for rapid growth and bio-assays with tobacco tissue cultures. Physiol Plant 15:473–497

Pink DAC (1992) Micropropagation of lettuce (*Lactuca sativa* L.). In: Bajaj YPS (ed) Biotechnology in agriculture and forestry, vol 19. High-Tech and micropropagation III. Springer, Berlin Heidelberg New York, pp 42–57

Pyrek JS (1977) Terpenes of Compositae plants. V. Sesquiterpene lactones of *Lactuca serriola* L. The structure of 8-deoxylactucin and the site of esterification of lactupicrin. Rocz Chem 51:2165–2170

Schripsema J, Verpoorte R (1995) Novel techniques for growth characterization and phytochemical analysis of plant cell suspension cultures. J Nat Prod 58:1305–1314

Schripsema J, Meijer AH, Van Iren F, Ten Hoopen HJG, Verpoorte R (1990) Dissimilation curves as a simple method for the characterization of growth of plant cell suspension cultures. Plant Cell Tissue Organ Cult 22:55–64

Seto M, Miyase T, Umehara K, Ueno A, Hirano Y, Otani N (1988) Sesquiterpene lactones from *Cichorium endivia* L. and *C. intybus* L. and cytotoxic activity. Chem Pharm Bull (Tokyo) 36:2423–2429

Song Q, Gomez-Barrios ML, Hopper EL, Hjortso MA, Fisher NH (1995) Biosynthetic studies of lactucin derivatives in hairy root culture of *Lactuca floridana*. Phytochemistry 40:1659–1665

Späth E, Lorenz R, Kuhn H (1951) Über das Lactucin (I. Mitteilung). Monatsh Chem 82:114–121

Stojakowska A, Malarz J, Kisiel W (1994) Sesquiterpene lactones in tissue culture of *Lactuca virosa*. Planta Med 60:93–94

Stojakowska A, Malarz J, Van der Heijden R, Verpoorte R (1997) Accumulation of ixerin F and activities of some terpenoid biosynthetic enzymes in a cell suspension culture of *Lactuca virosa* L. Acta Soc Bot Pol 66:185–188

Tamaki H, Robinson RW, Anderson JL, Stoewsand GS (1995) Sesquiterpene lactones in virus-resistant lettuce. J Agric Food Chem 43:6–8

Van Beek TA, Maas P, King BM, Leclercq E, Voragen AGJ, De Groot A (1990) Bitter sesquiterpene lactones from chicory roots. J Agric Food Chem 38:1035–1038

Weiss RF (1974) Lehrbuch der Phytotherapie. Hippokrates, Stuttgart, p 217

Wessely F, Lorenz R, Kuhn H (1951): Über das Lactucin (II. Mitteilung). Monatsh Chem 82:322–329

XVI *Marrubium vulgare* (White Horehound): In Vitro Culture, and the Production of Diterpene Marrubiin and Other Secondary Metabolites

W. Knöss

1 Introduction

Marrubium vulgare L. (Lamiaceae family) commonly known as white horehound is a perennial and reaches a height of up to 1 m (Fig. 1). Leaves are round-ovate and each node of the stem bears two opposing leaves. Calices, flowers and especially the leaves, are covered with different trichomes, amongst them numerous stellate trichomes, peltate glandular trichomes and capitate glandular trichomes (Smolen 1960). The plant is thought to have originated in the region between the Mediterranean Sea and Central Asia (Frohne 1987; Hänsel et al. 1993), and now grows on all continents, and is cultivated for example in south eastern Europe. In New South Wales (Australia) the plant is so widespread that control programs were initiated to restrict its growth (Sagliocco and Coupland 1995).

M. vulgare, was already known to have healing properties in Egypt and Greece some 2000 years ago. This is revealed by the plant's English name, white horehound, which is derived from Horus, an Egyptian goddess. This plant was mentioned in many antique and medieval herbal books (Frohne 1987). The use of medications containing extracts from *M. vulgare*, according to actual pharmacopoeias (e.g. from the United Kingdom, Austria, and Germany), is focused on treatment of stomach diseases and coughs. The German Commission E (Wichtl 1997) monograph describes the use of extracts in the case of lack of appetite, dyspeptic complaints and catarrhs in the respiratory tract. The effect of extracts from *M. vulgare* in antitussives and expectorants could not be assigned to a certain group of secondary metabolites. The treatment of stomach diseases may be explained by the diterpene marrubiin. This diterpene lactone with a furanic moiety is the drug's bitter principle. By opening of the lactone ring marrubiic acid is formed, which has been shown to have a choleretic effect in investigations with rats (Krejci and Zadina 1959). In folk medicine, the plant is used generally against gastrointestinal diseases (Weimann and Heinrich 1997) and also externally to treat skin lesions (Hänsel et al. 1993). Recently, alcoholic extracts of *M. vulgare* have been shown to have anti-spasmodic activity in preparations of smooth muscle cells (Schlemper et al. 1996).

Institute of Pharmaceutical Biology, University of Boon, Nussallee 6, 53115 Bonn, Germany

Biotechnology in Agriculture and Forestry, Vol. 43
Medicinal and Aromatic Plants XI (ed. by Y.P.S. Bajaj)

Fig. 1. *Marrubium vulgare* L.

While non-volatile furanic diterpenes are accumulated in large amounts, up to 4 mg/g fresh weight, *M. vulgare* produces only small amounts of essential oil. The main compounds were identified as monoterpenes, e.g. camphene, limonene, fenchene, pinene and sabinene. Other secondary metabolites found in *M. vulgare* are the alkaloids betonicin (0.3%), choline (0.2%) and stachydrin, which are widespread in the Lamiaceae familiy. The flavonoids described are mostly flavonglykosides and flavonolglykosides. Most of them are derivatives of apigenin and luteolin (Paudler and Wagner 1963; Frohne 1987; Duke 1992).

2 In Vitro Culture Studies

In vitro cultures of many plants of the Lamiaceae have been investigated because many species are known to accumulate either essential oils or non-volatile terpenoids (e.g. Banthorpe 1996; Callebaut et al. 1996). As far as *M. vulgare* is concerned, research was done only on growth and metabolites of plants (Bergeron et al. 1995). We initiated undifferentiated and differentiated culture systems of *M. vulgare* to investigate the prospects of production of terpenoids and to establish a model system to study regulation of biosynthesis of isoprenoids in vitro.

2.1 Establishment of Culture Systems

2.1.1 Callus Cultures

Callus cultures were initiated from stems, leaves or aseptically grown 3-week-old seedlings of *M. vulgare*. More than 80% callus formation was observed with explants from stems on LS medium (Linsmaier and Skoog 1965) or B5 medium (Gamborg et al. 1968) when initiation was performed without illumination. Normal growing light brown calli were subcultured. Some calli also formed small and hard green aggregates under illumination, but these aggregates did not grow further when subcultured. Mean growth of calli was about 10 g FW within 4 weeks (Knöss 1995).

2.1.2 Cell Suspension Cultures

Cell suspension cultures were established from callus material in liquid LS medium. Suspensions were cultivated in 300 ml Erlenmeyer flasks, either in the dark or under continuous light. Exponential growth started after a lag period of 3 to 4 days. After about 10 to 12 days, cell suspensions reached a stationary phase. The cell suspension cultures were subcultured every 14 days. The first cell suspension culture line has been maintained for more than 4 years (Wilhelm 1993).

2.1.3 Shoot Cultures

Shoot cultures may be a useful experimental system, if metabolic pathways need to be investigated in differentiated structures (Heble 1985; Hashimoto and Yamada 1991). Some callus cultures from *M. vulgare* formed shoots spontaneously. These shoots were cultivated on a modified MS medium (Murashige and Skoog 1962) without phytohormones and with only half the concentration of salts. Shoot cultures were maintained on agar media (Fig. 2), where they grew slowly and were subcultured every 6 to 8 weeks. Investigations of secondary metabolites were routinely done with fast-growing shoot cultures in liquid media. These liquid cultures were started from apices from shoot material growing on agar, normally consisting of the bud and a pair of leaves directly below the bud (Knöss 1994, 1995).

Leaves of shoot cultures were rather small compared with those of normally grown plants. While field grown plants develop leaves with blades up to

Fig. 2. Shoot cultures of *Marrubium vulgare* on MS/2 medium (continuous light)

4 cm, the maximum diameter of leaves from shoot cultures was 7 mm, a normal value being 5 mm. The petioles of leaves from shoot cultures were relatively long. Whereas similar shoot cultures from *Leonurus cardiaca* in liquid medium could be subcultured and maintained under continuous light for more than 1 year (Knöss 1994), shoot cultures of *M. vulgare* could only be subcultured 3 or 4 times under these conditions. The growth then stopped and no new buds were formed. However, shoot cultures in liquid medium in the darkness could be subcultured without such a degeneration.

2.2 Extraction and Structure of Furanic Labdane Diterpenes

Furanic diterpenes with a labdane skeleton are the most important secondary metabolites from *M. vulgare*. Such diterpenes are also known from other Lamiaceae, e.g. from other species of the genus *Marrubium* (Savona et al. 1979; Hatam et al. 1995) and the genera *Ballota* and *Leonurus* (Savona et al. 1976; Knöss 1994; Tasdemir et al. 1995). Marrubiin is the major diterpene in extracts of *M. vulgare* (Appleton et al. 1967). This compound was identified as the bitter principle of the drug in the 19th century, and its structure was elucidated in 1953 (Cocker et al. 1953). The structure and the conformation has been verified by X-ray studies (Busby et al. 1983), and it has been found that the genuine compound occurring in the plant is premarrubiin (Henderson and McCrindle 1969). The amount of diterpenes in field grown plants is up to 4 mg/g FW. During work-up procedure marrubiin is formed from premarrubiin, e.g. when extracts contain water or by heating. Premarrubiin can be isolated by careful extraction with acetone and preparative TLC. In preliminary studies we investigated acetone extracts of field grown plants of *M. vulgare* and the furanic labdane diterpenes premarrubiin, marrubiin, premarrubenol and marrubenol (Fig. 3) were isolated and identified by means of NMR and MS spectroscopy (Appleton et al. 1967; Henderson and McCrindle 1969).

The furanic labdane diterpenes of *M. vulgare* are soluble in $CHCl_3$, acetone, methanol, and to a limited extent even in water. Thus, we established a simple work-up procedure to separate furanic labdane diterpenes from other compounds using silica gel or reversed phase columns developed for solid phase extraction. First, the solvent from the extraction was evaporated and the residue dissolved in $CHCl_3$. By column chromatography on silica gel equili-

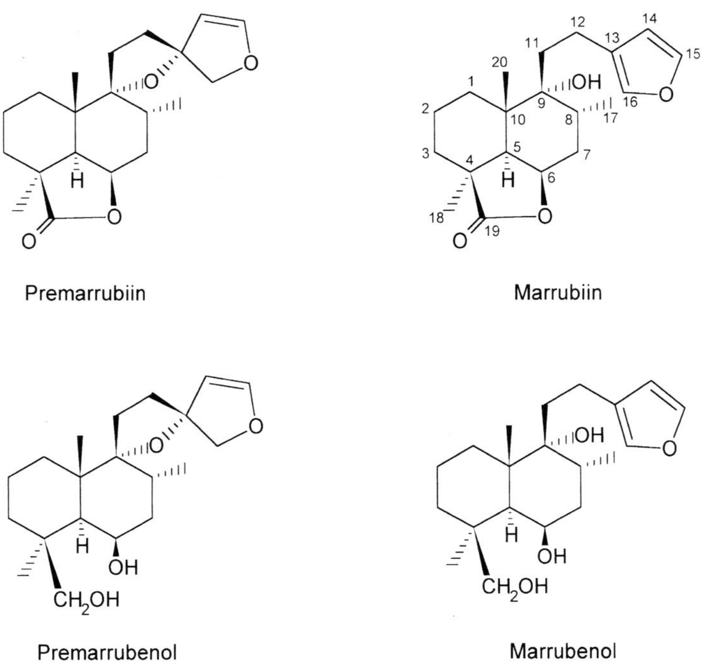

Fig. 3. Furanic labdane diterpenes from *Marrubium vulgare*

brated with $CHCl_3$, the furanic labdane diterpenes did not bind to the column material and were separated from more polar compounds (e.g. sugars, amino acids, glycosides). Subsequently, the fraction with furanic labdane diterpenes was again evaporated to dryness and the residue dissolved in a mixture of methanol/water (8:2 v/v). This mixture was subjected to column chromatography on RP18-material equilibrated with the same solvent system. In this step, furanic labdane diterpenes again did not bind to the column, while more apolar compounds bound to the RP-material (e.g. carotenoids, chlorophylls, sterols, aliphatic terpenes, diterpene alcohols). The amount of furanic labdane diterpenes obtained by this work-up procedure could be analyzed either by means of HPLC or by use of photometric estimations. The furan ring reacts with vanillin/sulfuric acid or 4-dimethylamino-benzaldehyde giving a specific red-violet color. The amount of furanic labdane diterpenes can be calculated by photometric estimation in solution or by means of TLC and evaluation with a TLC scanner after spraying (Fig. 4). With the latter method, the limit of detection is about 0.1 μg marrubiin, the linear range reaching from 0.2 to 2 μg. The estimation in solution is less sensitive.

These analytical procedures enabled us to analyze the furanic labdane diterpenes in small samples from in vitro cultures.

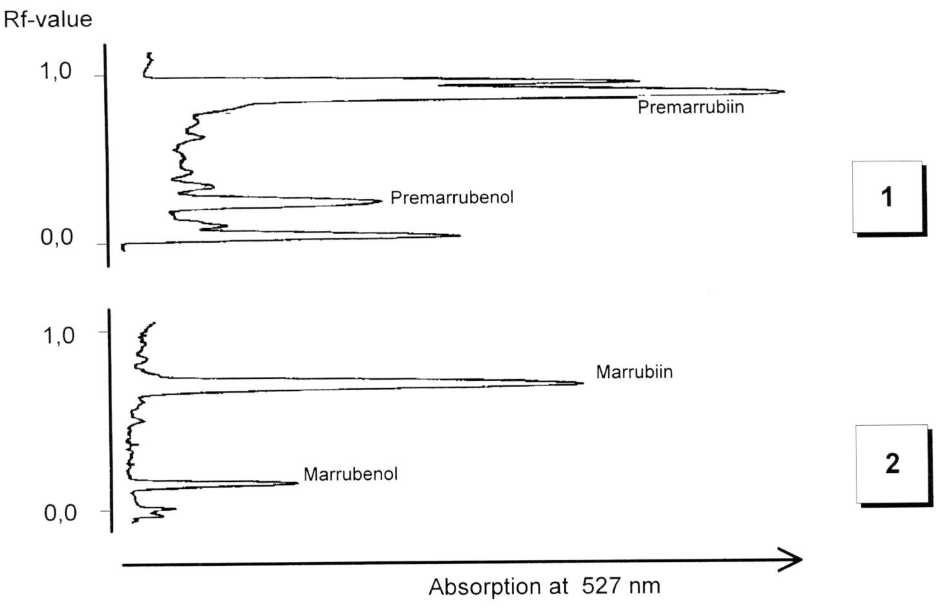

Fig. 4. Evaluation of extracts from *Marrubium vulgare* by TLC scanning (Silica gel: $CHCl_3$/MeOH (95:5 v/v); detection by spraying with vanillin/H_2SO_4 (1 g vanillin in 10 ml H_2SO_4 conc.) and subsequent heating for 5 min at 105 °C; scanning at 527 nm and calculation by using a calibration graph measured with authentic marrubiin). **1** TLC of fresh, carefully prepared acetone extract from leaves. **2** Fraction of furanic labdane diterpenes after the work-up procedure. (Knöss 1996)

2.3 Accumulation of Furanic Labdane Diterpenes

2.3.1 Undifferentiated Culture Systems

Production of terpenoids from secondary metabolism in callus or cell suspension cultures is often very low (Charlwood and Charlwood 1991). Nevertheless, there are examples which show formation of mono-, sesqui- or diterpenes in undifferentiated tissues (Mersinger et al. 1988; Miyasaka et al. 1989; Falk et al. 1990; Banthorpe et al. 1990; Paniego et al. 1993). In comparison with typical terpenoid compounds from essential oils, furanic labdane diterpenes are highly oxidized and relatively polar. This led us to the assumption that these compounds might be produced in cell suspension cultures like, for example, the labdane diterpene forskolin (Mersinger et al. 1988). However, during many investigations we did not find furanic labdane diterpenes in cell suspension cultures. Only in three samples of callus culture did we identify traces of marrubiin, but production was not stable when these calli were subcultured.

Figure 5 shows a typical growth curve from a cell suspension culture from *M. vulgare*. The amount of free sterols was estimated as an indicator of

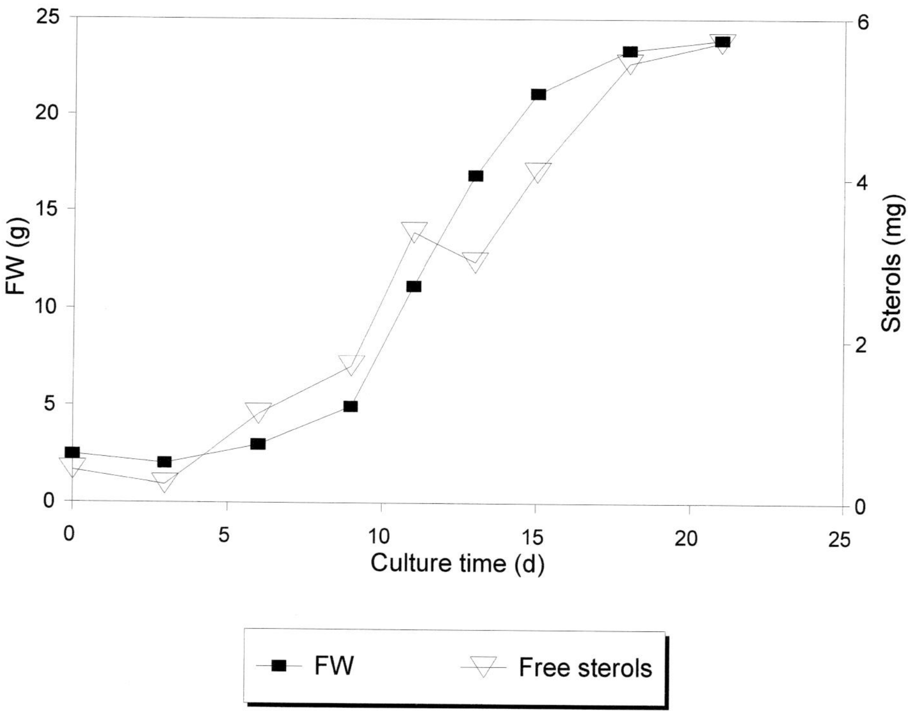

Fig. 5. Growth and free sterols in cell suspension cultures of *Marrubium vulgare* (inoculum 0.2 g FW; 60 ml LS medium in 300-ml Erlenmeyer flasks; dark-grown, 80 rpm, 25 °C). Data shown represents the means of three separate experiments. (Wilhelm 1993)

terpenoid metabolism; in this case it was parallel with the growth of the cell suspension, and only at the beginning of the exponential growth phase was there a significant additional maximum of sterols. In numerous experiments, the parameters of the cell suspension culture were varied (e.g. combination of phytohormones, light conditions, sugar content, nitrogen content, addition of acetate, elicitors like yeast extract, influence of heavy metals, jasmonic acid, and growth regulators), and cell suspensions derived from different callus lines were compared. In no case the induction of formation of furanic labdane diterpenes was achieved (Wilhelm 1993).

2.3.2 *Differentiated Systems*

Shoot cultures from *M. vulgare* produced and accumulated furanic labdane diterpenes. It was shown by TLC, that the main compound in shoot material was premarrubiin. The amount was normally more than 90% of the total furanic labdane diterpenes. When shoot cultures were subcultured with an inoculum of about 0.2g FW there was no definite lag phase as found with cell suspension cultures, and the shoots started growing immediately. From day 6 to day 14 of a culture period, shoots showed a linear growth curve. Under continuous light, maximum FW was about 2g (Fig. 6). In parallel with FW growth, the shoot cultures accumulated furanic labdane diterpenes. After 2 weeks the resources from the medium were exhausted, often tissues became partly brownish, and the FW and contents of furanic labdane diterpenes stagnated or even decreased. Depending on the starting material, the accumulation of furanic labdane diterpenes varied. Over a period of 3 years, values between 30μg and 500μg furanic labdane diterpenes per g FW were measured, mean production was about 120μg/g FW. It was not possible to select high producing lines, and perhaps the state of growth and differentiation of the starting material caused this variability. When homogeneous starting material was used, deviation was normally less than 10%.

Diterpenes are secondary metabolites which should be biosynthesized in plastids (Kleinig 1989; Chappell 1995; McGarvey and Croteau 1995). Therefore, formation of furanic labdane diterpenes was investigated under different light conditions. Biosynthesis and accumulation was not light-dependent. When shoot cultures were grown in the dark, furanic labdane diterpenes accumulated in the same range (Fig. 7). Growth of dark-grown etiolated cultures was somewhat slower, the leaves being smaller and the internodes elongated.

Production of furanic labdane diterpenes could not be increased by changing culture parameters (for variations see Sect. 2.3.1). All investigations showed that accumulation of furanic labdane diterpenes parallelled growth. When shoot cultures were effected by growth regulators (e.g. triazoles), accumulation of furanic labdane diterpenes also decreased. In stable, vital cultures no diterpenes or only small amounts (less than 3% of total furanic labdane diterpenes) were found in the culture medium. Under stress conditions (e.g.

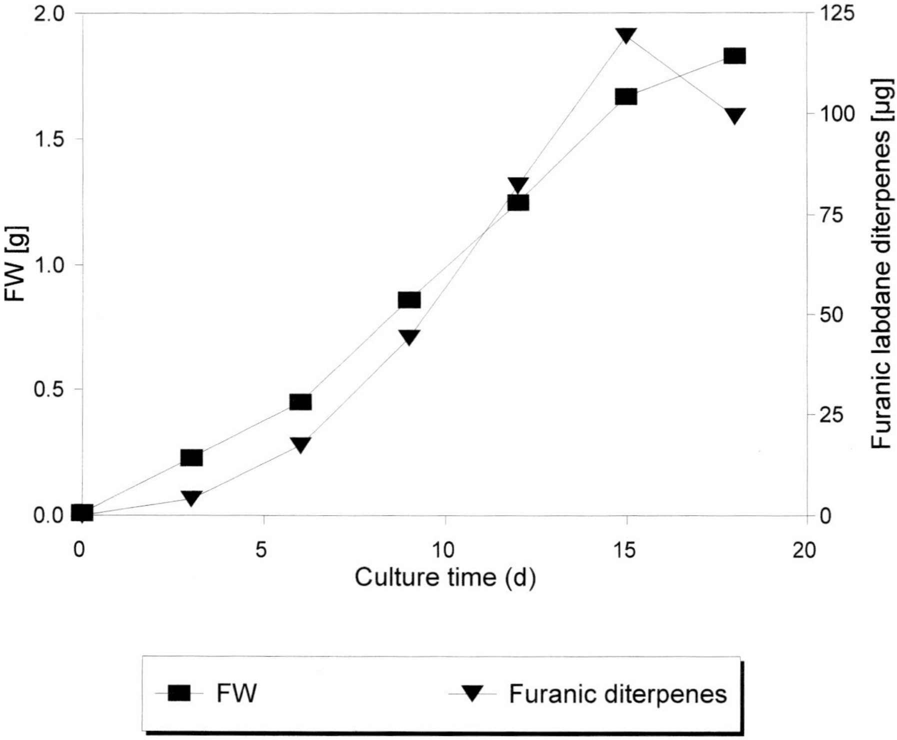

Fig. 6. Growth and accumulation of furanic labdane diterpenes in shoot cultures of *Marrubium vulgare* grown under continuous light (inoculum 0.2 g FW, 30 ml MS/2 medium in 100-ml Erlenmeyer flasks, 25 °C, 80 rpm). Data shown represents the means of three separate experiments. (Knöss 1996)

treatment with heavy metals) furanic labdane diterpenes were also extracted from the medium; maybe this was due to disruption of tissues.

Shoot cultures could be grown in alkaline media (Gormann 1994). They even grew at pH values of 7.5 and 8, the FW increased 65 and 30%, respectively, of that of the normal pH 5.8. The higher the pH value, the higher the amount of furanic labdane diterpenes, found in the medium (Fig. 8).

Thus, shoot cultures were a suitable system for producing furanic labdane diterpenes. The accumulation rate was normally about 10 to 20% that of field-grown plants, when compared with plants collected before flowering (i.e. about 3 months old). Considering the time which is necessary to accumulate furanic labdane diterpenes in shoot cultures (i.e. about 14 days), accumulation is nearly comparable. Additionally, the shoot cultures are grown under controlled conditions, without the risk of being influenced by climate or pathogens.

The production of furanic labdane diterpenes was also investigated in young seedlings (Knöss 1996). There was no accumulation of furanic labdane diterpenes in the first 4 to 5 weeks following germination. Then, correlated

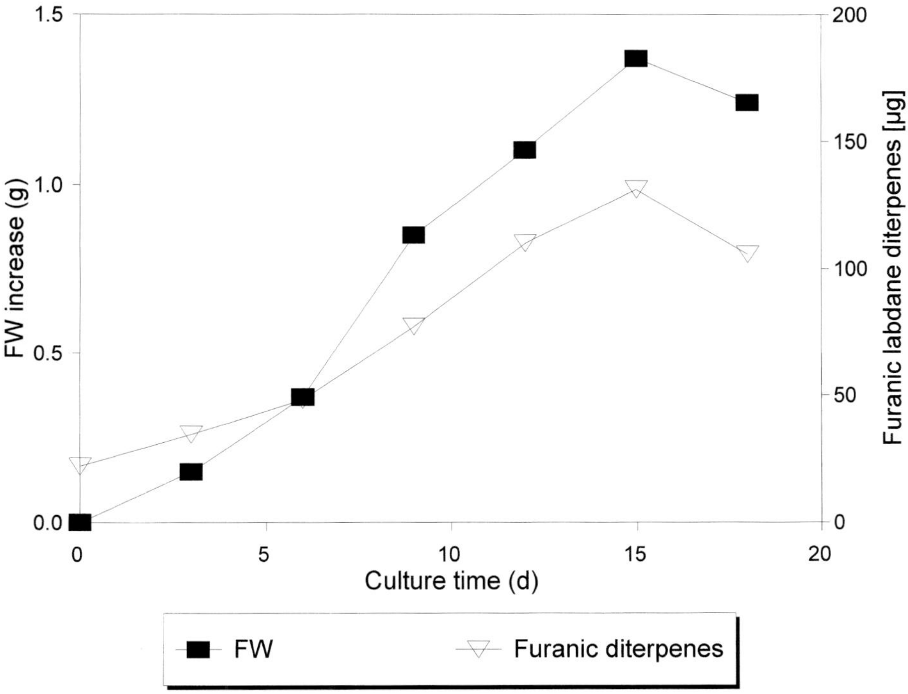

Fig. 7. Growth and accumulation of furanic labdane diterpenes in etiolated shoot cultures of *Marrubium vulgare* grown in the dark (inoculum 0.2 g FW, 30 ml MS/2 medium in 100-ml Erlenmeyer flasks, 25 °C, 80 rpm). Data shown represent the means of three separate experiments. (Knöss 1996)

with an observable progression in leaf development (i.e. form and shape as well as formation of trichomes) accumulation of furanic labdane diterpenes started. The plantlets showed highest production in the terminal bud and the pair of leaves directly below the bud. Therefore, the physiological state of the shoot cultures may represent a system that is comparable with the upper part of plantlets. Accumulation of furanic labdane diterpenes seems to be started by a developmental program, subsequently they are produced constitutively.

2.4 Biosynthetic Studies

2.4.1 Incorporation of ^{14}C-Labelled Precursors

The principal pathway of early steps in the biosynthesis of terpenoids was established more than 40 years ago (Ruzicka 1953). The sequence leading from acetyl-CoA to the biologically active isoprenes dimethylallyl diphosphate and isopentenyl diphosphate is known as the acetate/mevalonate pathway because a regulatory key step is the reduction of 3-hydroxy-3-methyl-

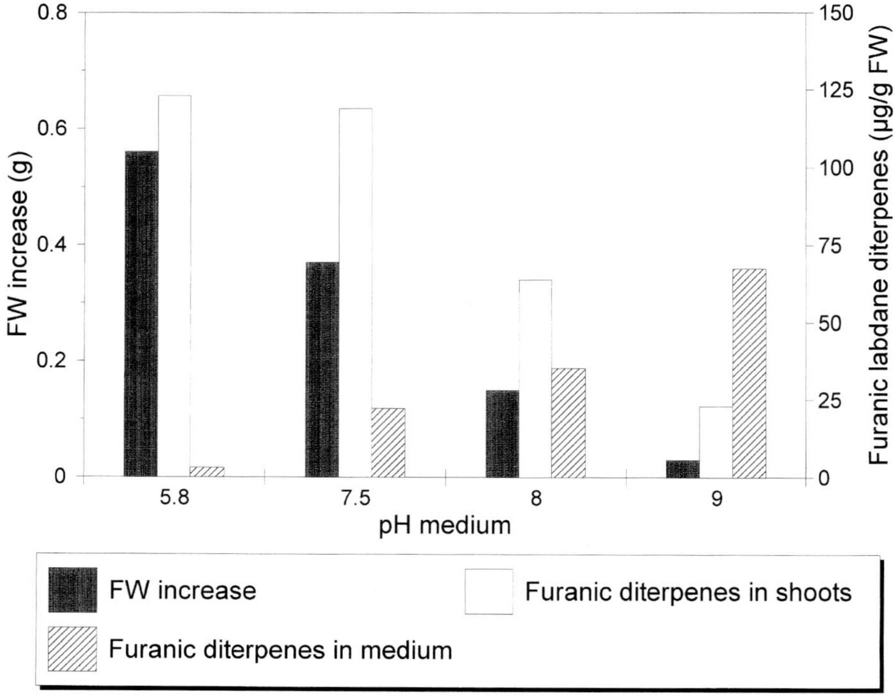

Fig. 8. Growth and accumulation of furanic labdane diterpenes in shoot cultures of *Marrubium vulgare* and in the medium grown under different pH-conditions (inoculum 0.2 g FW, culture time 7 days, 30 ml MS/2 medium in 100-ml Erlenmeyer flasks, 25 °C, 80 rpm, continuous light). Data shown represent the means of three separate experiments. (Gormann 1994)

glutaryl-CoA by HMGCoA-reductase to yield mevalonic acid. Preliminary investigations on the biosynthesis of marrubiin were made by Breccia and Badiello (1967). From incorporation studies with different ^{14}C-labelled precursors, e.g. acetate, mevalonic acid and different sugars, it was concluded that marrubiin is a diterpene, which is biosynthesized via the acetate/mevalonate pathway. The experiments were made with complete plants, and it was stated that incorporation was too low for further investigations.

The shoot cultures of *M. vulgare* were a suitable system to continue studies on the biosynthesis of the furanic labdane diterpenes because: (1) they produced and accumulated furanic labdane diterpenes, (2) they could be grown under defined conditions, and (3) the methods which had been established for quantification were so sensitive that it was possible to analyze the furanic labdane diterpenes in shoot samples of less than 0.5 g. The results of incorporation experiments with 2-[^{14}C]-acetate and 2-[^{14}C]-mevalonic acid were very puzzling. Incorporation of acetate into premarrubiin was very low and incorporation of mevalonic acid was beyond even the level of detection, when shoot cultures were supplied with about 1 µCi radioactive precursor. By extraction and subsequent scintillation counting of free sterols, it was shown that both substrates were taken up and used for biosynthesis of terpenoids. Moreover, radioactivity from mevalonic acid, taken up by plantlets or shoot cultures either by transpiration or by diffusion in submerged cultures, was homogeneously distributed over whole plants and shoots (Knöss 1996).

In contrast, when shoot cultures were incubated with [^{14}C]-GGPP, which was prepared enzymatically by using a prenyl transferase system (Knöss and Reuter 1996), radioactivity was detected in marrubiin. How the amphipolar molecule GGPP was taken up was not investigated.

Generally, the total radioactivity after incorporation experiments with [^{14}C]-GGPP found in shoot material was lower than with acetate or mevalonic acid, but the radioactivity found in marrubiin was higher with [^{14}C]-GGPP. This led us to the conclusion that either mevalonic acid is nearly completely excluded from biosynthesis of diterpenes by regulatory mechanisms, or there should be an alternative biosynthetic pathway.

2.4.2 Biosynthesis of Diterpenes Via a Non-Mevalonate Pathway – Incorporation Studies with 1-[^{13}C]-Glucose

In 1993 Rohmer et al. proposed a non-mevalonate pathway for the biosynthesis of triterpenoids from the hopane series and ubiquinon in some Eubacteria. The pathway starts with C_3-units and leads via a C_5-intermediate to the formation of activated isoprenic units. Up until now, it has been shown that a non-mevalonate pathway is also operating in the biosynthesis of ginkgolides (diterpenes from *Ginkgo biloba*, Schwarz 1994), taxane derivatives (diterpenes from *Taxus sinensis*, Eisenreich et al. 1996) and phytol, carotenoids and plastoquinone (from *Scenedesmus obliquus*, Chlorophyceae, and from *Hordeum vulgare*, *Lemna gibba* and *Daucus carota*, Schwender et al. 1996; Lichtenthaler et al. 1997). Recently, it has been demonstrated that isoprene, which is emitted by many plants, is also formed via a non-mevalonate pathway (Zeidler et al. 1997). By means of incubation experiments with etiolated shoot cultures from *M. vulgare* supplemented with 1-[^{13}C]-glucose, we demonstrated that furanic labdane diterpenes from this plant are also biosynthesized via a non-mevalonate pathway (Fig. 9). Because this example is a secondary metabolite from an angiosperm, it becomes evident that biosynthesis of special groups of terpenoids via an alternative pathway may be rather common.

The experiments used heterotrophically grown etiolated shoot cultures of *M. vulgare*. These are independent of CO_2 supply and all organic matter biosynthesized during incubation originates either from the starting material or from the medium. Shoot cultures were grown in darkness on medium containing glucose instead of sucrose. Glucose and 1-[^{13}C]-glucose were mixed in a 9:1 ratio. After 2 weeks, 22 g FW of shoots were harvested. About 2 mg ^{13}C-marrubiin were obtained after the work-up procedure and the positions labelled with ^{13}C were determined by means of NMR-spectroscopy (Knöss 1996; Knöss et al. 1997). The results showed that the labelling pattern of marrubiin could not be explained by the acetate/mevalonate pathway but was

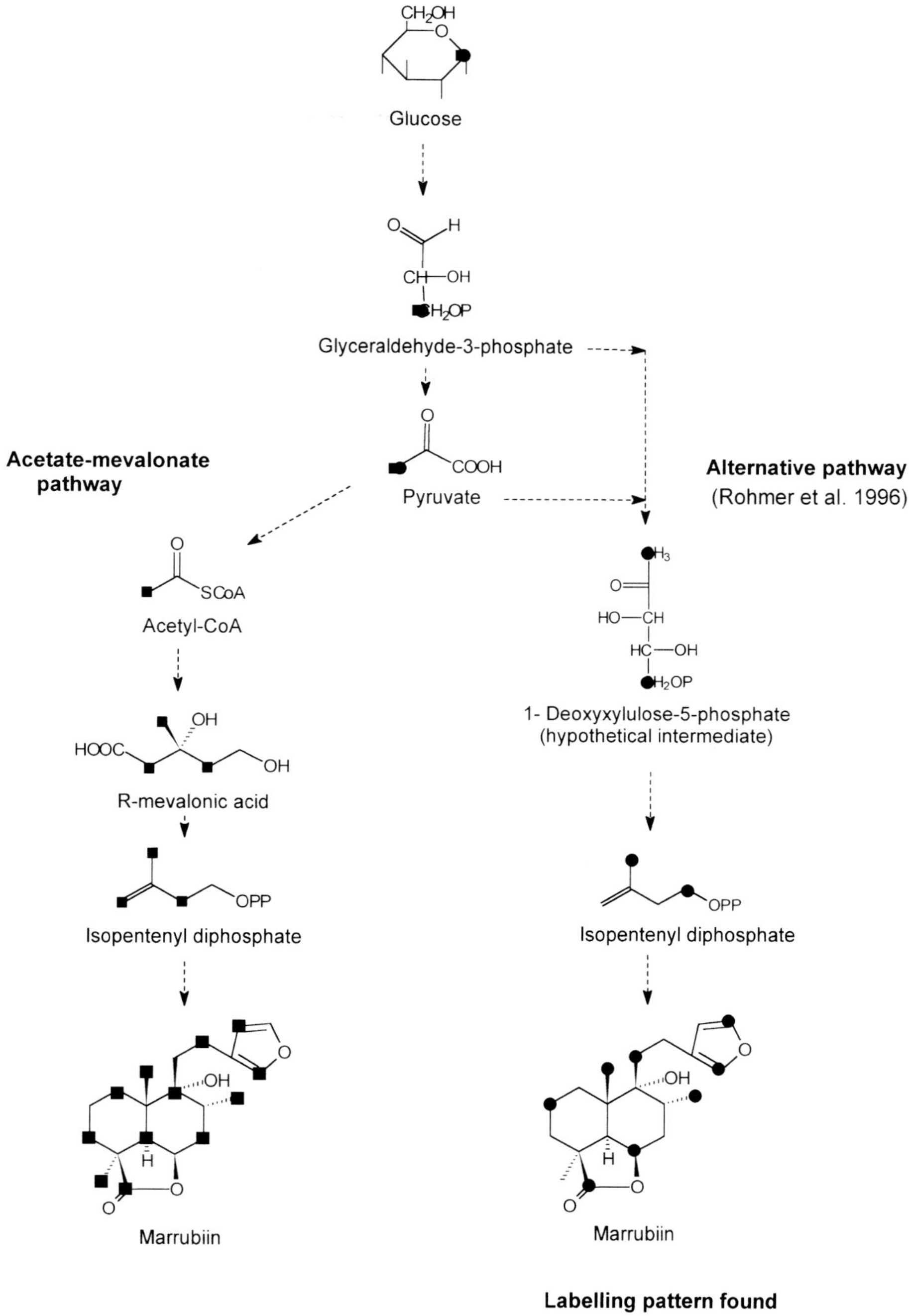

Fig. 9. Biosynthesis of marrubiin derived from incorporation experiments with etiolated shoot cultures of *Marrubium vulgare* growing on MS/2 medium supplied with 1-[^{13}C]-glucose. (Knöss 1996)

consistent with labelling results according to the alternative pathway proposed for bacteria by Rohmer et al. (1996). Actually, research is in progress to identify intermediates in this alternative pathway and to elucidate the principles of regulation.

3 Summary and Conclusion

Cell suspension cultures, callus cultures, and shoot cultures have been established from the *Marrubium vulgare*. All these systems could be initiated and cultivated on modified standard media and their growth was characterized (Knöss 1994, 1995). Cell suspension cultures and callus cultures did not accumulate furanic labdane diterpenes, which are typical secondary metabolites of *M. vulgare*, and they could not be induced by selection nor by elicitation or modification of the medium (Wilhelm 1993). Differentiated shoot cultures accumulated premarrubiin continuously, the amount being up to 20% that of field-grown plants. Production of furanic labdane diterpenes was not light-dependant. The shoot cultures were used to investigate the biosynthesis of the diterpene marrubiin, showing that the isoprenoid precursors of this diterpene are not formed via the acetate/mevalonate pathway (Knöss 1996; Knöss et al. 1997).

The market for the medicinal plant *M. vulgare* is rather limited, but there is a steady demand for extracts to prepare cough remedies and medications for treatment of gastrointestinal disorders. More interesting biological activities of diterpenes may be found in the near future (Buchbauer et al. 1990). To date, it is not of commercial relevance to produce shoot cultures of *M. vulgare* as a source of drug material. However, the in vitro system of differentiated and undifferentiated cultures of *M. vulgare* has been shown to be a suitable model for studying the biosynthesis of diterpenoids and the regulation of terpenoid metabolism.

References

Appleton RA, Fulke JWB, Henderson MS, McCrindle R (1967) The stereochemistry of marrubiin. J Chem Soc [C]: 1943–1947

Banthorpe DV (1996) *Mentha* species (Mints): in vitro culture and production of lower terpenoids and pigments. In: Bajaj YPS (ed) Biotechnology in agriculture and forestry, vol 37. Medicinal and aromatic plants IX. Springer, Berlin Heidelberg New York, pp 202–225

Banthorpe DV, Brown JT, Morris GS (1990) Accumulation of the anti-fungal diterpene sclareol by cell cultures of *Salvia sclarea* and *Nicotiana glutinosa*. Phytochemistry 29:2145–2148

Bergeron C, Charbonneau J, Desroches B, Gosselin A (1995) Influence of supplemental lighting and irrigation on mineral composition growth and premarrubiin content of horehound *Marrubium vulgare* L. J Herbs Spices 3:3–15

Breccia A, Badiello R (1967) The terpene marrubiin. Z Naturforsch [B] 22:44–49

Buchbauer G, Spreitzer H, Kiener G (1990) Biologische Wirkung von Diterpenen. Pharm Unserer Zeit 19:28–37
Busby MC, Day VW, Day RO, Wheeler DMS, Wheeler MM, Day CS (1983) The stereochemistry and conformation of marrubiin: an X-ray study. Proc R Ir Acad [B] 83:21–31
Callebaut A, Decleire M, Vandermeiren K (1996) *Ajuga reptans* (Bugle): in vitro production of anthocyanins. In: Bajaj YPS (ed) Biotechnology in agriculture and forestry, vol 37. Medicinal and aromatic plants IX. Springer, Berlin Heidelberg New York, pp 1–22
Chappell J (1995) Biochemistry and molecular biology of the isoprenoid biosynthetic pathway in plants. Annu Rev Plant Physiol Plant Mol Biol 46:521–547
Charlwood BV, Charlwood KA (1991) Terpenoid production in plant cell cultures. In: Harborne JB, Tomas-Barberan FA (eds) Ecological chemistry and biochemistry of plant terpenoids. Clarendon Press, Oxford, pp 95–104
Cocker W, Cross BE, Duff SR, Edward JT, Holley TF (1953) The constitution of marrubiin, part I. J Chem Soc:2540–2548
Duke JA (1992) Handbook of phytochemical constituents of grass herbs and other economic plants. CRC Press, Boca Raton
Eisenreich W, Menhard B, Hylands PJ, Zenk MH, Bacher A (1996) Studies on the biosynthesis of taxol. Proc Natl Acad Sci USA 93:6431–6436
Falk KL, Gershenzon J, Croteau R (1990) Metabolism of monoterpenes in cell cultures of common sage (*Salvia officinalis*). Plant Physiol 93:1559–1567
Frohne D (1987) Marrubii herba – Andornkraut. Kooperation Phytopharmaka, Bonn
Gamborg OL, Miller RA, Ojima K (1968) Nutrient requirements of suspension cultures of soybean root cells. Exp Cell Res 50:151–158
Gormann R (1994) Furanoditerpene in Sproßkulturen von *Leonurus cardiaca* und *Marrubium vulgare*. Diploma Thesis, University of Bonn
Hänsel R, Keller K, Rimpler H, Schneider G (1993) Hagers Handbuch der Pharmazeutischen Praxis. Drogen E-O, 5th edn. Springer, Berlin Heidelberg New York, pp 777–782
Hashimoto T, Yamada Y (1991) Organ culture and manipulation. Methods Plant Biochem 6:323–350
Hatam NAR, Porzell A, Seifert K (1995) Polyodonine a prefuranic labdane diterpene from *Marrubium polydon*. Phytochemistry 40:1575–1576
Heble MR (1985) Multiple shoot cultures: a viable alternative in vitro system for the production of known and new biologically active constituents. In: Neumann KH, Barz W, Reinhard E (eds) Primary and secondary metabolism of plant cell cultures. Springer, Berlin Heidelberg New York, pp 281–289
Henderson MS, McCrindle R (1969) Premarrubiin. A diterpenoid from *Marrubium vulgare*. J Chem Soc [C]: 2014–2015
Kleinig H (1989) The role of plastids in isoprenoid biosynthesis. Annu Rev Plant Physiol Plant Mol Biol 40:39–59
Knöss W (1994) Furanic labdane diterpenes in differentiated and undifferentiated cultures of *Marrubium vulgare* and *Leonurus cardiaca*. Plant Physiol Biochem 32:585–589
Knöss W (1995) Establishment of callus cell suspension and shoot cultures of *Leonurus cardiaca* and diterpene analysis. Plant Cell Rep 14:790–793
Knöss W (1996) Bildung der Furanolabdanditerpene in *Leonurus cardiaca* und *Marrubium vulgare*. *Habilitation*, University of Bonn
Knöss W, Reuter B (1996) Biosynthesis of [^{14}C]geranylgeranyl diphosphate by a prenyltransferase system from a mutant strain of *Gibberella fujikuroi*. Anal Biochem 239:208–212
Knöss W, Reuter B, Zapp J (1997) Biosynthesis of the labdane diterpene marrubiin in *Marrubium vulgare* via a non-mevalonate pathway. Biochem J 326:449–454
Krejci I, Zadina R (1959) Die gallentreibende Wikung von Marrubiin and Marrubiinsäure. Planta Med 7:1–7
Lichtenthaler HK, Schwender J, Disch A, Rohmer M (1997) Biosynthesis of isoprenoids in higher plant chloroplasts proceeds via a mevalonate-independent pathway. FEBS Lett 400:271–274
Linsmaier EM, Skoog F (1965) Organic growth factor requirements of tobacco tissue cultures. Physiol Plant 18:100–127
McGarvey DJ, Croteau R (1995) Terpenoid metabolism. Plant Cell 7:1015–1026

Mersinger R, Dornauer H, Reinhard E (1988) Formation of forskolin by suspension cultures of *Coleus forskohlii*. Planta Med 54:200–204

Miyasaka H, Nasu M, Yondea K (1989) *Salvia miltiorrhiza*: in vitro production of cryptotanshinone and ferruginol. In: Bajaj YPS (ed) Biotechnology in agriculture and forestry, vol 7. Medicinal and aromatic plants II. Springer, Berlin Heidelberg New York, pp 417–430

Murashige T, Skoog F (1962) A revised medium for rapid growth and bio-assays with tobacco tissue cultures. Physiol Plant 15:473–497

Paniego NB, Maligne AE, Giulietti AM (1993) *Artemisia annua* (Quing-Hao): in vitro culture and the production of artemisin. In: Bajaj YPS (ed) Biotechnology in agriculture and forestry, vol 24. Medicinal and aromatic plants V. Springer, Berlin Heidelberg New York, pp 64–78

Paudler WW, Wagner S (1963) The major alkaloid of *Marrubium vulgare*. Chem Ind:1693–1694

Rohmer M, Knani M, Simonin P, Sutter B, Sahm H (1993) Isoprenoid biosynthesis in bacteria: a novel pathway for the early steps leading to IPP. Biochem J 295:517–524

Rohmer M, Seemann M, Horbach S, Bringer-Meyer S, Sahm H (1996) Glyceraldehyde 3-phosphate and pyruvate as precursors of isoprenic units in an alternative non-mevalonate pathway for terpenoid biosynthesis. J Am Chem Soc 118:2564–2566

Ruzicka L (1953) The isoprene rule and the biogenesis of terpenic compounds. Experientia 9:357–367

Sagliocco JL, Coupland JB (1995) Biology and host specificity of *Chamaesphecia mysiniformis* (Lepidoptera: Sesiidae) a potential biological control agent of *Marrubium vulgare* (Lamiaceae) in Australia. Biocontrol Sci Technol 5:509–515

Savona G, Piozzi F, Hanson JR, Siverns M (1976) Structure of ballotinone, a diterpenoid from *Ballota nigra*. J Chem Soc Perkin I:1607–1609

Savona G, Piozzi F, Aránguez LM, Rodriguez B (1979) Diterpenes from *Marrubium sericeum*, *Marrubium supinum* and *Marrubium alysson*. Phytochemistry 18:859–860

Schlemper V, Ribas A, Nicolau M, Filho VC (1996) Antispasmodic effects of hydroalcoholic extract of *Marrubium vulgare* on isolated tissues. Phytomedicine 3:211–216

Schwarz M (1994) Terpen-Biosynthese in *Ginkgo biloba*: Eine überraschende Geschichte. PhD Thesis, University of Zurich

Schwender J, Seemann M, Lichtenthaler HK, Rohmer M (1996) Biosynthesis of isoprenoids (carotenoids, sterols, prenyl side-chains of chlorophylls and plastoquinone) via a novel pyruvate/glyceraldehyde 3-phosphate non mevalonate pathway in the green alga *Scenedesmus obliquus*. Biochem J 316:73–80

Smolen K (1960) Badania Farmakobotaniczne pieciu Gatunkow rodzaju *Marrubium* L. (Pharmacobotanical investigations of five species of the genus *Marrubium* L.). Diss Pharm 12(4):325–352

Tasdemir D, Wright AD, Sticher O, Calis I, Linden A (1995) Detailed ^{1}H- and ^{13}C-NMR investigations of some diterpenes isolated from *Leonurus persicus*. J Nat Prod 58:1543–1554

Weimann C, Heinrich M (1997) Indigenous medicinal plants in Mexico: the example of the Nahua (Sierra de Zongolica). Bot Acta 110:62–72

Wichtl M (1997) Teedrogen und Phytopharmaka. Wissenschaftliche Verlagsgesellschaft Stuttgart, pp 370–371

Wilhelm S (1993) Untersuchungen zum Sterolgehalt und zur Beeinflussung des Sekundärstoffwechsels in Zellkulturen von *Marrubium vulgare* L. (Lamiaceae). Diploma Thesis, University of Bonn

Zeidler JG, Lichtenthaler HK, May HU, Lichtenthaler FW (1997) Is isoprene emitted by plants synthesized via the novel isopentenyl pyrophosphate pathway? Z Naturforsch [C] 52:15–23

XVII *Menispermum* Species (Moonseed Vines): In Vitro Culture, and the Production of Dauricine and Other Secondary Metabolites

Y. SUGIMOTO

1 General Account

1.1 Distribution and Importance of the Plant

The genus *Menispermum* (Menispermaceae), a dicotyledonous woody vine, is composed of three species, *M. canadense*, *M. dauricum* (Fig. 1) and *M. mexicanum* (Hotta and Terabayashi 1989). These plants are commonly called moonseed because of the characteristic shape of their seeds (Fig. 2). The North American species, *M. canadense* (Canadian moonseed), is distributed widely in lowland forests in moist rich soils along streams in the eastern United States and West to Manitoba and Oklahoma (Purrington and Horn 1993). Its underground parts are used as a diuretic. The Asian species, *M. dauricum* (Siberian moonseed), is distributed from eastern Asia to China and Siberia. Rhizomes are used as a crude drug "Bian-Fu-Ge-Gen", in China, which is effective for the treatment of tonsillitis, diarrhea, gastroenteritis, etc. (Hu et al. 1993).

1.2 Alkaloids

Chemical investigations of the constituents of the members of the genus *Menispermum* have been concentrated mainly on the Asian species, *M. dauricum*. Twenty-five alkaloids have been identified from this species. On the other hand, *M. canadense* has received little attention since the systematic study of its alkaloids by Doskotch and Knapp (1971) over 25 years ago. Nine alkaloids were found in *M. canadense*, six of which are shared with *M. dauricum* (Table 1, Figs. 3–8). All the alkaloids identified in these plants are biosynthetically derived from the isoquinoline unit. Bisbenzylisoquinolines, aporphines and protoberberines are often found in Menispermaceae and related families. However, acutumine-type alkaloids, one of four types of chlorine-containing alkaloids found in higher plants (Engvild 1986), have been

Arid Land Research Center, Tottori University 1390 Hamasaka, Tottori 680-0001, Japan

Biotechnology in Agriculture and Forestry, Vol. 43
Medicinal and Aromatic Plants XI (ed. by Y.P.S. Bajaj)

Fig. 1. *Menispermum dauricum* DC

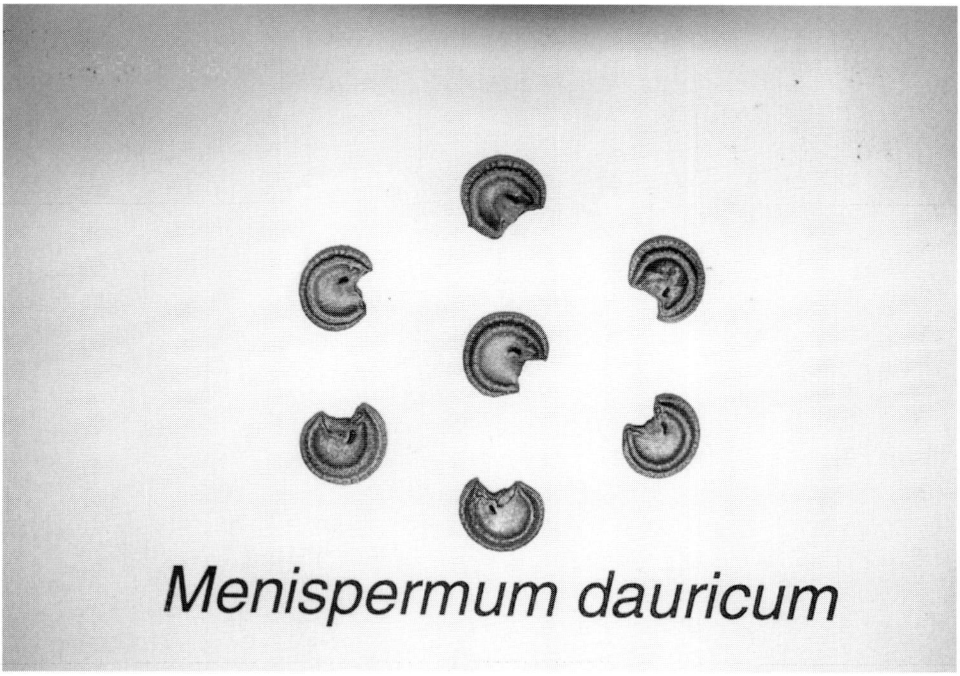

Fig. 2. Seeds of *M. dauricum*

Table 1. Alkaloids found in *M. canadense* and *M. dauricum*

Type of alkaloid	*M. canadense*	Reference	*M. dauricum*	Reference
Bisbenzylisoquinolines	Dauricine	Thornber (1970)	Dauricine	Thornber (1970)
			Dauricoline	Tomita et al. (1970a)
	Daurinoline	Doskotch and Knapp (1971)	Daurinoline	Thornber (1970)
			Dauricinoline	Tomita et al. (1970b)
			Daurisoline	Zheng et al. (1979)
			Dauriciline	Pang et al. (1991)
			N-Desmethyldauricine	Pan (1992)
	Tetrandrine	Thornber (1970)	Tetrandrine	Thornber (1970)
	N'-Desmethyldauricine	Doskotch and Knapp (1971)		
Hasubanans	Acutumine, acutumidine	Doskotch and Knapp (1971)	Acutumine, acutumidine	Thornber (1970)
			Acutuminine	Okamoto et al. (1969)
Aporphines			Menisperine	Thornber (1970)
	Magnoflorine, *N*-methyllindcarpine	Doskotch and Knapp (1971)	Magnoflorine	Tomita et al. (1970b)
Oxoisoaporphines			Menisporphine	Kunitomo and Sato (1982)
			Dauriporphine (bianfugenine)	Takani et al. (1983)
			Bianfugecine, bianfugedine	Hou and Xue (1985)
			Dauriporphinoline	Zhao et al. (1989)
			2,3-Dihydromenisporphine	Kunitomo et al. (1985)
			6-*O*-Demethylmenisporphine	Hu et al. (1993)
Protoberberines	Dehydrocheilanthifoline	Doskotch and Knapp (1971)	Cheilanthifoline, stepholidine	Okamoto et al. (1971)
			Dauricoside	Hu et al. (1993)
Miscellaneous bases			Sinomenine	Thornber (1970)
			Stepharine	Tomita et al. (1970b)

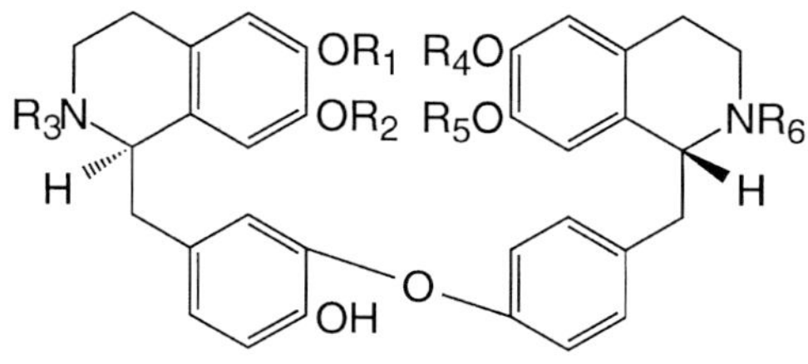

Alkaloid	R_1	R_2	R_3	R_4	R_5	R_6
Dauricine	Me	Me	Me	Me	Me	Me
N-Desmethyldauricine	Me	Me	H	Me	Me	Me
N'-Desmethyldauricine	Me	Me	Me	Me	Me	H
Daurinoline	Me	Me	Me	H	Me	Me
Dauricoline	H	Me	Me	H	Me	Me
Daurisoline	Me	H	Me	Me	Me	Me
Dauriciline	Me	H	Me	Me	H	Me
Dauricinoline	H	Me	Me	Me	Me	Me

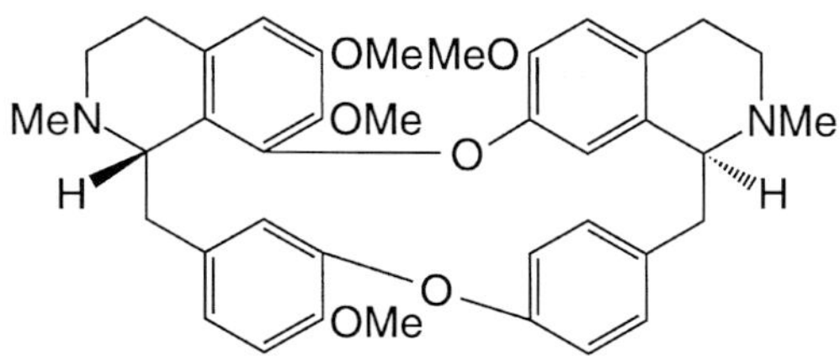

Tetrandrine

Fig. 3. Bisbenzylisoquinolines found in the genus *Menispermum*

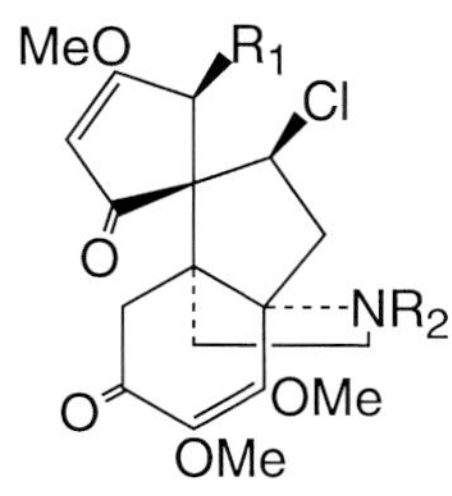

Alkaloid	R_1	R_2
Acutumine	OH	Me
Acutumidine	OH	H
Acutuminine	H	Me

Fig. 4. Hasubanans found in the genus *Menispermum*

found in *Sinomenium acutum* (Menispermaceae), *M. canadense* and *M. dauricum*. *M. dauricum* is the only source for oxoisoaporphines. There are several reports on tissue culture of *Menispermum* plants (Sugimoto et al. 1994, 1996, 1997).

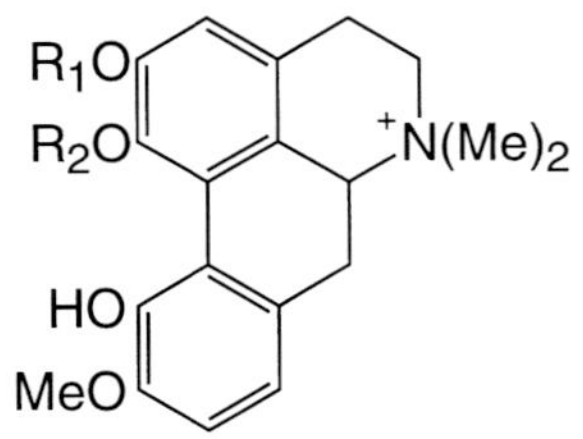

Alkaloid	R_1	R_2
Menisperine	Me	Me
Magnoflorine	Me	H
N,N-Dimethyllindcarpine	H	Me

Fig. 5. Aporphines found in the genus *Menispermum*

Alkaloid	R_1	R_2	R_3
Menisporphine	H	OMe	OMe
Dauriporphine	OMe	OMe	OMe
Dauriporphinoline	OMe	OMe	OH
6-*O*-Demethylmenisporphine	H	OMe	OH
Bianfugecine	H	OMe	H
Bianfugedine	H	OCH_2O	

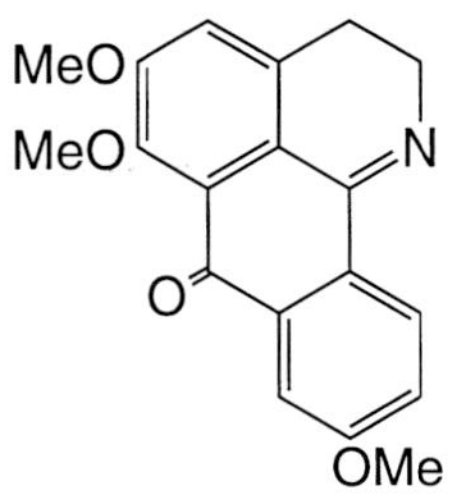

2,3-Dihydromenisporphine

Fig. 6. Oxoisoaporphines found in the genus *Menispermum*

2 In Vitro Studies

2.1 Root Cultures

Sterilized leaf segments of *M. canadense* and *M. dauricum* were inoculated on LS agar medium containing NAA and BA. For *M. dauricum*, callus

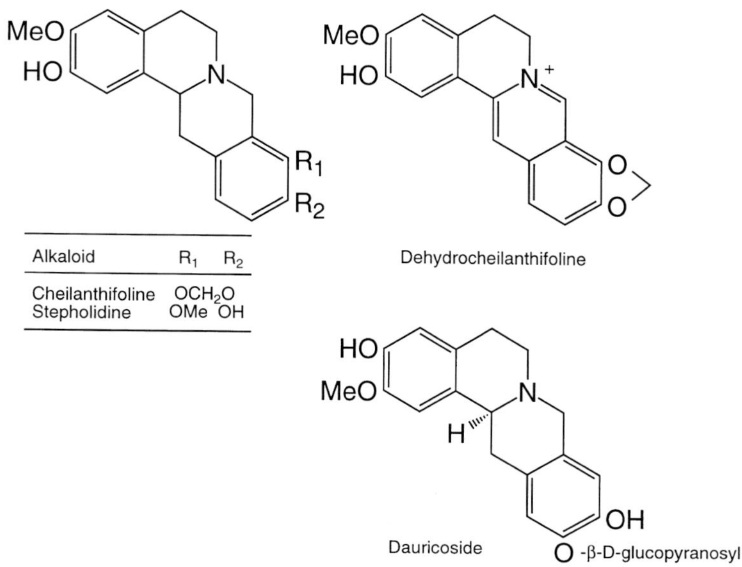

Fig. 7. Protoberberines found in the genus *Menispermum*

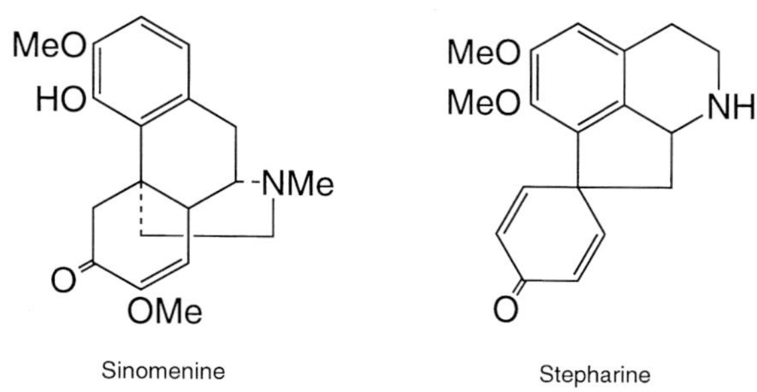

Fig. 8. Miscellaneous bases found in the genus *Menispermum*

was induced after 1 month of incubation in the dark at 27 °C. After another 1.5 months of incubation, adventitious roots had formed from the callus. For *M. canadense*, on the other hand, adventitious roots as well as callus were induced directly from the leaf segments. Adventitious roots thus obtained from *M. dauricum* and *M. canadense* were excised and cultured independently in liquid B5 media containing IBA. Calli of both species did not grow well.

TLC of methanol extracts from cultured roots of *M. canadense* and *M. dauricum* indicated that several bases were produced in the roots. The most abundant base was isolated as an amorphous solid and its ^{1}H-NMR, IR and mass spectra and specific rotation were studied. All these parameters matched literature values for dauricine (Manske et al. 1965; Baldas et al. 1972). The ^{13}C-NMR spectrum also showed this base to be dauricine, which has not been reported in the literature. This base was then treated with diazomethane and the ^{13}C-NMR of the product was measured, which matched that of *O*-methyldauricine (Jossang et al. 1986). Dauricine is a major secondary metabolite produced by *Menispermum* plants and this alkaloid has a variety of pharmacological activities including anti-inflammatory and cardiovascular effects (Schiff 1991).

In spite of a lot of effort, cultured roots of *M. canadense* did not grow well, therefore further investigation was concentrated on cultured roots of *M. dauricum*.

2.2 Effect of Cultural Parameters on Dauricine Production

Effects of cultural parameters, including plant growth regulators, nitrogen, phosphate and sugars, on growth and dauricine formation were studied. Several possible precursors of bisbenzylisoquinolines were also tested for their promotive effect on dauricine formation in *M. dauricum* root cultures.

2.2.1 Effect of Plant Growth Regulators

Three auxins (NAA, IAA and IBA), each at the concentrations of 0.1, 1, 10 or 100 μM, were tested for growth promotion in cultured roots (Fig. 9). Each auxin promoted root growth when present at 10 μM in the culture medium. At 10 μM NAA, cultured roots grew 28-fold in 40 days. With higher auxin concentrations, lateral roots were frequently induced and root elongation was inhibited. At the 10 μM auxin concentration, lateral root induction and root elongation were well balanced.

The effects of auxin concentrations on dauricine content in cultured roots were also studied (Fig. 9). At the optimum concentration of each auxin for root growth, roots produced the highest amount of dauricine. High concentrations of auxins (100 μM) drastically inhibited dauricine formation.

NAA was most effective for dauricine production (Fig. 9). The effects of NAA on root growth and dauricine content were examined in detail. As a result, cultured roots produced the highest amount of dauricine at 7.5 μM. At this concentration, root growth was 44-fold (40 days), containing 0.28% dauricine. Other growth regulators were also studied in combination with 7.5 μM NAA for their effects on the growth and dauricine content of *M. dauricum* root cultures. The growth regulators examined were 6-benzyladenine as cytokinin, gibberellic acid (GA_3) and abscisic acid. These regulators were studied at concentrations of 0.1, 1, 10 or 100 μM. None of these gave significant growth stimulation or increased alkaloid formation in cultured roots. At high concentrations (10 and 100 μM), each growth regulator completely inhibited root growth (data not shown).

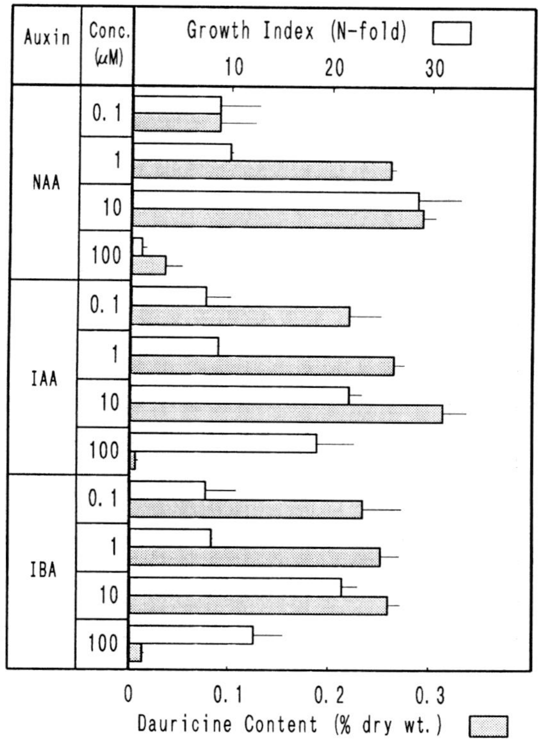

Fig. 9. Effects of auxin on growth and dauricine formation. Roots (0.1 g fresh weight) were cultured in 25 ml of medium with auxin for 40 days. *Thin lines* represent SE for three replicates. (Reprinted from Sugimoto et al. 1994, with kind permission from Elsevier Science Ltd., The Boulevard, Langford Lane, Kidlington OX5 1GB, UK)

2.2.2 Effects of Nitrogen, Phosphate and Carbohydrate Sources

Effects of the NH_4^+ to NO_3^- ratios in the culture medium on dauricine production were studied (Fig. 10A). The total amount of nitrogen was set equal to that in B5 medium, which has a ratio of NH_4^+ to NO_3^- of 1:14.6. Dauricine formation and root growth were inhibited by high concentrations of NH_4^+. The elimination of NH_4^+ from the medium did not reduce root growth significantly but dauricine formation was decreased. Changing the ratio of NH_4^+ to NO_3^- from half to twice that in B5 medium did not affect the dauricine production considerably. The original ratio in B5 medium gave the highest amount of dauricine.

Phosphate concentration was changed from 1/8 to twice that in the B5 medium. A low concentration (1/2) of phosphate gave a higher dauricine content, but slower root growth. Changing the phosphate concentration did not promote dauricine production significantly (Fig. 10B).

The effects of various sugars on root growth and dauricine formation were also studied. Two disaccharides, sucrose and lactose, and three

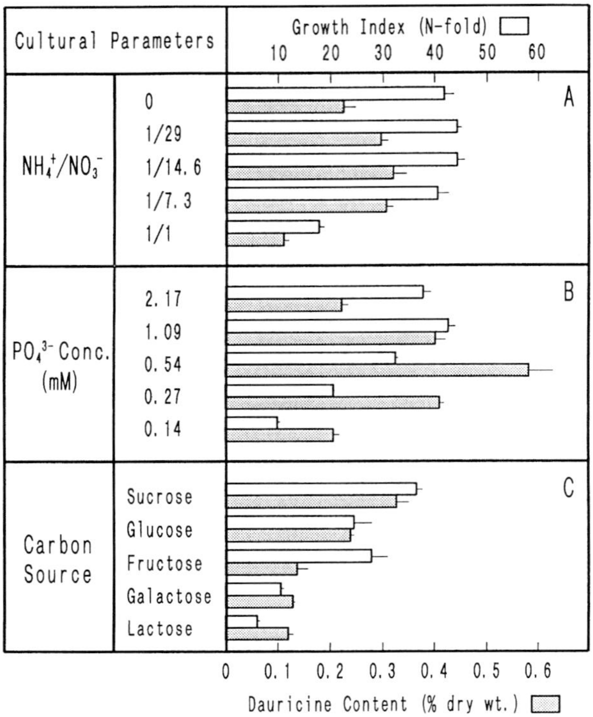

Fig. 10. A Effects of NH_4^+ to NO_3^- ratios; **B** phosphate concentration; **C** sugars on growth and dauricine formation. Roots (0.1 g fresh weight) were cultured in 25 ml of modified B5 medium for 40 days. *Thin lines* represent SE for three replicates. (Reprinted from Sugimoto et al. 1994, with kind permission from Elsevier Science Ltd., The Boulevard, Langford Lane, Kidlington OX5 1GB, UK)

monosaccharides, glucose, fructose and galactose, were used at 3%. Figure 10C shows that root growth in a medium containing sucrose, glucose or fructose is significantly higher than in a medium containing galactose or lactose. Dauricine contents were 0.33, 0.24 and 0.14% as a result of sucrose, glucose and fructose, respectively. Galactose and lactose decreased both root growth and dauricine formation. Since sucrose was the most suitable for dauricine production, the effects of sucrose concentration on dauricine production were studied. The optimum concentration of sucrose for root growth was 3%, on the other hand, the best concentration for dauricine content was 4%. Overall, the optimum sucrose concentration for total dauricine production was 3% (data not shown).

2.2.3 Dauricine Production Medium

We established the optimum medium for dauricine production in cultured roots of *M. dauricum*. This was found to be B5 medium with 7.5 μM NAA and 3% sucrose. Only auxin showed a promoting effect on dauricine production

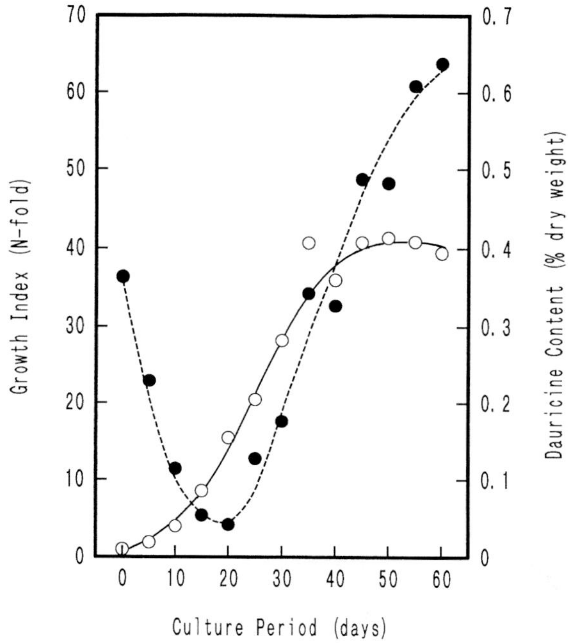

Fig. 11. Time course of growth and dauricine content in root cultures of *M. dauricum*. The *symbols* denote root growth (—○—) and dauricine content (---●---). (Reprinted from Sugimoto et al. 1994, with kind permission from Elsevier Science Ltd., The Boulevard, Langford Lane, Kidlington OX5 1GB, UK)

compared with other cultural parameters in the initial medium. A typical time course of root growth and dauricine content, for roots cultured in the optimized medium, is shown in Fig. 11. Cultured roots grew about 40-fold in 50 days (Fig. 12). Dauricine content was the lowest when roots were growing exponentially. In contrast, dauricine increased substantially at the late logphase and stationary phase. The dauricine content of the cultured roots reached more than 0.5% of dry weight. In the original plants, dauricine content in the aerial and underground parts was only about 0.03 and 0.2%, respectively.

2.2.4 *Effects of Precursors*

Effects of precursor feeding on dauricine formation were studied with the possible precursors of benzylisoquinoline biosynthesis; L-tyrosine, tyramine, L-dopa, dopamine, 4-hydroxyphenylpyruvic acid and *N*-methylcoclaurine (Fig. 13). The first four precursors were reported to be incorporated into the protoberberines. 4-Hydroxyphenylpyruvic acid is shown to be a building block of the benzylisoquinoline skeleton (Rueffer and Zenk 1987). L-Tyrosine, tyramine and *N*-methylcoclaurine were reported to be incorporated into bisbenzylisoquinolines (Stadler et al. 1988).

Fig. 12. Cultured roots of *M. dauricum*

Precursors were added to the root cultures at 0.1 or 1 mM at the beginning of incubation. Tyrosine and tyramine (0.1 mM) effectively increased dauricine formation, which indicates that the supply of these precursors might be the limiting factor for dauricine formation. Other precursors inhibited dauricine formation.

2.2.5 Relationship between Dauricine Content and Growth

The relationship between dauricine content and growth rate was analyzed for all the results described above. A statistically significant correlation was obtained. The correlation coefficient (r) was 0.584 ($p < 0.01$). The positive correlation means that faster-growing roots contained more dauricine. This character is very useful for selection to establish a high dauricine-producing culture.

2.3 Dauricine Biosynthesis in Cultured Roots of *M. dauricum*

Dauricine, a simple homodimer of (*R*)-armepavine with all the hydroxyl groups in the isoquinoline portions methylated, is an ideal alkaloid for studying bisbenzylisoquinoline biosynthesis. In this work, *M. dauricum* cultured roots were fed with ^{14}C-labeled tyrosine and ^{13}C-labeled tyrosine and tyramine,

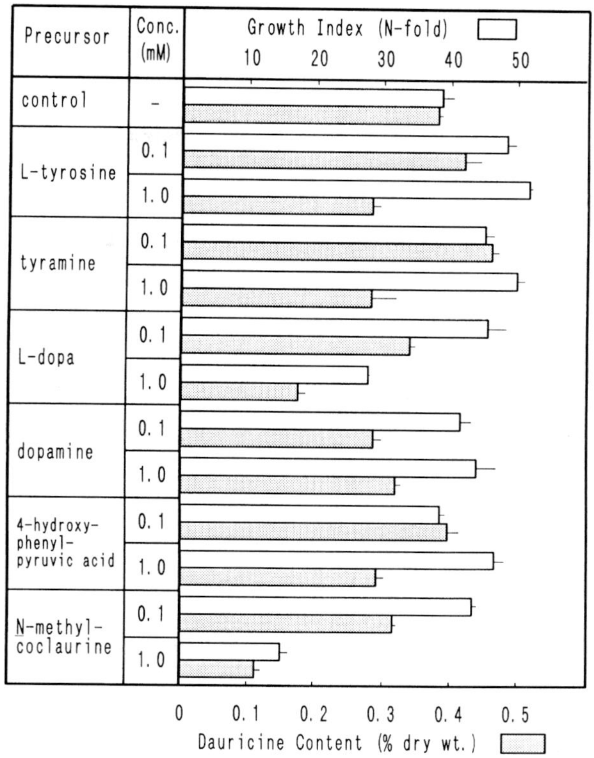

Fig. 13. Effects of possible precursors on growth and dauricine formation. Roots (0.1 g fresh weight) were cultured in 25 ml of medium with various precursors for 40 days. *Thin lines* represent SE for three replicates. (Reprinted from Sugimoto et al. 1994, with kind permission from Elsevier Science Ltd., The Boulevard, Langford Lane, Kidlington OX5 1GB, UK)

and their incorporation into dauricine was studied in an attempt to discover the exact role of these early precursors in bisbenzylisoquinoline biosynthesis.

Based on the remaining radioactivity in the medium, *M. dauricum* roots, harvested 20 days after applying L-[U-^{14}C] tyrosine (5.92×10^7 dpm), accumulated 92.2% of the total radioactivity. Of the radioactivity in the roots, 5.7% (33.8×10^5 dpm) of which 4.2×10^5 dpm was present in the crude alkaloids, was recovered in the methanol extract. HPLC analysis of the crude alkaloids showed that the distribution of radioactivity closely followed UV absorption (Fig. 14). Two main peaks with retention times (Rt) of 7.9 and 20 min, attributable to highly ^{14}C-labeled compounds, were present. One of the compounds, Rt 20 min, was identified as dauricine by comparison of its behavior on HPLC with authentic dauricine. The second compound, Rt 7.9 min, was identified based on physical and spectral data from a larger sample isolated in a separate experiment, as acutumine (Tomita et al. 1967). These results showed, unequivocally, that tyrosine is a building block of dauricine.

To obtain a more precise assessment of tyrosine incorporation into dauricine, feeding experiments of ^{13}C-labeled tyrosine and tyramine were

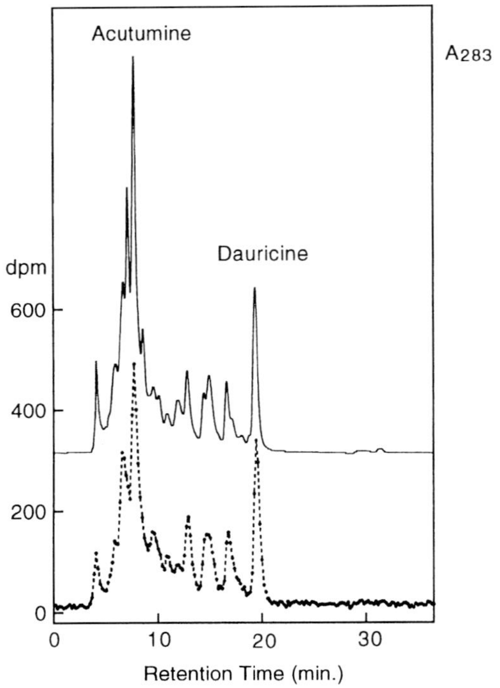

Fig. 14. HPLC analysis of the basic fraction extracted from cultured roots of *M. dauricum* administered L-[U-^{14}C] tyrosine.------radioactivity of the fractions;——absorbance at 283 nm. (Sugimoto et al. 1996)

carried out. ^{13}C-NMR of *O*-methyldauricine (OMD) obtained by feeding 0.5 mM L-[3-^{13}C] tyrosine showed no significant signal enhancement over that obtained from authentic OMD. In contrast, ^{13}C-NMR of OMD from cultures fed with 1 mM L-[3-^{13}C] tyrosine or 1 mM [2-^{13}C] tyramine displayed significant signal increases (Fig. 15). Comparing the spectra of OMDs prepared from dauricine isolated from roots fed with L-[3-^{13}C] tyrosine (Fig. 15A) or [2-^{13}C] tyramine (Fig. 15B) with that of unlabeled OMD (Fig. 15C), four signals (Fig. 15A) and two signals (Fig. 15B) were specifically increased. The ^{13}C-NMR spectrum of OMD was previously assigned by Jossang et al. (1986). According to their assignments, enriched signals correspond to the carbons at C-4 and 4′ (both δ 24.9) and C-α and α' (δ 39.5 and 39.7 or vice versa). The ^{13}C-enrichments of the signals calculated from peak-height analysis of ^{13}C-NMR spectra of OMD were as follows; for OMD from cultures fed with L-[3-^{13}C] tyrosine, C-4 and 4′ were 2.2 and 2.3% and C-α and α' were 0.8 and 0.9%. For OMD from cultures fed with [2-^{13}C] tyramine, C-4 and 4′ were both 1.3%.

These results showed that dauricine is composed of four molecules of tyrosine, and tyramine is incorporated exclusively into the isoquinoline portions. Therefore, tyrosine incorporation into the benzyl portion of the molecule proceeds not via tyramine, but via its corresponding α-keto acid, 4-hydroxyphenylpyruvic acid (Sugimoto and Yamada 1993).

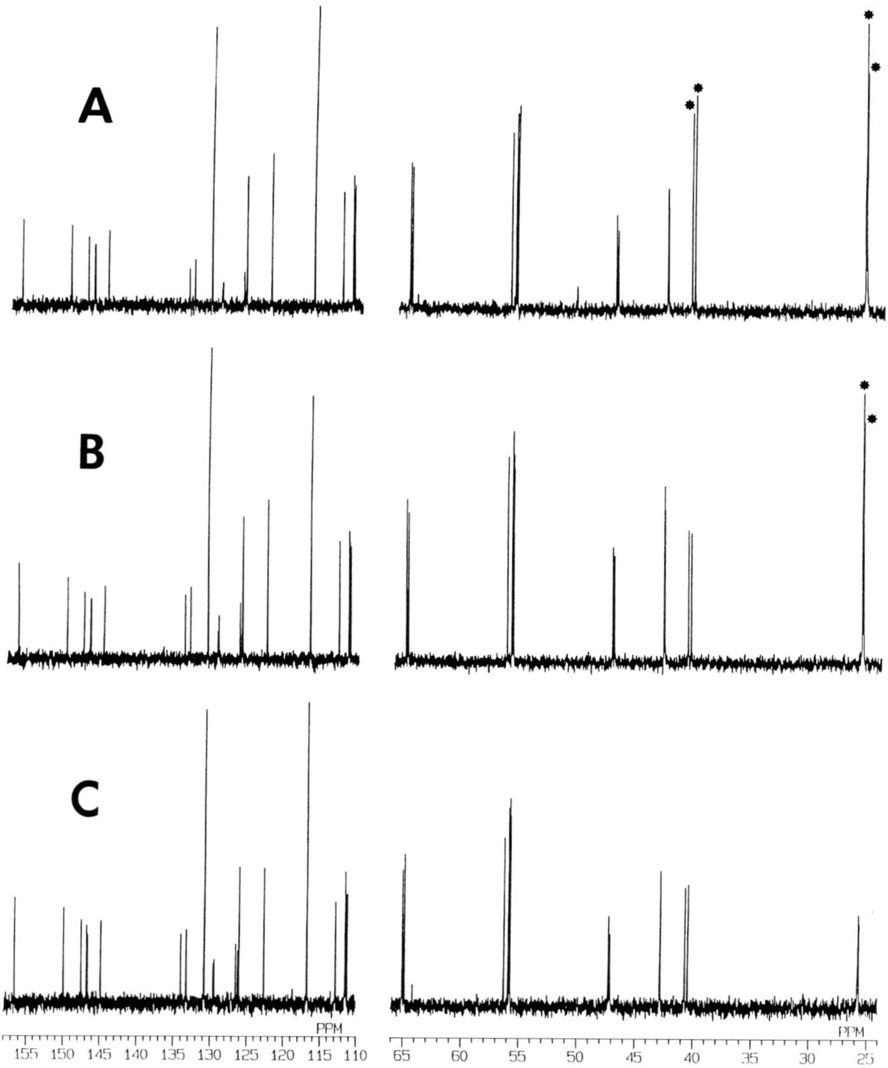

Fig. 15. ^{13}C-NMR spectra of OMD isolated from cultured roots of *M. dauricum*. The reference sample **C** has normalized peak heights for comparison with the OMDs labeled with **A** L-[3-^{13}C] tyrosine, and **B** [2-^{13}C] tyramine. *Enriched signal. (Sugimoto et al. 1996)

2.4 Other Metabolites Produced in Cultured Roots of *M. dauricum*

2.4.1 Alkaloids Induced by Ketoconazole, a Phytochrome P-450 Inhibitor

Cytochrome P-450 enzymes have been shown to mediate a number of biosynthetic reactions of secondary metabolites, including lipids, phenylpropanoids, flavonoids, terpenoids, and alkaloids, in higher plants

(Bolwell et al. 1994). These reactions require NADPH and molecular oxygen, and are inhibited by carbon monoxide and P-450 inhibitors (Coulson et al. 1984).

In this work, P-450 inhibitors were used to probe involvement of P-450 enzymes in the biosynthesis of bisbenzylisoquinolines and a related alkaloid, acutumine, in cultured roots of *M. dauricum*. It was demonstrated that both regio- and stereoselective intermolecular phenol oxidative couplings of benzylisoquinolines to yield bisbenzylisoquinolines are mediated by a P-450 enzyme (Kraus and Kutchan 1995). Furthermore, intramolecular C-C coupling of (*R*)-reticuline to salutaridine was reported to be catalyzed by a P-450 enzyme (Gerardy and Zenk 1993). Acutumine is a benzylisoquinoline alkaloid which arises via a complex biosynthetic pathway, involving intramolecular C-C coupling of a simple benzylisoquinoline (Barton et al. 1968).

Six P-450 inhibitors were chosen on the basis of water-solubility to add to liquid media. The effects of P-450 inhibitors on root growth and alkaloid contents are shown in Fig. 16. Root growth varied with the inhibitor used. Increasing concentrations of ancymidol and metyrapone promoted root growth while ketoconazole was inhibitory at all concentrations. The other inhibitors, however, had inconsistent effects. In contrast, the alkaloid contents showed no relationship to root growth and consistently displayed a progressive decline with increasing inhibitor concentration (Fig. 16).

Treatment of cultured roots with ketoconazole at 1 to 10 μM drastically reduced acutumine (Rt 7.8 min), dauricine (Rt 15.5 min) and some other components with Rts ranging from 7 to 14 min. However, accumulation of three bases with Rts of 9, 20.6 and 36 min was induced (Fig. 17). The peak with Rt of 9 min yielded a yellow powder that showed chromatographic behavior (TLC and HPLC), ^{1}H-, ^{13}C-NMR, and mass spectra identical to those of an authentic tyramine. The other two compounds with Rts of 20.6 and 36 min, tentatively named 1 and 2, respectively, showed positive reaction with Dragendorff's reagent. Their molecular masses were 353 and 456, respectively. Feeding L-[U-^{14}C] tyrosine to cultured roots in a medium supplemented with 2.5 μM ketoconazole resulted in radiolabeling of both compounds as well as tyramine.

Oikawa et al. (1992), working on microbes, showed that P-450 inhibitors resulted in a new generation of plausible precursors of secondary metabolites without serious growth inhibition. For *M. dauricum* roots, the effects of the inhibitors on alkaloid production and on root growth were at variance ($r = 0.50$ and 0.14 for dauricine and acutumine, respectively). At the concentration employed, the effect on root growth was dependent on the inhibitor used. It varied from stimulation with ancymidol and metyrapone to considerable suppression with ketoconazole. However, alkaloid production was, invariably, reduced by all inhibitors (Fig. 16). Therefore, a direct effect on alkaloid biosynthesis in *M. dauricum* roots is postulated, at least for some of the inhibitors.

Contrary to previous reports on biosynthesis of microbial secondary metabolites (Van Middlesworth et al. 1986; Oikawa et al. 1992), culturing of *M. dauricum* roots in media supplemented with P-450 inhibitors did

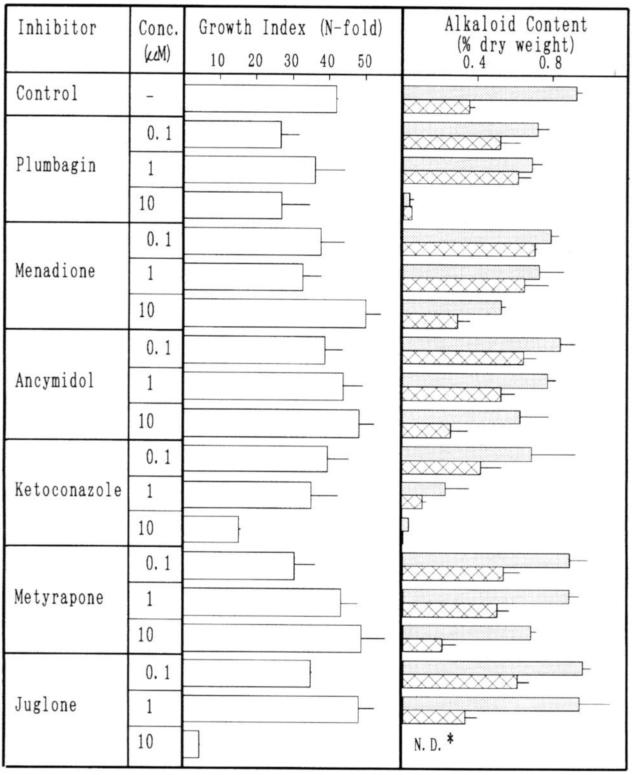

Fig. 16. Effects of P-450 inhibitors on growth and alkaloid contents in cultured roots of *M. dauricum*. Roots (0.1 g fresh weight) were cultured in 25 ml of medium with various P-450 inhibitors for 40 days. *Thin lines* represent SE for three replicates. ▨ dauricine, ⊠ acutumine, *not determined. (Sugimoto et al. 1997)

not lead to accumulation of immediate monomeric precursors of bisbenzylisoquinolines. However, treatment of *M. dauricum* roots with ketoconazole induced accumulation of three basic compounds (Fig. 17). One of these compounds was identified as tyramine, an early precursor of benzylisoquinoline alkaloids. This finding is consistent with an effective inhibition of oxidative coupling in the early steps of benzylisoquinoline biosynthesis. HPLC analysis of the basic fraction extracted from the roots showed that the distribution of radioactivity derived from L-[U-^{14}C] tyrosine closely followed UV absorption when roots were cultured in media with ketoconazole (data not shown) as well as when cultured without the inhibitor (Fig. 14). Incorporation of L-[U-^{14}C] tyrosine into the unidentified compounds 1 and 2 together with their positive reaction with Dragendorff's reagent showed that both compounds are tyrosine-derived alkaloids. Their molecular masses, 353 and 456, are too small for bisbenzylisoquinolines. Furthermore, judging from the chromatographic behavior on HPLC, the unidentified compounds 1 and 2 are

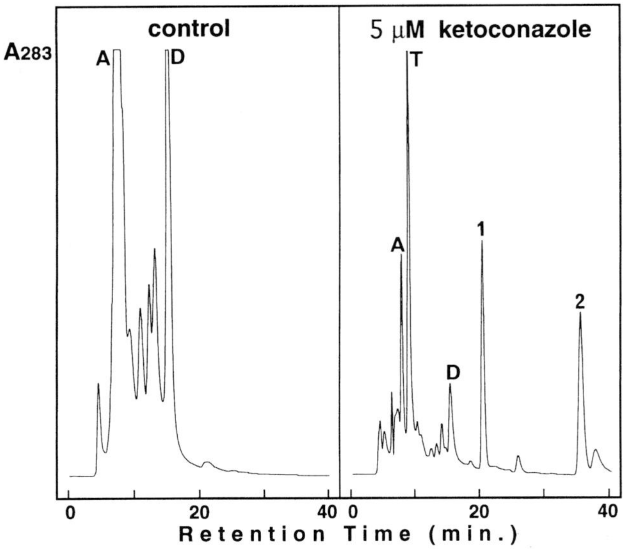

Fig. 17. HPLC analysis of the basic fraction extracted from control and ketoconazole treated roots of *M. dauricum*. Identification of peaks, *A* acutumine; *D* dauricine; *T* tyramine. Peaks with Rts of 20.6 and 36 min have not been identified. (Sugimoto et al. 1997)

too apolar for less oxidized precursors or intermediates in dauricine or acutumine biosynthesis. Based on the facts mentioned above, it is concluded that ketoconazole inhibited benzylisoquinoline biosynthesis at an early stage and activated another biosynthetic pathway, which leads tyrosine to the unidentified compounds 1 and 2, in *M. dauricum*. Structure elucidation of these compounds is in progress.

2.4.2 Chlorine-Containing Alkaloids

The finding of an unusual chlorine-containing alkaloid acutumine, as mentioned in the previous section, prompted us to investigate further chlorine-containing alkaloids in cultured roots of *M. dauricum*. The roots were cultured for 50 days in the media, in which Cl^- was substituted by radioactive $^{36}Cl^-$. Incorporation of $^{36}Cl^-$ was 12.5%, as determined by counting the remaining radioactivity in the medium. Crude alkaloids were isolated from methanol extracts of the roots. In the crude alkaloids, radioactivity derived from $^{36}Cl^-$ was recovered sufficiently for analysis by HPLC. HPLC analysis of the crude alkaloids showed that there were at least four bases, acutumine (Rt 11 min) and unidentified bases, labeled by ^{36}Cl, clear evidence that they are chlorine-containing alkaloids (Fig. 18). Radioactivity found in the four bases was equivalent to a quarter of that incorporated by roots. These bases may play a crucial role in localizing chlorine atoms in this plant.

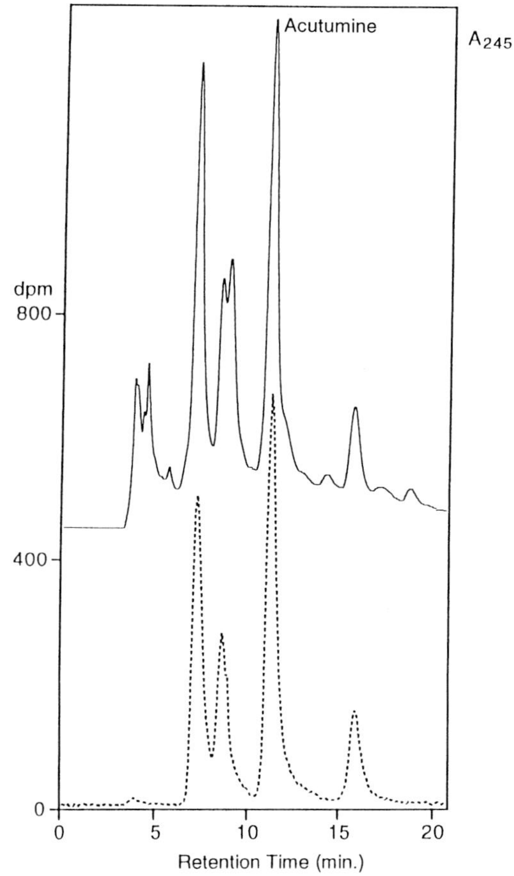

Fig. 18. HPLC analysis of the basic fraction extracted from cultured roots of *M. dauricum* administered $Na^{36}Cl$.------radioactivity of the fractions;——absorbance at 245 nm

3 Summary and Conclusion

Root cultures of *Menispermum dauricum* were established from adventitious roots formed on the leaf segments. These roots produced dauricine as a major constituent in the alkaloid fraction and the optimum conditions for dauricine production in cultured roots were investigated.

Feeding experiments with L-[U-^{14}C] tyrosine, L-[3-^{13}C] tyrosine, and [2-^{13}C] tyramine demonstrated that dauricine was composed of four molecules of tyrosine, and that tyramine was specifically incorporated into the isoquinoline portions of dauricine. The unusual chlorine-containing alkaloid acutumine, into which ^{14}C-labeled tyrosine was also incorporated, was identified as one of the main constituents in the alkaloid fraction from the roots. Further feeding experiments with $^{36}Cl^-$ showed that the roots produced at least four chlorine-containing alkaloids.

The effect of cytochrome P-450 inhibitors on biosynthesis of benzylisoquinoline alkaloids was studied. Production of dauricine and acutumine was curtailed by all inhibitors. Ketoconazole-treated roots accumulated tyramine, an early precursor of benzylisoquinoline, and two unidentified tyrosine-derived alkaloids with molecular masses of 353 and 426.

Cultured roots of *M. dauricum* are a rich source of a variety of benzylisoquinolines, and manipulation of culture parameters induces new constituents. These roots are also an excellent material for the study of biosynthesis. A complicated biosynthetic pathway from a simple isoquinoline to acutumine was proposed by Barton et al. in 1968, but so far there has been no experimental proof. Cultured roots of *M. dauricum* will make it possible to study acutumine biosynthesis in the near future. Mechanisms for biohalogenation remain poorly understood, though chlorinated products are sometimes found in higher plants. Cultured roots of *M. dauricum*, which produce more than 0.2% acutumine, must be a candidate for the study of mechanisms for biohalogenation in higher plants.

References

Baldas J, Bick IRC, Ibuka T, Kapil RS, Porter QN (1972) Alkaloids derived from coclaurine units joined tail-to-tail. J Chem Soc Perkin Trans I:592–596

Barton DHR, Kirby AJ, Kirby GW (1968) Investigations on the biosynthesis of sinomenine. J Chem Soc [C]:929–936

Bolwell GP, Bozak K, Zimmerlin A (1994) Plant cytochrome P450. Phytochemistry 37:1491–1506

Coulson CJ, King DJ, Wiseman A (1984) Chemotherapeutic and agrochemical applications of cytochrome P-450 ligands. Trends Biochem Sci 9:446–449

Doskotch RW, Knapp JE (1971) Alkaloids from *Menispermum canadense*. Lloydia 34:292–300

Engvild KC (1986) Chlorine-containing natural compounds in higher plants. Phytochemistry 25:781–791

Gerardy R, Zenk MH (1993) Formation of salutaridine from (*R*)-reticuline by a membrane-bound cytochrome P-450 enzyme from *Papaver somniferum*. Phytochemistry 32:79–86

Hotta M, Terabayashi S (1989) Menispermaceae. In: Hotta M (ed) Useful plants of the world. Heibonsha, Tokyo, pp 673–674

Hou CY, Xue H (1985) Studies on the chemical constituents of *Menispermum dauricum* DC. Acta Pharm Sin 20:112–117

Hu SM, Xu SX, Yao XS, Cui CB, Tezuka Y, Kikuchi T (1993) Dauricoside, a new glycosidal alkaloid having an inhibitory activity against blood-platelet aggregation. Chem Pharm Bull (Tokyo) 41:1866–1868

Jossang A, Leboeuf M, Cave A (1986) Alcaloides de *Popowia pisocarpa*, premiere partie: nouvelles bisbenzylisoquinoleines. J Nat Prod 49:1018–1027

Kraus PFX, Kutchan TM (1995) Molecular cloning and heterologous expression of a cDNA encoding berbamunine synthase, a C-O phenol-coupling cytochrome P450 from the higher plant *Berberis stolonifera*. Proc Natl Acad Sci USA 92:2071–2075

Kunitomo J, Satoh M (1982) Structure of menisporphine: a new type of isoquinoline alkaloid. Chem Pharm Bull (Tokyo) 30:2659–2660

Kunitomo J, Kaede S, Satoh M (1985) The structure of 2,3-dihydromenisporphine and the synthesis of dauriporphine, oxoisoaporphine alkaloids from *Menispermum dauricum* DC. Chem Pharm Bull (Tokyo) 33:2778–2782

Manske RHF, Tomita M, Fujitani K, Okamoto Y (1965) Dauricine from *Menispermum canadense* L. Chem Pharm Bull (Tokyo) 13:1476–1477

Oikawa H, Murakami Y, Ichihara A (1992) Useful approach to find the plausible biosynthetic precursors of secondary metabolites using P-450 inhibitors: postulated intermediates of chaetoglobosin A. J Chem Soc Perkin Trans I:2949–2953

Okamoto Y, Yuge E, Nagai Y, Katsuta R, Kishimoto A, Kobayashi Y, Kikuchi T, Tomita M (1969) Acutuminine, a new alkaloid from the leaves of *Menispermum dauricum* DC. Tetrahedron Lett: 1933–1935

Okamoto Y, Tanaka S, Kitayama K, Isomoto M, Masaishi M, Yanagawa H, Kunitomo J (1971) Alkaloids of *Menispermum dauricum* DC (8). Yakugaku Zasshi 91:684–687

Pan XP (1992) A new alkaloid from *Menispermum dauricum* DC. *N*-demethyldauricine. Acta Pharm Sin 27:788–791

Pang XP, Chen YW, Li XJ, Long JG (1991) A new alkaloid of *Menispermum dauricum* DC. dauriciline. Acta Pharm Sin 26:387–390

Purrington FF, Horn DJ (1993) Canada moonseed vine (Menispermanceae): host of four roundheaded wood borers in central Ohio (Coleoptera: Cerambycidae). Proc Entomol Soc Wash 95:313–320

Rueffer M, Zenk MH (1987) Distant precursors of benzylisoquinoline alkaloids and their enzymatic formation. Z Naturforsch [C] 42:319–332

Schiff PL Jr (1991) Bisbenzylisoquinoline alkaloids. J Nat Prod 54:645–749

Stadler R, Loeffler S, Cassels BK, Zenk MH (1988) Bisbenzylisoquinoline biosynthesis in *Berberis stolonifera* cell cultures. Phytochemistry 27:2557–2565

Sugimoto Y, Yamada Y (1993) Involvement of tyramine in bisbenzylisoquinoline biosynthesis in cultured roots of *Stephania cepharantha*. J Nat Prod 56:1229–1233

Sugimoto Y, Yoshida A, Uchida S, Inanaga S, Yamada Y (1994) Dauricine production in cultured roots of *Menispermum dauricum*. Phytochemistry 36:679–683

Sugimoto Y, Uchida S, Inanaga S, Kimura Y, Hashimoto M, Isogai A (1996) Early steps of dauricine biosynthesis in cultured roots of *Menispermum dauricum*. Biosci Biotechnol Biochem 60:503–505

Sugimoto Y, Uchida S, Inanaga S, Isogai A (1997) Effect of P-450 inhibitors on benzylisoquinoline alkaloid biosynthesis in cultured roots of *Stephania cepharantha* and *Menispermum dauricum*. J Plant Physiol 150:376–380

Takani M, Takasu Y, Takahashi K (1983) Constituents of the vines of *Menispermum dauricum* DC. (2). Chem Pharm Bull (Tokyo) 31:3091–3093

Thornber CW (1970) Alkaloids of the Menispermaceae. Phytochemistry 9:157–187

Tomita M, Okamoto Y, Kikuchi T, Osaki K, Nishikawa M, Kamiya K, Sasaki Y, Matoba K, Goto K (1967) X-ray analysis of acutumine. Tetrahedron Lett: 2421–2424

Tomita M, Okamoto Y, Nagai Y, Kitayama K, Yanagawa H (1970a) The structure of a new tertiary phenolic biscoclaurine type alkaloid "dauricoline". Yakugaku Zasshi 90:1178–1181

Tomita M, Okamoto Y, Nagai Y, Tanaka S, Hayata T (1970b) Basic components of Siberian *Menispermum dahuricum* DC. Yakugaku Zasshi 90:1182–1186

Van Middlesworth F, Desjardins AE, Taylor SL, Plattner RD (1986) Trichodiene accumulation by ancymidol treatment of *Gibberella pulicaris*. J Chem Soc Chem Commun: 1156–1157

Zheng XW, Min ZD, Zhao SX (1979) A new alkaloid of *Menispermum dauricum*. Daurisoline. K'o Hsueh T'ung Pao 24:285–288

Zhao S, Ye W, Tan N, Zhao H, Xia Z (1989) A novel oxoisoaporphine alkaloid from the rhizome of *Menispermum dauricum*. Zhongguo Yaoke Daxue Xuebao 20:312

XVIII *Ornithopus sativus* Brot. (Serradella): In Vitro Culture, Phytochemical Studies, and Biotransformation

A. Kolbe[1], G. Krauss[2], and B. Schneider[3]

1 Introduction

Ornithopus sativus Brot. (serradella, Portuguese for small saw, due to the shape of the leaves) (Fig. 1) belongs to the family Fabaceae. The leaves are 3–7 cm in length with imparipinnate sessile side leaves. The pedunculate umbel-like flowers are 5–8 mm in length, typically papillionaceous in shape, and of rose to white color. Like lupins, serradella prefers sandy soils but requires rather humid conditions. *O. sativus* is native to south-western Mediterranean Europe. Depending on the growth conditions, it is an annual or, as in a markedly mediterranean climate, a perennial species (Gladstones and McKeown 1977; Geister 1991). This plant has been cultivated since the middle of the 19th century in central Europe and later on to some extent in eastern Europe and in temperate regions of the Southern Hemisphere as a fodder plant. It easily becomes wild in areas of cultivation. In contrast with the lupins, serradella obviously is lacking in alkaloids and, therefore, is more compatible with cattle pasture. However, it has a rather small yield, and is competitive with clover only on poor sandy soils. Frequently, it is grown as an intermediate crop between cereals and root crops and, recently, it has been used as a green fertilizer. However, overall its economic importance is relatively limited.

Phytochemical studies have revealed a variety of secondary products which exhibit different pharmacological activities per se or serve as starting materials for the synthesis of steroidal drugs. Thus, β-sitosterol was found to be the major sterol component in *O. sativus* and the triterpene lupeol also occurred in relatively high concentrations (Spengler 1995; Schmidt et al. 1993a). Furthermore, natural products have been isolated which are involved in defense against plant pathogens (Seifert et al. 1993) or play a role in plant growth and development (Schmidt et al. 1993b; Spengler et al. 1995).

[1] Institut für Pflanzenbiochemie, Weinberg 3, 06120 Halle, Germany
[2] UFZ – Umweltforschungszentrum Leipzig-Halle GmbH, Sektion Hydrogeologie, Hallesche Straße 44, 06246 Bad Lauchstädt, Germany
[3] Max-Planck-Institut für Chemische Ökologie, Tatzendpromenade 1a, 07745 Jena, Germany

Biotechnology in Agriculture and Forestry, Vol. 43
Medicinal and Aromatic Plants XI (ed. by Y.P.S. Bajaj)

Fig. 1. *Ornithopus sativus* plants

2 In Vitro Culture

Cell culture systems of *O. sativus* have been established at our laboratory (Seifert et al. 1993). The hard-coated seeds were washed with distilled water followed by removal of the coat. The scarified seeds were submerged for 10 min in 70% ethanol. After washing with distilled water, seeds were submerged in saturated sodium hypochlorite for 20 min and rinsed 3 times in sterile water. Seeds were placed on agar prepared only with water in 100-ml Erlenmeyer flasks and were kept at 25 °C in the dark.

For the induction of callus slices from cotyledones, stems or roots of the seedlings were placed on a modified MS agar (Murashige and Skoog 1962) supplemented with sucrose (3%), Ca-pantothenate (5 mg/l), inositol (80 mg/l), 2,4-dichlorophenoxyacetic acid (1 mg/l) and kinetin (0.2 mg/l), and were kept

Fig. 2. Callus culture obtained from stem tissue of *Ornithopus sativus*

at 25 °C in the dark for 1 month. The calli appear pale yellow or light brown, and tend to aggregate to form clusters (Fig. 2).

The cell suspension cultures obtained from stem-derived callus cultures were initiated by transfering callus cultures (ca. 5 g fresh mass) to 250-ml Erlenmeyer flasks containing 30 ml of a modified MS medium as described above. The cells were grown in this medium on rotary shakers (80 rpm) at 27 °C in the dark. The pH of the medium was adjusted to 5.6–5.8. Transfer of cells (10 ml) to fresh medium (30 ml) was performed every 10–11 days.

For biochemical studies, a cell suspension culture which had been subcultured for a long time was used, containing finely dispersed cells without large cell aggregates. Figure 3 shows a typical growth curve of the *O. sativus* cell suspension culture.

3 Phytochemical Studies

Relatively little research has been carried out on the phytochemistry of *O. sativus*, Table 1 summarises previous investigations. A variety of triterpenes and 4α-methylsterols have been obtained in different ratios from various parts of intact plants (shoots, roots, seeds) and cell cultures (Spengler 1995). β-Amyrine, lupeol and taraxarol were identified as the major components in roots. In shoots, lanosterol and 24-methylenlophenol also occurred in signifi-

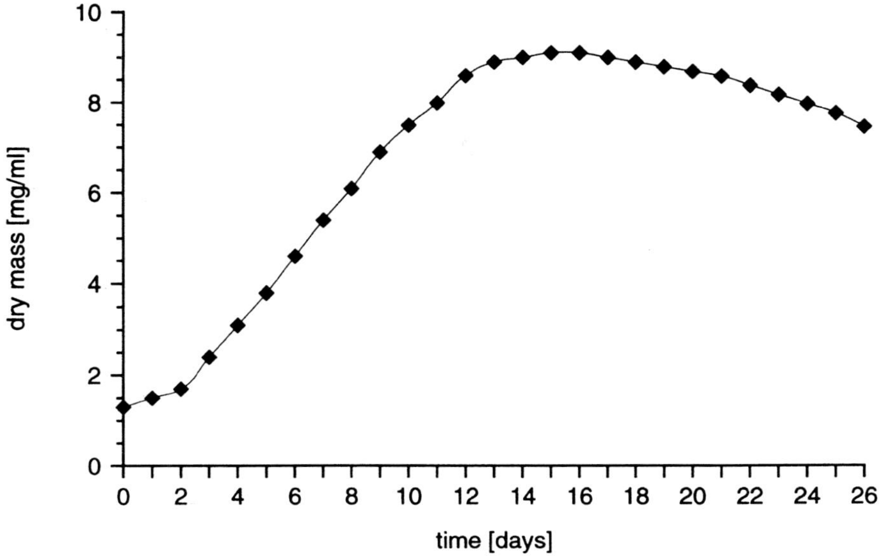

Fig. 3. Growth curve of *Ornithopus sativus* cell suspension (means $n = 4$)

Table 1. Summary of in vitro culture, and secondary metabolite studies on *Ornithopus sativus*

References	System studied	Observation/Remarks
Seifert et al. (1993)	Suspension cultures, plants	Phytoalexin accumulation
Schmidt et al. (1993a)	Seeds	Phytosterols
Schmidt et al. (1993b)	Seeds	Brassinosteroids
Kolbe et al. (1994)	Suspension cultures	Brassinosteroid metabolism, pregnanes
Kolbe et al. (1995)	Suspension cultures	Brassinosteroid metabolism, acyl conjugates
Spengler 1995	Suspension cultures, plants	Phytochemical investigation
Spengler et al. (1995)	Plants	Brassinosteroids
Kolbe et al. (1996)	Suspension cultures	Brassinosteroids, metabolic pathways

cant amounts, while only traces of taraxerol were present. Lanost-7-en-3β-ol, 24-methylene-24-dihydroparkeol and 24-methylenecycloartanol (roots) and 24-methyllophenol (shoots) were found in minor amounts. In seeds, β-amyrine was found as the only triterpene. 24-Methylenecycloartanol was the predominant component of the triterpene/4α-methylsterol fraction in cell cultures which also showed significant amounts of β-amyrine and minor amounts of 24-methylene-24-dihydroparkeol and 24-methylene-24-dihydrolanosterol. The latter compound did not accumulate to detectable concentrations in intact plants.

The sterols of *O. sativus* have been investigated with regard to their role as biosynthetic precursors of brassinosteroids (Spengler et al. 1995). Sitosterol

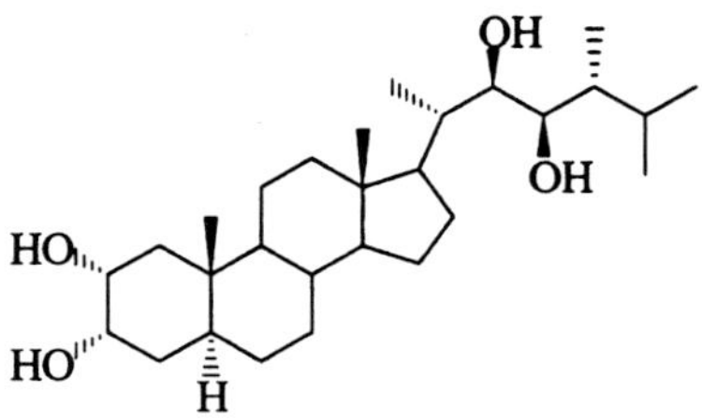

6-Deoxo-24-epicastasterone

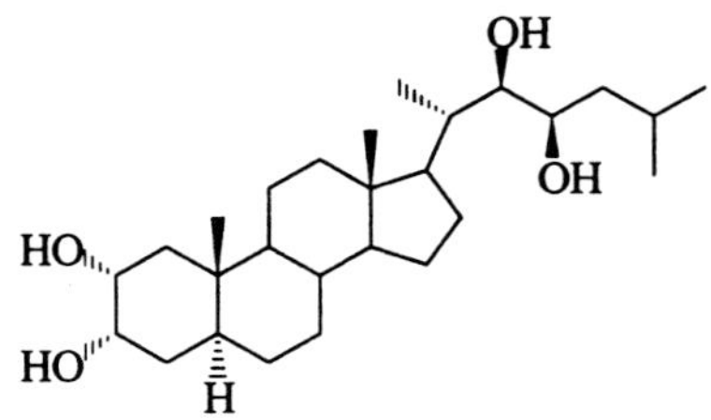

6-Deoxo-28-norcastasterone

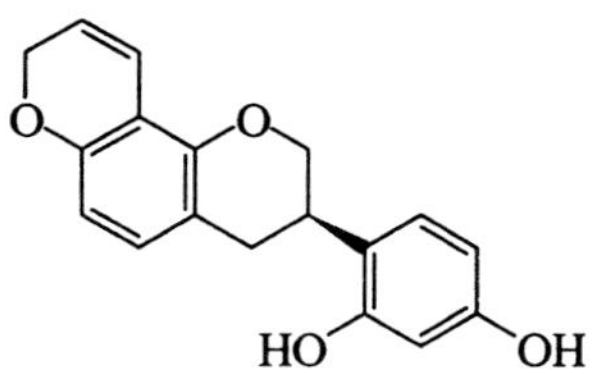

Glabridin

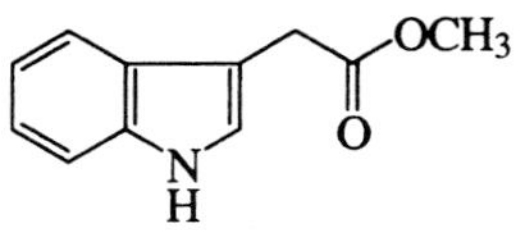

Indolyl-3-acetic acid methyl ester

Fig. 4. Structures of natural products isolated from *Ornithopus sativus*

was found to be the predominant phytosterol in all analysed plant parts (seeds, roots, shoots) and in cell cultures. Major portions of stigmasterol were identified in roots and in cell cultures while relatively small concentrations were detected in shoots and seeds. Campesterol, cholesterol, campestanol, stigmastanol, isofucosterol, stigmasta-5,7-dien-3β-ol and stigma-22-en-3β-ol were minor components in cell cultures (Schmidt et al. 1993a; Spengler 1995).

Five brassinosteroids; castasterone, 24-epicastasterone, 6-deoxocastasterone, 6-deoxo-24-epicastasterone, and 6-deoxo-28-norcastasterone, were found in shoots of *O. sativus* (Spengler et al. 1995). The last two compounds are not yet known from any other species (Fig. 4). The highly active plant hormone castasterone is widespread in plants and 24-epicastasterone was found in a variety of species (Adam et al. 1996). The occurrence of 6-deoxobrassinosteroids suggests that the so-called late C-6 oxidation pathway is operative in brassinosteroid biosynthesis (Choi et al. 1996). Castasterone and 24-epicastasterone were also detected in seeds (Schmidt et al. 1993b). Interestingly, with regard to brassinosteroid biosynthesis, the side chain structures and configurations of these brassinosteroids do not directly correlate with the predominant sterol in that species, sitosterol. Furthermore, no sterol bearing a side chain with 24*R* methyl was found which could serve as a direct precursor for 24-epicastasterone and 6-deoxo-24-epicastasterone. Due to these findings, more detailed investigations on brassinosteroid biosynthesis are required.

Sterols bearing an oxygen function at C-6 (7-oxo-, 7α-hydroxy and 7β-hydroxy derivatives of campesterol, stigmasterol and sitosterol) have been identified in seeds, roots, shoots and cell cultures (Schmidt et al. 1993a; Spengler 1995). In each case, the sitosterol derivative was the most abundant component in the 7-oxo and 7-hydroxy fractions and the ratios corresponded with the sterol pattern. The occurrence of oxygenated sterols might be due to autoxidation of the Δ^5-sterols. Otherwise such compounds are biosynthesized by many microorganisms, plants and several cell cultures (Parish 1991). Five 3-oxo-Δ^4-sterols have been identified in roots, stigmast-4-en-3-one, stigmast-4,22-dien-3-one, campest-4-en-3-one, cholest-4-en-3-one, and stigmast-4,6-dien-3-one. These compounds, except the last one, were also found in shoots.

The isolation of brassinosteroids is usually carried out by bioactivity-guided fractionation, using the rice lamina inclination test. Depending on the activity and concentration of the compound to be tested, brassinosteroids induce leaf lamina joint inclination in etiolated rice seedlings. This bioassay was originally developed as an auxin text (Maeda 1965) which, in higher concentrations, causes the same effect. While searching for brassinosteroids in cell cultures of *O. sativus*, besides a fraction due to brassinosteroid activity, the methyl ester of indolyl-3-acetic acid (Fig. 4) was isolated and identified by GC-MS analysis (Spengler 1995). To exclude an artifact, extraction was carried out with ethanol instead of methanol. Since, in these experiments, *O. sativus* cell cultures were grown in MS medium, which does not contain IAA

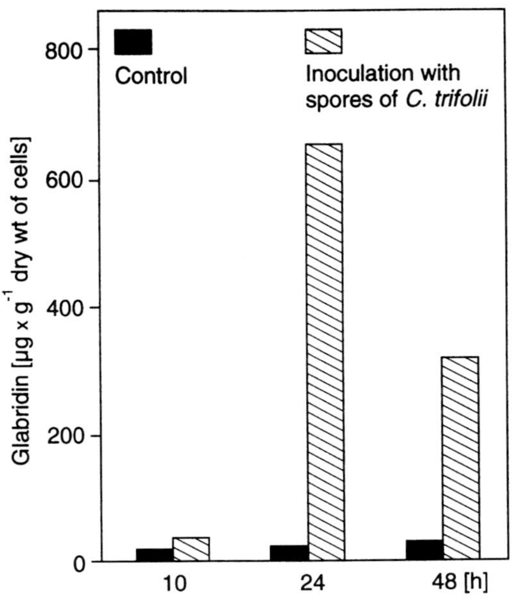

Fig. 5. Accumulation of glabridin in the cell suspension culture of *Ornithopus sativus* inoculated with spores of *C. trifolii*. (Seifert et al. 1993)

but 2,4-D as an auxin, this methyl ester must be considered as an endogenous metabolite.

Elicitor treatment of plants and cell cultures of *O. sativus* resulted in a significant increase in phytoalexin concentration (Seifert et al. 1993). An isoflavan derivative, glabridin (Fig. 4), has been isolated from aerial parts of the plants and from cell suspension cultures as the fungitoxic component. Glabridin was shown to completely inhibit the germination of *Cladosporium cucumerinum* conidia. The plants and cell suspensions responded to treatment with $CuSO_4$, yeast polysaccharides, or spores of *Colletotricum trifolii* Bain and Ess. (Deuteromycetes), which is responsible for yield and quality reductions in field cultures, with considerable accumulation of glabridin. For example, the highest concentration of glabridin was found in cell cultures 24h after application of *C. trifolii* spores (Fig. 5). It has been further demonstrated that glabridin formation is associated with a large transient increase of the enzymes involved in isoflavan biosynthesis, phenylalanine ammonia lyase (PAL) and chalcone synthase (CHS, Fig. 6).

4 Biotransformation

A variety of brassinosteroids have been found in *Ornithopus sativus*. In addition to phytochemical investigations, cell cultures were employed to study the fate of exogenously applied brassinosteroids in plant cells. The objective of these investigations was to elucidate the pathways of transformation, conjugation and degradation of 24-epicastasterone and 24-epibrassinolide, which are highly active plant hormones (Adam et al. 1996; Clouse 1996). Prior to metabolic studies, the influence of brassinosteroids on the growth of cultured cells was investigated. Compared with control experiments, both fresh weights and dry weights of *O. sativus* cells were markedly enhanced by increasing the concentration of 24-epibrassinolide (Fig. 7). This effect was clearly higher with 24-epibrassinolide than with 24-epicastasterone.

In further experiments, both compounds were administered in radiolabelled form to enable quantification of precursor uptake and transformation. Interestingly, the distribution of radioactivity between the cells and the medium did not significantly change during the entire incubation period of 7 days after tritium-labelled 24-epicastasterone and 24-epibrassinolide, respectively, were applied. After application of [5,7,7-^{3}H]24-epibrassinolide, about 40% of the radioactivity and after application of [5,7,7-^{3}H]24-epicastasterone about 25% of the radioactivity were found in the medium, and the remainder was associated with the cells (Kolbe et al. 1995). There was also a clear distribution of the different types of metabolites between the cells and the medium. In the cells, a major fraction of lipophilic and a minor fraction of hydrophilic conjugates were found, while the medium contained almost solely nonconjugated metabolites. The hydrophilic components of the cells are probably related to glucosides which have been found in metabolism experiments using cell cultures of *Lycopersicon esculentum* (Hai et al. 1996).

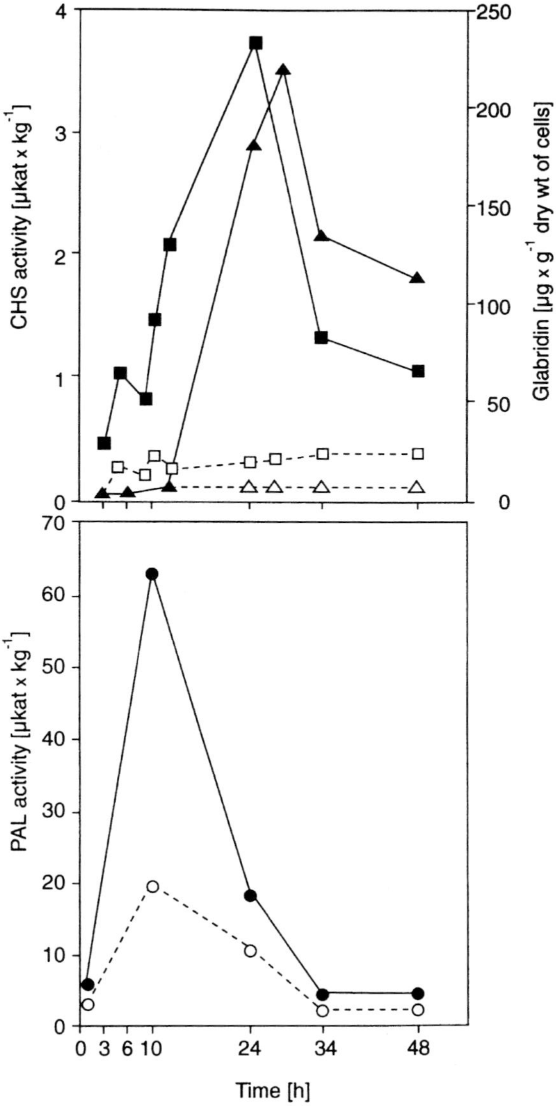

Fig. 6. Time course of elicitor-induced changes in the activity of PAL (●), CHS (■) and accumulation of glabridin (▲) in the cell suspension culture of *Ornithopus sativus*. Open symbols are controls (○), (□), (△). (Seifert et al. 1993)

The lipophilic fractions of the cell extracts after administration of 24-epicastasterone and 24-epibrassinolide, respectively, showed similar HPLC patterns. Each chromatogram exhibited three radioactive peaks of about the same quantity (Fig. 8). Structure elucidation revealed these metabolites to be brassinosteroid acyl conjugates, namely 3β-laurate, 3β-myristate and 3β-palmitate of 3,24-diepibrassinolide and 3,24-diepicastasterone, respectively.

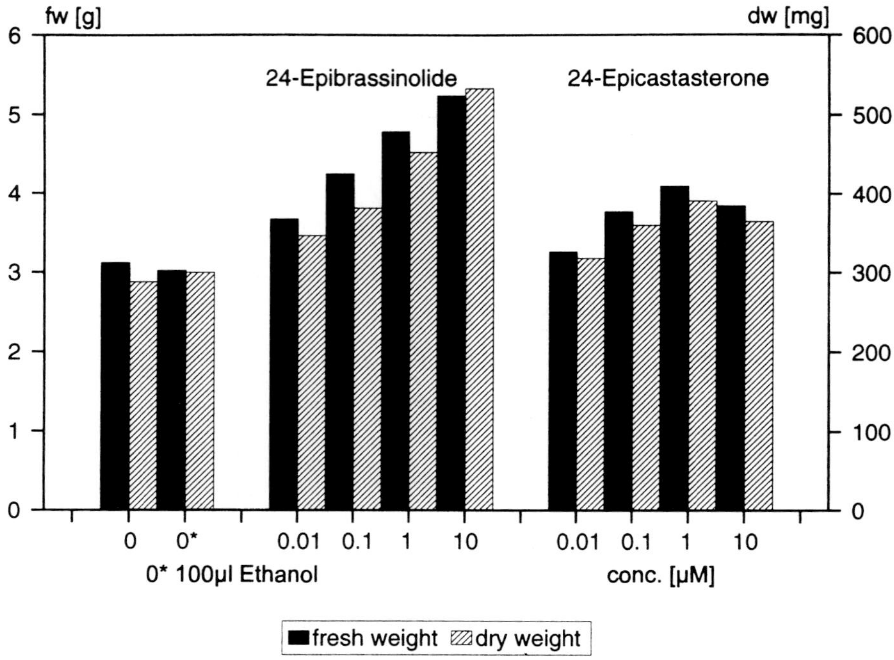

Fig. 7. Influence of 24-epibrassinolide and 24-epicastasterone on the growth behavior of *Ornithopus sativus* cells in suspension culture (fresh weights and dry weights were determined 7 days after administration of 24-epibrassinolide and 24-epicastasterone, respectively)

Obviously, the parent compounds underwent epimerization at C-3 prior to acyl conjugation. Acyl conjugates of brassinosteroids hitherto have been described only from lily pollen, also exhibiting 3β-configuration (Asakawa et al. 1996). The function of these fatty acid conjugates is not yet known. As discussed for phytosterol acyl conjugates (Wojchiechowski 1991) they might be involved in compartmentalization of membrane structures.

From the medium of the cell suspension cultures of *O. sativus*, supplied with radiolabelled 24-epicastasterone and 24-epibrassinolide, respectively, pregnane-type metabolites were isolated (Kolbe et al. 1994). 24-Epicastasterone was transformed to 2α,3β-dihydroxy-5α-pregnane-6,20-dione and 2α,3β,6β-trihydroxy-5α-pregnane-20-one, and after exogenous application of 24-epibrassinolide, the metabolic product was 2α, 3β-dihydroxy-B-homo-6a-oxa-5α-pregnane-6,20-dione (Fig. 9). These pregnane-type metabolites seem to be the final products of catabolism of brassinosteroids in cell suspension cultures of *O. sativus* which are expected to be formed by side chain cleavage. Although occurring in very low concentrations, in either case these putative intermediates of the metabolic sequence between the parent brassinosteroids and the pregnane derivatives could be isolated from the cell culture medium and identified by NMR and mass spectroscopy. Thus, complete metabolic pathways including 3,24-diepibrassinolide and

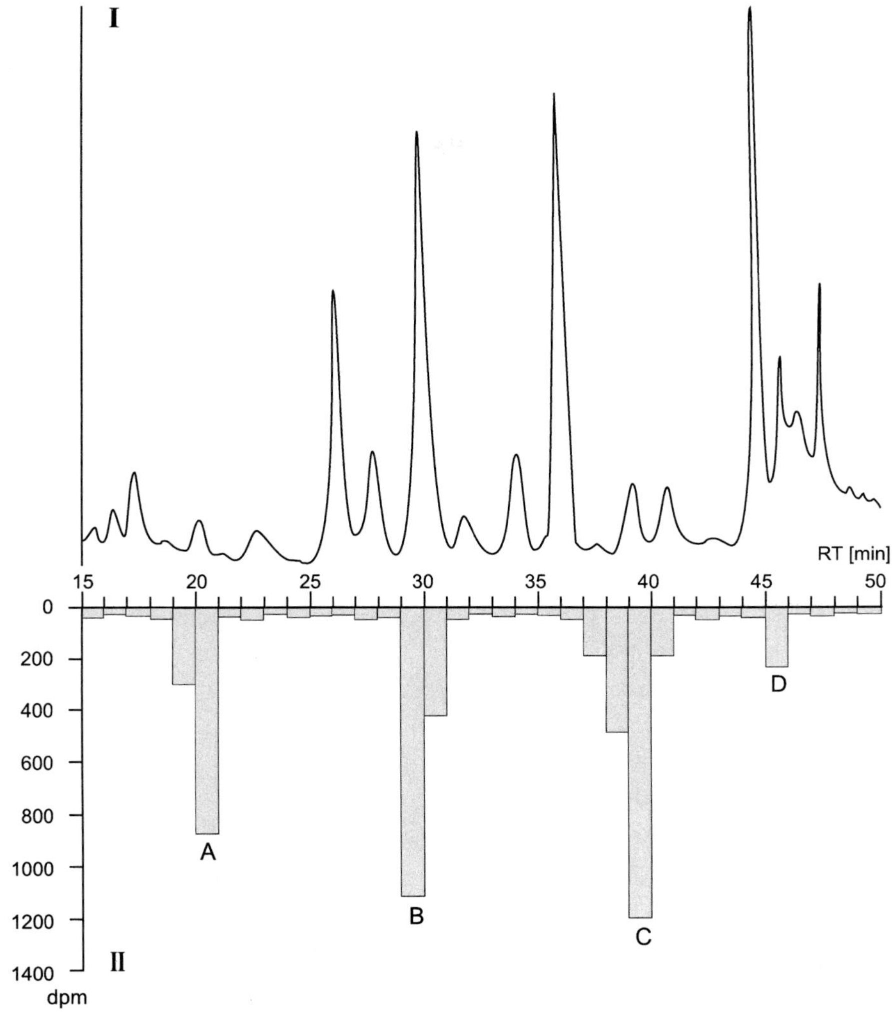

Fig. 8. HPLC elution profile (RP-8; acetonitrile-water 4:1 for 20 min, to 9:1 in 20 min, to 1:0 in 2 min) of an extract obtained from *Ornithopus sativus* suspended cells after application of tritium labelled 24-epibrassinolide (**I**, UV 204 nm; **II**, radioactivity, determined by liquid scintillation counting of 1 min interval fractions; **A** 3,24-bisepibrassinolide-3β-laurate; **B** 3,24-bisepibrassinolide-3β-myristate; **C** 3,24-bisepibrassinolide-3β-palmitate; **D** unknown metabolite)

(20R)-hydroxy-3,24-diepibrassinolide as metabolites of 24-epibrassinolide, and 3,24-diepicastasterone and (20R)-hydroxy-3,24-diepicastasterone as metabolites of 24-epicastasterone, could be established (Kolbe et al. 1996).

Inversion of the 3α hydroxyl group to the equatorial 3β hydroxyl is the initial step of the brassinosteroid metabolism in *O. sativus*. This epimerization is probably catalyzed by an oxidoreductase which might also be involved in the transformation of teasterone to typhasterol (Suzuki et al. 1994) and 24-

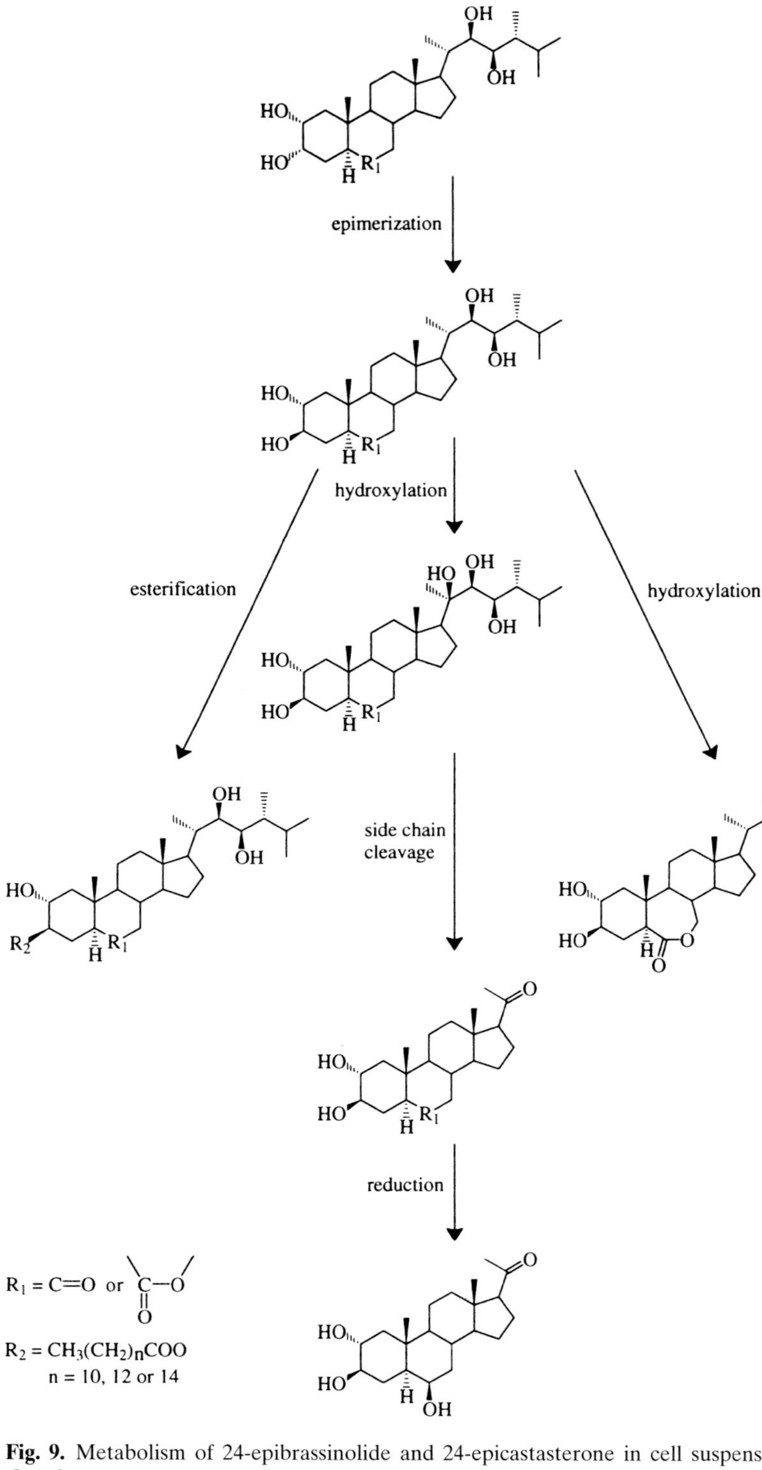

Fig. 9. Metabolism of 24-epibrassinolide and 24-epicastasterone in cell suspension cultures of *Ornithopus sativus*

epiteasterone to 24-epityphasterol (Kolbe et al., unpubl. results) during brassinosteroid biosynthesis. This assumption has been supported by the finding of 3-dehydro-24-epicastasterone as an intermediate in the metabolism of 24-epicastasterone in tomato cell cultures (Hai et al. 1996).

The next metabolic step, hydroxylation of the 3,24-diepimeric brassinosteroids at C-20, affords a novel type of pentahydroxylated steroid. Obviously, the C-20/C-22 bond is destabilized by hydroxylation at C-20 and, consequently, becomes accessible to enzymatic attack resulting in bond cleavage. Typical dominant fragments in the mass spectra of the 20-hydroxy compounds confirmed the weakness of the C-20/C-22 bond. The mechanism of the side chain cleavage is assumed to be analogous to that proposed for the metabolism of 20-hydroxyecdysone in insects (Galbraith et al. 1969) and in crustaceans (Lachaise and Lafont 1984).

Although a number of plants do contain pregnanes of various structural types which in part show significant pharmacological potential and are key intermediates in the biosynthesis of cardenolides and bufadienolides (Deepak et al. 1989), their biosynthesis in plants hitherto is little understood. Thus, the evidence for the involvement of 20-hydroxy intermediates in the formation of pregnanes from steroid precursors in cell cultures of *O. sativus* is of more general interest.

The presence of hydroxypentanoic acids as side chain fragments could be an argument for the mechanism via 20-hydroxylation. 4-Hydroxy-4-methylpentanoic acid, for example, was formed by side chain cleavage of 20-hydroxyecdysone (Galbraith et al. 1969; Lachaise and Lafont 1984). In *O. sativus*, a hypothetical pentanoic acid moiety, metabolically derived from ring-labelled 24-epibrassinolide and 24-epicastasterone was not detectable due to lack of label within the side chain of the parent compounds. However, the proposed mechanism via 20-hydroxylation seems to be operative only with steroids without a double bond between C-20/C-22. An alternative pathway of the formation of pregnanes from Δ^{22}-sterols via 22,23-epoxides has recently been established in a marine sponge (Kerr et al. 1995).

Besides hydroxylation at C-20, introduction of an additional hydroxyl group occurs also in the terminal part of the side chain at C-25 of 3,24-diepibrassinolide (Fig. 9). Hydroxylation at C-25 is well known for a variety of other steroid compounds, e.g. ecdysones (Rees 1989) and was observed also for brassinosteroids in tomato cell cultures (Schneider et al. 1994; Hai et al. 1996).

5 Summary and Conclusions

Ornithopus sativus, a legume with some agricultural importance, has turned out as a valuable tool for studying aspects of plant resistance against fungal pathogens and metabolism of brassinosteroids, a novel type of plant hormone. These investigations have been carried out mainly by using a well-growing, stable cell culture strain, which provides a number of advantages.

The activity of the enzymes of the phenylpropane pathway was shown to increase in response to elicitor treatment and is followed by accumulation of an antifungal isoflavonoid, glabridin, in these cell cultures. Interestingly, a deuteromycete, *Colletotricum trifolii*, which is pathogenic to *O. sativus*, was also able to induce glabridin formation.

Cell suspension cultures of serradella have been established as one of the most used plant systems for metabolic studies on brassinosteroids. Since interesting transformations were found which have not been observed in other systems, this cell culture strain will surely be used for future investigations in this field. In particular, *O. sativus* cell cultures might serve as an object to further study the pathway of pregnane biosynthesis in plants via vicinal hydroxylation at C-20 and C-22.

References

Adam G, Porzel A, Schmidt J, Schneider B, Voigt B (1996) New developments in brassinosteroid research. In: Atta-ur-Rahman (ed) Studies in natural products chemistry, vol 18. Elsevier, Amsterdam, pp 495–549

Asakawa S, Abe H, Nishikawa N, Natsume M, Koshioka M (1996) Purification and identification of new acyl-conjugated teasterones in lily pollen. Biosci Biotechnol Biochem 60:1416–1420

Choi Y-H, Fujioka S, Harada A, Yokota T, Takatsuto S, Sakurai A (1996) A brassinolide biosynthetic pathway via 6-deoxocastasterone. Phytochemistry 43:593–596

Clouse SD (1996) Molecular genetic studies confirm the role of brassinosteroids in plant growth and development. Plant J 19:1–8

Deepak D, Khare A, Khare MP (1989) Plant pregnanes. Phytochemistry 28:3255–3263

Galbraith MN, Horn DHS, Middleton EJ, Thomson JA, Siddall JB, Hafferl W (1969) The catabolism of crustecdysone in the blowfly *Calliphora stygia*. J Chem Soc Chem Commun:1134–1135

Geister G (1991) Farbatlas Landwirtschaftlicher Nutzpflanzen. Ulmer, Stuttgart

Gladstones JS, McKeown NR (1977) Botany and origins of serradella. J Agric (West Aust) 18:10–27

Hai T, Schneider B, Porzel A, Adam G (1996) Metabolism of 24-*epi*-castasterone in cell suspension cultures of *Lycopersicon esculentum*. Phytochemistry 41:197–201

Kerr RG, Kelly K, Schulman A (1995) A novel biosynthetic route to pregnanes in the marine sponge *Amphimedon compressa*. J Nat Prod 58:1077–1080

Kolbe A, Schneider B, Porzel A, Voigt B, Krauss G, Adam G (1994) Pregnane-type metabolites of brassinosteroids in cell suspension cultures of *Ornithopus sativus*. Phytochemistry 36:671–673

Kolbe A, Schneider B, Porzel A, Schmidt J, Adam G (1995) Acyl-conjugated metabolites of brassinosteroids in cell suspension cultures of *Ornithopus sativus*. Phytochemistry 38:633–636

Kolbe A, Schneider B, Porzel A, Adam G (1996) Metabolism of 24-*epi*-castasterone and 24-*epi*-brassinolide in cell suspension cultures of *Ornithopus sativus*. Phytochemistry 41:163–167

Lachaise F, Lafont R (1984) Ecdysteroid metabolism in a crab: *Carcinus maenas*. Steroids 43:243–259

Maeda E (1965) Rate of lamina inclination in excised rice leaves. Plant Physiol 18:813–827

Murashige T, Skoog F (1962) A revised medium for rapid growth and bio-assays with tobacco tissue cultures. Physiol Plant 15:473–497

Parish EJ (1991) The biosynthesis of oxysteroids in plants and microorganisms. In: Patterson GW, Nes WD (eds) Physiology and biochemistry of sterols. American Oil Chemists' Society, Champaign, Illinois, pp 324–336

Rees HH (1989) Pathways of biosynthesis of ecdysone. In: Koolman J (ed) Ecdysone – from chemistry to mode of action. Thieme, Stuttgart pp 152–160

Schmidt J, Spengler B, Adam G, Budzikiewicz H (1993a) Sterol constituents in seeds of *Ornithopus sativus*. Phytochemistry 33:506–507

Schmidt J, Spengler B, Yokota T, Adam G (1993b) The co-occurrence of 24-epi-castasterone and castasterone in seeds of *Ornithopus sativus*. Phytochemistry 32:1614–1615

Schneider B, Kolbe A, Porzel A, Adam G (1994) A metabolite of 24-*epi*-brassinolide in cell suspension cultures of *Lycopersicon esculentum*. Phytochemistry 36:319–321

Seifert K, Härtling S, Porzel A, Johne S, Krauss G (1993) Phytoalexin accumulation in *Ornithopus sativus* as a response to elicitor treatment. Z Naturforsch [C] 48:550–555

Spengler B (1995) Brassinosteroide und Sterole aus den europäischen Kulturpflanzen *Ornithopus sativus*, *Raphanus sativus* und *Secale cereale*. PhD Thesis, University of Halle

Spengler B, Schmidt J, Voigt B, Adam G (1995) 6-Deoxo-28-norcastasterone and 6-deoxo-24-epicastasterone – two new brassinosteroids from *Ornithopus sativus*. Phytochemistry 40:907–910

Suzuki H, Inoue T, Fujioka S, Takatsuto S, Yanagisawa T, Yokota T, Murofushi N, Sakurai A (1994) Possible involvement of 3-dehydroteasterone in the conversion of teasterone to typhasterol in cultured cells of *Catharanthus roseus*. Biosci Biotechnol Biochem 58:1186–1188

Wojciechowski ZA (1991) Biochemistry of phytosterol conjugates. In: Patterson GW, Nes WD (eds) Physiology and biochemistry of sterols. American Oil Chemists' Society, Champaign, Illinois, pp 361–395

XIX *Petroselinum crispum* (Mill.) Nyman (Parsley): In Vitro Culture, Production and Metabolism of Volatile Constituents

A.A. GBOLADE[1] and G.B. LOCKWOOD[2]

1 Introduction

Fresh leaves of Parsley, *Petroselinum crispum* (Mill.) Nyman are widely used whole as a garnish, and used chopped, either fresh or dried, to give a strong characteristic flavour to foods throughout the world. Parsley (Fig. 1) is listed by the Council of Europe as a natural source of food flavouring (category N2; Council of Europe 1981). It is an annual or biennial herb indigenous to the eastern Mediterranean, but naturalised in most parts of Europe, and cultivated widely through temperate to tropical areas of the rest of the World. As a member of the family Umbelliferae, parsley has ridged seeds and distinctive leaves, and a white to yellowish root. A range of cultivars have been developed, with leaves available from plain to curled. All parts of the plant have historic use in herbal medicine (British Herbal Pharmacopoeia 1990). All plant parts, as well as infusions and tinctures are used for carminative, antispasmodic, diuretic, emmenagogue, expectorant, antirheumatic, and antimicrobial activities. Traditionally, it has been used for flatulent dyspepsia, colic, cystitis, dysuria, bronchitic cough in the elderley, dysmenorrhoea, functional amenorrhoea, myalgia, and specifically for flatulent dyspepsia with intestinal colic (Newall et al. 1996).

2 Constituents

Macleod et al. (1985) identified 45 volatile aroma constituents from the leaves, and found the major constituents to be myristicin, apiole, and β-phellandrene (Fig. 2). Apiole, in particular, had the characteristic parsley aroma. Another group of researchers later found nine compounds not previously identified, and pronounced that apiole was not responsible for the parsley flavour (Jung et al. 1992). Volatile oil from the leaf (0.05%) contains lesser amounts of 1,3,8-ρ-menthatriene, 4-isopropenyl-1-methylbenzene, methyl disulphide, α- and

[1] Department of Pharmacognosy, Faculty of Pharmacy, Obafemi Awolowo University, Ile-Ife, Nigeria
[2] School of Pharmacy and Pharmaceutical Sciences, University of Manchester, Manchester M13 9PL, UK

Biotechnology in Agriculture and Forestry, Vol. 43
Medicinal and Aromatic Plants XI (ed. by Y.P.S. Bajaj)

Fig. 1. *Petroselinum crispum* L. in flower. (Photograph S. Edwards 1987)

β-pinene, β-myrcene, β-ocimene, ρ-terpinene, α-terpineol, α-copaene, carotol, caryophyllene, and other mono- and sesqui-terpenes. The volatile oil from the seed (2–7% in content) contains apiole, myristicin, tetramethoxyallylbenzene, and a range of mono- and sesqui-terpenes (Newall et al. 1996). In addition to the volatile constituents, flavonoids, furanocoumarins, fixed oil, oleo-resin, proteins, carbohydrates, and vitamins have been identified. The flavonoids are apiin and glycosides of apigenin and luteolin in the main, and a range of furanocoumarins have been identified, including bergapten and oxypeucedanin as the major examples (AGIS PhytochemDB 1996).

The presence of certain of these constituents has been found to cause adverse effects in patients taking parsley. Individual constituents such as

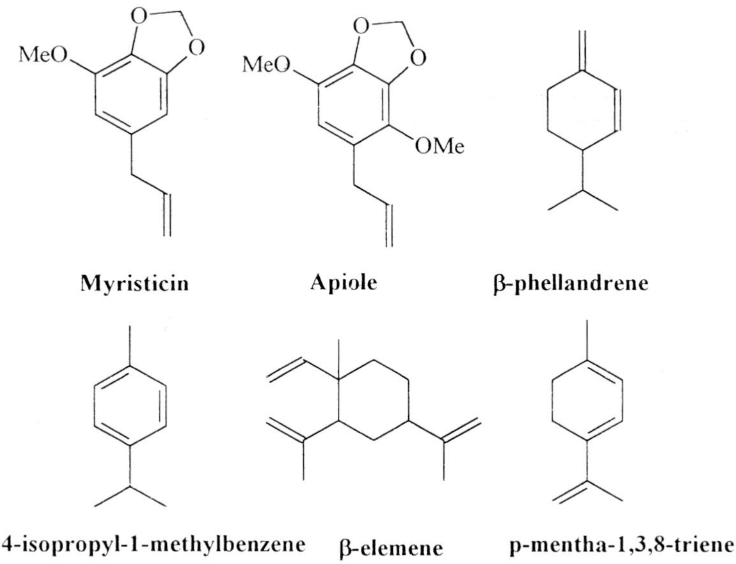

Fig. 2. Major volatile aroma constituents

apiole, which in large doses can cause hepatic dysfunction, acute haemolytic anaemia, and abortion, and myristicin, which can cause hallucinations and giddiness, followed by liver and kidney degeneration are the probable causative agents. Although photodermatitis resulting from ingestion of the furanocoumarins is possible, there appear to be relatively safe levels of the phototoxic psoralen derivitives. Parsley seed oil has GRAS status in the USA and it has an acute oral LD_{50} estimated to be 277 g/70-kg man (Opdyke 1975).

3 In Vitro Culture Studies

3.1 Review

A number of reports have appeared on the tissue culture of Parsley (see Table 1). Extensive studies have been carried out using parsley cell cultures as an experimental model for investigation of flavonoid and phenylpropanoid metabolism. Most of the enzymes of flavonoid glycoside biosynthesis were isolated from parsley cultures, and the literature has been reviewed (Hahlbrock 1981). Surprisingly, few morphological studies have been published with parsley, although callus formation and somatic embryogenesis have been reported (Vasil and Hildebrandt 1966; Masuda et al. 1977). Vasil and Hildebrandt produced highly friable calli, and in several media nodular structures with negatively geotropic roots were formed. In MS medium (Murashige and Skoog 1962), supplemented with 15–20 mg/l adenine sulphate, embryoids were

Table 1. In vitro culture studies on *Petroselinum*

Reference	Study/observations/remarks
Vasil and Hildebrandt (1966)	Embryoids/callus/plantlets
Hahlbrock (1975)	Flavonoid biosynthetic pathway
Spiess and Seitz (1975)	Suspension morphology
Masuda et al. (1977)	Embryogenic callus
Ellenbracht et al. (1980)	Unusual fatty acids
Hahlbrock (1981)	Elucidation of flavonoid biosynthesis
Hauffe et al. (1986)	Elicitor stimulated furanocoumarins
Bruns et al. (1986)	Light stimulated chalcone synthesis
Gbolade and Lockwood (1989a)	Callus and suspension volatiles
Gbolade and Lockwood (1989b)	Geraniol metabolism in suspensions
Conrath et al. (1989)	Elicitor stimulation of callose
McCue and Conn (1989)	Enzyme stimulation by elicitor
Gbolade and Lockwood (1990a)	Monoterpene aldehyde biotransformation
Gbolade and Lockwood (1990b)	Biotransformation in immobilised cells
McCue and Conn (1990)	Enzyme stimulation by light
Bollman and Hahlbrock (1990)	Elicitor stimulation of protein synthesis
Gbolade and Lockwood (1991)	Callus morphology
Pakusch et al. (1991)	Elicitor inducible enzyme
Dittrich et al. (1992)	Elicitor induction of phytoalexin
Kauss et al. (1993)	Phytoalexin elicitor conditioning
Dusek et al. (1992)	Herbicide effect on metabolites
Kauss et al. (1994)	Methyl jasmonate/elicitor effects
Conn and McCue (1994)	Shikimate elicitors
Appert et al. (1994)	PAL enzymes
Kauss and Jeblick (1995)	Salicylic acid elicitation

formed, which finally developed into plantlets (Vasil and Hildebrandt 1966). Masuda et al. (1977) also formed embryogenic calli using MS medium, but without addition of any growth substances, and the callus was seen to consist of approximately triploid cells whereas diploid ones predominated in the embryos originating from this callus. Friable callus was produced when grown on both MS and B5 media (Gbolade and Lockwood 1989a, 1991) and suspensions were initiated from calli of three cultivars "Paramount", "Champion Moss – curled", and "Plain-leaved" (Gbolade and Lockwood 1989b, 1990a). Monoterpene biotransformation was studied in suspension cultures which had been immobilised in polyurethane foam (Gbolade and Lockwood 1990b).

By carrying out conductivity measurements (Hahlbrock 1975), biochemical monitoring of nitrate depletion from the medium was used for the production of growth curves and the ultrastructure of freely suspended callus cells was monitored and found to be growth dependent (Spiess and Seitz 1975).

Apart from the phenylpropanoid metabolites reported above, few other categories of secondary metabolites have been studied. An unusual fatty acid, vaccenic acid, was isolated from cultures along with the expected lipids, petroselinic and octadecadienoic acids, but no desaturated derivatives of these lipids could be found (Ellenbracht et al. 1980). The effects of a herbicide,

Potablan D 85, on therapeutically useful metabolites produced in the cultures was shown to be insignificant (Dusek et al. 1992). The cultures have been widely used in investigations of the effects of light on the activities of enzymes involved in the biosynthesis of phenylpropanoids (Bruns et al. 1986; McCue and Conn 1990; Conn and McCue 1994).

Most commonly, recent research involves the use of elicitors to stimulate production of both primary and secondary metabolites. The furanocoumarin, isopimpinellin was produced from 5-hydroxyxanthotoxin, and bergapten and xanthotoxin were accumulated in the medium after treatment with an elicitor from the soybean pathogen *Phytophora megasperma glycinea*. The activities of S-adenosyl methionine O-methyltransferases showed transient increases upon elicitor application (Hauffe et al. 1986). Induction of coumarin synthesis was reported after elicitation with chitosan (Conrath et al. 1989), with *Phytophora megasperma glycinea* (Pakusch et al. 1991), with two non-pathogenic strains of fungi (Tietjen et al. 1983), and with 2,6-dichloroisonicotinic acid or salicylic acid derivatives. Kauss et al. (1993) conditioned parsley suspension cells prior to treatment with an elicitor from *Phytopthora megasperma glycinea*. In this last case where pretreatment was employed, greatly increased phenylpropanoid levels were reported when pre-incubation for 1 day was the protocol, and low elicitor concentrations were employed. It was suggested that salicylic acid derivatives were most likely to improve signal induction leading to activation of genes encoding phenylalanine ammonia lyase and 4-coumarate:CoA ligase (Kauss et al. 1993). Salicylic acid, and a range of elicitors were also found to stimulate production of hydrogen peroxide (Kauss and Jeblick 1995), and methyl jasmonate enhanced the production of an unidentified active oxygen species (Kauss et al. 1994). The elicitor from the soybean pathogen (McCue and Conn 1989), or actinomycin or cycloheximide (McCue and Conn 1990) also induced or altered levels of shikimic acid pathway enzymes, and the jasmonate precursor, 12-oxo-phytodienoic acid was found to stimulate accumulation of the flavonoid apiin (Dittrich et al. 1992). Interestingly, soybean pathogen elicitor was shown to inhibit sterol biosynthesis at the same time as stimulating accumulation of furanocoumarins (Haudenschild and Hartmann 1995).

Another application of parsley cultures has been in the investigation of protein synthesis. Using elicitors, both total protein levels and those of enzymes involved in the synthesis of furanocoumarin phytoalexins were shown to be altered in suspension cells (Bollman and Hahlbrock 1990). Shikimic acid pathway enzymes were shown to be affected by both light and elicitors by a number of researchers (Kombrink and Hahlbrock 1986; Hauffe et al. 1986; McCue and Conn 1989, 1990; Pakusch et al. 1991; Dittrich et al. 1992; Conn and McCue 1994; Appert et al. 1994).

Cell wall constituents such as 1,3-beta-glucan callose were shown to exhibit rapid deposition on treatment with chitosan elicitors, at the same time as coumarin synthesis was stimulated, possibly indicating common initial events in their synthesis, in spite of the fact that two fully independent pathways are involved (Conrath et al. 1989).

3.2 Establishment and Growth of Callus and Suspension Cultures

Cell cultures from three cultivars of parsley were established from aseptically germinated seedlings. The cultures acquired different growth characteristics on both B5 and MS medium. The B5 medium encouraged profuse growth of fresh creamy-yellow, highly friable and root-differentiated callus (Table 2). Growth on MS medium was similar, but much slower. The differences observed in callus growth on both media can be attributed to the differences in the nutritional composition of the media. Growth of callus cultures of cultivars "Paramount", "Champion Moss – curled", and "Plain Leaved" followed the typical sigmoid growth curve, and had a regular subculture interval of 6 weeks. The cultivar "Champion Moss – curled" consisted of a slightly slower growing, less friable and more compact callus tissue than the other cultivars (Table 3). Various concentrations of hormones stimulated both chlorophyll production and root differentation, which later disappeared on 2,4-D supplemented cultures over the period of seven generations (Table 4; Gbolade and Lockwood 1991). Chlorophyll-containing cultures of *P. crispum* were found to be no

Table 2. Influence of growth regulators on growth of parsley callus cultures[a]

Treatment (mg/l)	Callus growth	Morphological differentiation		Greening
		Plant/shoot	Root	
0.1 K + 1 2,4-D	+/2+	±/+	2+/3+	+
0.1 K + 0.1 2,4-D	+/2+	±	2+/3+	±
1 K + 0.1 2,4-D	3+	+	+/2+	±
0.5 K only	+	2+	3+	+
0.5 K + 0.1 NAA	±/+	+/2+	2+/3+	+
0.1 K + 0.5 NAA	±/+	+/2+	2+/3+	+
0.5 K + 0.5 NAA	±/+	+/2+	2+/3+	+
0.1 K + 0.1 NAA	±/+	+/2+	2+/3+	+

± = Ill-defined growth.
[a] Growth medium B5.

Table 3. Influence of different cultivars on growth of parsley callus cultures

Parameter	Paramount	Parsley cultivars	Plain-leaved
		Champion Moss-curled	
Growth	2+/3+	+/2+	3+
Consistency			
Friability	3+	+	3+
Looseness	3+	−	3+
Compactness	+	3+	+
Granular nature	±/+	3+	±
Morphological differentiation			
Root formation	2+	2+	2+
Shoot formation	±	+	±
Greening	+	2+	+

Table 4. Influence of subculture on growth of parsley callus cultures

Parameter	Young cultures 1st/2nd generations	Old cultures 6th/7th generations
Growth	3+	3+
Morphological differentiations		
Greening	±	±
Root formation	3+	−/±
Green shoot formation	+	−
Consistency		
Friability	3+	2+/3+
Granular nature	−	±/+
Yellow pigmentation	+	2+

better than those lacking this pigment in their biochemical productivities of volatile constituents including monoterpenes. Previously, Banthorpe and Wirz-Justice (1972) had indicated that production of monoterpenes in cultures was independent of chlorophyll development.

3.3 Production of Volatile Constituents by Callus and Suspension Cultures

The first report on the identification of volatile constituents in parsley cultures (Gbolade and Lockwood 1989a), described the aroma constituents found in both calli and suspensions of the "Paramount" cultivar, which had been harvested at the onset of the stationary phase of growth. Suspension cells were found to accumulate 4-isopropenyl-1-methylbenzene, ρ-menthe-1,3,8-triene, β-elemene, and myristicin (Fig. 2), while calli contained γ-terpinene in addition. Although these cultures exhibited a commonly reported (Charlwood et al. 1986) qualitative similarity in the secondary product spectrum with the parent plant, the terpenes α- and β-pinene, β-myrcene, and β-phellandrene, and the coumarin apiole, were absent. These cultures were completely undifferentiated, and showed a total absence of glandular or secretory structures or shoots. Although reported increased terpenoid accumulation in certain cultures has been associated with an increase in tissue differentation (Brown and Charlwood 1986), the formation of detectable amounts of mono- and di-terpenes by undifferentiated suspension cultures of *Thuja occidentalis* (Berlin et al. 1984), having different composition from those of the leaves of the intact plant, shows that differentiation may not be a prerequisite.

Later work on the volatile constituents of the "Paramount" cultivar and two others showed β-elemene, myristicin, and apiole to be the major metabolites (Gbolade and Lockwood 1990a). The influence of culture age on levels of these products shows that levels generally fall away after the fourth generation (Table 5), and by the tenth generation only β-elemene and apiole were detected. There were large fluctuations in levels of products, particularly between generations. Myristicin and apiole were the major constituents of the cultivar "Plain-leaved", but no products were detected after the fourth

Table 5. Yield (μg/g dry weight) of the major volatile constituents in callus cultures of different parsley cultivars

Generation	β-Elemene			Myristicin			Apiole		
	CMC	PPL	PP	CMC	PPL	PP	CMC	PPL	PP
1	0.4	+	–	39.0	8.4	–	–	13.3	–
2	0.5	+	–	13.0	2.2	–	8.2	13.8	–
3	0.7	0.2	–	4.1	0.8	–	0.4	3.5	–
4	0.3	2.6	–	0.3	–	–	–	–	–
6	0.2	–	–	–	–	–	–	–	–
8	0.2	–	0.6	–	–	–	–	–	1.4
9	0.1	–	0.7	–	–	–	–	–	23.3
10	0.2	–	1.2	–	–	–	–	–	+

+ = trace (≤0.1); – = not detected;
CMC = "Champion Moss – curled"; PPL = "Plain-leaved"; PP = "Paramount".

generation. While myristicin was absent in older generations of "Paramount", β-elemene and higher levels of apiole were accumulated. Apiole and β-elemene were continuosly present in low yields from "Champion Moss – curled" up to the third and tenth generations respectively, and the cultures showed the greatest myristicin accumulation up to the fourth generation. Apart from the concentration of β-elemene and apiole in "Paramount", levels of metabolites generally decreased with age, a phenomenon previously reported for some species of culture (Charlwood et al. 1986). As in the case of previous workers' experience, these particular callus cultures accumulated higher yields of metabolites than their corresponding suspensions (Charlwood et al. 1986). Removal of 2,4-D (1 mg/l) from the medium, resulted in increased product levels, but much reduced growth. Modification of the medium sucrose content had dramatic effects on the growth rate of cultures of the same cultivars, but little effect on the β-elemene yield, except in the "Plain-leaved" cultivar, where 5–10% sucrose supplementation resulted in dramatic increases in levels.

3.4 Biotransformation of Exogenous Monoterpenoids

Feeding of citronellal (30 mg/l) to suspensions of the two cultivars, "Paramount" and "Plain-leaved", resulted in conversion exclusively into the alcohol, citronellol (Gbolade and Lockwood 1990a). The kinetics of the conversion differed; in "Plain-leaved" there was a steady increase for the first 8 h, followed by a sharp decline, but in "Paramount" there was dramatic degradation or loss of precursor in the first hour, and negligible conversion into citronellol. These particular cultures were shown to be unable to perform a reverse conversion (alcohol to aldehyde), possibly as a consequence of preferential metabolism. Again, using suspensions of the same cultivars, exclusive conversion of geraniol into nerol was demonstrated over the first 24 h after feeding (Gbolade and Lockwood 1989b). The kinetics of disappearance

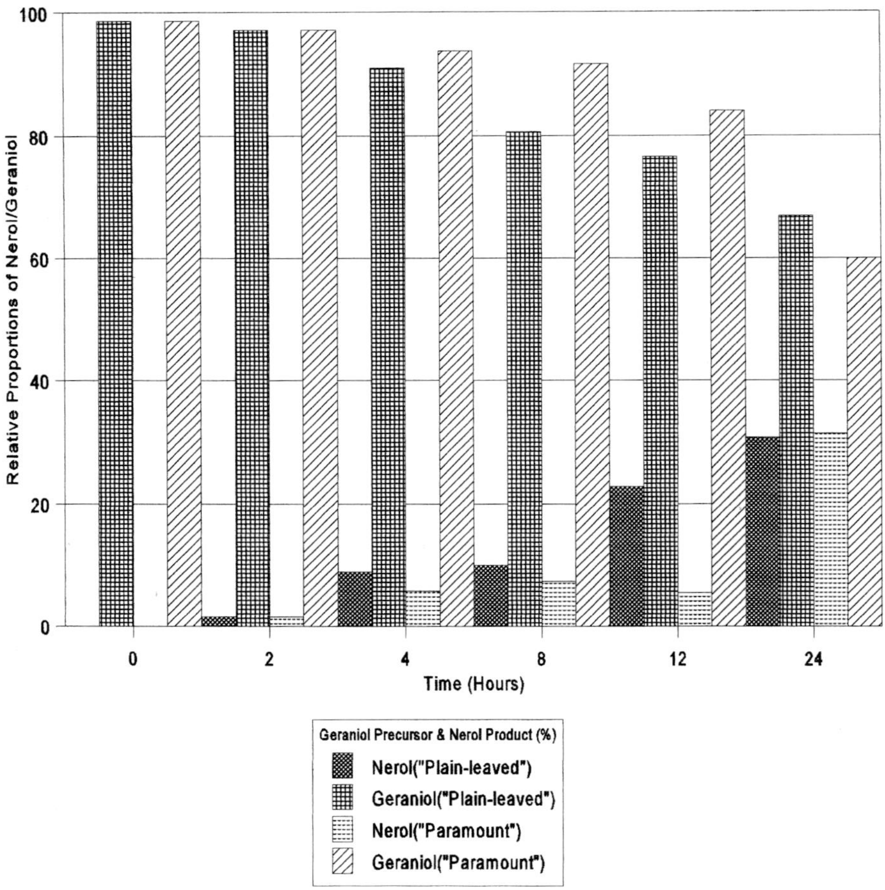

Fig. 3. Biotransformation of geraniol by cultures of two parsley cultivars, during the first 24 h after innoculation with 30 mg/l geraniol. (Redrawn from, Gbolade and Lockwood 1989b)

of geraniol from the cultures and the concomitant formation of nerol, followed a characteristic pattern which is frequently described for the biotransformation of monoterpenes in plant cell cultures (Fig. 3). Experiments employing growth-limiting protocols (elimination of sucrose from the medium) showed greatly increased rates of biotransformation of geraniol Similarly increased rates of biotransformation of citronellol in sucrose-supplemented suspensions of the two cultivars was reported (Gbolade and Lockwood 1990b).

3.5 Biotransformation of Monoterpenes Using Immobilised Cells

Immobilisation is one of the potential techniques for improving the productivity of plant cell cultures. Usually, cells are immobilised by entrapment in

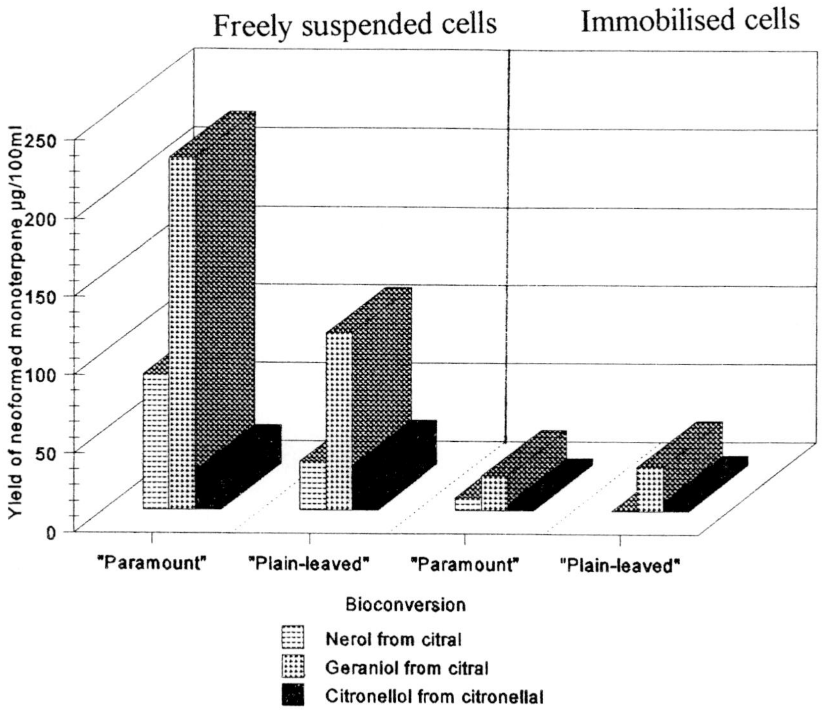

Fig. 4. Biotransformation of monoterpenes by freely suspended and immobilised cells of the two cultivars, "Paramount" and "Plain-leaved" (Redrawn from, Gbolade and Lockwood 1990b)

calcium alginate, agar, or porous (reticulate) polyurethane foam matrices, but only work with the last medium has been published (Gbolade and Lockwood 1990b). One of the desired characteristics of immobilised cell suspensions is the release of the desired metabolic products into the medium. However, with parsley suspensions this did not occur, and lower levels of bioconversion products were found than with freely suspended cells (Fig. 4).

4 Summary and Conclusion

Both callus and suspension cultures of *Petroselinum crispum* produced volatile constituents characteristic of leaves of the whole plant, but the spectrum of compounds was at a reduced level, and the more volatile monoterpenoids and apiole were absent from undifferentiated cultures. This phenomenon is frequently reported for all types of volatile constituents (Everitt and Lockwood 1989), and one of the causes could be that the conditions for cell culture, namely warmth and lack of differentiated storage structures, are ideal for volatilisation of low boiling point constituents.

Levels of volatile constituents decreased over a period of ten subcultures, when using three different cultivars. Exclusion of 2,4-D from culture media was shown to yield higher levels of β-elemene, myristicin, and apiole, but reduction of sucrose to 1% reduced growth dramatically. Increase of sucrose to 10% had variable effects on cell growth and production of volatiles.

There was evidence for the presence of enzymic systems involved in the biosynthesis of volatile substances, as certain monoterpenoids were metabolised to identified end products. Although immobilisation of various plant cell cultures has had beneficial effects on levels of secondary metabolites, this has not been shown with parsley cultures.

5 Protocol

Callus and Suspension Cultures. Callus cultures were established from aseptically germinated seedlings on B5 medium (Gamborg et al. 1968), supplemented with 3% sucrose, 1% agar, 5% coconut water, and 0.1 mg/l kinetin and 1 mg/l 2,4-D or NAA. Calli were incubated at 27 °C under 12-h intermittent light/dark conditions, and subcultured every 6 weeks. Suspensions were initiated from 10–14 g of third generation callus, into 50 ml of liquid devoid of agar, and incubated as above at a shaking rate of 60 rpm.

Analytical Procedures. Vacuum-filtered cells were freeze-dried for 24–36 h. Dichloromethane (redistilled), 15 ml/g cells dry weight, was used for extraction, and extracts concentrated under nitrogen to 200 μl. One microlitre samples were chromatographed on a Carbowax 20 M (25 m × 0.32 mm i.d.) column fitted in a 5890 Hewlett-Packard GC (temperature programme 50–180 °C at 5 °C/min; carrier gas, nitrogen, 2 ml/min; FID detector held at 185 °C). GC-MS was carried out on the same column, ionisation potential 70 eV; ionisation current 1 A, source temperature 250 °C, resolution 600; scan speed 1 s/decade; carrier gas helium. Identification was based on retention data, and EI-MS plus CI spectra.

Precursor Feeding. Erlenmeyer flasks (250 ml) containing 100 ml of suspension (packed cell volume 53%) were innoculated with 30 mg/l substrate.

References

AGIS PhytochemDB (1996) http://probe.nalusda.gov:8300/cgi-bin/dbrun/phytochemdb?find+Genus

Appert C, Logeman E, Hahlbrock K, Schmid J, Amrhein N (1994) Structural and catalytic properties of the four phenylalanine ammonia-lyase isoenzymes from parsley (*Petroselinum crispum* Nym.). Eur J Biochem 225:491–499

Banthorpe DV, Wirz-Justice AJ (1972) Terpene biosynthesis. I. Preliminary tracer studies on terpenoid and chlorophyll of *Tanacetum vulgare* L. J Chem Soc Perkin Trans 1:1769–1772

Berlin J, Witte L, Schubert W, Wray V (1984) Determination and quantification of monoterpenoids secreted into the medium of cell cultures of *Thuja occidentalis*. Phytochemistry 23:1277–1279

Bollman J, Hahlbrock K (1990) Timing of changes in protein synthesis pattern in elicitor-treated cell suspension cultures of parsley (*Petroselinum crispum*). Z Naturforsch [C] 45:1011–1020

British Herbal Pharmacopoeia (1996) 4th edn. British Herbal Medicine Association, Bournemouth

Brown JT, Charlwood BV (1986) The control of callus formation and differentiation in scented Pelargoniums. J Plant Physiol 123:409–417

Bruns B, Hahlbrock K, Schaefer E (1986) Fluence dependence of the ultraviolet-light-induced accumulation of chalcone synthase mRNA and the effects of blue and far-red light in cultured parsley cells. Planta 169:393–398

Charlwood BV, Hegarty PK, Charlwood KA (1986) The synthesis and biotransformation of monoterpenes by plant cells in culture. In: Morris P, Scragg AH, Stafford A, Fowler MW (eds). Secondary metabolism in plant cell cultures. Cambridge University Press, Cambridge, pp 15–34

Conn EE, McCue KF (1994) Regulation of the shikimic acid pathway in suspension cultured cells of parsley (*Petroselinum crispum* L.) Stud Plant Sci 4:95–102

Conrath U, Domard A, Kauss H (1989) Chitosan-elicited synthesis of callose and of coumarin derivatives in parsley cell suspension cultures. Plant Cell Rep 8:152–155

Council of Europe (1981) Flavouring substances and natural sources of flavourings, 3rd edn. Maisonneuve, Strasbourg

Dittrich H, Kutchan TM, Zenk M (1992) The jasmonate precursor, 12-oxo-phytodienoic acid, induces phytoalexin synthesis in *Petroselinum crispum* cell cultures. FEBS Lett 309:33–36

Dusek J, Duskova J, Sicha J, Tumova L (1992) A study of the effect of the herbicide Potablan D 85 on the production of selected metabolites in the in vivo and in vitro cultures of *Petroselinum crispum* (Mill) Nyman and *Foeniculum vulgare* Mill. Folia Pharm Univ Carol 16:89–98

Ellenbracht F, Barz W, Mangold HK (1980) Unusual fatty acids in the lipids from organs and cell cultures of *Petroselinum crispum*. Planta 150:114–119

Everitt Z, Lockwood GB (1989) Recent advances in the production of volatile constituents by plant cell cultures. Plant Physiol (Life Sci. Advances) 8:75–79

Gamborg OL, Miller RA, Ojima K (1968) Nutrient requirements of suspension cultures of soybean root cells. Exp Cell Res 50:151–158

Gbolade AA, Lockwood GB (1989a) Volatile constituents from parsley cultures. Flavour Fragrance J 4:69–71

Gbolade AA, Lockwood GB (1989b) Selective biotransformation of monoterpenoids by cell suspensions of *Petroselinum crispum*. Z Naturforsch [C] 44:1066–1068

Gbolade AA, Lockwood GB (1990a) Metabolic studies of volatile constituents in tissue cultures of *Petroselinum crispum* (Mill) Nyman. J Plant Physiol 136:198–202

Gbolade AA, Lockwood GB (1990b) Biotransformation of monoterpenes by polyurethane foam-immobilized cells of *Petroselinum crispum* (Mill) Nyman. Z Naturforsch [C] 45:245–248

Gbolade AA, Lockwood GB (1991) Growth and production of volatile substances by *Melissa officinalis* and *Petroselinum crispum* cultures. Fitoterapia 62:237–242

Hahlbrock K (1975) Further studies on the relationship between the rates of nitrate uptake, growth and conductivity changes in the medium of plant cell suspension cultures. Planta 124:311–318

Hahlbrock K (1981) Flavonoids. In: Conn E (ed) The biochemistry of plants, vol 7. Secondary plant products. Academic Press, New York, pp 425–456

Haudenschild C, Hartmann MA (1995) Inhibition of sterol biosynthesis during elicitor-induced accumulation of furanocoumarins in parsley cell suspension cultures. Phytochemistry 40:1117–1124

Hauffe KD, Hahlbrock K, Scheel D (1986) Elicitor stimulated furanocoumarin biosynthesis in cultured parsley cells: S-adenosyl-L-methionime:bergaptol and S-adenosyl-L-methionine:xanthotoxol O-methyltransferases. Z Naturforsch [C] 41:228–239

Jung HP, Sen A, Grosch W (1992) Evaluation of potent odorants in parsley leaves *Petroselinum crispum* (Mill.) Nym. *spp. crispum* by aroma extract dilution analysis. Lebens – Wiss Technol 25:55–60

Kauss H, Jeblick W (1995) Pretreatment of parsley suspension cultures with salicylic acid enhances spontaneous and elicited production of hydrogen peroxide. Plant Physiol 108:1171–1178

Kauss H, Franke R, Kraus K, Conrath U, Jeblick W, Grimmig B, Matern U (1993) Conditioning of parsley (*Petroselinum crispum* L.) suspension cells increases elicitor-induced incorporation of cell wall phenolics. Plant Physiol 102:459–466

Kauss H, Jeblick W, Ziegler J, Krabler W (1994) Pretreatment of parsley (*Petroselinum crispum* L.) suspension cultures with methyl jasmonate enhances elicitation of activated oxygen species. Plant Physiol. 105:89–94

Kombrink E, Hahlbrock K (1986) Responses of cultured parsley cells to elicitors from phytopathogenic fungi. Timing and dose dependency of elicitor-induced reactions. Plant Physiol 81:216–221

Macleod AJ, Snyder CH, Subramanian G (1985) Volatile aroma constituents of parsley leaves. Phytochemistry 24:2623–2627

Masuda K, Kodo Y, Okazawa Y (1977) Callus formation and embryogenesis of endosperm tissues of parsley seed cultured on hormone-free medium. Physiol Plant 41:135–138

McCue K, Conn EE (1989) Induction of 3-deoxy-D-arabino-heptulosonate-7-phosphate synthase activity by fungal elicitor in cultures of *Petroselinum crispum*. Proc Natl Acad Sci USA 86:7374–7377

McCue K, Conn EE (1990) Induction of shikimic acid pathway enzymes by light in suspension cultured cells of parsley (*Petroselinum crispum*). Plant Physiol 94:507–510

Murashige T, Skoog F (1962) A revised medium for rapid growth and bio-assays with tobacco tissue cultures. Physiol Plant 15:473–497

Newall CA, Anderson LA, Phillipson JD (1996) Herbal medicines: a guide for health-care professionals. Pharmaceutical Press, London

Opdyke DLJ (1975) Parsley seed oil. Food Cosmet Toxicol 13:897–898

Pakusch AE, Matern U, Schlitz E (1991) Elicitor-inducible caffeoyl-coenzyme A 3-O-methyltransferase from *Petroselinum crispum* cell suspensions. Purification, partial sequence, and antigenicity. Plant Physiol 95:137–143

Spiess E, Seitz U (1975) Quantitative determination of growth-dependent changes in the ultrastructure of freely suspended callus cells of *Petroselinum crispum* (Mill) A. W. Hill. Ber Dtsch Bot Ges 88:319–328

Tietjen KG, Hankler D, Matern U (1983) Differential response of cultured parsley cells to elicitors from two non-pathogenic strains of fungi. Eur J Biochem 131:401–407

Vasil IK, Hildebrandt AC (1966) Variations of morphogenetic behaviour in plant tissue cultures II *Petroselinum hortense*. Am J Bot 53:869–874

XX *Phellodendron amurense* (Cork Tree): In Vitro Culture, Micropropagation, and the Production of Berberine

Y.G. PARK[1] and M.S. CHOI[2]

1 Introduction

1.1 Distribution and Importance of the Plant

The genus *Phellodendron* (family Rutaceae) comprises ten species of cork trees indigenous to the Far East region, including Korea, China, and Japan. *P. amurense* (Amur cork tree) is the best known and almost widely grown of the genus. It grows rapidly and has a broad crown. *P. sachalinensis* (Sakhalin cork tree) is slightly more upright in habit and has a more rounded crown. The bark is brown and similar to *P. amurense*, which is a larger tree. The cork tree is dioecious, 10m high, and cork tissue of the outer bark is developed as indicated in Fig. 1C. The grey-brown bark becomes fissured and corky with age (Fig. 1B). Flowers are greenish and very conspicuous. They are borne in panicles of generally five to eight flowers on spurs or lateral branchlets; the ovary contains five carpels. Female trees produce large clusters of fleshy black fruit which remain on the tree into the winter (Fig. 1A). This tree grows well in the humid climates.

Trees harvested for medicinal purposes are usually felled in the spring or autumn, and the bark is removed and dried in the sun. It is normally used to make aqueous extracts. Extracts of the wood and root bark are used as dye for wool and leather. The crude extract from the bark of cork trees was well known as an antibiotic. Hong et al. (1988) reported that bark extract was very effective in inhibition of Japanese apple canker caused by *Valsa ceratosperma*. The active constituent of bark extract contained berberine derivatives. This result indicated possibilities as a biopesticide. This species is frequently planted as an ornamental or a roadside tree because it is relatively resistant to air pollution. It is also utilized for furniture woods, and attracts bees for honey production.

[1] Dept. of Forestry, Kyungpook National University, Taegu, 702-701 Korea
[2] Dept. of Biotechnology, Forest Genetics Research Institute, Suwon 441-350, Korea

Biotechnology in Agriculture and Forestry, Vol. 43
Medicinal and Aromatic Plants XI (ed. by Y.P.S. Bajaj)

Fig. 1. Mature tree of *Phellodendron amurense* growing in Korea, 20 years old. **A** leaves, and fruits, **B** outer bark, **C** mature cork tree

1.2 Conventional Propagation, and the Production of Alkaloids and Medicinal Components

The cork tree can be propagated from seeds, stem cuttings, and roots. Propagation by seeds is possible, but germination in the field under natural conditions is poor, due to weak dormancy. Stem cuttings have a low rooting rate. More than 90% of stem cuttings from young seedlings root, but for mature trees, this drops to below 50%. The shoot survival rate of a 28-year-old tree cutting decreased markedly to below 10%, although treatment with indolebutyric acid (IBA) somewhat improves the survival and rooting rate (Park et al. 1988).

The secondary metabolites include protoberberine alkaloids such as berberine, palmatine, columbamine, and magnoflorine, flavonoids and terpenoid (Fig. 2). Most recently, research has focused mainly on berberine which is found in the bark, root, and fruit. This is most abundant in bark tissues, and this single compound can account for up to 5% of dry weight (Tosa et al. 1989).

Berberine is well known for its anti-inflamatory, antimicrobial, antileukemic, and antineoplastic effects (Otsuka et al. 1981). It has been used as a fluorescent marker in several areas of medicinal research and as a stomach drug in the Orient. Berberine is also found in *Berberis* spp. (Hint and Zenk 1981), *Thalictrum minus* (Nakagawa et al. 1984), and *Coptis* (Ikuta and Itokawa 1989). The demand for drugs from the cork tree is increasing. In 1992 the demand for the crude material in Korea was 6000 kg/year; 5000 kg

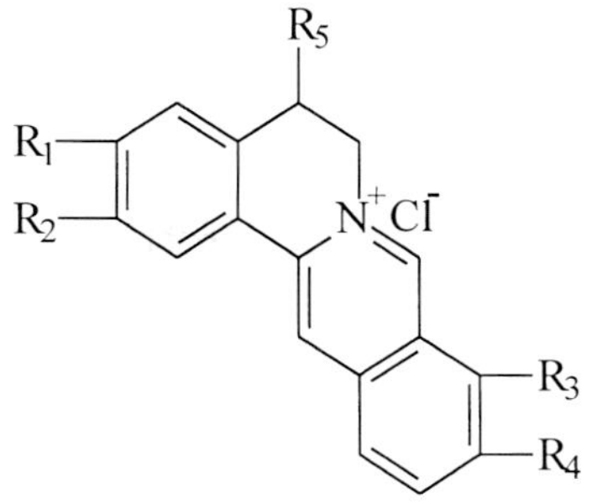

	R_1	R_2	R_3	R_4	R_5
Berberine	O (—CH_2—)	O	OMe	OMe	H
Palmatine	OMe	OMe	OMe	OMe	H
Jateorrhizine	OH	OMe	OMe	OMe	H
Columbamine	OMe	OH	OMe	OMe	H

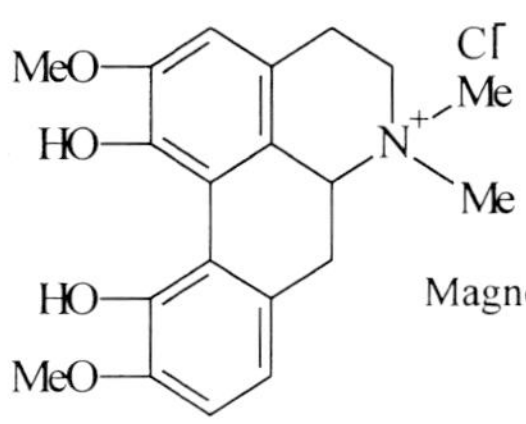

Magnoflorine

Fig. 2. Protoberberine alkaloids from *Phellodendron* sp

imported from Japan and China, and 1000 kg self-produced in Korea (Seong 1992).

2 In Vitro Approaches

2.1 Review of Tissue Culture Studies

There are a few publications concerning in vitro culture of cork trees which are related to mass propagation by axillary bud culture and berberine production by cell cultures (Table 1). Kunio (1986) reported successful micropropagation using intrapetiolar buds formed after cutting mature cork trees. Best growth was obtained on liquid WPM (Lloyd and McCown 1980) containing 0.23–0.64 mg/l BA or 0.02 mg/l IBA. Multiple shoots formed from lateral buds in the nodal segments on the same medium. Individual shoots were excised and rooted successfully in 30–40% of cases by the fourth generation. Rooting efficiency was very low, but increased after several generations of in vitro plantlets. Kim et al. (1992) reported mass propagation by using axillary buds from 15-year-old trees. Over 90% of in vitro proliferated shoots survived in the greenhouse.

The production of secondary metabolites by cell cultures of the cork tree has been reported (Park et al. 1992b; Choi et al. 1996). Park et al. (1992a)

Table 1. Summary of various in vitro studies on *Phellodendron* species

Reference	Purpose of study	Results/remarks
1. Kunio (1986)	Mass propagation of mature tree	Shoot multiplication, in vitro rooting
2. Kim et al. (1992)	Mass propagation of mature tree	Shoot multiplication, in vitro rooting, soil acclimatization
3. Park et al. (1992)	Berberine production by cell culture	Immobilization, berberine secretion
4. Choi et al. (1996)	Berberine production by cell culture	Two-stage culture

reported that immobilized cells produced about 70% of the berberine produced by free cells, and 50%–70% berberine was secreted into the medium from the Ca-alginate beads containing the cells. Continuous light, addition of DMSO, and XAD-2 resin stimulated release of berberine from cells. Choi et al. (1996) concluded that optimization of nutrient components can improve yields of berberine and palmatine.

2.2 Establishment of Tissue Culture and Cell Suspension for Berberine Production

2.2.1 Micropropagation

For in vitro proliferation of cork tree axillary buds and juvenile shoots were surface sterilized with 70% ethanol for 30s, 2% sodium hypochlorite solution for 3min, and rinsed with sterile distilled water. The coat of the bud was peeled off and then the bud was cultured on MS medium (Murashige and Skoog 1962) with various growth regulators. Axillary buds cultured on MS medium started to form new shoots within 1–2 weeks. After 4 weeks in culture, axillary shoots were cultured on MS medium containing 0.05μM NAA and 0.44μM BA, which promoted development of axillary shoots (Table 2 and Fig. 3A). 0.05μM NAA, the lowest concentration tested, promoted shoot bud initiation and shoot growth, but concentrations exceeding 0.5μM reduced the shoot growth.

Excised shoot tips exhibited multiple shoot formation independent of the BA concentration (Fig. 3B). The addition of high concentrations of BA in the medium promoted multiple shoot formation. Approximately eight shoots were obtained in MS medium supplemented with 44.39μM BA and 3% sucrose (data not shown). During subculturing of plants, BAP concentrations of 4.5 to 9μM produced callus in the shoot base and led to abnormal growth of shoots (Grusselle et al. 1987). However, high levels of BA in the culture medium promoted shoot multiplication, but the size of the shoot became smaller.

Culture conditions such as concentrations of salt and auxins in media were investigated. Suitable conditions for root induction on various media (1/16 MS, 1/8 MS, 1/4 MS, 1/2 MS, MS, GD, ACM, WPM, and DKW, respectively)

Table 2. Effects of different concentrations of NAA and BA on the number and length of shoots derived from axillary buds of *Phellodendron amurense* after 4 weeks in culture. (Kim et al. 1992)

Growth regulators (μM)		Number of shoots/explant	Length of shoot (mm)
NAA	BA		
0.54	0.44	2.3 ± 0.6[a]	2.7 ± 0.8
	0.22	4.0 ± 0.8	5.3 ± 1.5
	4.44	–[b]	–
0.05	0.44	8.7 ± 1.5	1.9 ± 1.0
	2.22	4.3 ± 1.5	4.3 ± 1.2
	4.44	3.0 ± 1.0	4.7 ± 1.5

[a] Each value represents the mean S.D. of 3 replications.
[b] No response.

Fig. 3A–D. Micropropagation of *Phellodendron amurense*. **A** Shoots regenerated from axillary buds in MS medium with 0.05 μM BA after 4 weeks of culture. **B** Multiplication of shoots in MS medium with 44.39 μM BA after 4 weeks of culture. **C** Roots initiated by dipping treatment in 500 μM IAA solution after 4 weeks of culture. **D** Potted plantlets cultivated for 2 months in a greenhouse. (Kim et al. 1992)

Table 3. Ex vitro rooting of cork tree shoots by dipping in various auxin solutions after 4 weeks of culture. (Kim et al. 1992)

Treatment (μM)		Survival rate (%)[a]	Rooting rate (%)[a]	Growth
IAA	50	87.5	62.5	+++[b]
	500	87.5	75.5	+++
	5000	62.5	12.5	++
IBA	50	75.0	25.0	++
	500	25.0	–	+
	5000	25.0	–	++
2,4-D	50	50.0	–	++
	500	50.0	–	++
	5000	12.5	–	+
NAA	50	37.5	12.5	++
	500	50.0	–	++
	5000	–	–	+

[a] Nine shoots were used for each treatment.
[b] Visual estimation, – no response, + poor, ++ moderate, +++ good.

were investigated. It was found that the low nutrient media such as 1/16, 1/8, and 1/2 MS stimulated in vitro rooting. The 1/4 MS medium containing 25μM IAA showed a higher rooting rate (66.7%) and better quality roots than media with other auxins.

For in vitro rooting, microshoots were cultured on MS medium containing auxins (IAA, IBA, 2,4-D, and NAA.). The addition of IAA, IBA, and NAA did not inhibit the survival rate of plantlets. The addition of 2,4-D into the culture medium, formed a callus at the base of the plantlet. Shoot growth was also promoted in media with IAA, IBA, and NAA and without 2,4-D, especially in the medium containing 25μM IAA. Root growth was promoted in a medium with IAA and IBA (Fig. 3C). Success of in vitro rooting of microshoots excised from proliferating cultures depended on the species and clone used, on media composition, and on the concentration of the auxin used (Chalupa 1992). In general, low salt media stimulated root number and length in *Phellodendron* (Maene and Derbergh 1985).

Ex vitro rooting was achieved by dipping the base of excised microshoots in 500μM IAA solution for 30s (Table 3). Rooting rate ex vitro was similar to that in vitro but shoot length was higher in ex vitro rooting. When cultured in rooting medium for 20 days, survival rate after transplantation was 100% (data not shown). In vitro rooted shoots were easily acclimated in potting media containing vermiculite under high humidity, and resulted in over 90% survival rate (Fig. 2D).

2.2.2 *Establishment of Cell Cultures*

Primary callus cultures were established from stem cambial tissues of cork tree cultured on LS (Linsmaier and Skoog 1965) liquid medium with 0.5mg/l 2,4-D,

0.1 mg/l BA, and 3% sucrose. Suspension culture was initiated by transferring callus to liquid MS medium supplemented with 3% sucrose, 1 mg/l 2,4-D and 0.1 mg/l benzylamino purine (BA). Cultures were placed in the dark at 25 °C, and were subcultured every 4 weeks. The effect of various macronutrients was tested by transferring duplicate samples of 1.5 g callus of 10-day old suspension into 250-ml Erlenmeyer flasks, each containing 50 ml of test medium. The test media for studying the effect of nitrate, phosphate, sucrose, and growth regulators (2,4-D, IAA and BAP) were prepared by modifying their concentrations in the LS basal medium. All cultures were maintained in the dark on a gyratory shaker at 25 °C.

2.2.3 *Extraction and Analysis of Berberine*

Samples for berberine analysis were collected by taking 0.5 g of cells (fresh weight), and were analyzed using the method described by Nakagawa et al. (1984). Dried cells were then extracted with 10 ml methanol in a sonicating bath at 50 °C for 1 h. Quantitative determinations of berberine and palmatine were made by the calibration curve obtained with authentic compounds using a Spectra Physics SP8800 HPLC system equipped with a UV detector. Separation was accomplished using a μ-Bondapak C18 (300 × 3.9 mm) column; solvent 1 mM tetrabutylammonium phosphate (pH 2.0):acetonitrile (40:60); flow rate 0.5 ml/min; wavelength 265 nm. Berberine was tentatively identified based on retention time and UV spectra (Fig. 4). The qualitative analysis of berberine was done by EI mass spectrometry. The mass spectrum of berberine purified from the cultured cells was essentially the same as that of the authentic sample. The mass spectrum of berberine showed an abundant protonated molecule at m/z 337 suitable for identification (Fig. 5).

Berberine was detected in extracts of cultured cells, but the amounts varied depending on cell lines and culture conditions (data not shown).

The effect of a nitrogen source on both cell growth and berberine production was examined by varying concentrations of NH_4NO_3 and KNO_3 (Fig. 6). High nitrate concentration in the medium resulted in a marked increase in berberine production, whereas low level nitrate promoted cell growth. For example, high berberine production was obtained in the medium containing 80 mM nitrate, while cell growth was shown to increase at 40 mM nitrate. Nitrate has been known to play an important role in the production of secondary metabolites (De-Eknamkul and Ellis 1988). It was also reported that high nitrate concentration suppresses cell growth in a hybrid poplar, but promoted anthocyanin production (Park et al. 1992a,b).

Cell growth was promoted by phosphate in the medium up to the concentration of 54.8 mM. Higher phosphate levels than this did not promote cell growth. However, cell growth increased exceptionally at 2.25 mM phosphate (Fig. 7). The optimal concentration for berberine production was 8.98 mM. This optimal initial phosphate concentration is half that of normal LS medium. The internal phosphate level seemingly played a key role in cell growth and secondary metabolism (De-Eknamkul and Ellis 1988).

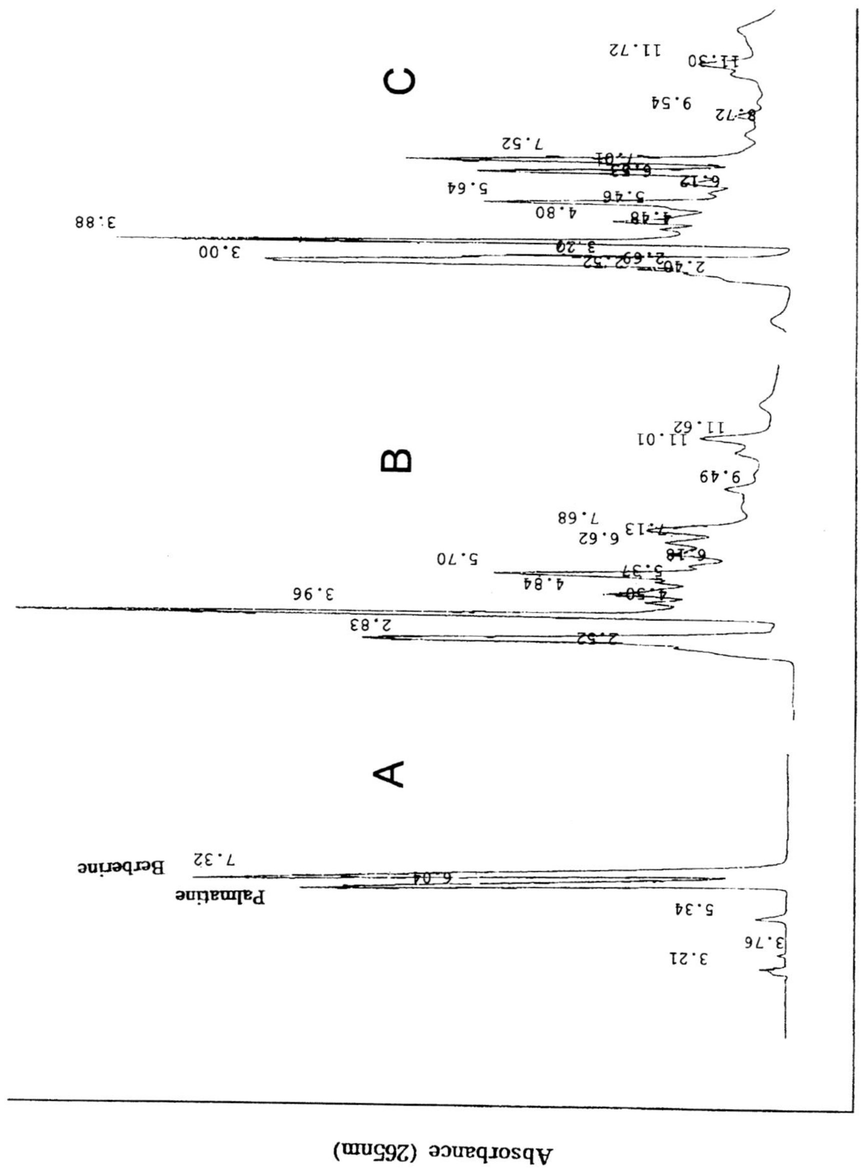

Fig. 4. HPLC separation of authentic berberine compounds and extracts of suspension cultured cells of *Phellodendron amurense* Rupr. *A* authentic berberine and palmatine, *B* cell extracts, *C* co-chromatography of authentic compounds and cell extracts (*A* + *B*). (Choi et al. 1996)

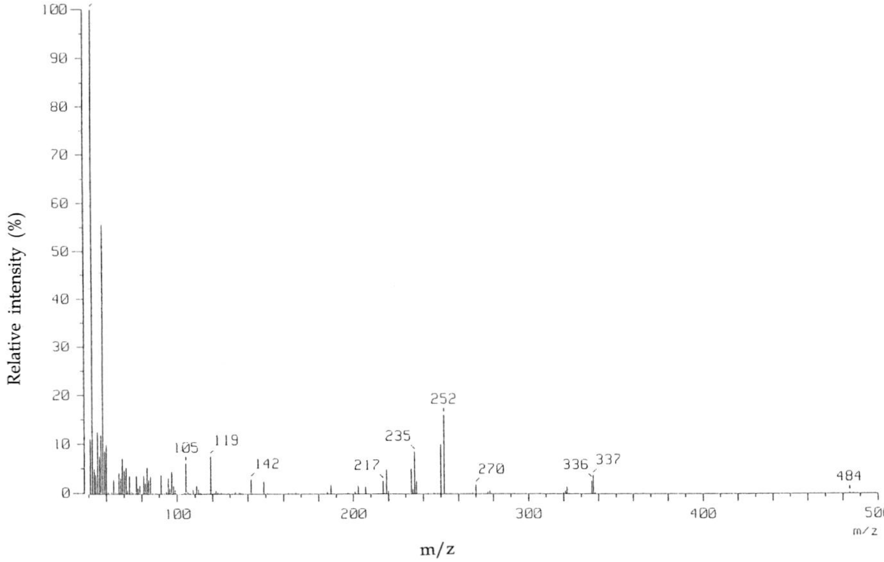

Fig. 5. EI mass spectrum of endogenous berberine isolated from cultured cells

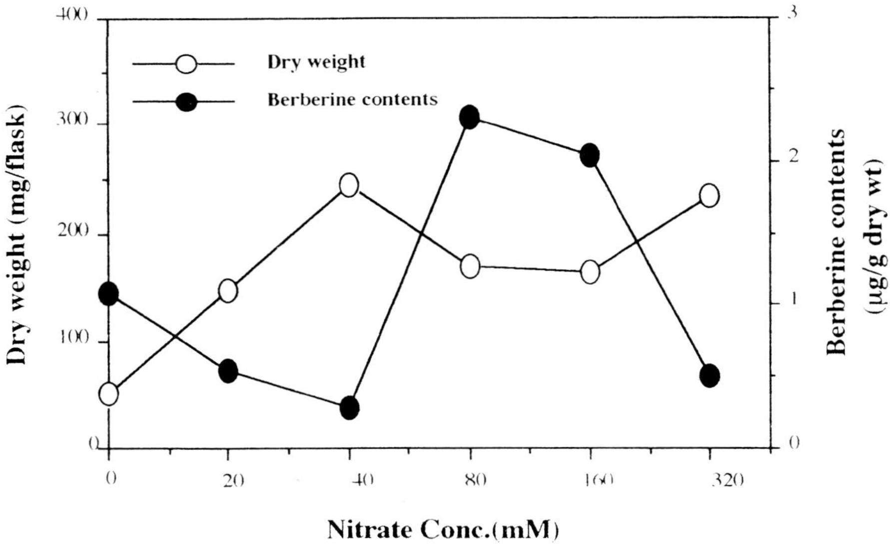

Fig. 6. Effect of nitrate concentration on cell growth and berberine production. (Choi et al. 1996)

Cell growth as well as berberine yield increased with sucrose concentration (Fig. 8), the optimal concentration of sucrose for both was 7%. Sucrose has been known to support a high growth rate and biomass yield in plant cell cultures (De-Eknamkul and Ellis 1988).

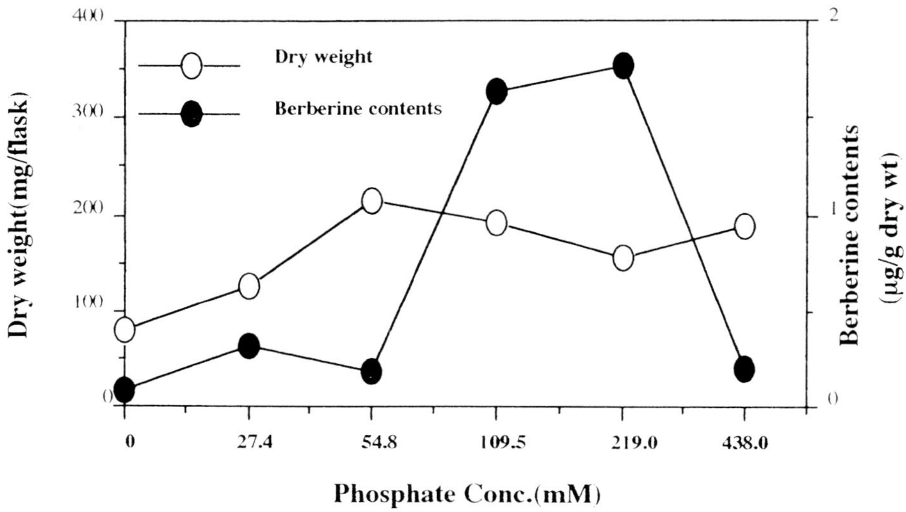

Fig. 7. Effect of phosphate concentration on cell growth and berberine production

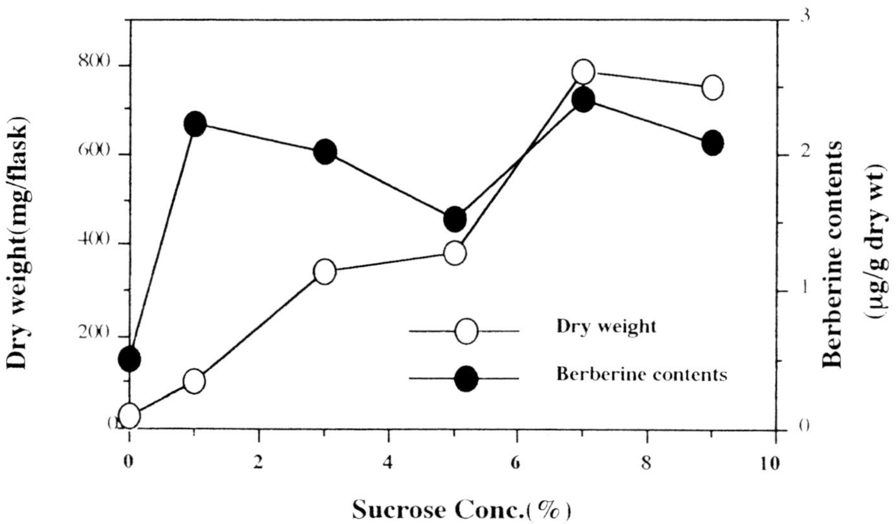

Fig. 8. Effect of sucrose concentration on cell growth and berberine production (Choi et al. 1996)

In cork tree cells, cell growth and berberine production needed different growth regulators. 2,4-D and IAA promoted growth of cells (Table 4). Although cell growth in a medium with high concentrations of 2,4-D and IAA was markedly superior to low concentrations and controls, berberine production was inhibited by a high level of auxin. Addition of 2,4-D has been known

Table 4. Effect of growth regulators on cell growth and berberine production. (Choi et al. 1996)

Growth regulators (mg/l)			Cell growth (g dry wt./flask)	Berberine content (mg/g dry wt.)
2,4-D	IAA	BA		
1	1	–	0.098	0.056
1	5	–	0.116	0.340
1	10	–	0.074	0.254
–	1	–	0.075	3.170
–	5	–	0.091	0.051
–	10	–	0.120	0.750
–	5	0.5	0.103	0.056
–	10	0.5	0.398	0.056
5	–	0.5	0.090	0.690

to be responsible for suppressing formation of secondary metabolites in many plant cell culture systems (Fowler and Stephan-Sarkissian 1985; Stafford et al. 1985; Park et al. 1992b).

Berberine production in cork tree cells was stimulated when IAA was incorporated with cytokinin (BA). This observation suggested an important role of BAP as a promoter of berberine production. Hara et al. (1993) reported that BA activated enzymatic reactions after the formation of amines in the biosynthesis of berberine. The highest berberine yield was obtained from LS basal medium containing 1 mg/l IAA.

2.2.4 *Two Stage Cultures*

Based on the results obtained in suspension culture, growth medium (GM) and production medium (PM) were formulated for two stage cultures. Twenty-five days after culture, GM was replaced with fresh PM. Cells were harvested by suction filtration after a 10-day incubation in the test medium. Dry weight was measured after desiccation at 60 °C for 24 h.

When the optimal condition for cell growth is different from that for synthesis of secondary metabolites, a two stage culture system can improve productivity by optimizing conditions separately for cell growth and product formation. Growth medium and production medium were formulated based on the previous results for optimal nutrient constituents. It was noteworthy that several conditions, such as concentrations of nitrate, phosphate, carbon source, and growth regulators were changed. Thus, we attempted to find conditions for two stage cultures to ensure both maximum growth in GM and maximum production in PM. Ten days after two stage culture, the berberine content in cork tree cells increased to 5.06 mg/g dry weight (Table 5). This result differed from a previous report by Knobloch and Berlin (1980), who reported that transfer of *Catharanthus roseous* hairy root from GM to PM reduced the growth rate.

Table 5. Comparison between one-stage culture and two-stage culture on cell growth and berberine production. (Choi et al. 1996)

Cultures	Cell growth D.W. (mg/flask)	Berberine content (μg/g dry wt.)
One stage cultures (GM → GM)	250 ± 30[a]	2350 ± 359
Two stage cultures (GM → PM)	290 ± 45	5065 ± 698

[a] The values represent the mean ± SD.

3. Conclusion and Prospects

The cork tree (*Phellodendron amurense* Rupr.) is principally grown in the Far East as a medicinal and ornamental tree, particularly in places, where it withstands pollution well. In Korea, extracts of cork tree bark are used as an antimicrobial, anticancer, and anti-inflammatory agent.

Mass propagation by conventional breeding methods appears difficult. Propagation by in vitro culture techniques is more easily achieved than that of cutting and root sprouting. The results presented here suggest potential applications for fast multiplication and high yields of desired products.

The major secondary metabolites from cork tree are berberine; yellow alkaloids. Callus and suspension cultures of cork tree accumulate berberine and palmatine. Optimal conditions for the increase of both cell growth and product yield can be created by changing the chemical components such as nitrate, phosphate and phytohormones. Two stage cultures by growth medium and production medium increased the dry weight to 5.06 mg/g after 10 days of culture.

Acknowledgement. This work was made possible by financial support from the Institute of Genetic Engineering, Kyungpook National University, Taegu, Korea

References

Aitken-Christe J, Connett M (1992) Micropropagation of forest trees. In: Kurata K, Kozai T (eds) Transplant production systems. Kluwer, Dordrecht, pp 163–194

Chalupa V (1992) Micropropagation of european mountain ash (*Sorbus acuparia* L.) and wild service tree [*Sorbus torminalis* (L.)Cr.], In: Bajaj YPS (ed) Biotechnology in agriculture and forestry, vol 18. High tech and micropropagation II. Springer, Berlin Heidelberg New York, pp 211–216

Choi MS, Shin DI, Park YG (1996) Berberine production by cell suspension cultures of cork tree (*Phellodendron amurense* Rupr). Korean J Pharmacogn 27(1):32–36

De-Eknamkul W, Ellis BE (1988) Rosmaric acid production in plant cell culture. In: Bajaj YPS (ed) Biotechnology in agriculture and forestry, vol 4. Medicinal and aromatic plants I. Springer, Berlin Heidelberg New York, pp 310–329

Fowler MW, Stephan-Sarkissian G (1985) Carbohydrates source, biomass productivity, and natural product yield in cell suspension cultures. In: Neumann (ed) Primary and secondary metabolism of plant cell cultures. Springer, Berlin Heidelberg New York, pp 66–73

Gruselle R, Badia N, Boxus P (1987) Walnut propagation: first results. Acta Hortic 212:511–515
Hara M, Toshihito T, Fukui H, Tabata M (1993) Induction of berberine biosynthesis by cytokinins in *Thalictrum minus* cell suspension cultures. Plant Cell Rep 112:70–73
Hinz H, Zenk MH (1981) Production of protoberberine alkaloids by cell suspension cultures of *Berberis* species. Naturwissenschaften 67:620–621
Hong MK, Jeong YH, Hong JU (1988) Development of botanical fungicide on Japanese apple canker. I. Isolation of antifungal compound from plants. Res Rep Rural Der Admin (CP) 30(3):24–30
Ikuta A, Itokawa H (1989) *Coptis*: in vitro regeneration of plants and production of berberine. In: Bajaj YPS (ed) Biotechnology in agriculture and forestry, Vol 7. Medicinal and aromatic plants II. Springer, Berlin Heidelberg New York, pp 110–128
Kim JH, Goo GH, Choi MS, Park YG (1992) Micropropagation and soil adjustment of cork tree (*Phellodendron amurense* Rupr.) through in vitro culture. Korean J Plant Tissue Cult 19(1):37–42
Knobloch KH, Berlin J (1980) Influence of medium composition on the formation of secondary compounds in cell suspension cultures of *Catharanthus roseous* (L.) Don. Z Naturforsch [c] 35:551–556
Kunio A (1986) Plantlet formation by intrapetiolar bud culture of *Phellodendron amurense* Rupr. For Tree Breed 138:24–26
Linsmaier E, Skoog F (1965) Organic growth factor requirement of tobacco tissue cultures. Physiol Plant 18:100–127
Lloyd G, McCown B (1980) Commercially feasible micropropagation of mountain laurel, *Kalmia latifolia* by use of shoot tip culture. Proc Int Plant Prop Soc 30:421–427
Maene L, Debergh P (1985) Liquid medium additions to established tissue cultures to improve elongation and rooting in vivo. Plant Cell Tissue Organ Cult 5:23–33
Misawa M, Nakanishio TM (1988) Antitumor compounds. Production by plant cell culture In: Bajaj YPS (ed) Biotechnology in agriculture and forestry, vol 4. Medicinal and aromatic plants I. Springer, Berlin Heidelberg New York, pp 99–207
Murashige T, Skoog F (1962) A revised medium for rapid growth and bio-assays with tobacco tissue cultures. Physiol Plant 15:473–497
Nakagawa K, Konagai A, Fukui H, Tabata M (1984) Release and crystallization of berberine in the liquid medium of *Thalictrum minus* cell suspension cultures. Plant Cell Rep 3:254–257
Otsuka H, Fujimura H, Sawada T, Goto M (1981) Studies on anti-inflammatory agents. II. Anti-inflammatory constituents from rhizome of *Coptis japonica*. Makino Yakugaku Zasshi 101:883–890
Park YK, Song WS, Park HS, Keon OU (1988) Breeding of useful plants. Research Report of the Forest Genetic Research Institute (FGRI) Suwon, Korea, pp 89–111
Park JC, Kim LS, Choi YW (1992a) Production of protopberberine-type alkaloid from cell cultures. II. Culture methods for effective berberine production. Korean J Plant Tissue Cult 19(6):323–329
Park YG, Choi MS, Son SH (1992b) Anthocyanin synthesis in cell cultures of *Populus alba* X *P. glandulosa*. Korean J For Soc 81(2):183–190
Seong NS (1992) High quality production and competitive improvement strategy of medicinal crops. In: Seoung NS (ed) Prospects of high quality production and strategy of medicinal crops. Cheju ADI, pp 235–240
Stafford A, Smith L, Fowler MW (1985) Regulation of product synthesis in cell cultures of *Catharanthus roseous* (L) G. Don. Growth-related indole alkaloid accumulation in batch cultures. Plant Cell Tissue Organ Cult 4:83–94
Tosa S, Ishihara S, Nose Y, Usikawa T, Yoshida S, Nakazawa H, Tomimatsu T (1989) Studies on quality control of crude drugs. I. Analysis and distribution of alkaloids in *Phellodendron amurense*. Shoyakugaku Zasshi 43(1):28–34

XXI *Physochlaina* Species: In Vitro Culture and the Production of Physochlaine and Other Tropane Alkaloids

N.I. GORINOVA[1], A.I. ATANASSOV[2], and M.P. VELCHEVA[3]

1 General Account

A large number of genera of the Solanaceae contain alkaloids derived from ornithine via pyrrolizidine intermediates (Seigler 1977). Some species of the genus *Physochlaina* G. Don (family Solanaceae) contain tropane alkaloids. One proposal for tropine biosynthesis is an outgrowth of studies with tissue cultures (Elze and Teuscher 1972). A cell suspension culture of *Datura ferrox* L., when supplied with *dl*-[2-^{14}C] ornithine, yielded radioactive α-keto-δ-amino-valeric acid (1) among other products. However, none of tropane alkaloids produced was radioactive. It has been proposed that this in vitro cell culture lacks the enzyme that catalyzed the reaction between Δ^1-pyrroline-2-carboxylic acid (2) and acetoacetylcoenzyme A. This condensation product can ultimately yield hygrine and tropine, as illustrated in Fig. 1.

The study of the alkaloid content of the cell suspension culture of *P. orientalis* shows biosynthesis of the tropane alkaloids hygrine, tropinone and tropine (Gorinova et al. 1994). The enzyme catalyzing the reaction between Δ^1-pyrroline-2-carboxylic acid and acetoacetylcoenzyme A in this culture may be present, according to the above results. To confirm this, further experiments with *dl*-[2-^{14}C] ornithine in *P. orientalis* cell culture need to be performed.

The genus *Physochlaina* G. Don (family Solanaceae) comprises six species: *Physochlaina orientalis, P. physaloides* (Fig. 2), *P. semenowii*, *P. alaica*, *P. praealta* and *P. officinalis*. These plants are perennial herbs (Academy of Science USSR 1955). The first four species contain tropane alkaloids, mainly apoatropine, apohyoscine, hyoscyamine, scopolamine, 6-hydroxyatropine and physochlaine (Table 1).

Physochlaina orientalis occurs across the Caucasus and Middle Asia (Academy of Science USSR 1955). Levels of total alkaloids, tropane alkaloids and scopolamine were studied in this plant. Tropane alkaloids represented about 90% of the total alkaloid mixture. The maximum content of alkaloids in the underground parts during flowering was found to be 0.718%, while the

[1] Institute of Genetic Engineering, 2232 Kostinbrod – 2, Bulgaria
[2] De Montfort University, Norman Borlaug Center for Plant Science, Institute of Genetic Engineering, 2232 Kostinbrod – 2, Bulgaria
[3] Institute of Organic Chemistry with Centre of Phytochemistry, Bulgarian Academy of Sciences, 1113 Sofia, Bulgaria

Biotechnology in Agriculture and Forestry, Vol. 43
Medicinal and Aromatic Plants XI (ed. by Y.P.S. Bajaj)

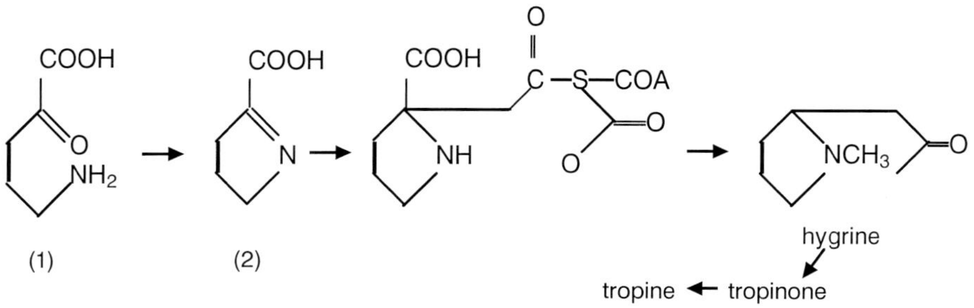

Fig. 1. Biochemical pathway leading to the production of tropine

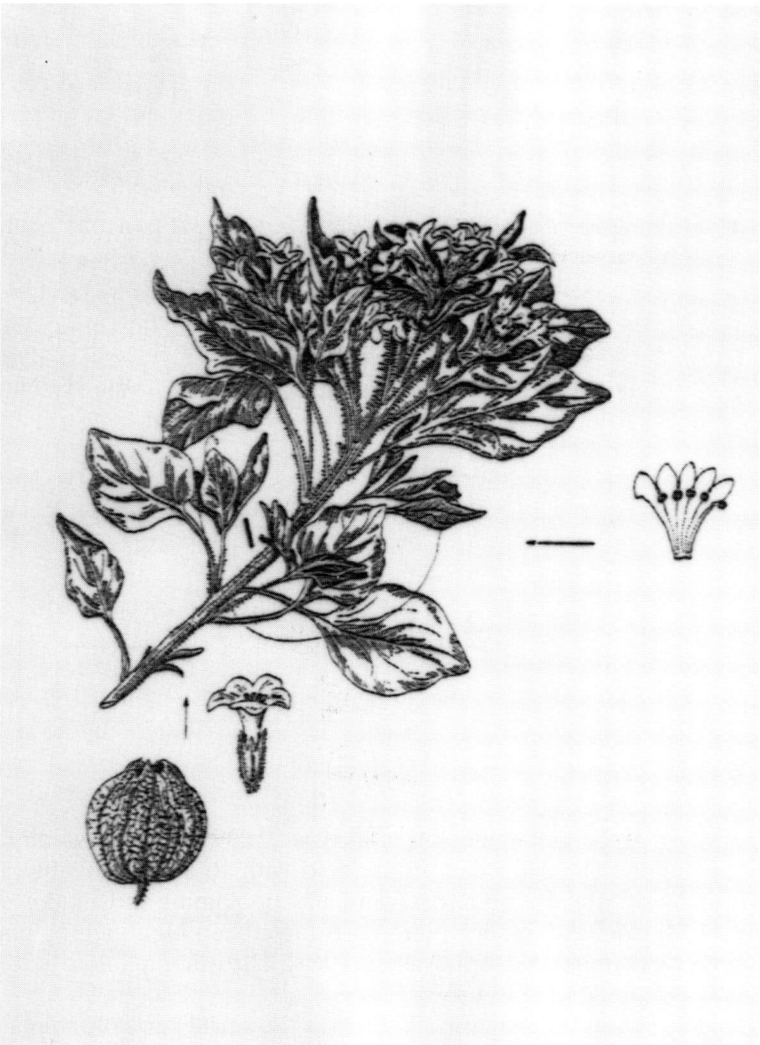

Fig. 2. *Physochlaina physaloides* (L.) G. Don; part of the plant

Table 1. Structure of the tropane alkaloids of *Physochlaina* species

Alkaloid	R_1	R_2	R_3	R_4	Reference
Apoatropine	$O—CO—C(=CH_2)—C_6H_5$	H	H	H	Yunusov (1981), Glasby (1975b)
Apohyoscine	$O—CO—C(=CH_2)—C_6H_5$	H	—O—		Mirzamatov et al. (1974a), Glasby (1975b)
Hyoscyamine	$O—CO—C(CH_2OH)—C_6H_5$	H	H	H	Yunusov (1981), Glasby (1975a)
6-Hydroxyatropine (*dl*-6-hydroxyhyoscyamine)	$O—CO—CH(CH_2OH)—C_6H_5$	H	OH	H	Mirzamatov et al. (1974b)
6-Hydroxyhyoscyamine	$O—CO—CH(CH_2OH)—C_6H_5$	H	OH	H	Mirzamatov et al. (1974a)
Physochlaine	$O—CO—CH_2—C_6H_4—OCH_3$	H	OH	H	Mirzamatov et al. (1974a), Manske (1977)
Scopolamine	$O—CO—CH(CH_2OH)—C_6H_5$	H	—O—		Yunusov (1981), Glasby (1975c)

content of scopolamine in stalks is 0.239%. Hyoscyamine (R_f 0.73), scopolamine (R_f 0.63) and an alkaloid with R_f 0.86 (probably apoatropine) from the aerial parts, and hyoscyamine, scopolamine and alkaloids with R_f 0.82 and 0.20 from the underground parts have been identified by paper chromatography (Minina and Aduevskaya 1965). The other alkaloid isolated from the plant is tropine (Table 1, Fig. 3).

The content of tropane alkaloids in the aerial parts of the plant was found to be 0.16–0.22% and in the roots, – 0.32–1.03% according to the results of Yunusov (1981).

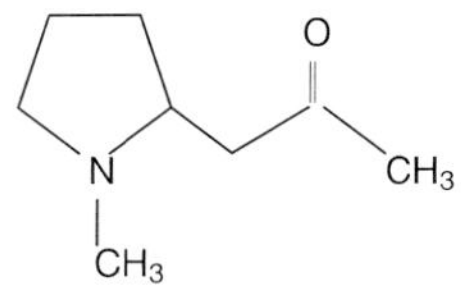

2 Hygrine

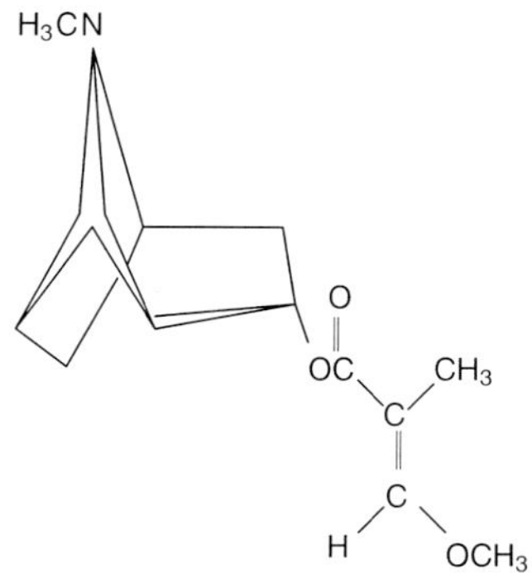

4 3a-(3-methoxymethacryloyloxy)-tropane

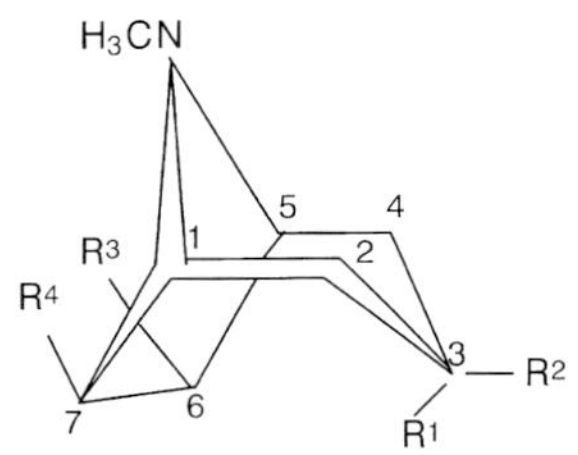

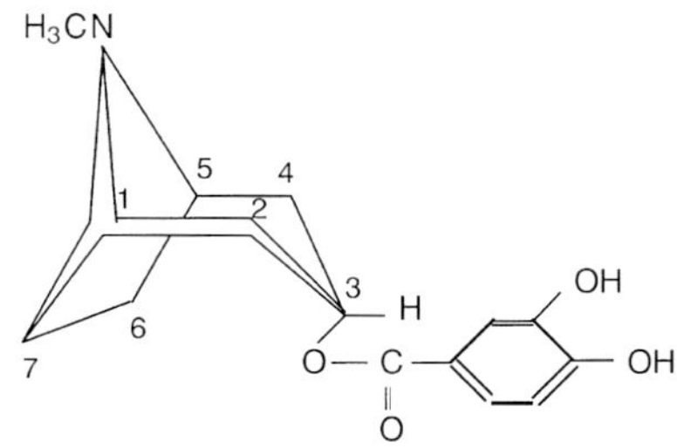

5 Demethylphyllalbine

No of alkaloids	R^1	R^2	R^3	R^4
1	–O–		H	H
3	OH	H	H	H
6	O–CO–C–C_6H_5 ‖ CH_2	H	–O–	
7	O–CO–C = CH │ │ CH_3 CH_3	H	O–CO–CH_3	H
8	O–CO–C = CH │ │ CH_3 CH_3	H	O–CO–CH–$(CH_3)_2$	H

Fig. 3. Structure of the alkaloids in cell suspension cultures of *P. orientalis*. *Numbers* are identical to the numbers in Table 2

Physochlaina physaloides is found in the mountains of Siberia, Middle Asia, Mongolia, Japan and China. The concentration of total alkaloids in *P. physaloides* was highest during flowering (0.78, 0.61, 0.57 and 0.54% in stem, leaves, roots and rootstock, respectively) and decreased gradually toward the end of plant growth (Peshkova 1964). All parts of the plant contain tropane alkaloids: 0.25% in roots, 0.22% in aerial parts together with the fruits and 0.18% in the seeds (Kuvaev and Blinova 1960).

Physochlaina alaica E. Korot is the endemic species for the Altai region in the former USSR (Mirzamatov et al. 1973b). The Russian *P. alaica* contains 1.1% alkaloids in the roots and 0.5% in the aerial parts (Yunusov 1981). Twelve tropane alkaloids have been isolated from this plant. The main alkaloid in the different parts of the plant and during different phases of its individual development is hyoscyamine – 60–74.3% (Mirzamatov et al. 1973b; Yunusov 1981; Table 1). The amount of alkaloids from the roots of *P. alaica*, collected when the aerial parts have died back, are divided into three parts in benzol, chloroform and water (Mirzamatov et al. 1974a). The chloroform extract contains 6-hydroxyhyoscyamine after treatment with benzol. The new base, physochlaine [3α-(*p*-methoxyphenylacetoxy)-tropane-6β-ol] (Table 1), is separated from the chloroform and water extracts, which are not soluble in benzol. Two other alkaloids, apohyoscine and 6-hydroxyatropine, have been isolated from the roots (Table 1; Mirzamatov et al. 1974b). The dimeric alkaloids α- and β-belladonine (Mirzamatov et al. 1973a) have been found together with *l*-scopolamine (Table 1) in the chloroform extract of the roots. The structures of these compounds (Fig. 4) have been established by chemical analysis and spectral data. It has been suggested that the biogenetic formation of dimeric tropane alkaloids is due to a photochemical reaction in the plant under certain conditions (temperature, light, enzyme action). The photochemical reaction is a common and unique method of obtaining compounds with a different structure (Mirzamatov et al. 1973a).

Physochlaina semeniwii is found in Middle Asia (Academy of Science USSR 1955). Scopolamine is one of the main tropane alkaloids in *Physochlaina* species. Expression of the hydroxylation gene (hyoscyamine 6 β-hydroxylase gene) from *Hyosciamus niger* has been examined (Herbert 1996). This gene, fused to a reporter gene, has been expressed in hairy roots of *Atropa belladona*. It was concluded from the results that it is controlled by some genetic regulation specific to scopolamine-producing plants (6-β-hydroxyhyoscyamine is a precursor for scopolamine).

The following typical physiological effects exerted by tropane bases: (1) mydriatic, i.e., parasympathetic blocking, anesthetic, (2) stimulating the central nervous system, and (3) ganglion blocking, i.e., courare-like activity, have been detected in some synthetic tropeines. Effects are based on tropanes with a C-3 α-placed esterified hydroxyl, e.g., atropine and hyoscyamine, as contrasted with (1) which seems to be specific for 3β-tropanol esters only, such as tropacocaine, cocaine, whereas activities (2) and (3) seem to depend to a greater extent on the substituents of nitrogen and the structure of the esterifying acid than on the α- or β-location of the hydroxyl group at C-3 (Fodor 1960).

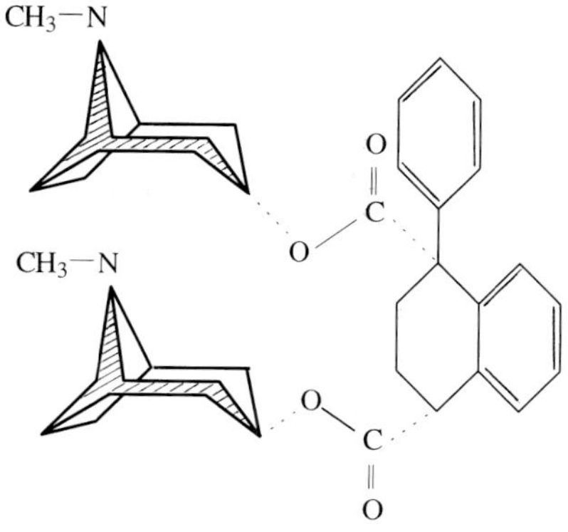

1

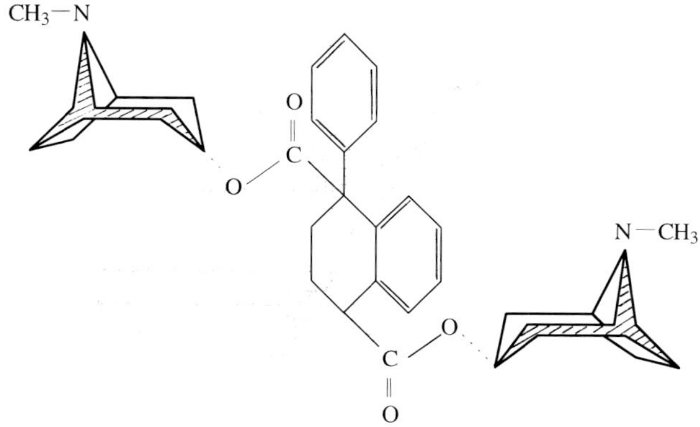

2

Fig. 4. Structures of the dimeric alkaloids. *1* α-Belladonine; *2* β-belladonine

The *Physochlaina* herbs, named "hun horse" in Tibet and "garag chig tav" in Mongolia, and those occurring in the Mongolian and Gobi Altai, and Hanghai have been used for many years in traditional medicine against acute and chronic infections for wound healing, because of its anti-inflammatory action (Khaidav and Menschikova 1978).

Preparations of species of the genus *Physochlaina* have been used against skin and venereal diseases, headache, and nervous disorders. The preparations have also been applied during the time of infectious epidemic diseases (Khaidov and Menschikova 1978).

(−)Hyoscyamine is the most widely spread of the tropane alkaloids. The optically inactive form is atropine. The (−) form of the alkaloid acts more powerfully upon the peripheral nerves than the (+) form. Both closely resemble atropine in their bitter taste and mydriatic action (Glasby 1975a).

Scopolamine has a similar pharmacological action to that of atropine upon the peripheral, cholinergic, autonomic nervous system, but one which is more transitory. The alkaloid is used in medicine as a sedative, in particular for treatment of motion sickness. Although there may be some preliminary excitement in small doses, the normal reaction is one of fatigue passing into sleep (Glasby 1975a).

The biological activity of dimeric tropane alkaloids is less known. The quaternary derivatives of belladonines possess courare-like activity. That is why the study of such tropane alkaloids is necessary. It is possible that different pharmacological activities may be expressed by new types of tropane alkaloids (Aripova and Yunusov 1991).

Tissue cultures of alkaloid-producing plants would be an ideal system for studying biosynthetic routes, since these systems can be so well controlled. Tissue cultures of *Physochlaina* species producing tropane alkaloids are not known. Callus and suspension cultures of *P. orientalis* were obtained in order to examine the production of tropane alkaloids (Gorinova et al. 1994). The structures of many tropane alkaloids have been successfully elucidated by examining their mass spectra. The mass spectral fragmentation behavior of these alicyclic class alkaloids – tropane alkaloids has been studied by Blossey et al. (1964). Capillary gas liquid chromatography (GLC) combined with mass spectrometry (MS) has been used successfully for the separation and identification of more than 30 tropane alkaloids from *Datura innoxia* (Witte et al. 1987). The more powerful GLC-MS technique using capillary columns was successfully applied for the separation and structure elucidation of alkaloids produced by cell suspension culture of *P. orientalis* (Gorinova et al. 1994). The results are presented here.

2 Techniques

GLC-MS: Jeol JG 20K equipped with fused silica capillary column 30 × 0.2 mm i.d. coated with SPB-1 phase and coupled with Jeol JMS-D-300 MS was used. Carrier gas: He. Programmed temperature: 50–280 °C (10 min). Ionization energy: 70 eV. Volume injected: 1 μl. GLC-CI (chemical ionization) mass spectra, gas reagent: isobutane.

The suspension culture was obtained from callus initiated from leaf segments of in vitro plants of *P. orientalis* (Botanical Garden, Marburg, Germany). The callus was cultivated in MS medium (Murashige and Skoog 1962) containing sucrose (30 g/l), agar (7 g/l), NAA (0.1 mg/l) and kinetin (0.05 mg/l). The pH of the culture medium was adjusted to 5.8 before autoclaving.

The suspension culture was subcultured in the same MS liquid medium every 3 weeks at 25 ± 1 °C on a rotary shaker at 100 rpm. The culture was illuminated with a 16-h photoperiod and at light intensity of 2000 lx.

Harvested cells of the suspension culture, cultivated for 3 months in liquid medium, were freeze dried (18.4 g dry wt.), powdered, extracted in EtOH: 25% NH_4OH (11 : 1) and homogenized during 20 h at room temperature. The extract was filtered, evaporated to dryness and the residue was dissolved in 0.1 M HCl. The acidic aqueous solution was extracted with $CHCl_3$ (3 × 25 ml), filtered and adjusted with 25% NH_4OH to pH 10. The alkaline solution was extracted with $CHCl_3$ (3 × 25 ml), the combined $CHCl_3$ extracts were evaporated for dryness, yielding 0.065 g of crude alkaloids. The alkaloid mixture was dissolved in $CHCl_3$ and examined by TLC (Si-gel GF_{60} plates, light petrol:$CHCl_3$:Me_2CO:MeOH 4:4:1:1, Dragendorff reagent).

3 Tropane Alkaloids of *P. orientalis* Cell Suspension Culture

Nine alkaloids (Table 2, Fig. 3) were detected in the total alkaloid mixture obtained from the cell suspension culture of *P. orientalis*. Although most of them are known constituents of other genera of the Solanaceae, only tropine among the detected alkaloids has been found earlier in this plant (Yunusov 1981).

The mass spectral fragmentations of the known alkaloids are identical to those published in the literature as follows: (numbers refer to compounds in Table 2): tropinone (Blossey et al. 1964; Hesse and Bernard 1975) **1**; hygrine (Hesse and Bernard 1975; Witte et al. 1987) **2**; tropine (Blossey et al. 1964) **3**; aposcopolamine (Witte et al. 1987) **6**; 3α-tigloyloxy-6-acetoxytropane (Hesse and Bernard 1975) **7**; and 3α-tigloyloxy-6-isobutyryloxytropane (Witte et al. 1987) **8**.

The ester moiety at carbon 3 of the tropane nucleus can be determined easily by MS from ions formed by losses of the ester group (Hesse and Bernard 1975). In the mass spectrum of compounds **4** and **5**, the loss of the ester group gave a low intensive ion at *m/z* 140 and a more intensive ion at *m/z* 124.

The mass spectrum of compound **4** (Fig. 5A) shows a molecular ion at *m/z* 239 and base peak at *m/z* 83, accompanied by similarly strong ions at *m/z* 224 (M^+-15), 123, 124, 82, 42. The molecular weight of compound **4** was confirmed by the presence of an adduct ion at *m/z* 240 in the CI mass spectrum (CIMS). The MS fragmentation of **4** can be ascribed to 3α-(3-methoxymethacryloyloxy)-tropane. A similar structure, containing two hydrogens less due to the presence of a double bond at C-3 and C-4, has been obtained as an additional product during the synthesis of bellendine (Bick et al. 1971, 1973). Neither a synthetic compound with a molecular weight of 237 nor alkaloid **4** have been found as natural products to date.

Compound **5**, isolated by PTLC in a sufficient amount (3 mg), has a mass spectral fragmentation (Fig. 5B) very close to that of phyllalbine (Parrello et al. 1963). The molecular ion peak of **5** appears at *m/z* 277, being 14 units lower than that of phyllalbine, *m/z* 291. The acid moiety of phyllalbine has been characterized using the fragment at *m/z* 151. The same fragment with 14 units less at *m/z* 137 can be seen in the spectrum of alkaloid **5**. A low intensive ion fragment at *m/z* 140, coming from O-acyl cleavage, is present in the MS

Table 2. GLC-MS data for alkaloids in cell suspension cultures of *P. orientalis*

Compound	R_t (min)	M+	Characteristic ions (abundance %)							
(**1**) Tropinone	9.05	139	112	110	97	96	83	82	56	41
		(15)	(2)	(4)	(4)	(5)	(36)	(100)	(11)	(20)
(**2**) Hygrine	10.01	141	124	98	85	84	82	70	55	42
		(4)	(20)	(2)	(5)	(100)	(7)	(8)	(4)	(25)
(**3**) Tropine	10.20	141	124	113	96	83	82	67	55	42
		(2)	(20)	(10)	(40)	(50)	(100)	(9)	(10)	(30)
(**4**) 3α-(3-Methoxymetacryloyloxy)-tropane	18.12	239	224	168	124	83	82	69	55	42
		(13)	(18)	(11)	(36)	(100)	(53)	(75)	(76)	(11)
(**5**) Demethylphyllalbine	19.34	277	168	151	137	140	124	95	82	55
		(22)	(2)	(2)	(6)	(6)	(36)	(78)	(100)	(36)
(**6**) Aposcopolamine	20.15	285	154	138	136	108	103	94	81	42
		(25)	(41)	(2)	(25)	(40)	(40)	(100)	(20)	(30)
(**7**) 3α-Tigloyloxy-6-acetoxytropane	21.23	281	182	181	138	122	95	94	55	43
		(1)	(10)	(5)	(10)	(46)	(90)	(100)	(25)	(30)
(**8**) 3α-Tigloyloxy-6-isobutyryloxy-tropane	22.17	309	210	209	138	122	95	94	55	43
		(5)	(8)	(4)	(7)	(40)	(100)	(98)	(20)	(25)
(**9**) Physochlainoorientaline	23.12	256	239	195	163	112	94	84	82	55
		(3)	(36)	(6)	(7)	(100)	(7)	(29)	(25)	(56)

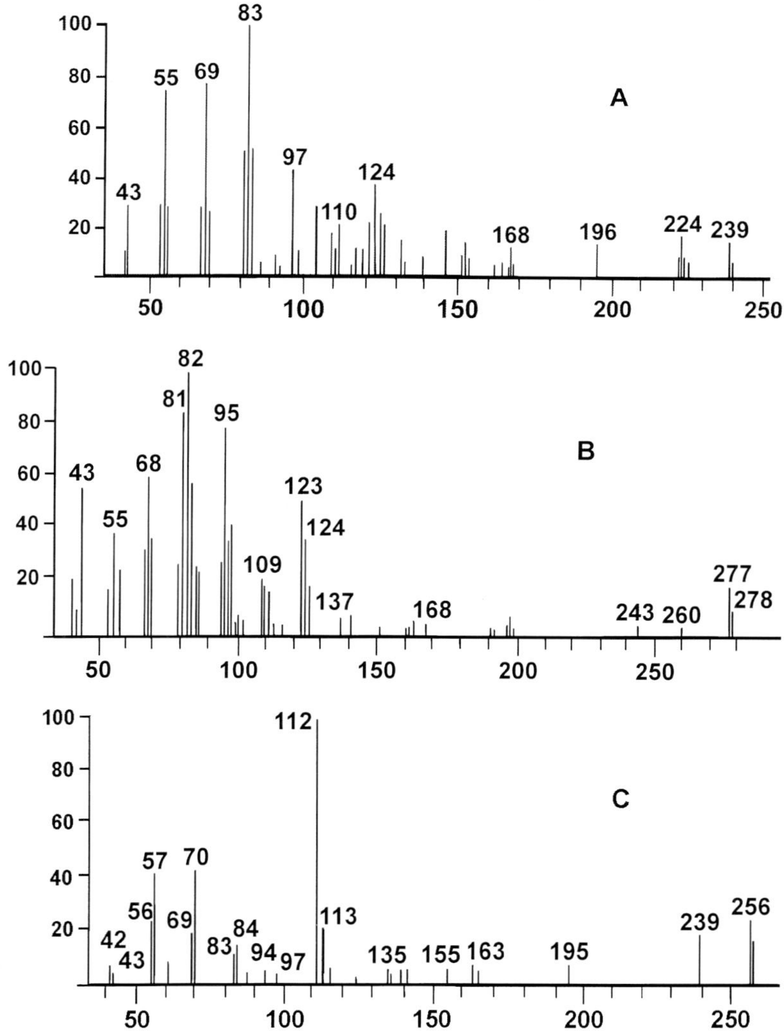

Fig. 5. Mass spectra of the alkaloids isolated from cell suspension cultures of *P. orientalis*. **A** 3α-(3-Methoxymethacryloyloxy)-tropane; **B** demethylphyllalbine; **C** physochlainoorientaline

spectrum of phyllalbine (Parrello et al. 1963) and of **5**. The mass spectrum of demethylphyllalbine showed relatively slight characteristic peaks of its acidic part like phyllalbine. The fragment $[C_6H_5O_2]^+$ showed a peak at *m/z* 109, but a peak at *m/z* 137 was due to an acyclic fragment $[COC_6H_5O_2]^+$. The other fragment $[C_8H_{14}NO]^+$ showed a peak at *m/z* 140. After the separation of the acidic part, the acyclic moiety had undergone changes in the arrangement of a hydrogen atom in the skeleton and obtained an olefine at *m/z* 123, $[C_8H_{13}N]^+$.

The cleavage of the bond 1–7 gave an ion at *m/z* 95 and loss of one molecule of ethylene. This ion could lose one hydrogen atom and give the methyl-pyridine ion at *m/z* 94.

An intensive peak for the molecular ion adduct at *m/z* 278 can be seen in the CI mass spectrum of alkaloid **5**. The IR spectrum of **5** shows IR maxima (film) at 3307 (OH), 1728, 1713 (C = O), 1666 (C = C), 760 and 802 (aromatic) cm^{-1}. The UV spectrum showed maxima in EtOH at 221 (3.15), 265 (3.01), 295 (2.77) nm. The ^{1}H-NMR spectrum showed signals at δ 2.35 (*s*, 3H, NMe), 3.49 (*m*, 2H, H-1 and H-5), 5.37 (*t*, 1H, H-3), 6.63 (*s*, 1H), 6.76 (*s*, 1H) and 6.99 (*s*, 1H). The signal at δ 7.97 (*m*, 2H, 2 × OH) was exchangeable with D_2O. On the basis of these spectral data, compound **5** was identified as the new alkaloid demethylphyllalbine.

The electron impact mass spectrum (EIMS) of compound **9** shows the base peak at *m/z* 112 together with prominent peaks at *m/z* 155, 97, 84, 43, 42 (Fig. 5C). The same base peaks have been registered in EI mass spectra of two pseudotropane alkaloids, named product C and knightalbinol, isolated from endemic plants *Knightia deplanchei* and *K. strobilina* (Protaceae) of New Caledonia (Fan-Kan and Lounasmaa 1973; Lounasmaa et al. 1980). The base peak of knightalbinol is due to the combination of an acetyl group at C-7 and hydroxyl group at C-6 in the tropane moiety of the alkaloid (Lounasmaa et al. 1980). The product C contains a benzylic substituent at C-2 and C-6 (Fan-Kan and Lounasmaa 1973). The molecular ion peak of **9** was supposed to be at *m/z* 256 according its EIMS (Gorinova et al. 1994). It was found later that the molecular ion peak of **9** has to be at *m/z* 257, because in its CI mass spectrum in NH_3 the adduct was registered at *m/z* 258. The molecular ion peak at *m/z* 257 corresponds to the formula $C_{13}H_{23}NO_4$.

It is known that the major fragment ion of the tropane ring system after EIMS can be rationalized by assuming primary cleavage of one of the carbon-carbon bonds adjacent to nitrogen with further bond fission (with or without hydrogen rearrangement) then being governed by the presence of other substituents. These additional functional groups influence the relative abundance of different fragments (Blossey et al. 1964). On the basis of EIMS and CIMS data, we propose the structure shown in Fig. 6 for alkaloid **9**. The possible EIMS fragmentation of **9** is shown in Fig. 7. The positions of acetyl and hydroxyl groups at C-6 and C-7 may be interchangeable. The fragment at *m/z* 77, typical for the benzyl substituent, was not registered in the EIMS of **9** (Fig. 5C). All fragments shown in Fig. 7 can be found in the EIMS spectrum of **9** together with the fragmentation at *m/z* 239 due to the loss of water from the molecular ion peak. The EIMS fragmentation of **9** is in good accordance with the proposed structure shown in Fig. 6. The new tropane alkaloid was named physochlainoorientaline. In conclusion, the structure of physochlainoorientaline (**9**) was tentatively elucidated by its EI and CI mass spectral data. Final confirmation of the proposed structure can be obtained if the additional quantity from the alkaloid is isolated in order to measure its ^{1}H and ^{13}C NMR, IR and UV spectra. Such a quantity can be obtained either from cell culture of *P. orientalis* or from the intact plant.

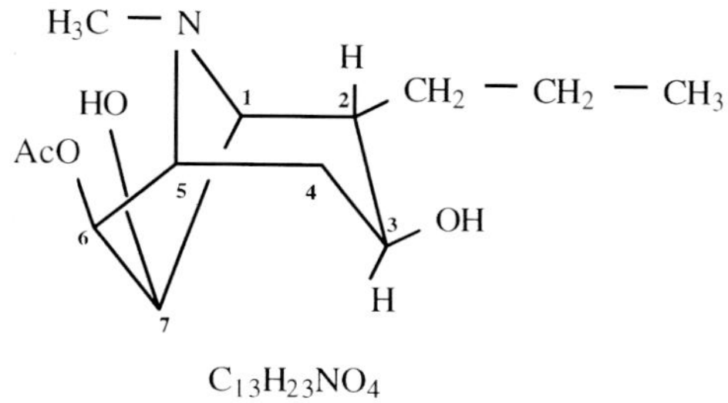

Fig. 6. Structure of physochlainoorientaline

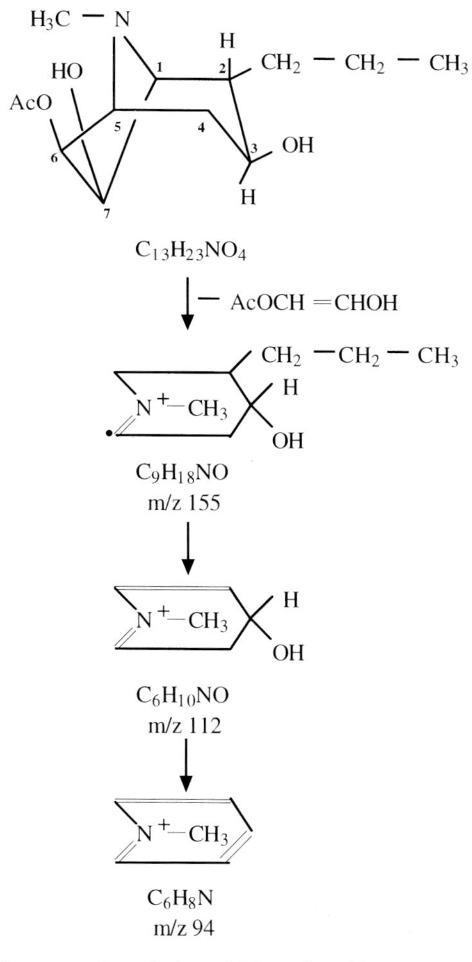

Fig. 7. EIMS (electron impact mass spectral) fragmentation of physochlainoorientaline

4 Summary and Conclusion

Solanaceae is one of the richest families with regard to the number of species containing tropane alkaloids. The genus *Physochlaina* comprises six species: *Physochlaina alaica, P. orientalis, P. physaloides, P. semenowii, P. praealta* and *P. officinalis*. They are perennial herbs.

The Russian *P. alaica* contains 12 tropane alkaloids, the most important of which is hyoscyamine (60–74.3%). The new base physochlaine was isolated from the aerial part of the plant. Two dimeric tropane alkaloids, α- and β-belladonine, were found in the roots of the plant.

Physochlaina orientalis, found in the Caucasus and Middle Asia, contains tropane alkaloids in the aerial part (0.16–0.22%) and in the roots (0.32–1.03%). Its tropane alkaloids represent about 90% of the total alkaloid mixture. The tropane alkaloids in this species are: scopolamine, hyoscyamine, tropine and apoatropine.

Tropane alkaloids, found in *Physochlaina* species, include some medicinally important secondary metabolites. *Physochlaina* herbs have been used and are still in use in the traditional medicine of Tibet and Mongolia because of their powerful anti-inflammatory action against skin and venereal diseases and nervous disorders.

Tissue cultures of *Physochlaina* species producing tropane alkaloids are not known. Callus and cell suspension cultures of *P. orientalis* were obtained in order to examine the production of tropane alkaloids. Capillary GLC combined with mass spectrometry was successfully applied for the separation of the isolated alkaloids and elucidation of their structure. The new tropane bases demethylphyllalbine, 3α-(3-methoxymethacryloyloxy)-tropane and physochlainoorientaline together with the known tropane alkaloids tropinone, hygrine, tropine, aposcopolamine, 3α-tigloyloxy-6-acetoxytropane and 3α-tigloyloxy-6-isobutyryloxytropane were detected.

The cell suspension culture of *Physochlaina orientalis* contains six known and three new tropane alkaloids. One of the main tropane alkaloids is tropine which is the source of biosynthesis for other tropane alkaloids. This shows that the cell culture of *P. orientalis* may possess the enzyme responsible for the synthesis of tropine. Therefore, this cell suspension culture is a good system for further investigation of the biosynthesis of tropane alkaloids.

References

Academy of Science USSR (1955) Genus *Physochlaina* G. Don. In: Flora USSR, vol 22. Academy of Science, Moscow, pp 103–105

Aripova S, Yunusov S (1991) Dimernie tropanovie alkaloidi. Khim Prir Soed 4:447–454

Bick I, Bremner J, Gillard J (1971) Methyl (p-hydroxybenzoyl) acetate and an alkaloid, bellendine from *Bellendena montana*. Phytochemistry 10:475–477

Bick I, Bremner J, Gillard J (1973) The synthesis of bellendine. Tetrahedron Lett 51:5099–5100

Blossey E, Budzikiewicz H, Ohashi M, Fodor G, Djerassi C (1964) Mass spectrometry in structural and stereochemical problems. Tetrahedron 20:585–595

Elze H, Teuscher E (1972) Ornitine metabolism by *Datura* tissue cultures. In: Mothes K (ed) Biochem Physiol Alkaloide, Int Symp 1969. Academie-Verlag, Berlin, pp 239–244

Fodor G (1960) Some new physiological aspects of natural tropane bases and of their synthetic derivatives. In: Manske R (ed) The alkaloids, vol 6. The tropane alkaloids. Academic Press, New York, pp 171–172

Glasby J (1975a) Encyclopedia of the alkaloids, vol 1. Plenum Press, New York, p 730

Glasby J (1975b) Encyclopedia of the alkaloids, vol 1. Plenum Press, New York, pp 164–165

Glasby J (1975c) Encyclopedia of the alkaloids, vol 1. Plenum Press, New York, pp 729–730

Gorinova N, Velcheva M, Dyulgerov A, Atanassov A (1994) Tropane alkaloids in cell cultures of *Physochlaina orientalis*. Fitoterapia 5:452–456

Herbert R (1996) The biosynthesis of plant alkaloids and nitrogenous microbal metabolites. Nat Prod Rep 1:45–46

Hesse M, Bernard H (1975) Pyrrol- und Pyridin-Derivate. Tropane und verwandte Basen. In: Budzikiewicz H (ed) Progress in mass spectrometry, vol 3. Alkaloide außer Indole-, Triterpen- und Steroidalkaloide. Verlag Chemie, Weinheim, pp 40–47

Kan-Fan C, Lounasmaa M (1973) Sur les alkaloids de *Knightia deplanchei* Vieille. e X Brongn. et Gris (Proteacees). Acta Chem Scand 27:1039–1052

Khaidav Z, Menschikova T (eds) (1978) Lekarstvenie rastenia v Mongolskoi medizine. Academia Nauk MNR, Ulan Bator

Kuvaev V, Blinova K (1960) Predvaritelnaya khimicheskaya otzenka lekarstvenich rastenii Tibetskoi medicini, proizrastaustih v Zabaikalie. Voprosi Pharmacognosii 12:213–262

Lounasmaa M, Pusset J, Sevenet T (1980) Benzyl-2 pseudotropanes et dihydro γ-pyranotropanes, alcaloides de *Knightia Strobilina*. Phytochemistry 19:953–955

Manske R (1977) Plants and their contained alkaloids. In: Manske R (ed) The alkaloids, vol 16. Alkaloids unclassified and of unknown structure. Academic Press, New York, p 541

Minina S, Aduevskaya G (1965) Alkaloidnoe soderjanie v *Physochlaina orientalis*. Aptechn Delo 2:66–68

Mirzamatov R, Luftullin K, Mallikov V, Yunusov S (1973a) Skopolamin i belladonin iz *Physochlaina alaica*. Khim Prir Soedin 5:680

Mirzamatov R, Mallikov V, Luftullin K, Hakimov O, Yunusov S (1973b) Dinamika nakuplenia alkaloidov v rastenii *Physochlaina alaica*. Khim Prir Soedin 4:566

Mirzamatov R, Luftullin K, Mallikov V, Yunusov S (1974a) Physochlaine – novi alkaloid iz *Physochlaina alaica*. Khim Prir Soedin 3:415–416

Mirzamatov R, Luftullin K, Mallikov V, Yunusov S (1974b) Videlenie apohyoscina i 6-oksiatropina iz *Physochlaina alaica*. Khim Prir Soedin 3:416–417

Murashige T, Skoog F (1962) A revised medium for rapid growth and bioassays with tobacco tissue cultures. Physiol Plant 15:473–497

Parrello J, Longevialle P, Vetter W, McCloskey J (1963) Structure de la phyllalbine. Application de la resonance magnetique nucleaire et de la spectrometrie de masse a l'etude des derives du tropane. Bull Soc Chim Fr 461:2787–2793

Peshkova V (1964) Biological characteristics and dynamics of the accumulation of alkaloids of *Physochlaina physaloides*. Farm Zh 4:23–28

Seigler D (1977) The Solanaceae. In: Manske R. (ed) The alkaloids, vol 16. Plant systematics and alkaloids. Academic Press, New York, p 61

Witte L, Muller K, Arfmann HA (1987) Investigation of the alkaloid pattern of *Datura innoxia* plants by capillary gas-liquid chromatography – mass spectrometry. Planta Med 49:192–197

Yunusov S (1981) Family Solanaceae. In: Fan UzUSSR (ed) The alkaloids. Fan UzUSSR, Tashkent, pp 328–331

XXII *Solanum malacoxylon* Sendt. (*Solanum glaucophyllum* Desf.): In Vitro Culture and the Production of Sterols and Vitamin D_3 Metabolites

L.A. MANZOCCHI[1] and M.L. RACCHI[2]

1 Calcinogenic Plants, Vitamin D and Vitamin D-Like Substances

The interest in plants exhibiting vitamin D-like activity (calcinogenic plants) is primarily due to their possible use as a low-cost source of $1\alpha,25(OH)_2D_3$ (calcitriol) for human and veterinary use. Among calcinogenic plants, *Solanum malacoxylon* Sendt. (also identified as *Solanum glaucophyllum* Desf.; Fig. 1), known in South America as "Duraznillo blanco", has been the most extensively studied, in relation both to its biological activity in animals and to the identification of the chemical compounds with vitamin D activity (Wasserman et al. 1976; Esparza et al. 1982).

The presence of vitamin D or D-like substances in plants was first suspected because of the occurrence of "calcinosis" in cattle and other grazing livestock, described by Grant (1951). Affected animals are usually hypercalcemic and hyperphosphatemic with extensive soft tissue calcification, resulting in stiffness of limbs, emaciation, and possibly death. The disease therefore appears to be related to disruption of the regulation of Ca^{2+} homeostasis. Anti-rachitic properties have been observed in mature plant dry tissues, suggesting the occurrence of vitamin D precursors in fresh material, but attempts to demonstrate the presence of such compounds were unsuccessful. Later, Raoul et al. (1953) indicated the presence of a substance with anti-rachitic activity in an extract of an unidentified species of grass and finally, following extensive analysis, the presence of vitamin D_3 was demonstrated in extracts from the grass *Dactylis glomerata* (Raoul et al. 1970).

Calcinogenic plants are widespread all over the world and have been recorded in North and South America and in Europe. The description of a disease called "enzootic calcinosis" in the Alpine region of Germany and Austria resulted in the identification of a new calcinogenic plant, the grass *Trisetum flavescens*. The similarity of the symptoms to those of other calcinogenic plants hinted at the search for $1\alpha,25(OH)_2$ vitamin D_3 in this plant (Rambeck et al. 1979).

[1] Istituto Biosintesi Vegetali, C.N.R., via Bassini 15, 20133 Milan, Italy
[2] Genetics Unit, Institute of Silviculture, University of Florence, via San Bonaventura 13, 50145 Florence, Italy

Biotechnology in Agriculture and Forestry, Vol. 43
Medicinal and Aromatic Plants XI (ed. by Y.P.S. Bajaj)

Fig. 1. *Solanum malacoxylon* Sendt. (*Solanum glaucophyllum* Desf.). (Photograph by M.L. Racchi)

In *T. flavescens*, anti-rachitic activity has been shown to depend on the amount of UV radiation to which the plant has been exposed (Zucker et al. 1980). This result was confirmed by Rambeck et al. (1981) who showed that a very low intensity radiation is effective in the photo conversion, but they failed to detect any pro-vitamin D_3. Since neither $1\alpha,25(OH)_2D_3$ nor any other known metabolite of vitamin D_3 (cholecalciferol) was present, cholecalciferol itself must account for at least the major part of the vitamin D-like activity in *Trisetum flavescens*. Another group of workers, studying the same plant, found water soluble vitamin D-like substances, one of which had a biological activity similar to that of $1\alpha,25(OH)_2$ vitamin D_3. Probably the state of the plant material and the methods of isolation accounted for the differences in the compounds found (Morris and Levak 1982).

Solanum malcoxylon is the most studied calcinogenic plant which grows in Argentina, and is known to be the cause of a disease of cattle (locally called "enteque seco") with the same characteristics as the calcinosis described above. The plant contains the glycosides of $1\alpha,25(OH)_2$ vitamin D_3 (Wasserman et al. 1976), of 25(OH) vitamin D_3, and of vitamin D_3 (Esparza et al. 1982; Vidal et al. 1985; Skliar et al. 1992). The calcinogenic activity, reported at first primarily in leaves, was later detected in berries, stems and roots, in both the aglycone and glycoside forms (Weissenberg et al. 1989). The values reported for the concentrations of vitamin D-like substances depends on the assay used: for leaf tissues, it varies between 0.15 and 120 μg vitamin D_3 per g dry weight (Boland 1986).

A disease similar to *S. malacoxylon* poisoning, this time associated with the intake of *Cestrum diurnum* (a wild species of the Solanaceae family), was described in horses and cattle in Florida. The occurrence of vitamin D_3 and its metabolites was investigated in leaves of this plant. The results showed that the apical leaves contained concentrations of free vitamin D_3 much higher than those of their corresponding glycosides. This result suggested that the apical region may be the site of synthesis, while glycosilation yields both transport and storage forms. Alternatively, these metabolites could have a role in the differentiation of the apical region of the plant (Huges et al. 1977; Prema and Raghuramolu 1994).

The use of more sensitive analytical techniques has allowed the detection of both vitamin D_2 and D_3 in different tissues of *Medicago sativa*. Stem, flowers, seeds and roots of this plant showed low concentration levels of these substances, while high concentrations were observed in young leaves (Horst et al. 1984).

Most of the plants so far investigated belong to either the Solanaceae or to the Gramineae. The tomato, *Lycopersicum esculentum*, also belonging to the Solanaceae, was therefore tested for the presence of vitamin D-like activity. The results demonstrated that the tomato is also a calcinogenic plant, and that the activity is confined to the leaves, but absent in the fruit. The plant contains vitamin D_3, 25(OH) vitamin D_3 and 1α,25$(OH)_2$ vitamin D_3 and their glycosidic forms. Free vitamin D_3 was observed to be the main active principle and free forms of the metabolites were present in concentrations higher than those of the corresponding glycosides (Prema and Raghuramulu 1996). Table 1 summarizes the quantitative data on the distribution of vitamin D-like substances in plant tissues of calcinogenic species.

Scarce data are available for physiological roles of vitamin D-related compounds in plants. The observation that vitamin D_3, or secosterols in general, promote root differentiation has suggested a role as plant growth regulators for these compounds (Buchala and Pythoud 1988). Results obtained by different authors pointed out effects of vitamin D-like substances on adventitious rooting. Buchala and Schmid (1979) reported an enhancement in the number of adventitious roots on green cuttings of *Populus tremula*; this effect was exerted both by vitamin D_3 and by vitamin D_2 (a reduced derivative of dihydrotachysterol), which acted synergically with indol butyric acid (IBA) in promoting rooting (Pythoud et al. 1986). Similar results were obtained in mung bean *Vigna radiata* (Jarvis and Booth 1981), in which vitamin D_2 was demonstrated to stimulate rooting of cuttings. The mechanism of stimulation is unknown, though a possible effect on RNA metabolism has been postulated (Jarvis et al. 1985). Vitamin D_2 alone and in combination with α-naphtyl acetic acid improves root formation on *Cynara scolimus* shoots (Moncousin and Gaspar 1983); a possible interaction of vitamin D_2 with the IAA metabolism is discussed by the authors, although no variation in peroxidase activity has been observed.

Table 1. Summary of literature on the distribution of Vitamin D-like substances in different tissues of calcinogenic plants

Plant	Tissue	Compound	Amount	References
Cestrum diurnum	Leaf	Vitamin D_3	102 $\mu g\,kg^{-1}$ dry matter	Prema and Raghuramulu (1994)
		25-OH D_3	102 " "	
		$1\alpha,25(OH)_2D_3$ (free and glycosides)	1000 " "	
Lycopersicum esculentum	Leaf	Vitamin D_3	778 $\mu g\,kg^{-1}$	
		25-OH D_3	22 " "	Prema and Raghuramulu (1996)
		$1\alpha,25(OH)_2D_3$ (free and glycosides)	100 " "	
Trisetum flavescens	Leaf	cholecalciferol	0.01 ppm/dry matter	Rambeck et al. (1979)
Medicago sativa	Leaf	Vitamin D_2	48.00 $ng\,g^{-1}$	Horst et al. (1984)
		Vitamin D_3	0.63 " "	
Solanum malacoxylon	Leaf	$1\alpha,25(OH)_2D_3$	2.7 $\mu g\,mg^{-1}$ protein	
	Stem	(free and glycosides)	13.0 " "	Weissenberg et al. (1989)
	Berry		4.4 " "	
	Root		4.2 " "	

2 Calcitriol

2.1 Structure and Physiological Role

Toxicity in calcinogenic plants has been identified as due to 1α,25$(OH)_2$ vitamin D_3 (calcitriol, Fig. 2), which is the hormonally active form of vitamin D_3 in higher animals (DeLuca and Schones 1983). The functions and biosynthetic pathways of vitamin D_3 and its hydroxylated metabolites have been largely studied in mammals and birds, while very few data are available for their metabolism and possible physiological role in higher plants (Vega et al. 1985; Buchala and Pythoud 1988).

Vitamin D_3 and its metabolites are structurally related to cholesterol and other steroid hormones. Their biosynthetic pathway is summarized in Fig. 3. In the rumen of grazing animals, cholesterol undergoes a dehydrogenation to 7-dehydrocholesterol, the provitamin D with its characteristic Δ 5,7 conjugated double-bond system in the B ring. The resulting diene is transported to the skin where, in a photochemical reaction (Mac Laughlin et al. 1982), the B ring is cleaved, yielding provitamin D_3, which then promptly converts to vitamin D_3. The introduction of two hydroxyl groups is required for biological activity. The hydroxylation in position 25 is effected in the liver by a microsomal vitamin D_3-25 hydroxylase; the subsequent introduction of a hydroxyl group in position 1α of 25-hydroxy vitamin D_3 is catalyzed, in the kidney, by a mitochondrial enzyme; both enzymes are highly specific cytochrome P-450 monooxygenases (Henry 1992).

The biologically active form 1α, 25-dihydroxy vitamin D_3 (1α,25$(OH)_2D_3$) can get through the cell membranes and binds to the vitamin D_3 receptors (VDREs) within the target cells. In animals, it is involved in a number of essential biological processes: intestinal absorption of calcium and phosphorus and their mobilisation from bones; participation in the regulation of the immune system; involvement in the biosynthesis of polyamine, lymphokine and Ca-binding proteins; transcription of oncogenes; regulation of cell differentiation and proliferation (DeLuca and Schones 1983; Minghetti and Norman

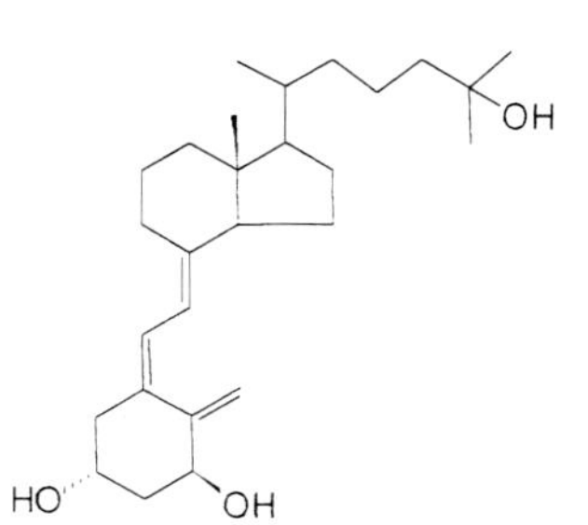

1α,25 $(OH)_2D_3$

Fig. 2. The structure of calcitriol

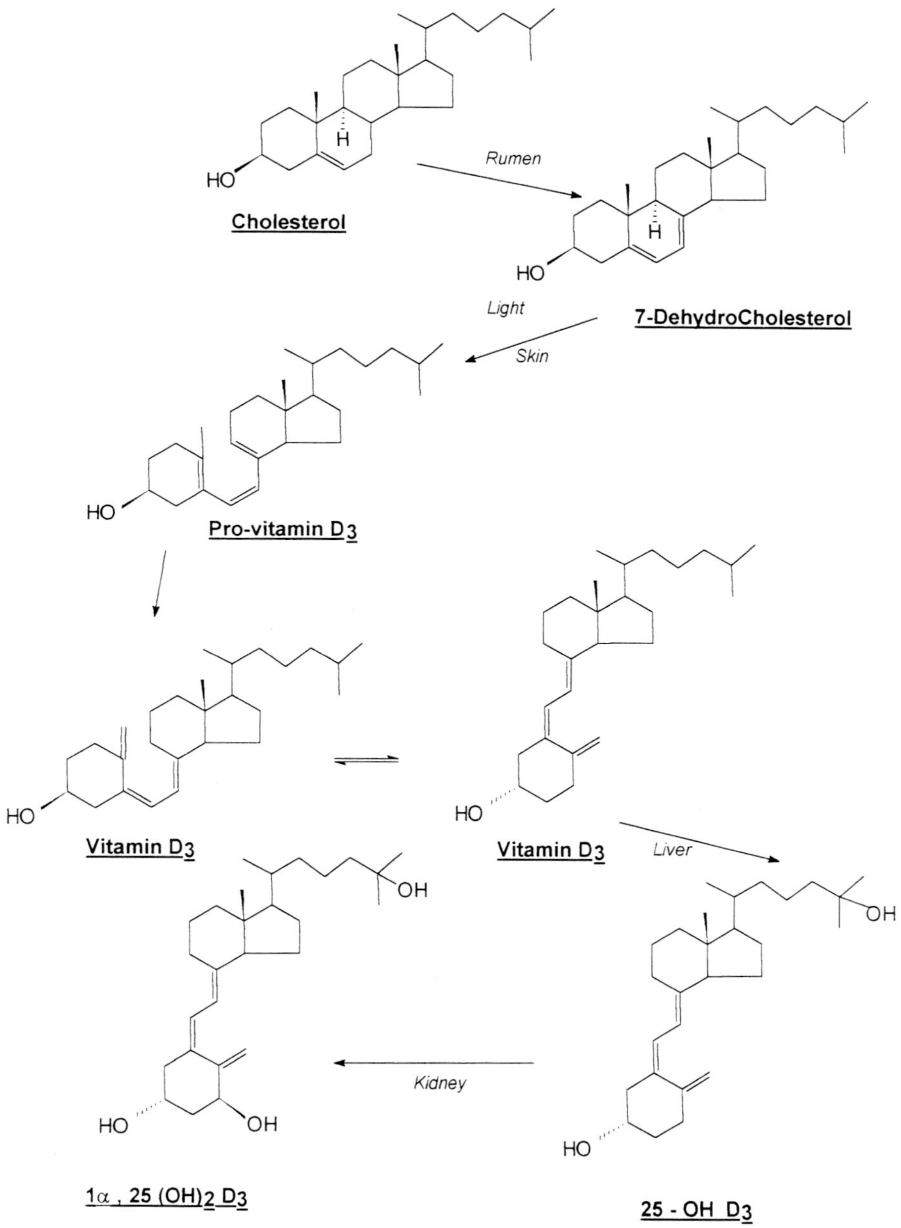

Fig. 3. Biosynthesis of vitamin D_3, $25(OH)D_3$ and $1\alpha,25(OH)_2D_3$ in mammals. (Adapted from Henry 1992)

1988). The molecular function of calcitriol consists of a positive transcriptional activation, through binding to nuclear receptors (Ferrara et al. 1994; Whitfield et al. 1995).

2.2 Medicinal Importance and Demand on the World Market

Calcitriol is employed clinically for chronic renal failure, hypoparathyroidism, osteoporosis and osteodistrophy. Moreover, calcitriol plays an important role in the therapy of some cancers, immunodeficiency and skin diseases, or in cases in which the metabolic pathway of D_3 is blocked (review in Sasaki et al. 1992). Although the pro-die dose is not high (0.25 μg, Seino et al. 1987), the market requirement of calcitriol for therapeutic use is high.

The commercial production of calcitriol is constrained by a very high cost, due to a number of limiting factors. Extraction from kidneys is not possible, as the compound, though produced there, does not accumulate. The chemical synthesis of calcitriol is a complicated procedure requiring many steps (Vanmaele et al. 1985; Baggiolini et al. 1986; Kametani and Furuyama 1987). The drug can be produced by semi-synthesis, i.e. chemical modifications of naturally occurring compounds; this approach is complex, and requires several steps with very low yields (Andrews et al. 1986; Dai and Posner 1994). The most critical step is the introduction of the hydroxyl group in the 1α position of $25(OH)D_3$, which is regio- and stereo-specific and is accomplished in kidney cells by a specific mitochondrial cytochrome P-450 hydroxylase.

Therefore, obtaining calcitriol from biological sources would offer a valuable alternative to chemical synthesis. Microorganisms that can hydroxylate vitamin D_3 have been recently found (Sasaki et al. 1992) and their exploitation is being studied. An alternative supply of calcitriol may come from calcinogenic plants; the amounts that can be extracted from plant tissues are, however, very low. Plant cell culture technology offers a feasible option for the production of secondary metabolites in cases in which supply from natural sources, or chemical synthesis does not meet the commercial demand.

2.3 Occurrence in *S. malacoxylon*

In higher plants, sterols belong to the large family of isoprenoid compounds, whose biosynthetic pattern and multiple physiological roles are shown in Fig. 4 (Chappell 1995). In plants as in animals, the key and rate-limiting step in the biosynthetic pathway is the first passage, catalyzed by 3-hydroxy-3-methylglutaryl coenzyme A reductase (HMGR); subsequent steps lead to the synthesis of the specific classes of isoprenoid compounds, through pathways whose control is complex and largely unknown (Chappell et al. 1995; Schaller et al. 1995). The ability of many wild *Solanum* species to produce steroidal alkaloids (such as solasodine, solasonine and solamargine) through the general pathway of steroid biosynthesis is well known (Balandrin et al. 1985; Cham 1994; Valkonen et al. 1996); in contrast, the ability of plant tissues

```
  acetyl- CoA
     ↓↓↓                          CLASSES          EXAMPLES
   HMG-CoA
     ↓ a
  mevalonate
     ↓↓↓
  isopentenyl
  diphosphate          →       prenyl group       cytokinin
     IPP
     ↓ b
   geranyl             c
  diphosphate          →       monoterpenes        menthol
     GPP
     ↓ b
                               sesquiterpenes      capsidiol
   farnesyl          d ↗
  diphosphate        e →          sterols         stigmasterol
     FPP               ↘
                       f        prenylated        ras-like GTP
                                 proteins         binding prot.
     ↓ b               ↗
 geranylgeranyl      g →        diterpenes      gibberellic ac.
  diphosphate        h ↘
     GGPP                       polyprenols       chlorophyll
```

Fig. 4. Biosynthetic pathway of isoprenoids in plants, with respect to the types of end products. Adapted from Chappell (1995). *Multiple arrows* indicate multiple steps or reactions. **a** HMGR (3-hydroxy-3-methylglutaryl coenzyme A reductase); **b** phenyltransferases; **c** monoterpene synthases; **d** sesquiterpene synthases; **e** squalene synthase; **f** prenyl:protein transferases; **g** diterpene synthases; **h** phytoene synthase

to biosynthesize steroid molecules of the vitamin D-like group of compounds has been identified and studied in a limited group of species (calcinogenic plants) only.

The presence of substances with $1\alpha,25(OH)_2D_3$-like activity in extracts from plant tissues is tested by biological methods, based on in vivo stimulation of Ca^{2+} mobilization or induction of Ca^{2+} binding protein in chicks fed on a high-strontium, low-calcium diet. This diet inhibits kidney 1α-hydroxylase responsible for the formation of $1\alpha,25(OH)_2D_3$. Administration of a source of $1\alpha,25(OH)_2D_3$ to strontium-inhibited chicks restores the capacity of the intestine to synthesize CaBP, and this thereby serves as an indicator of the presence of a $1\alpha,25(OH)_2D_3$-like derivative in the test sample (Wasserman et al. 1976; Weissenberg et al. 1989).

The *S. malacoxylon* calcinogenic factor was first chemically identified in biologically active extracts from leaves. The compound was water-soluble and became lipophylic after enzymatic cleavage with a mixed glycosidase preparation: this strongly suggested that the naturally occurring form of the factor was a glycoside. The sterol moiety was purified by column chromatographies on Sephadex LH-20, silicic acid and celite, and identified by combined gas chromatography and mass spectrometry as $1\alpha,25(OH)_2D_3$ (Wasserman et al. 1976). Usually, vitamin D_3-like substances are extracted from plant tissues in two steps. A first extraction with a non polar solvent (methylene chloride, chloroform) and a subsequent extraction of the remaining material with a polar solvent (water, aqueous ethanol, n-butanol) allows a selective separation of aglycones (in the former extract) from hydrophilic glycosides (in the latter); following hydrolysis of this fraction with glycosidases, the aglycone moieties are extracted with an organic solvent (Esparza et al. 1982; Weissenberg et al. 1989). Steroids from both organic extractions are separated by column chromatography (Sephadex LH-20), TLC and HPLC. Chemical identification is made by comparing Rfs on TLC and Rts on HPLC of standard products, and by UV, mass and ^{1}H NMR spectra (Wasserman et al. 1976; Esparza et al. 1982; Aburjai et al. 1996a). In *S. malacoxylon* plant tissues, the prevalence of glycosydic forms of vitamin D_3, 25(OH)D_3 and of 1α,25$(OH)_2D_3$, and the presence of minor amounts of the free aglycones is usually reported (Wasserman et al. 1976; Esparza et al. 1982; Vidal et al. 1985; Skliar et al. 1992). In other calcinogenic plants, such as *Cestrum diurnum* and tomato, on the contrary, high concentrations of the algycone, prevailing over the glycoside forms, have been observed (Prema and Raghuromulu 1994, 1996). Calcitriol activity is present in *S. malacoxylon*, in leaves, berries, stems and roots (Weissenberg et al. 1989). The concentrations reported for vitamin D_3-like compounds are in a wide range (between 0.15 and 120 μg vitamin D_3 per g dry weight of leaves), depending on extraction and assay methods (Boland 1986).

3 In Vitro Approaches

For the wild *Solanum* species known to synthesize steroidal glycoalkaloids (SGAs), abundant literature on tissue culture is available (Bhatt et al. 1983; Macek 1989; Tsoulpha and Doran 1991; Alvarez et al. 1994; Drewer and van Staden 1995; El Badaoui et al. 1996); cell cultures are actually employed for part of the commercial production of anticancer SGAs. The study of *S. malcoxylon* tissue culture and production of vitamin D_3 metabolites has, by contrast, received relatively little attention. The references on the subject are summarized in Table 2. As the amounts of calcitriol that can be extracted from plant tissues are negligible, and the growing demand cannot be met by chemical synthesis, in vitro culture may offer a valuable source of pharmaceutically useful calcitriol.

3.1 Callus Cultures

Callus cultures have been obtained under axenic conditions from seeds (Jain and Sahoo 1986), hypocotyls, leaves and roots detached from germinating seeds, and from leaves from micropropagated plantlets (Weissenberg et al. 1989; Suardi et al. 1994). Suardi et al. (1994) obtained calli from primary

Table 2. Summary of literature on the production of sterols and vitamin D_3 metabolites by in vitro cell cultures of *S. malacoxylon*

Primary explant	Medium and growth regulators	Sterols produced	References
Seeds	Revised MS medium 2,4-D 1 mg/l	Sitosterol, diosgenin, solasodine	Jain and Sahoo (1986)
Hypocotyls	MS medium 2,4-D: 1 mg/l kinetin: 0.3 mg/l	Calcitriol activity	Weissenberg et al. (1989)
Hypocotyls, leaves, roots	B5 medium 2,4-D: 1 mg/l kinetin: 0.3 mg/l	Campesterol, β-sitosterol, stigmasterol, cholesterol, 28-isofucosterol	Suardi et al. (1994)
Leaves	B5 medium 2,4-D: 1 mg/l kinetin: 0.3 mg/l	7-Dehydro-cholesterol (dark), vitamin D_3, 25-OH D_3 and 1α,25(OH)2D_3(light)	Aburjai et al. (1996a) Aburjai et al. (1997)

MS = Murashige and Skoog (1962); B5 = Gamborg et al. (1968), 2,4-D = 2,4-dichlorophenoxyacetic acid.

explants cultured both on MS medium (Murashige and Skoog 1962) or B5 medium (Gamborg et al. 1968) supplemented with 3% sucrose, 1 mg/l 2,4-dichlorophenoxy-acetic acid (2,4-D) and 0.3 mg/l kinetin, in the dark at 25 °C. After a week, white, friable calli appeared on the cut surfaces of the explants and, within 2–3 weeks, the initial explant became covered by a proliferating mass, from which established cultures could be obtained in 6 weeks. The growth rates of calli derived from different primary explants are shown in Fig. 5.

Callus cultures were maintained by routine subcultures every 3 weeks; if calli were shifted to the light in a culture medium deprived of hormones, organogenesis and plantlet regeneration could be induced. Established calli maintained in the dark on media containing 1 mg/l 2,4-D were soft and friable, and could be cultivated for years (Aburjai 1996). Figure 6 shows both organogenic and friable calli, derived from leaf primary explants. Friable calli from leaf explants were the best source for long term cultures (Aburjai 1996), and have been employed by Aburjai et al. (1996b) for vitamin D_3 biosynthesis and biotransformation experiments.

3.2 Cell Suspensions-Cell Immobilization

Liquid cultures were obtained by Aburjai et al. (1996a) by dispersing friable calli in liquid MS medium, containing 2 mg/l 2,4-D and cultivating them on a rotatory shaker. Suspensions were subcultured with fresh medium every 2 weeks; cell viability was tested by the fluoresceine diacetate (FDA) method (Widholm 1972), and cell growth was measured as packed cell volume (PCV). A logarithmic growth phase was reached at day 2–3, and maintained for

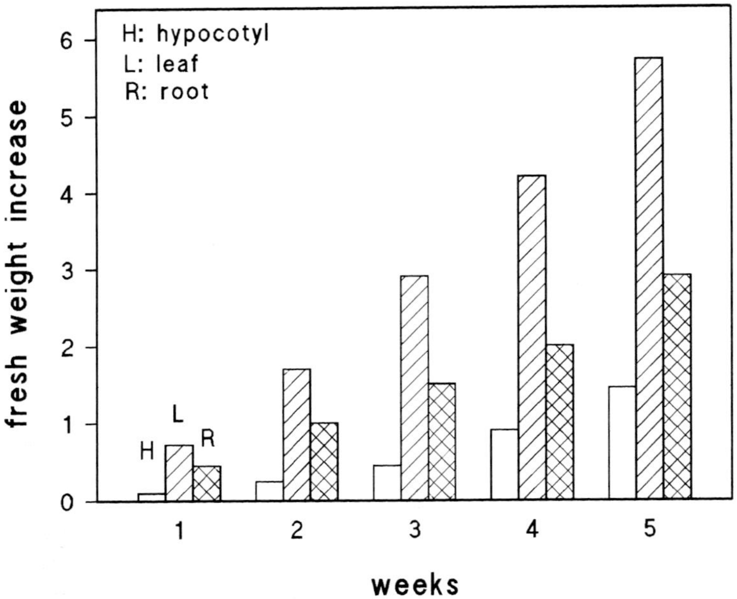

Fig. 5. Growth rates on B5 medium of calli derived from different primary explants. (Adapted from Suardi et al. 1994)

several days. A sixfold increase in PCV was observed in 2 weeks, starting with an inoculum of 3 g cells in 50 ml culture medium.

Cell immobilization in calcium alginate beads according to the method of Brodelius et al. (1979) has been recently described by Aburjai et al. (1997). Cell growth of immobilized cells (measured by packed cell volume after dissolving the alginate gel beads in sodium citrate) was strongly reduced compared with the growth of free suspended cells: typical growth curves are shown in Fig. 7. Cell viability as tested by fluorescein diacetate staining, however, was maintained at good levels (over 70%) for up to 30 days in culture.

3.3 Production of Vitamin D_3 Metabolites by Cell Cultures

In their first paper on *S. glaucophyllum* tissue culture, Jain and Sahoo (1986), reported the detection of free steroids and steroidal glycoalkaloids in methylene chloride and ethanol extracts of long-term callus cultures induced from seeds. Their data show that in in vitro cultures the synthesis of diosgenin is enhanced, while sitosterol and solasodine are produced at reduced levels compared with the amounts produced in the seeds.

Cell cultures induced from hypocotyls of *S. glaucophyllum* (Weissenberg et al. 1989) showed calcitriol activity, measured as stimulation of the synthesis of duodenal Ca-binding protein in strontium-inhibited chickens. The activity was present at variable levels in different cell cultures; the cell line with the

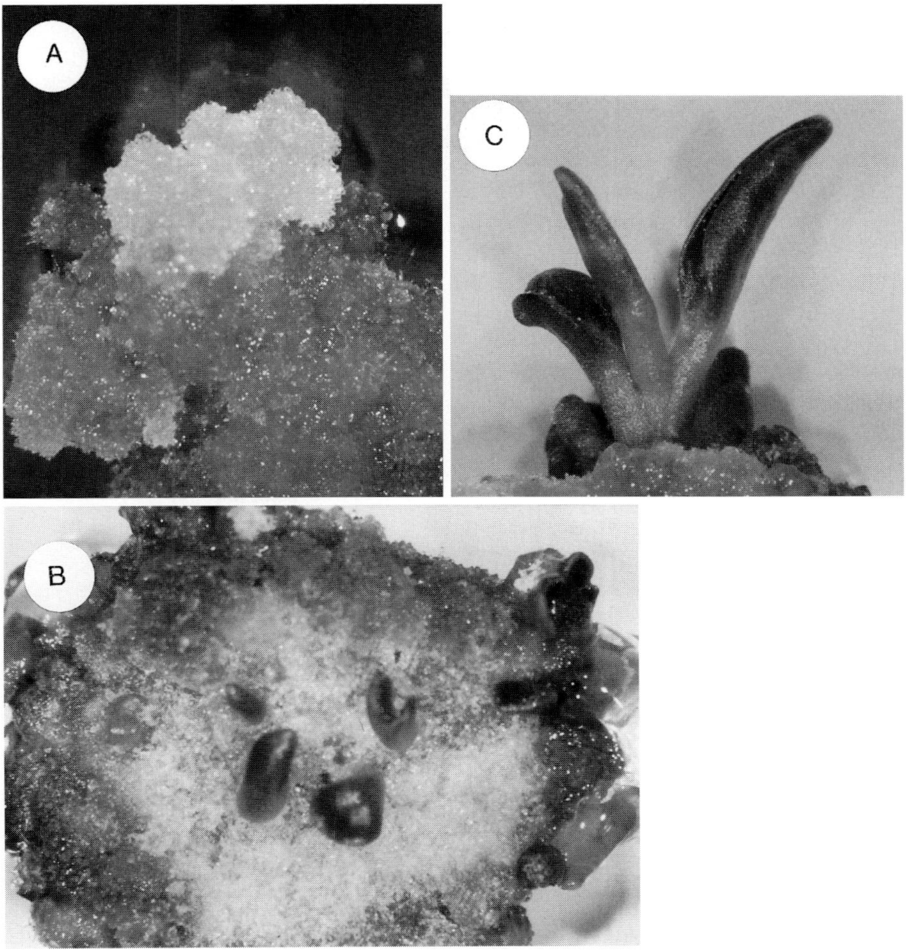

Fig. 6. Calli induced from leaf explants. **A** established friable callus tissue, subcultured 10 months in the dark on MS medium (×60). **B,C** Organogenic calli, subcultured 6 weeks in the light on culture medium without hormones (×60). (Photographs by L.A. Manzocchi, and T. Aburjai)

highest production showed a higher calcitriol activity than that of leaf, stem and root tissues, and comparable to that found in berries. The activities reported were obtained in cultures incubated in the dark; however, the authors suggest that exposure to light or to UV irradiation may enhance the production of calcitriol activity. The ability of *S. malacoxylon* tissue cultures to produce steroidal metabolites has been further investigated in systematic studies of induction from hypocotyl, leaf and root explants (Bernasconi et al. 1993; Suardi et al. 1994). Callus cultures grown in the dark have been shown to produce campesterol, β-sitosterol, stigmasterol, cholesterol, 28-isofucosterol and 24-methylencycloartenol. The concentration levels of sterol compounds were different in cultures derived from different explants, as shown in Fig. 8.

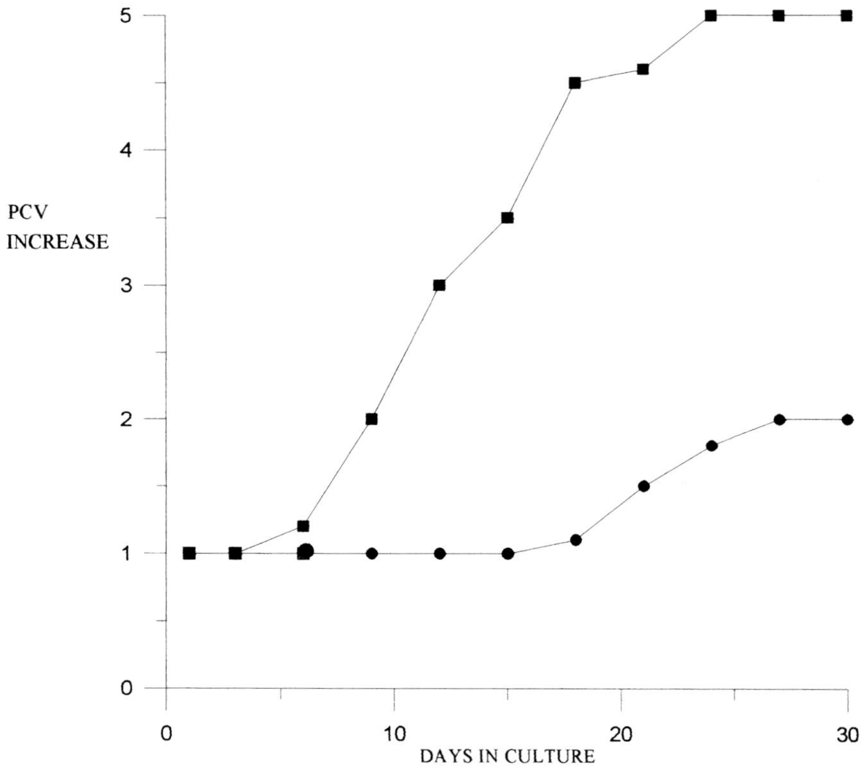

Fig. 7. Growth curves of *S. malacoxylon* cell cultures. ■, liquid suspension cultures; ●, Ca alginate-immobilized cells. (Adapted from Aburjai et al. 1996a)

No metabolite of vitamin D_3 was detected; however, significant levels of cholesterol were present in cultures derived from leaves (Suardi et al. 1994).

The level of sterol production can be changed by altering culture parameters. Establishment of callus cultures from different explants is usually performed on MS based media supplemented with 2,4-D, capable of substaining a good cell growth (Jain and Sahoo 1986; Weissenberg et al. 1989; Suardi et al. 1994). Shifting of the cultures to media with lower concentrations of macro- and micronutrients and vitamins (such as B5 medium) results in a reduction of the growth rate of the callus, but also in a significant increase in the levels of steroidal metabolites, as shown in Fig. 9 (Suardi et al. 1994). The induction of liquid suspension cultures (Aburjai et al. 1996b) from mesophyll cells has allowed further investigations on the vitamin D_3 biosynthetic pathway in *S. malacoxylon*. In addition to the already detected cholesterol, 7-dehydrocholesterol was found to accumulate in dark-grown cells; shifting the cultures to the light caused a rapid decline in the level of 7-dehydrocholesterol and an increase of vitamin D_3 and its hydroxylated metabolites (25-OH-D_3 and $1\alpha,25(OH)_2D_3$); the chemical identities of the

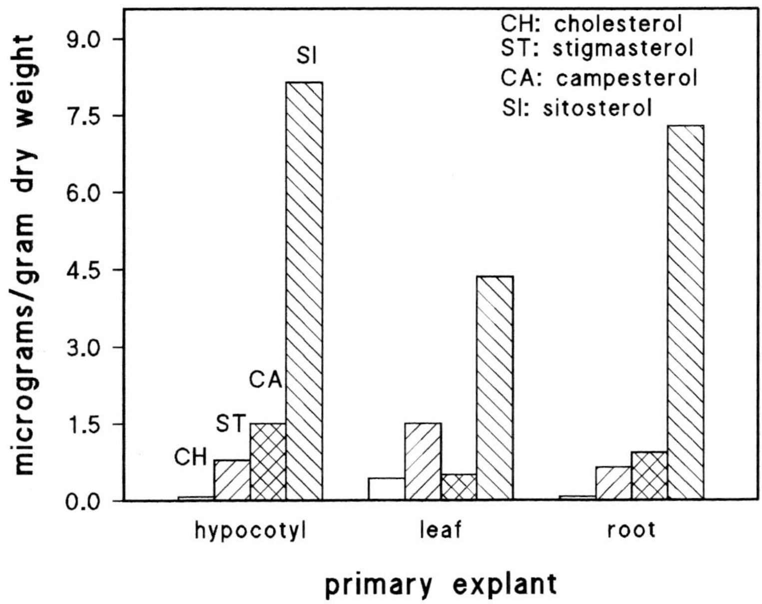

Fig. 8. Sterol content of established calli from different primary explants, grown on B5 medium. (Adapted from Suardi et al. 1994)

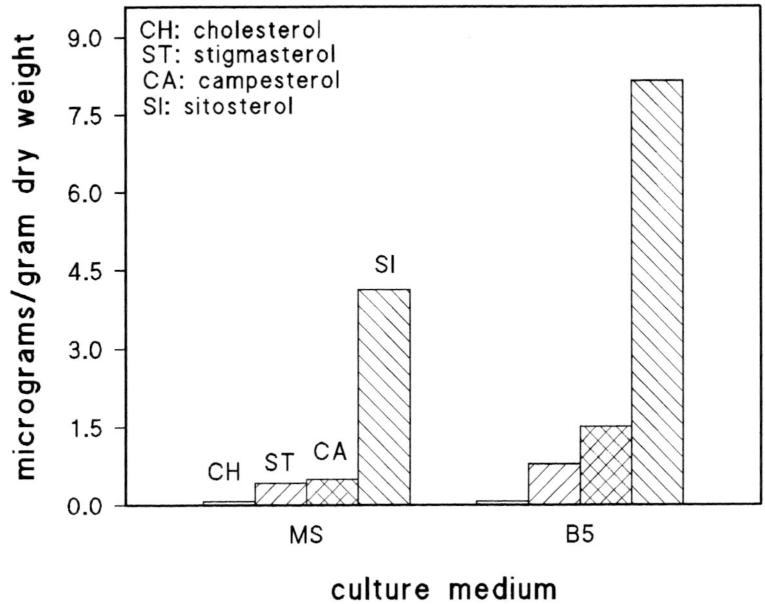

Fig. 9. Effect of culture medium on sterol content of established calli from hypocotyl. (Data from Suardi et al. 1994)

compounds were confirmed by ^{1}H NMR and mass spectrometry. As for other steroids, the levels of metabolites related to vitamin D_3 was higher in B5 than in MS medium (Aburjai et al. 1996a). The data suggest that vitamin D_3 may be occurring in *S. malacoxylon* cells by means of the photolytic activation of an intermediate, 7-dehydrocholesterol, following a pathway similar to the one present in the skin of vertebrates.

The effect of calcium on the level of metabolites related to vitamin D_3 in cell cultures of *S. malacoxylon* has been studied by Aburjai et al. (1996a, 1997). Cell suspensions obtained from leaf-derived callus cultures were cultivated in a standard liquid B5 medium containing different concentrations of $CaCl_2$. The modification of the Ca^{++} content in the culture medium had no significant effect on the viability and growth of the cells during this period. The chloroform extracts (containing free sterols) of both cells and culture medium were analysed by TLC and HPLC: vitamin D_3 metabolites reached their maximum level after 48h in culture and then decreased very rapidly. The metabolites were distributed between cells and culture medium in the ratio 60%–40%. Low levels of vitamin D_3 were detected in cell cultures cultivated in 10mM Ca^{++} medium. In cultures grown in 1mM Ca^{2+} media, vitamin D_3, 25-OH D_3 and 1α,25(OH)$_2$D$_3$ (the latter at very low levels: 0.1 μg/g fresh weight) were detected. In calcium-free cultures, vitamin D_3 was present in similar amounts to those found in cells cultured in 1mM Ca^{2+} medium, while the level of 25-OH D_3 was strongly increased: data are shown in Fig. 10. The results obtained by Aburjai et al. (1997) therefore show a stimulating effect of calcium

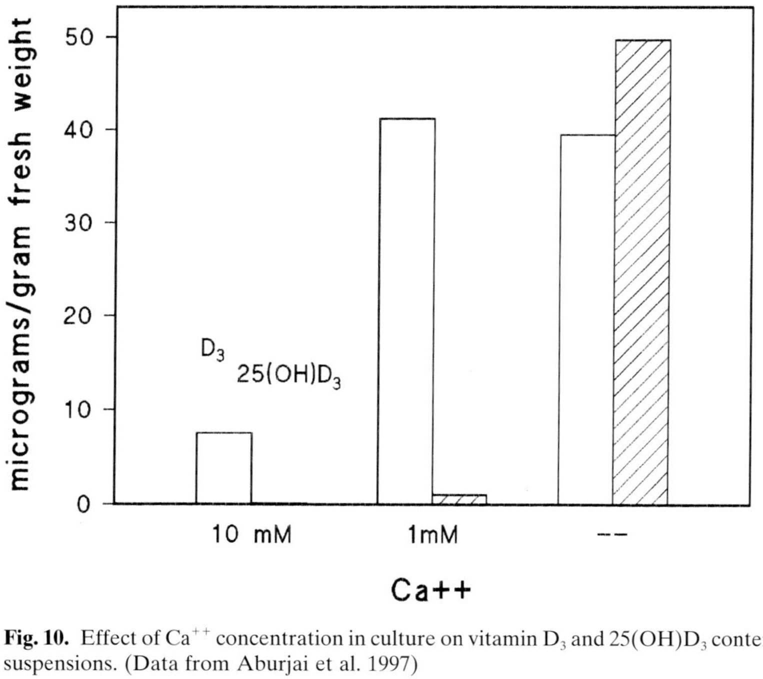

Fig. 10. Effect of Ca^{++} concentration in culture on vitamin D_3 and 25(OH)D_3 content of liquid cell suspensions. (Data from Aburjai et al. 1997)

deprivation on vitamin D_3 hydroxylation. Similarly, a low calcium diet and hypocalcemia in animals results in a marked increase in vitamin D_3 hydroxylated metabolites (DeLuca and Schones 1983).

The residues of chloroform extraction were treated with n-butanol, to investigate the presence of the glycosydic forms of the metabolites. After hydrolysis with mixed glycosidases, the butanol extracts were analyzed by TLC and HPLC. Vitamin D_3 and 25-OH D_3 were detected in much lower amounts, but in the same proportions, as the corresponding chloroform extracts. In *S. malacoxylon* cell cultures, therefore, the concentration of free vitamin D_3 metabolites appear to predominate compared with the corresponding glycosides; in the plant tissues, on the contrary, glycosidic forms predominate. Among several hypotheses that may account for this difference, we can recall the suggestion of Prema and Raghuramulu (1994) that glycosylation in the plant may represent the transport and storage form of vitamin D_3 metabolites.

In cultures of calcium alginate-immobilized cells, vitamin D_3 and 25-OH D_3 accumulate in the same proportions as in liquid suspensions, but for much longer periods (20 days) and in higher amounts (levels up to fivefold; Aburjai et al. 1997). This report of enhanced levels of vitamin D_3 metabolites is in accordance with the already reported stimulation of secondary metabolite production in alginate-immobilized cells (Lindsey and Yeoman 1985). Although the levels of 25-OH D_3 were significantly enhanced by the immobilized cell technique, the recovery of 1α,25$(OH)_2$ D_3 did not increase.

The ability of established cell lines to hydroxylate vitamin D_3, added as an exogenous precursor, in positions 25 and 1α, has been studied in biotransformation experiments. Addition of vitamin D_3 to suspension cultures (Suardi et al. 1994; Aburjai et al. 1995, 1996a) led to the isolation of some unknown products and also of small amounts of calcitriol, both free and in the glycosidic form. The presence in cultured cells of an enzymatic system able to hydroxylate vitamin D_3 in the 25- position was thus demonstrated; 1α hydroxylation is also present, although at much lower levels. In experiments with Ca alginate-immobilized cells (Aburjai et al. 1997), exogenous vitamin D_3 was enclosed within the calcium alginate matrix together with the cells. The results show the efficiency of the *S. malacoxylon* immobilized cell system in long term biotransformation experiments. The maximum levels of 25-OH D_3 obtained 9 days after the beginning of the experiment were over 60 times greater than those detected in the absence of the added precursor. 1α,25$(OH)_2D_3$ was also detected, although at low levels.

4 Conclusions and Prospects

The data reported above suggest that the synthesis of calcitriol (from the dehydrogenation of cholesterol to 7-dehydrocholesterol, to the synthesis of vitamin D_3, to the hydroxylation of the latter in 25- and 1α- positions) can be accomplished by cell cultures derived from leaves of *Solanum malacoxylon*, in a pathway similar to the one operating in the animals. Yields are low, but the alteration of some culture parameters (nutrients, Ca^{++} level, cell immobilization) appears to be effective in increasing metabolite production. A deeper understanding of the role of hydroxylated vitamin D_3 metabolites in plant cells could help to devise better conditions to elicit their biosynthesis.

In contrast to intact plant tissues, where glycosides are the prevalent forms, free sterols are predominantly recovered from cell cultures. We have no data to support explanations for this, due to lack of information on the cellular compartmentalization of synthesis, storage and degradation of the metabolites. A deeper knowledge of these subjects might help to develop methodologies aimed at increasing the recovery of the desired compounds. Moreover, further studies are necessary on the enzymes involved in the biosynthetic pathway of calcitriol. The most critical steps are the hydroxylations in the 25- and 1α- positions of vitamin D_3. In *S. malacoxylon*, two hydroxylase activities have been detected, in microsomes and mitochondria respectively, in leaf extracts (Esparza et al. 1982); no data are available for cultured cells. These hydroxylases belong to the P450 family of enzymes: a gene encoding a 25-D_3 hydroxylase has been cloned from *Amycolata antitrophica*, an actinomycete strain (Kawauchi et al. 1994), and very recently the cloning of 1α-25$(OH)_2$ D_3 hydroxylase from kidney of mice has been reported (Takeyama et al. 1997).

Acknowledgments. A fundamental aspect (the chemical part) of the researches of the Italian group has been performed in the Department of Organic and Industrial Chemistry, University of Milan, under the supervision of Prof. Francesca Pelizzoni. The authors wish to acknowledge her, together with S. Bernasconi and T. Aburjai, for valuable research, collaboration and stimulating discussion. Research during 1996–1997 has been consistently supported by MIRAAF, Piano Nazionale "Biotecnologie Vegetali".

References

Aburjai T (1996) Metaboliti secondari di origine vegetale – studi di biosintesi e biotrasformazione con colture cellulari. PhD Thesis, University of Milan

Aburjai T, Bernasconi S, Manzocchi L, Pelizzoni F (1995) Biotrasformation of vitamin D_3 with *Solanum malacoxylon* cell cultures. European Research Conference on Plant Cell Biology and Biological Applications – Dourdan, Paris, Feb 14–19

Aburjai T, Bernasconi S, Manzocchi L, Pelizzoni F (1996a) Isolation of 7-dehydrocholesterol from cell cultures of *Solanum malacoxylon*. Phytochemistry 43:773–776

Aburjai T, Bernasconi S, Manzocchi L, Pelizzoni F (1996b) Effetti del calcio su biosintesi e biotransformazione della vitamina D_3 in colture cellulari di *Solanum malacoxylon*. Conv. Biocatalisi e sintesi organica, Società Chimica Italiana, Acquafredda di Maratea May 3–5, 1996; P18

Aburjai T, Bernasconi S, Manzocchi L, Pelizzoni F (1997) Effect of calcium and cell immobilization in *Solanum malacoxylon* cell cultures. Phytochemistry 46:1015–1018

Alvarez MA, Rodriguz Talou J, Paniego NB, Giulietti AM (1994) Solasodine production in transformed organ cultures (roots and shoots) of *Solanum eleagnifolium* Cav. Biotechnol Lett 16:393–396

Andrews DR, Barton DHR, Hesse RH, Pechet MM (1986) Synthesis of 25-hydroxy- and 1,25-dihydroxy vitamin D_3 from vitamin D_2 (calcifero). J Org Chem 51:4819–4828
Baggiolini EG, Jacobelli JA, Hennessy M, Batcho AD, Sereno JF, Uskokovic MR (1986) Stereo-controlled total synthesis of 1,25 dihydroxycholecalciferol and 1,25 dihydroxyergocalciferol. J Org Chem 51:3098–3108
Balandrin MF, Klocke JA, Wurtele ES, Bollinger WH (1985) Natural plant chemicals: sources of industrial and medicinal materials. Science 228:1154–1160
Bernasconi S, Manzocchi L, Pelizzoni F (1993) Studi su colture cellulari di *Solanum malacoxylon*. XXI Congr Naz Div Chim Org, Società Chimica Italiana, Terrasini (Pa), Sept 28–Oct. 2, P118, 186
Bhatt PN, Bhatt DP, Sussex I (1983) Studies on some factors affecting solasodine content in tissue cultures of *Solanum nigrum*. Physiol Plant 57:159–162
Boland RL (1986) Plants as a source of vitamin D metabolites. Nutr Rev 44:1–8
Brodelius P, Deus B, Mosbach K, Zenk MH (1979) Immobilized plant cells for the production and transformation of natural products. FEBS Lett 103:93–97
Buchala AJ, Pythoud F (1988) Vitamin D and related compounds as plant growth substances. Physiol Plant 74:391–396
Buchala AJ, Schmid A (1979) Vitamin D and its analogues as a new class of plant growth substances affecting rhizogenesis. Nature 280:230–231
Cham BE (1994) Solasodine glycosides as anticancer agents: pre-clinical and clinical studies. Asia Pac J Pharmacol 9:113–118
Chappell J (1995) Biochemistry and molecular biology of the isoprenoid biosynthetic pathway in plants. Annu Rev Plant Physiol Plant Mol Biol 46:521–547
Chappell J, Wolf F, Proulx J, Cuellar R, Saunders C (1995) Is the reaction catalyzed by 3-hydroxy-3-methylglutaryl coenzyme A reductase a rate limiting step for isoprenoid biosynthesis in plants? Plant Physiol 109:1337–1343
Dai H, Posner JH (1994) Synthetic approaches to vitamin D. Synthesis 1994:1383–1398
DeLuca HF, Schones HK (1983) Vitamin D: recent advances. Annu Rev Biochem 52:411–439
Drewer FE, van Staden J (1995) Attempts to produce solasodine in callus and suspension cultures of *Solanum mauritianum* Scop. Plant Growth Regul 17:21–25
El Badaoui H, Muguet B, Henry M (1996) Production of solamargine by in vitro cultures of *Solanum paludosum*. Plant Cell Tissue Organ Cult 45:123–127
Esparza MS, Vega M, Boland RL (1982) Synthesis and composition of vitamin D_3 metabolites in *Solanum malacoxylon*. Biochim Biophys Acta 719:633–640
Ferrara J, McCuaig K, Hendy GN, Uskokovic M, White JH (1994) Highly potent transcriptional activation by 16-ene derivatives of 1,25-dihydroxyvitamin D_3. J Biol Chem 269:2971–2981
Gamborg OL, Miller RA, Ojima K (1968) Nutrient requirements of suspension cultures of soybean root cells. Exp Cell Res 50:151–158
Grant AB (1951) Antivitamin D factor. Nature 268:347–349
Henry HL (1992) Vitamin D hydrolases. J Cell Biochem 49:4–9
Horst RL, Reinhardt TA, Russell JR, Napoli JL (1984) The isolation and identification of vitamin D_2 and vitamin D_3 from *Medicago sativa* (alfalfa plant). Arch Biochem Biophys 231:67–71
Huges MR, McCain TA, Chang SY, Haussler MR, Villareale M, Wasserman R (1977) Presence of 1,25-dihydroxyvitamin D_3-glycoside in the calcinogenic plant *Cestrum diurnum*. Nature 268:347–349
Jain SC, Sahoo S (1986) Isolation of steroids and glycoalkaloids from *Solanum glaucophyllum* Desf. Pharmazie 41:820–821
Jarvis B, Booth A (1981) Influence of indol-butyric acid, boron, myo-inositol, vitamin D_2 and seedling age on adventitious root development in cuttings of *Phaseolus aureus*. Physiol Plant 53:213–218
Jarvis B, Yasmin S, Coleman MT (1985) RNA and protein metabolism during adventitious root formation in stem cuttings of *Phaseolus aureus*. Physiol Plant 64:53–59
Kametani T, Furuyama H (1987) Synthesis of vitamin D_3 and related compounds. Med Res Rev 7:147–171
Kawauchi H, Sasaki J, Adachi T, Hanada K, Beppu T, Horinouchi S (1994) Cloning and nucleotide sequence of a bacterial cytochrome P-450 VD25 gene encoding vitamin D_3 25-hydroxylase. Biochim Biophys Acta 1219:179–183

Lindsey K, Yeoman MM (1985) Immobilised plant cell culture systems. In: Neumann KH, Barz W (eds) Primary and secondary metabolism of plant cell cultures. Springer, Berlin Heidelberg New York, pp 304–315

Macek TE (1989) *Solanum aviculare* Forst., *Solanum laciniatum* Ait. (Poroporo): In vitro culture and the production of solasodine. In: Bajaj YPS (ed) Biotechnology in agriculture and forestry, vol 7. Medicinal and aromatic plants II. Springer, Berlin Heidelberg, New York, pp 443–467

Maclaughlin JA, Anderson RR, Holick MF (1982) Spectral character of sunlight modulates photosynthesis of previtamin D_3 and its photoisomers in human skin. Science 216:1001–1003

Minghetti PP, Norman AW (1988) $1,25(OH)_2$-vitamin D_3 receptors: gene regulation and genetic circuitry. FASEB J 2:3043–3053

Moncousin CH, Gaspar TH (1983) Peroxidase as a marker for rooting improvement of *Cynara scolimus* cultured in vitro. Biochem Physiol Pflanz 178:263–271

Morris KML, Levak VM (1982) Evidence for aqueous soluble vitamin D-like substances in the calcinogenic plant *Trisetum flavescens*. Life Sci 30:1255–1262

Murashige T, Skoog F (1962) A revised medium for rapid growth and bio-assays with tobacco tissue cultures. Physiol Plant 15:473–497

Prema TP, Raghuramulu N (1994) Free vitamin D_3 metabolites in *Cestrum diurnum* leaves. Phytochemistry 37:677–681

Prema TP, Raghuramulu N (1996) Vitamin D_3 and its metabolites in the tomato plant. Phytochemistry 42:617–620

Pythoud F, Buchala AJ, Schmid A (1986) Adventitious root formation in green cuttings of *Populus tremula* L.: characterization of the effect of Vit D_3 and indolylbutyric acid. Physiol Plant 68:93–99

Rambeck W, Oesterhelt W, Vecchi M, Zucker H (1979) Occurrence of cholecalciferol in the calcinogenic plant *Trisetum flavescens*. Biochem Biophys Res Commun 87:743–749

Rambeck W, Kreutzberg O, Bruns Droste C, Zucker H (1981) Vitamin D_3 in the grass *Trisetum flavescens*. Z Pflanzenphysiol 104:9–16

Rambeck W, Weiser H, Zucker H (1987) A vitamin D_3 steroid hormone in the calcinogenic grass *Trisetum flavescens*. Z Naturforsch [C] 42:430–434

Raoul Y, Le Buolch N, Chopin J, Meunier P, Guerillot-Veniet A (1953) Données nouvelles sur les substances antirachitiques naturelles; leur présence dans le végétaux. C R Acad Sci Paris [D] 237:439–444

Raoul Y, Le Buolch N, Gounelle JC, Marnay-Gulat C, Ourisson G (1970) Substances antirachitiques des végétaux. Présence du cholécalciférol. Bull Soc Chim Biol 52:641–657

Sasaki J, Miyakazi A, Saito M, Adachi T, Mitzue K, Hanada K, Omura S (1992) Transformation of vitamin D_3 using *Amycolate* sp. strains. Appl Microbiol Biotechnol 38:152–157

Schaller H, Grausem B, Benveniste P, Chye ML, Tan CT, Song YH, Chua NH (1995) Expression of the *Hevea brasiliensis* (H.B.K.) Mull. Arg. 3-hydroxy-3-methylglutaryl coenzyme A reductase 1 in tobacco results in sterol overproduction. Plant Physiol 109:761–770

Seino Y, Tanaka H, Yamaoka K, Yabuuchi H (1987) Circulating 1α,25 dihydroxyvitamin D levels after a single dose of 1α,25 dihydroxyvitamin D_3 or 1α-hydroxyvitamin D_3 in normal men. Bone Miner 2:479–458

Skliar MI, Boland RL, Mourino A, Tojo G (1992) Isolation and identification of vitamin D_3, 25-hydroxyvitamin D_3, 1,25-dihydroxyvitamin D_3 and 1,24,25-trihydroxyvitamin D_3 in *Solanum malacoxylon* incubated with ruminal fluid. J Steroid Biochem Mol Biol 43:677–682

Suardi ML, Bernasconi S, Pelizzoni F, Racchi ML (1994) In vitro cultures of *Solanum malacoxylon* Sendt.: nutritional requirements and sterol production. Plant Cell Tiss Org Cult 36:9–14

Takeyama K, Kitanaka S, Sato T, Kobori M, Yanagisawa J, Kato S (1997) 25. Hydroxyvitamin D_3 1α-hydroxylase and vitamin D synthesis. Science 277:1827–1830

Tsoulpha P, Doran PM (1991) Solasodine production from self-immobilized *Solanum aviculare* cells. J Biotechnol 19:99–110

Valkonen JPT, Keskitalo M, Vasara T, Pietila L (1996) Potato glycolalkaloids: a burden or a blessing? Crit Rev Plant Sci 15:1–20

Vanmaele LP, DeClercq PG, Vandewalle M (1985) An efficient synthesis of 1,25-dihydroxy vitamin D_3. Tetrahedron 41:141–144

Vega MA, Santamaria EC, Morales A, Boland RL (1985) Vitamin D_3 affects growth and Ca^{2+} uptake by *Phaseolus vulgaris* roots cultured in vitro. Physiol Plant 65:423–426

Vidal MC, Lescano W, Advolov R, Puche RC (1985) Partial structure elucidation of the carbohydrate moiety of 1,25-dihydroxycholecalciferol glycoside isolated from *Solanum glaucophyllum* leaves. Turrialba 35:65–70

Wasserman RH, Henion JD, Haussler MR, McCain TA (1976) Calcinogenic factor in *Solanum malacoxylon*: evidence that it is, 1,25-dihydroxyvitamin D_3-glycoside. Science 194:853–855

Weissenberg M, Levy A, Wasserman RH (1989) Distribution of calcitriol activity in *Solanum glaucophyllum* plants and cell cultures. Phytochemistry 28:795–798

Whitfield GK, Hsieh JC, Jurutka PW, Selznick SH, Haussler CA, McDonald PN, Haussler MR (1995) Genomic actions of 1,25-dihydroxyvitamin D_3. J Nutr 125:S1690–S1694

Widholm JM (1972) The use of fluorescein diacetate and phenosafranine for determining viability of cultured plant cells. Stain Technol 47:189–194

Zucker JH, Stank H, Rambeck WA (1980) Light dependent synthesis of cholecalciferol in a green plant. Nature 283:68–69

XXIII *Solidago* L. Species (Goldenrod): In Vitro Regeneration and Biologically Active Secondary Metabolites

L. SKRZYPCZAK, M. WESOŁOWSKA, B. THIEM, and J. BUDZIANOWSKI

1 Distribution and Morphology of the Plants

The genus *Solidago* L. (family Asteraceae, Compositae), comprises some 120 species, mainly distributed throughout North America. From the USA, about 62 goldenrod species are reported, usually found in the north-western territories (Gleason 1952). This genus is also common in Canada, where the species have been classified into three groups on the basis of the morphological features of leaves and stems (Scoggan 1979).

North American species have been transferred to the European flora (McNeill 1976). These are: *Solidago canadensis* L.Sp. Pl.878 (1753), *S.gigantea* Aiton, Hort. Kew. 3:211 (1789) (=*S.serotina* Ait.), and *S.graminifolia* (L.) Salisb. Prodr. 199 (1796) (=*S.lanceolata* L.). The first two species are abundant in Europe, and in Poland they are considered to be the most rampant plants of foreign origin (Guzikowa and Maycock 1986). The species *S.canadensis* is found in two varieties in Poland: var. *canadensis* and var. *scabra* (Muhl.) Torr. et Gray (Rostański 1971). The variety *canadensis*, common in Europe, is very similar to *S.altissima* L.Sp. Pl.878 (1753) (*S.canadensis* var. *scabra* Torr. et Gray). The range of occurrence of the American species *S.graminifolia*, in turn, covers restricted areas of central Europe. In Poland this plant can be found in Lower Silesia and East Pomerania (Rostański 1971). A widespread European species is *S.virgaurea* L.Sp. Pl.880 (1753) (incl. *S.taurica* Juz.), which is only absent from the islands of Crete, Sicily, Iceland, the Azores and the Faeroes. Rostański (1971) divides *S.virgaurea* into two subspecies: *virgaurea*, including the forms *angustifolia* C.Kocha, *folisa* Fiori., and *obtusisquamea* Kem.Nat., and the subspecies *alpestris* (W. et K.) Reichenb. *Flora Europaea* (McNeill 1976), in turn, distinguishes the subspecies *minuta* (L.) Arcangeli, *Comp.Fl.Ital.* 339 (1882) [subsp. *alpestris* (Walds. and Kit. ex Willd.) Hayek, *S.lapponica* With., *S.jailarum* Juz.]. The goldenrod often referred to this subspecies is *S.macrorrhiza* Lange in Willk. and Lange, *Prodr.Fl.Hisp.*2,39 (1865). This goldenrod occurs in south-western France and north-western Spain. The species known from Italy is *S.litoralis* Savi, *Due Cent. Piante Etrusc.* 182 (1804), and from north Bulgaria, *S.virgaurea* subsp.*centiflora* Velen., Fl.Bulg.278 (1891).

Department of Pharmaceutical Botany, K.Marcinkowski University of Medical Sciences in Poznań, Św.Marii Magdaleny 14, 61-861 Poznań, Poland

Biotechnology in Agriculture and Forestry, Vol. 43
Medicinal and Aromatic Plants XI (ed. by Y.P.S. Bajaj)

Flora Europaea also lists a North American species, *S.sempervirens* L.,Sp.Pl. 878 (1753), whose occurrence is restricted to the Azores and is termed *S.sempervirens* var. *azorica* (Hochst.) St. John.

In the former USSR, 17 goldenrod species occur, including *S.virgaurea* and *S.canadensis* (Sziszkin 1959).

Solidago species are perennial herbs with rhizomes, single stems, and simple entire or serrate leaves, sometimes forming rosettes. Capitula, small or medium, in a paniculate or corymbose inflorescence. Receptacles are flat with several rows of imbricate involucral bracts. In a capitulum, florets are yellow. Outer florets are ligulate, female; inner florets are five-lobed, tubular, hermaphrodite. Pistil with two stigmas. Fruits are achenes, many veined, shortly ciliate, with the pappus hairs in one to two rows. The number of chromosomes is $2n = 18$ (Rostański 1971; McNeill 1976). Figure 1 shows an example of *Solidago virgaurea* L. obtained from in vitro cultures.

2 Chemical Components and Their Significance (Table 1)

Chemical compounds occurring in goldenrod species determine their medicinal uses. The plants are listed in a variety of the world's pharmacopoeias (Penso 1983). The most frequently mentioned, however, are *S.virgaurea* L. (syn. *Amphiraphis leiocarpa* DC., *A. pubescens* DC., *Dectis decurrens* Raf.var. *leiocarpa* Benth., *Solidago cantonensis* Lour., *S.decurrens* Lour., and *S.virga-aurea* L.).

Most commonly used in phytotherapy are the apical shoots of goldenrod, known as *Herba Solidaginis*. In Europe, as a result of multi-year quality control of the medicinal material, it has been shown that this material contains *S.gigantea* (50–100%) and *S.canadensis* (10–50%), with only a negligible addition of *S.virgaurea*. The declared commercial material *Herba Virgaureae* has also turned out to be a mixture of the three species mentioned (Schilcher 1965; Schilcher and Bornschein 1986). Among these three species, *S.virgaurea* seems to be most important, and its main therapeutical application is in treatment of diseases of the urinary tract, where herbs of *S.gigantea* and *S.canadensis* are considered to be equally valuable (Schilcher 1965).

Species of the genus *Solidago* synthesize numerous compounds from various chemical groups, of which flavonoids, saponins and, in *S.virgaurea* only, a phenolic glucoside-leiocarposide are responsible for diuretic effects.

2.1 Flavonoids

This group of compounds has been the object of much interest. Guthrie et al. (1944) isolated quercitrin and its aglycone from the species *S.leavenworthii* T.G., while Fuchs and Iliev (1949) obtained this glucoside from leaves of *S.virgaurea*, *S.canadensis* and *S.serotina*. The presence of phenyl chromones in

Fig. 1. *Solidago virgaurea* L. in ground, from in vitro culture

goldenrods has been reported in many publications (Skrzypczakowa 1961, 1962a,b; Borkowski and Skrzypczakowa 1962; Schilcher 1965; Batjuk and Kowalewa 1985; Tamas 1986; Batjuk et al. 1988). It has also been shown that flavonoids present in goldenrods are all flavonol 3-0-glycosides (Budzianowski et al. 1987, 1988) and they differentiate the varieties *S.canadensis* var. *canadensis* from *S.canadensis* var. *scabra* (Table 2). The chemical data were in agreement with the results of caryological studies which showed 2n = 18 for var. *canadensis* and 2n = 18 + 1 − 2B for var. *scabra* (Małecka 1988). On the basis of further chemical analyses, var. *scabra* was found to be distinguished by the presence of acylated glucosides of quercetin, kaempferol and isorhamnetin (Budzianowski et al. 1990). The compounds that turned out to be new for the genus *Solidago* were flavonoid C-glycosyls: schaftoside and isoschaftoside

Table 1. Summary of various compounds isolated from species of *Solidago* L.

Solidago species	Compounds reported	References
1. *S. altissima*	Diterpenes	Bohlmann et al. (1985)
	Phytosterol	Tada et al. (1989)
	cis-Dehydromatricaria Ester	Matsunaga et al. (1990)
	Mineral elements	Hosoyama et al. (1994)
2. *S. canadensis*	Flavonoids	Fuchs and Iliev (1949) Skrzypczakowa (1961, 1962a,b) Borkowski and Skrzypczakowa (1962) Schilcher (1965) Batjuk and Kowalewa (1985) Tamas (1986) Batjuk et al. (1988) Budzianowski et al. (1987, 1988, 1990)
	Saponosides	Murah et al. (1975) Hiller et al. (1987a) Bader et al. (1990c) Reznicek et al. (1991, 1992)
	Phenolic acids	Skrzypczakowa (1962b) Borkowski and Skrzypczakowa (1962) Gerlach (1972) Kalemba (1992) Poetsch et al. (1997) Budzianowski (unpubl.)
	Coumarines	Gerlach (1972)
	Diterpenes	Krepinsky and Herout (1962) Lu et al. (1993) Herz (1996)
	Sesquiterpenes	Melching et al. (1997)
	Polyacetylenes	Lam et al. (1992) Lu et al. (1993)
	Polysaccharides	Gerlach (1968)
	Essential oil	Kalemba et al. (1990a, 1993)
	Tannins	Hałas et al. (1991)
	Mineral elements	Rebele (1989)
3. *S. graminifolia*	Flavonoids C-glycosyls (schaftoside, isoschaftoside)	Budzianowski (1990)
	Phenolic acids	Kalemba (1992) Budzianowski (unpubl.)
	Essential oil	Kalemba et al. (1994)
4. *S. gigantea*	Flavonoids	Fuchs and Iliev (1949) Skrzypczakowa (1961, 1962a,b) Borkowski and Skrzypczakowa (1962) Schilcher (1965) Batjuk and Kowalewa (1985) Tamas (1986) Batjuk et al. (1988) Budzianowski (1990)
	Saponosides	Jurenitsch et al. (1986) Hiller et al. (1987b) Reznicek et al. (1989a,b, 1990)

Table 1. *Continuous*

Solidago species	Commpounds reported	References
	Phenolic acids	Skrzypczakowa (1962b) Borkowski and Skrzypczakowa (1962) Gerlach (1972) Kalemba (1992) Poetsch et al. (1997) Budzianowski (unpubl.)
5. *S. leavenworthii*	Flavonoids	Guthrie et al. (1944)
6. *S. microglossa*	Mineral elements	Vaz et al. (1995)
7. *S. missouriensis*	Diterpenes	Le Quesne et al. (1985)
8. *S. rigida*	Diterpenes	Le Quesne et al. (1985)
9. *S. rugosa*	Diterpenes	Lu et al. (1995) Saraswathy et al. (1996)
10. *S. sempervirens*	Diterpenes	Lu et al. (1995) Saraswathy et al. (1996)
11. *S. virgaurea*	Flavonoids	Fuchs and Iliev (1949) Skrzypczakowa (1961, 1962a,b) Borkowski and Skrzypczakowa (1962) Schilcher (1965) Batjuk and Kowalewa (1985) Tamas (1986) Batjuk et al. (1988) Budzianowski et al. (1990)
	Saponosides	Gnekow (1938) Hiller et al. (1975, 1987b,c, 1991) Hiller and Gil-Rjong (1979) Bader and Hiller (1992) Bader et al. (1992, 1995) Inose et al. (1991, 1992) Miyase et al. (1994)
	Phenolic glycosides (leiocarposide, virgaureoside)	Skrzypczakowa (1962) Hiller and Gil-Rjong (1979) Hiller et al. (1979, 1985) Gründemann et al. (1979) Skrzypczak and Ellnein-Wojtaszek (1981) Fötsch et al. (1988) Budzianowski et al. (1990)
	Phenolic acids	Skrzypczakowa (1962b) Borkowski and Skrzypczakowa (1962) Gerlach (1972) Bohlmann et al. (1981) Kalemba (1992, 1993) Poetsch et al. (1997) Budzianowski (unpubl.)
	Diterpenes	Goswami et al. (1984)
	Polyacetylenes	Knütter and Pohloudek-Fabini (1962) Lam et al. (1992) Lu et al. (1993)
	Polysaccharides	Gerlach (1968)
	Essential oil	Kalemba and Góra (1995)

Table 2. Distribution of leiocarposide and flavonoids in the genus *Solidago* L.

Compound	*S. virgaurea*	*S. gigantea*	*S. canadensis var. canadensis*	*S. canadensis var. scabra*
1. Leiocarposide	+	–	–	–
2. Quercetin [Q]	+	+	+	+
3. Q-3-β-Glucoside	+	+	+	+
4. Q-3-β-Galactoside[a]	+	+	+	–
5. Q-3-α-L-Arabinopyranoside[a]	+	+	–	–
6. Q-3-α-Rhamnoside	–	+	–	–
7. Q-3-/6″ -O-Acetyl/-β-glucopyranoside	–	–	–	+
8. Q-3-/X″ -O-Hydroxycinnamoyl ?/-glucoside[a]	–	+	–	–
9. Q-3-β-Rutinoside	+	+	+	+
10. Q-3-β-Robinobioside[a]	+	+	+	+
11. Q-3-β-Apiosyl-β-glucoside[a]	–	+	–	–
12. Kaempferol [K]	+	+	+	+
13. K-3-β-Glucoside	+	+	+	+
14. K-3-β-Galactoside[a]	+	+	+	+
15. K-3-Rhamnoside	–	+	–	–
16. K-3-Arabinoside[a]	+	+	–	–
17. K-3-/6″ -O-Acetyl/-glucopyranoside[a]	–	–	–	+
18. K-3-β-Rutinoside	+	–	+	+
19. K-3-β-Robinobioside[a]	+	–	+	+
20. Isorhamnetin [I]	+	+	+	+
21. I-3-β-Glucoside	+	+	+	+
22. I-3-β-Galactoside[a]	+	+	–	–
23. I-3-β-Arabinoside[a]	–	+	–	–
24. I-3-/6″ -O-Acetyl/-β-glucopyranoside	–	–	–	+
25. I-3-β-Rutinoside	+	–	+	+
26. Dillenitin (?)-3-glucoside[a]	+	+	–	–
27. D(?)-3-Galactoside[a]	+	+	–	–
28. D(?)-3-Arabinoside[a]	–	+	–	–

+ = Present, – = Absent.
[a] New for the species (Budzianowski et al. 1990).

(Fig. 2) isolated from *S.graminifolia* (Budzianowski 1990). Some studies also determined the quantitative amount of flavonoids in *Solidago* species (Pietta et al. 1991; Kalemba et al. 1993; Bongartz and Hesse 1995).

2.2 Saponosides

As Hegnauer (1964) states, *S.virgaurea* has been defined as a saponin plant. The presence of these compounds in the herb was reported by Gnekow (1938). The first data about structure of the saponosides were supplied by Hiller and his coworkers (1975). From the herb of *S.virgaurea* they isolated a mixture of saponosides of high haemolytic activity. Hydrolysis released eight genins, among which polygalacic and oleanolic acids were identified. From the species

FLAVONOIDS

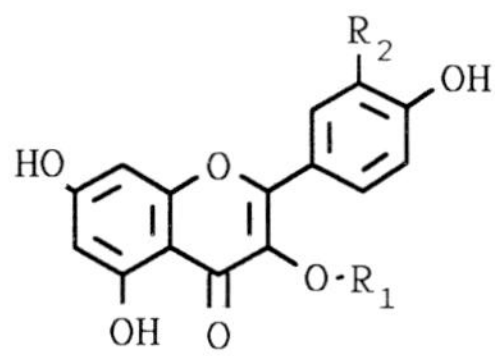

R_1	R_2
6"-O-acetylglucose	H
6"-O-acetylglucose	OH
6"-O-acetylglucose	OCH_3

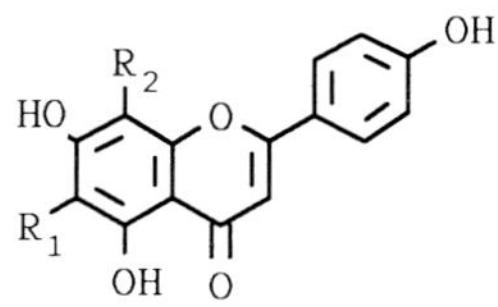

R_1	R_2	
glucose	arabinose	schaftoside
arabinose	glucose	isoschaftoside

SAPONINS

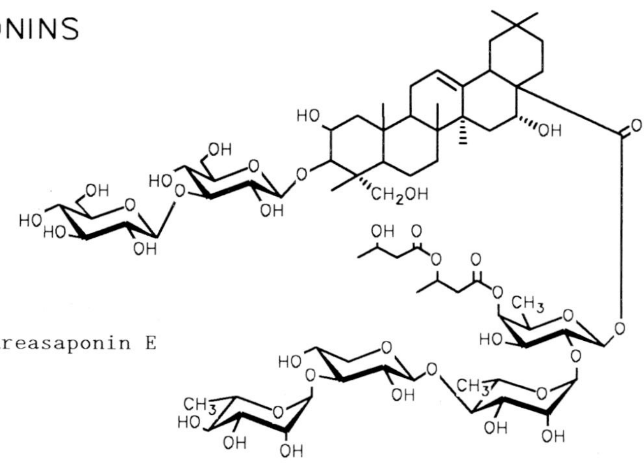

Virgaureasaponin E

PHENOL GLYCOSIDES

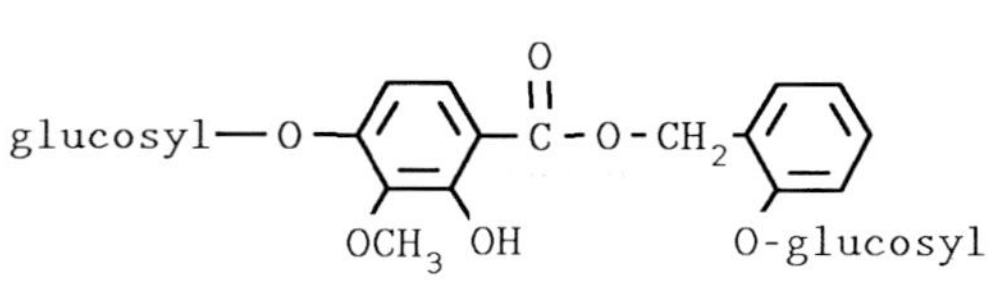

leiocarposide

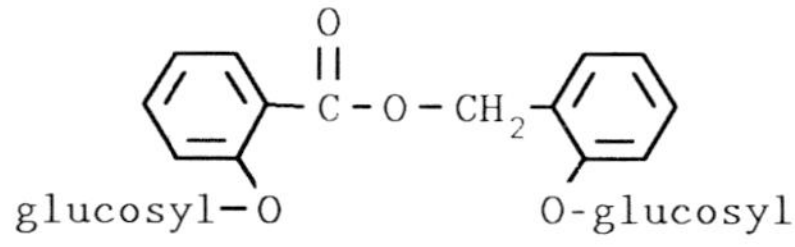

virgaureoside A

Fig. 2. Chemical structures of various compounds in *Solidago* species

S.canadensis, saponosides were isolated that were derivatives of oleanolic acid and bayogenin. The latter compound had not been found in the family Asteraceae before (Murach et al. 1975).

Other plants which were analysed for saponosides represented the species *S.virgaurea* var. *leiocarpa* (Benth.) A. Gray (Hiller and Gil-Rjong 1979), *S.gigantea* (Jurenitsch et al. 1986; Reznicek et al. 1989a,b, 1990), and *S.canadensis* (Hiller et al. 1987a; Bader et al. 1990c; Reznicek et al. 1991, 1992). As a result of further studies, ester triterpenoid saponosides with a bisdesmosidic structure were isolated from *S.virgaurea*. These compounds, which are derivatives of polygalacic acid, have been termed virgaureasaponin 1 (Hiller et al. 1987b), virgaureasaponin 2 (Hiller et al. 1987c) and virgaureasaponin 3 (Bader and Hiller 1992; Bader et al. 1992). An inventory of major saponosides found in the genus *Solidago* has been given in Hiller et al. (1991).

Japanese authors (Inose et al. 1991, 1992) isolated 22 saponoside glycosides, derivatives of polygalacic acid, from fresh *S.virgaurea*. Also other saponosides, derivatives of the oleanane type, were obtained (Miyase et al. 1994). Two new acylated bisdesmosidic triterpenoid glycosides were obtained from dry aerial parts and roots of the species *S.virgaurea* subsp. *virgaurea* (Bader et al. 1995). Tamas et al. (1988) searched for saponins in the genus *Solidago* from Romania.

2.3 Phenolic Glucosides, Phenolic Acids, and Other Compounds

From *S.virgaurea*, Skrzypczakowa (1962a) isolated a crystalline compound (m.p. 186–188 °C) which she defined as a new phenolic glucoside. Fuller chemical data were given in Skrzypczak and Ellnain-Wojtaszek (1981). When presenting the results of their research on saponosides in *S.virgaurea* var. *leiocarpa* at a congress in Budapest, Hiller and Gil-Rjong (1979) told of the isolation of a phenolic bisglycoside of a benzoic acid ester. A direct comparison of samples of the two glycosides showed them to be identical. The compound was termed leiocarposide (Hiller et al. 1979; Gründemann et al. 1979) and its complete structure (Fig. 2) was given (Fötsch et al. 1988).

During research on saponosides in *S.virgaurea*, a new phenolic bisdesmosidic glycoside was isolated as a by-product, viz. virgaureoside A (Fig. 2, Hiller et al. 1985). This compound has also been discovered in *Prunus grayana* (Shimomura et al. 1989). Studies have also been made on the quantitative contents of phenolic glycosides (Hiller and Fötsch 1986; Bader et al. 1990a), and of their biotransformation in animals (Fötsch et al. 1989). After revision of earlier studies, the presence of leiocarposide was found to be restricted to the species *S.virgaurea* (Budzianowski et al. 1990). A communication appeared (Poetsch et al. 1997) on the presence of three caffeoylquinic acids, namely, chlorogenic acid (5-0-caffeoylquinic acid), neochlorogenic acid (4-0-chlorogenic acid) and 3,5-di-0-caffeoylquinic acid in all three investigated species: *S.canadensis*, *S.gigantea* and *S.virgaurea*, only differing in quantitative distribution.

Independently, at the same time, we isolated chlorogenic acid, 3,5-di-0-caffeoylquinic acid and 4,5-di-0-caffeoylquinic acid from the flowers of *S.canadensis var. scabra* and we detected the same compounds in *S.canadensis var. canadensis*, *S.gigantea*, *S.virgaurea* and *S.graminifolia* (unpubl.). We eludidated the structure by comparing UV, ^{1}H and ^{13}C NMR spectral analysis with the published data (Cheminat el al. 1988). Chlorogenic acid is very common in the dicotyledons while isomeric dicaffeoylquinic acids are restricted to the Asteraceae (Molgaard and Ravn 1988).

Derivatives of benzoic acid are commonly found in the family Asteraceae. From the aerial parts and roots of a Chinese species *S.decurrens* Lour. (synonymous with *S.virgaurea* L.), several new compounds have been isolated: benzyl benzoates and cinnamyl angelates (Bohlmann et al. 1981). Other compounds discovered in goldenrod include chlorogenic and caffeic acids (Borkowski and Skrzypczakowa 1962) as well as isochlorogenic acid, esculin and esculetin (Gerlach 1972). The contents of phenolic compounds were analysed in goldenrod species (Kalemba 1992; Kalemba et al. 1993).

Derivatives of diterpenes have also been found in *S.canadensis* (Krepinsky and Herout 1962), *S.virgaurea* (Goswami et al. 1984), *S.altissima* (Bohlmann et al. 1985), *S.missouriensis* and *S.rigida* (Le Quesne et al. 1985), and *S.sempervirens* and *S.rugosa*, (Lu et al. 1995; Saraswathy et al. 1996). The occurrence of diterpenes in Asteraceae in general has been discussed by Hegnauer (1989) and Herz (1996). Melching et al. (1997) describe sesquiterpene from *S.canadensis* as an enantiometric mixture of germacrene D.

Polyacetylene compounds are common in Astereae. Their source is the roots of various *Solidago* species (e.g. Knütter and Pohloudek-Fabini 1962; Lam et al. 1992; Lu et al. 1993).

During the flowering of goldenrods, their underground organs can contain up to 20% inulin (Gerlach 1968) and they also produce other polysaccharides.

From the aerial parts of *Solidago* species, essential oils have been isolated and subjected to qualitative and quantitative analysis (Kalemba et al. 1990a; Weyerstahl et al. 1993, Kalemba et al. 1994; Kalemba and Góra 1995). Oil components have also been examined in material from various parts of Japan (Fujita 1990). Other issues of interest have been the accumulation of elements, e.g. Pb in *S.canadensis* (Rebele 1989), Cd in *S.altissima* (Hosoyama et al. 1994), and Hg and Se in *S.microglossa*, as well as other elements of therapeutic value (Vaz et al. 1995). *S.altissima* has proved to be a source of an active phytosterol – brassinolide (Tada et al. 1989). The characteristics and ways of obtaining tanning agents from *S.canadensis* (Hałas et al. 1991), among other species, have also been described.

2.4 Application in Phytotherapy and Biological Properties

In the Middle Ages the roots and aerial parts of goldenrods were used in kidney conditions, hydrops and nephrolithiasis, while their flowers and leaves were used as a natural yellow dye (Hayek 1908–1931). The plant was then forgotten, only to find its place once again in modern phytotherapy (e.g.

Schilcher 1987; Schilcher et al. 1988; Schilcher and Rau 1988; Weiss 1988; Eberwein and Vogel 1993; Hiller and Bader 1996).

The *Rote Liste* (1996) includes 66 herbal specifics containing extracts of goldenrod, applicable in diseases of the urinary tract, nephrolithiasis, and prostate conditions. The diuretic properties of goldenrods are mainly ascribed to flavonoid and saponoside compounds (Schilcher 1984; Schilcher et al. 1988; Schilcher and Rau 1988; Batjuk et al. 1988; Chodera et al. 1991). The ester phenolic glucoside – leiocarposide – present in *S.virgaurea* is indispensable for the material to have a diuretic effect (Chodera et al. 1985, 1986, 1988). It has also been found to have analgesic and anti-inflammatory properties (Metzner et al. 1984). There have also been reports of hypotensive and sedative effects from extracts of *S.virgaurea* subsp. *alpestris* (Wet K.) Goud. (Racz-Kotilla et al. 1977) and antimycotic activities of *S.virgaurea* and *S.gigantea* (Pepeljnjak et al. 1997).

The known preparation Phytodolor N, containing extracts of *Populus tremula*, *S.virgaurea* and *Fraxinus excelsior*, has anti-inflammatory effects (El-Ghazaly et al. 1992) and antirheumatic properties (Kruedener et al. 1995). This natural medication also shows an antioxidant effect (Meyer et al. 1995) and inhibits the activity of dihydrofolate reductase (Strehl et al. 1995). Also, extract of *S.virgaurea* has been demonstrated to have antioxidant effects (Dite et al. 1996).

Diuretic, spasmolytic and anti-irradiation effects have been observed in *S.gigantea* extracts (Leuschner 1995). The chlorogenic and caffeic acids present in goldenrods have bacterostatic and cholagogic properties (Czok and Land 1961a,b). Both compounds also show an immunostimulating activity. The effect of leiocarposide has been observed to be much weaker (Sawicka et al. 1994).

The saponosides present in the *Solidago* species are ascribed to have activities against pathogenic yeasts (Bader et al. 1987, 1990b) and tumor cells using virgaurea-saponins (Bader et al. 1996a,b; Fig. 2). The cytotoxicity and the antifungal activity of the triterpenoid saponins depend on the structure of the carbohydrate units (Bader et al. 1997). The polysaccharides in *Solidago* have been studied for their anticarcinogenic properties (Kraus et al. 1986; Franz and Kraus 1987), and the polyacetylene compound called cis-dehydromatricaria, an ester isolated from *S.altissima*, has been demonstrated to inhibit the development of mammal cells (Matsunga et al. 1990). Japanese authors (Matsumoto et al. 1991) have obtained a patent on an anticarcinogenic polyacetylene extract of *S.virga-aurea*. These compounds isolated from *S.gigantea* have been shown to have insecticidal effects (Polos et al. 1993). Experiments on larvae of grass-eating insects have revealed the chemical compounds of *S.altissima* to have a deadly effect (Bosio et al. 1990). The essential oil from *S.canadensis* has proved to be highly toxic, paralysing and acting as an antifeedant with respect to some insects – pests of food stores (Kalemba et al. 1990b). These last properties have also been discovered in compounds from various plant species of the family Compositae, including *S.canadensis* (Mullin et al. 1991). There has also been research on the effect of *S.canadensis* pollen allergens on workers exposed to lead (Shoji et al. 1993).

3 In Vitro Culture Studies

Earlier cell cultures of *S.altissima* were used for the biotransformation of sesquiterpenes (Sakamoto et al. 1994). The callus tissue was obtained from young leaves of the plant. The subculture was grown for over a year on MS agar medium (Murashige and Skoog, 1962). Next, the tissue was transferred into a liquid MS medium supplemented with 2,4-D and BA (0.1–0.2; 3–0.2 mg/l), and 2,4-D and kinetin (1–0.2 mg/l). Twenty milligrams of a substrate, germacrone, dissolved in 0.5 cm^3 EtOH and was added to 200 cm^3 of the suspension. The cultures were incubated for 10 days. The separated media and cells were used to isolate the products of biotransformation.

3.1 Establishment of Tissue Cultures and Micropropagation

In our studies, seeds of four species: *S.canadensis*, *S.gigantea*, *S.graminifolia* and *S.virgaurea*, after previous stratification at about 4 °C for a minimum of 10 days, were placed on a solidified MS medium with no plant growth regulators. An MS medium supplemented with GA_3 (1 mg/l), or GA_3 (1 mg/l) + kinetin (0.5 mg/l) was also suitable for seedlings, resulting in the germination of about 50% of seeds.

Micropropagation of the *Solidago* species from shoot tips was carried out on a solidified MS medium supplemented with kinetin (2 mg/l) and IAA (2 mg/l), or only BAP (1 mg/l). The first variant of the medium yielded better results, since lateral buds grew on it and then developed into normal shoots. Some shoots developed roots in vitro directly on the same medium, while others required the addition of IAA (1 mg/l; Fig. 3). When transferred to pots with earth, plantlets grew successfully, and they reached the blooming and fruiting stage (Fig. 1).

In parallel studies, callus tissues were obtained from the three *Solidago* species which were derived both from fragments of seedlings, roots, hypocotyls and cotyledons, and from leaves. The callus tissue was first established on MS medium with 2,4-D (1 mg/l) and BAP (0.5 mg/l; Fig. 4). The undifferentiated calli obtained were subcultured on MS medium supplemented with BAP (1 or 0.2 mg/l) and 2,4-D (0.25 or 0.2 mg/l). On this medium callus was friable and yellowish with red centers.

3.2 Characterisation of Compounds in Plants from In Vitro Cultures

Plants from in vitro cultures were transferred to soil; material was collected during the flowering stage. The species identity was confirmed in the Department of Plant Taxonomy, Adam Mickiewicz University, Poznań.

Qualitative analyses were made using thin-layer chromatography (TLC). For this purpose, methanol extracts were prepared, from which chloroform, butanol and water fractions were derived. The butanol fractions were

Fig. 3. Rooted shoots of *Solidago canadensis* on MS mediuum with 1 mg/l IAA, after 1 month

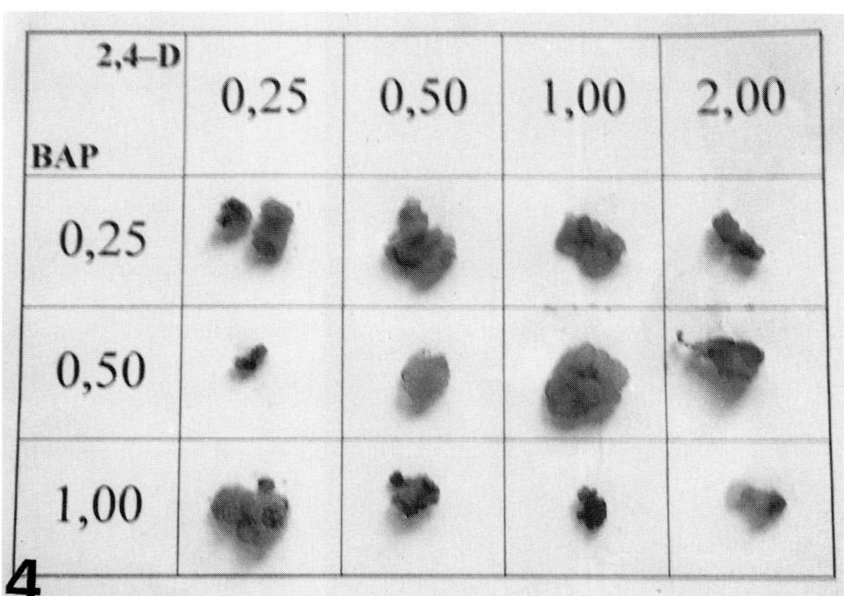

Fig. 4. The effect of 2,4-D and BA (mg/l medium) on callus formation from hypocotyls of *S.virgaurea* . (Wesołowska et al. 1997a,b)

subjected to two-dimensional chromatography (2D-TLC) on plates covered with polyamide (DC-6, Macherey Nagel) in the following developing phases: dimension I: $CHCl_3$-MeOH-MeCOEt-Ac_2CH_2 (9:4:2:1); dimension II: H_2O-EtOH-Ac_2CH_2 (4:2:1) (Budzianowski et al. 1990). Alongside the extracts, solutions of selected standards were developed. A sample chromatogram is presented in Fig. 5.

Preliminary analyses of *S. virgaurea* extracts from in vitro cultures showed the presence of a phenolic glucoside-leiocarposide. This compound was isolated for full identification. Extracts were obtained from 100 g of dried inflorescence by boiling in methanol three times. Extracts were combined, dried under lower pressure, and dissolved in hot water. After cooling, the extract was shaken with chloroform and the water phase separated on a polyamide column. The purified leiocarposide was chromatographically (TLC) and spectroscopically (^{1}H and ^{13}C NMR) identical to the sample isolated from naturally growing plants in our previous investigations. The comparative presentation of ^{13}C NMR spectra of both samples is given in Fig. 6. Moreover, we confirmed the chemical structure and checked the correctness of assignments of ^{1}H and ^{13}C NMR resonances for this compound (Fötsch et al. 1988) using two-dimensional NMR (2D-NMR) techniques like ^{1}H^{1}H-COSY, HMQC and HMBC. Extracts were also prepared from rooted, non-flowering shoots from in vitro cultures. Chromatographic analysis showed that flavonoids and the phenolic glucoside were synthesised as early as the shoot stage. The data on

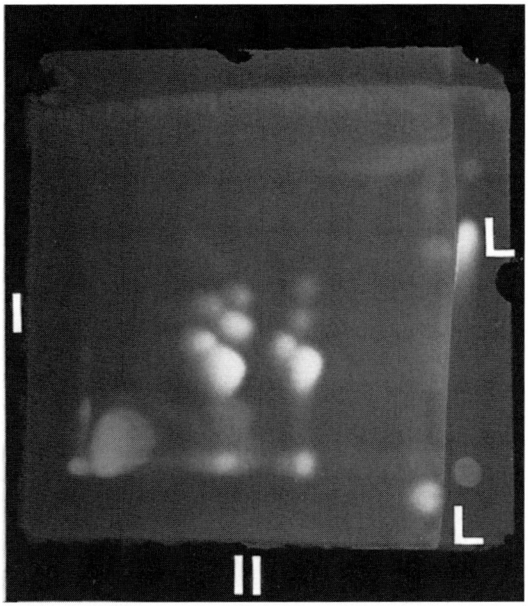

Fig. 5. 2D-TLC of butanolic fraction from *S. virgaurea*. *L* leiocarposide, *I* $CHCl_3$-MeOH-MeCOEt-Ac_2CH_2 (9:4:2:1), *II* H_2O-EtOH-Ac_2CH_2 (4:2:1), $AlCl_3$-UV_{365}. (Wesołowska et al. 1997a,b)

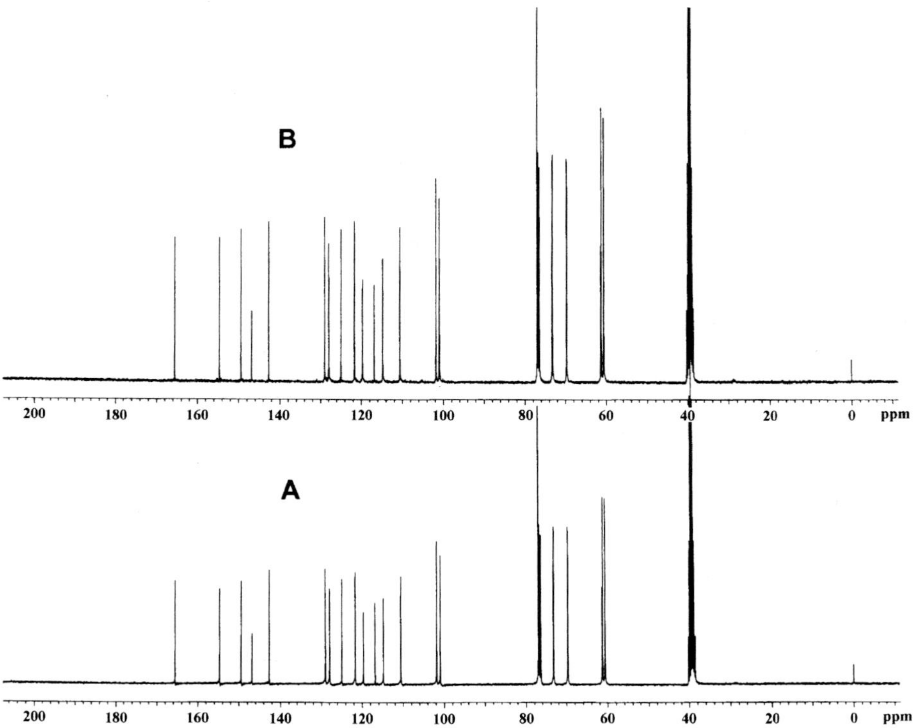

Fig. 6. ^{13}CNMR spectra (75 MHz in DMSO-d_6) of leiocarposide isolated from naturally growing *S. virgaurea* (**A**) and from in vitro culture (**B**). The structure of leiocarposide is given in Fig. 2

micropropagation of the *Solidago* species and chemical analyses were discussed at various congresses (Wesołowska et al. 1997a,b). Analysis of volatile compounds in the essential oils of *Solidago* species from in vitro culture, was done by GC and GC-MS. The content of essential oils and their composition was comparable to that of the naturally growing plants (unpubl.).

4 Summary

The genus *Solidago* L., typical of the family Asteraceae, embraces many species synthesizing secondary metabolites of medicinal importance. Therapeutic activity in diseases of the urinary tract is ascribed to such chemical constituents as flavonoids and saponins and, in the case of *Solidago virgaurea*, particularly to the unique phenolic ester glucoside named leiocarposide.

We micropropagated *Solidago canadensis*, *S. gigantea*, *S. graminifolia* and *S. virgaurea* through proliferation of axillary shoots on MS medium with IAA and kinetin. After in vitro rooting, plantlets were transferred to the soil where they followed the regular course of development, reaching the stage of blooming. Inflorescences and aerial parts were subjected to phytochemical analysis to check the presence of some compounds (phenolic glucoside-leiocarposide, flavonoids, essential oils). The analysis showed composition of these compounds to be the same as in naturally growing plants. The presence of leiocarposide in *S. virgaurea* was confirmed by its isolation and identification by NMR analysis.

A well-established callus tissue culture was set up that could be utilised in cell suspension culture. The best results were obtained when the callus tissue was derived from seedling roots and leaves on MS medium enriched with 2,4-D and BAP.

Acknowledgements. The authors are grateful to Prof. K. Latowski (Department of Plant Taxonomy, A. Mickiewicz University, Poznań) for the identification of plants, and to M.S. Mariola Pawlik for her assistance in the preparation of this manuscript.

References

Bader G, Hiller K (1992) New ester saponins from *Solidago virgaurea*. 40th Annu Congr Med Plant Res, Trieste, Abstr, pp 134–135

Bader G, Binder K, Hiller K, Ziegler-Böhme H (1987) Zur antimyzetischen Wirksamkeit von Triterpensaponinen aus *Solidago virgaurea* L. Pharmazie 42:140

Bader G, Janka M, Hannig HJ, Hiller K (1990a) Zur quantitativen Bestimmung von Leiocarposid in *Solidago virgaurea* L. Pharmazie 45:380–381

Bader G, Kulhanek Y, Ziegler-Böhme H (1990b) Zur antimyzetischen Wirksamkeit von Polygalsäureglycosiden. Pharmazie 45:618–620

Bader G, Hiller K, Ehwald R, Guempel C, Rathgen K (1990c) Purification of pharmaceutical triterpene glycosides. Ger (East) Patent DD 276, 287. Chem Abstr 1991, 114:30111b

Bader G, Wray V, Hiller K (1992) Virgaurea-saponin 3, a 3, 28-bisdesmosidic triterpenoid saponin from *Solidago virgaurea*. Phytochemistry 31:621–623

Bader G, Wray V, Hiller K (1995) The main saponins from the aerial parts and the roots of *Solidago virgaurea* subsp. *virgaurea*. Planta Med 61:158–161

Bader G, Plohmann B, Franz G, Hiller K (1996a) Saponins from *Solidago virgaurea* L. – possible agents for therapy of cancer? 44th Annu Congr Soc Med Plant Res and Joint Meet Czech Biotech Soc Prague, Abstr, pp 21–22

Bader G, Plohmann B, Hiller K, Franz G (1996b) Cytotoxicity of triterpenoid saponins. Pharmazie 51:414–417

Bader G, Melzig M, Plohmann B, Franz G, Hiller K (1997) Biological effects of different triterpenoid saponins in selected cell systems. 45th Annu Congr Med Plant Res, Regensburg, Germany, Abstr 015

Batjuk WS, Kowalewa SN (1985) Flavonoids of *Solidago canadensis* and *S. virgaurea*. Khim Prir Soedin: 566 (in Russian)

Batjuk WS, Vasilchenko EA, Kowalewa SN (1988) Flavonoids of *Solidago virgaurea* L. and *Solidago canadensis* L. and their pharmacological properties. Rast Res 24:92–99 (in Russian)

Bohlmann F, Chen Z-L, Schuster A (1981) Aromatic esters from *Solidago decurrens*. Phytochemistry 20:2601–2602

Bohlmann F, Singh P, Singh R, Joshi KC, Jakupowic J (1985) A diterpene with a new carbon skeleton from *Solidago altissima*. Phytochemistry 24:1114–1115

Bongartz D, Hesse A (1995) Selective extraction of quercetin in vegetable drugs and urine by off-line coupling of boronic acid affinity chromatography and high-performance liquid chromatography. J Chromatogr B: Biomed Appl 673:223–230; Chem Abstr 1996, 124:44594e

Borkowski B, Skrzypczakowa L (1962) Zwiaozki wielofenolowe w krajowych gatunkach nawłoci *Solidago* L. Acta Pol Pharm 19:491–495

Bosio CF, McCrea KD, Nitao JK, Abrahamson WG (1990) Defense chemistry of *Solidago altissima*: effects on the generalist herbivore *Trichoplusia ni* (Lepidoptera: Noctuidae). Environ Entomol 19:465–468; Chem Abstr 1991, 114:182010f

Budzianowski J (1990) Two-Di-C-glycosylfavones from *Solidago graminofolia*. Sci Pharm 58:413–416

Budzianowski J, Skrzypczak L, Wesołowska M (1987) Flavonoid patterns in the genus *Solidago* Pharm. Weekblad Sci Ed 9:230

Budzianowski J, Skrzypczak L, Wesołowska M (1988) Studies on the genus *Solidago* L II. Phenolic compounds in taxonomy of *Solidago canadensis* L. Acta Biol Cracov Ser Bot 30:33–37

Budzianowski J, Skrzypczak L, Wesołowska M (1990) Flavonoids and leiocarposide in four *Solidago* taxa. Sci Pharm 58:15–23

Cheminat A, Zawatzky R, Becker H, Brouillard R (1988) Caffeoyl conjugates from *Echinacea* species: structures and biological activity. Phytochemistry 27:2787–2794

Chodera A, Dąbrowska K, Seńczuk M, Wasik-Olejnik A, Skrzypczak L, Budzianowski J, Ellnain-Wojtaszek M (1985) Badania nad działaniem diuretycznym estrowego glukozydu z rodzaju *Solidago* L. Acta Pol Pharm 42:199–203

Chodera A, Dąbrowska K, Skrzypczak L, Budzianowski J (1986) Dalsze badania nad działaniem diuretycznym lejokarpozydu. Acta Pol Pharm 43:499–503

Chodera A, Dąbrowska K, Bobkiewicz-Kozłowska T, Tkaczyk J, Skrzypczak L, Budzianowski J (1988) Wpływ lejocarpozydu na doświadczalną kamicę moczową u szczurów. Acta Pol Pharm 45:181–186

Chodera A, Dąbrowska K, Śloderbach A, Skrzypczak L, Budzianowski J (1991) Wpływ frakcji flawonoidowych rodzaju *Solidago* L. na diurezę i stężenie elektrolitów. Acta Pol Pharm-Drug Res 48:35–37

Czok G, Lang K (1961a) Zur erregenden Wirkung von Chlorogensaure. Arzneim-Forsch 11:448–450

Czok G, Lang K (1961b) Chlorogensäure-Wirkungen am Magen-Darmkanal. Arzneim-Forsch 11:545–549

Dite J, Kubinova M, Pizova M, Suchy V (1996) Antioxidative activity of extracts of ninety-six plant species. 44th Annu Congr Soc Med Plant Res and Joint Meet Czech Biotech Soc, Prague, Abstr, pp 61–62

Eberwein E, Vogel G (1993) Arzneipflanzen in der Phytotherapie. Translated by Borkowski et al. (1994) Wyd Inst Rośl Przetw Ziel, Poznań, pp 496–497, 554–555

El-Ghazaly M, Khayyal MT, Okpanyi SN, Arens-Corell M (1992) Study of anti-inflammatory activity of *Populus tremula*, *Solidago virgaurea* and *Fraxinus excelsior*. Arzneim-Forsch 42:333–336

Fötsch G, Gründemann E, Pfeifer S, Hiller K, Salzwedel D (1988) Zur Struktur von Leiocarposid. Pharmazie 43:278–280

Fötsch G, Pfeifer S, Bartoszek M, Franke P, Hiller K (1989) Biotransformation der Phenolglycoside Leiocarposid und Salicin. Pharmazie 44:555–558

Franz G, Kraus J (1987) Pflanzliche Polysaccharide mit Antitumorwirkung. Z Phytother 8:114–118

Fuchs L, Iliev V (1949) Isolierung von Quercitrin aus *Solidago virga aurea* L., *S.serotina* Ait. und *S.canadensis* L. (Eine alte Heilpflanze in neuem Licht). Sci Pharm 17:128–131

Fujita S (1990) Miscellaneous contributions to the essential oils of plants from various territories. Components of the essential oils of *Solidago virgaurea* Linn.ssp. Nippon Nogei Kagaku Kaishi 64:1729–1732 (Jpn). Chem Abstr 1991, 114:98252d

Gerlach H (1968) *Solidago* Arten als Arzneipflanzen und ihre Inhaltsstoffe. 2. Mitt.Das Inulin als Hauptbestandteil in Wurzeln verschiedener *Solidago*-Arten. Sein qualitative und quantitative Bestimmung. Pharm Zentralhalle 107:584–592

Gerlach H (1972) Über einige Inhaltsstoffe aus verschiedenen *Solidago*-Arten. Herba Pol 18:155–159

Gleason HA (1952) Illustrated Flora of the northeastern United States and adjacent Canada, vol 3. Lancaster Press, Lancaster, Pennsylvania, pp 413–439

Gnekow R (1938) Die Goldrute, *Solidago virga-aurea* L., Monographien alter Heilpflanzen, heft 1. Hansischer Gildenverlag, Hamburg cited in Hegnauer R (1964)

Goswami A, Robindra NB, Sharma RP, Baruah JN, Kulanthaivel P, Herz W (1984) Clerodanes from *Solidago virgaurea*. Phytochemistry 23:837–841

Gründemann E, Gil-Rjong K, Hiller K (1979) ^{1}H-und ^{13}-C-NMR-spektroskopische Untersuchungen an Leiocarposid, einem neuen Phenologlykosid. Pharmazie 34:430–431

Guthrie JD, O'Connor RT, Stansbury MF, Savich TR (1944) Isolation of quercitrin and quercetin from goldenrod material. J Am Chem Soc 66:1794–1795

Guzikowa M, Maycock PF (1986) The invasion and expansion of three North American species of goldenrod (*Solidago canadensis* L. sensu lato, *S.gigantea* Ait. and *S.graminifolia* (L.) Salisb.) in Poland. Acta Soc Bot Pol 55:367–384

Hałas J, Krawiecki C, Góra J, Kalemba D, Kurowska A (1991) Możliwości pozyskiwania i charakterystyka garbników z roślin zielarskich. Przegl Skórzany 46:63–64

Hayek A (1908–1931) DCCXIX. *Solidago* L. Goldrute. In: Hegi G (ed) Illustrierte Flora von Mittel-Europa, vol 6, t 4. Lehmann, München, pp 409–413

Hegnauer R (1964) Chemotaxonomie der Pflanzen, vol 3. Birkhäuser, Basel, pp 486, 488

Hegnauer R (1989) Chemotaxonomie der Pflanzen, vol 8. Birkhäuser, Basel, pp 276–279

Herz W (1996) Terpenoid chemistry of the Astereae. In: Hind DJN, Beentje HJ (eds) Compositae: systematics. Proc Int Compositae Conf, Kew, 1994, vol 1. Royal Botanic Gardens, Kew, pp 261–293

Hiller K, Bader G (1996) Goldruten-Kraut. Z Phytother 17:123–130

Hiller K, Fötsch G (1986) Zur quantitativen Verteilung der Phenolglycoside Virgaureosid A und Leiocarposid in *Solidago virgaurea* L. Pharmazie 41:415–416

Hiller K, Gil-Rjong R (1979) Structures of saponins and sapogenins in species of genus *Solidago*. Planta Med 36:267–268

Hiller K, Genzel S, Murach M, Franke P (1975) Zur Kenntnis der Saponine der Gattung *Solidago*. 1 Mitteilung: Über die Saponine von *Solidago virgaurea* L. Pharmazie 30:188–190

Hiller K, Gil-Rjong R, Franke P, Gründemann E (1979) Über die Saponine von *Solidago virgaurea* L. var.*leiocarpa* (Benth) A. Gray. Pharmazie 34:360–361

Hiller K, Dube G, Zeigan D (1985) Virgaureosid A – ein neues, bisdesmosidisches Phenolglycosid aus *Solidago virgaurea* L. Pharmazie 40:795–796

Hiller K, Bardella H, Schulten H-R (1987a) Beiträge zur Struktur der Saponine von *Solidago canadensis* L. Pharmazie 42:622–625

Hiller K, Bader G, Schulten H-R (1987b) Virgaureasaponin 1, ein neues bisdesmosidisches Triterpensaponin aus *Solidago virgaurea* L. Pharmazie 42:541–543

Hiller K, Bader G, Dube G (1987c) Zur Struktur von Virgaureasaponin 2. Pharmazie 42:744–745

Hiller K, Bader G, Reznicek G, Jurenitsch J, Kubelka W (1991) Die Hauptsaponine der arzneilich genutzten Arten der Gattung *Solidago*. Pharmazie 46:405–408

Hosoyama Y, Daiten S, Sakai MI (1994) Histochemical demonstration of cadmium in plant tissues. Kenkyu Kiyo-Nihon Daigaku Bunrigakubu Shizen Kagaku Kenkyusho 29:107–110; Chem Abstr 1994,121:51842a

Inose Y, Miyase T, Ueno A (1991) Studies on the constituents of *Solidago virga-aurea* L. I Structural elucidation of saponins in the herb. Chem Pharm Bull 39:2037–2042

Inose Y, Miyase T, Ueno A (1992) Studies on the constituents of *Solidago virga-aurea* L. II. Structures of solidagosaponins X–XX. Chem Pharm Bull 40:946–953

Jurenitsch J, Lichtenberger E, Robien W, Jentzsch K (1986) Über die Inhaltsstoffe der Riesengoldrute. Planta Med 52:236–238

Kalemba D (1992) Phenolic acids in four *Solidago* species. Pharmazie 47:471–472

Kalemba D, Góra J (1995) Chemical composition of the essential oil of *Solidago virgaurea* L. 26th Int Symp on Essential oils, Hamburg, Abstr, p 44

Kalemba D, Góra J, Kurowska A (1990a) Analysis of the essential oil of *Solidago canadensis*. Planta Med 56:222–223

Kalemba D, Góra J, Kurowska A, Majda T (1990b) Badanie olejków eterycznych w aspekcie ich działania na owady. Zesz Nauk Pl Technol Chem Spoż 47:91–97

Kalemba D, Góra J, Kurowska A, Lis A, Lasocka I (1993) Polyphenolic compounds in four *Solidago* species. Acta Pol Pharm – Drug Res 50:441–442

Kalemba D, Weyerstahl P, Marschall H (1994) Constituents of the essential oil of *Solidago graminifolia* (L.) Salisb. Flavour Fragr J 9:269–274

Knütter S, Pohloudek-Fabini R (1962) Versuche zur chemischen Identifizierung und Bestimmung des Matricariaesters aus Wurzeln von *Solidago virgaurea*. Pharmazie 17:456–462

Kraus J, Schneider M, Franz G (1986) Antitumorpolysaccharide aus *Solidago* sp. Dtsch Apoth Ztg 126:2045–2049

Krepinsky J, Herout V (1962) Plant substances, XVIII. Isolation of terpenic compounds from *Solidago canadensis* L. Collection Chechoslov Chem Commun 27:2459–2462

Kruedener S, Schneider W, Elstner EF (1995) A combination of *Populus tremula*, *Solidago virgaurea* and *Fraxinus excelsior* as an anti-inflammatory and antirheumatic drug: a short review. Arzneim-Forsch 42:169–171

Lam J, Christensen LP, Färch T, Thomasen T (1992) Acetylenes from the roots of *Solidago* species. Phytochemistry 31:4159–4161

Le Quesne PW, Honkan V, Onan KD, Morrow PA, Tonkyn D (1985) Oxidized kaurene derivatives from leaves of *Solidago missouriensis* and *S.rigida*. Phytochemistry 24:1785–1787

Leuschner J (1995) Anti-inflammatory, spasmolytic and diuretic effects of commercially available *Solidago gigantea* herb extract. Arzneim-Forsch 45:165–168

Lu T, Menelaou MA, Vargas D, Fronczek FR, Fischer NH (1993) Plyacetylenes and diterpenes from *Solidago canadensis*. Phytochemistry 32:1483–1488

Lu T, Vargas D, Franzblau SG, Fischer NH (1995) Diterpenes from *Solidago rugosa*. Phytochemistry 38:451–456

McNeill J (1976) *Solidago* L. In: Tutin TG, Meywood YH, Burges NA (eds) Flora Europaea, vol 4. Cambridge University Press, Cambridge, pp 109–111

Małecka J (1988) Studies on the genus *Solidago* L. I Karyotype analysis of *Solidago canadensis* L. sl. Acta Biol Cracov Ser Bot 30:143–151

Matsumoto A, Katsuya H, Matsumoto T, Tokuda H (1991) Antitumor polyacetylene extraction from plants. Jpn Kokai Tokkyo Koho JP03, 287, 532, (91, 287, 532); Chem Abstr 1992, 116:181120w

Matsunaga H, Katano M, Tasaki M, Yamamoto H, Mori M, Takata K (1990) Inhibitory effect of *cis*-dehydromatricaria ester isolated from *Solidago altissima* on the growth of mammalian cells. Chem Pharm Bull 38:3483–3484

Melching S, Bülow N, Wihstutz K, Jung S, König WA (1997) Natural occurrence of both enantiomers of cadina-3,5-diene and δ-amorphene. Phytochemistry 44:1291–1296

Metzner J, Hirschelmann R, Hiller K (1984) Antiphlogistische und analgetische Wirkungen von Leiocarposid einem phenolischen Bisglucosid aus *Solidago virgaurea* L. Pharmazie 39:869–870

Meyer B, Schneider W, Elstner EF (1995) Antioxidative properties of alcoholic extracts from *Fraxinus excelsior*, *Populus tremula* and *Solidago virgaurea*. Arzneim-Forsch 45:174–176

Miyase T, Inose Y, Ueno A (1994) Studies on the constituents of *Solidago virga-aurea* L. III. Structures of solidagosaponins XXI-XXIX. Chem Pharm Bull 42:617–624

Molgaard P, Ravn H (1988) Evoluutionary aspects of caffeoyl ester distribution in dicotyledons. Phytochemistry 27:2411–2421

Mullin CA, Alfatafta AA, Harman JL, Serino AA, Everett SL (1991) Corn rootworm feeding on sunflower and other Compositae. Influence of floral terpenoid and phenolic factors. ACS Symp Ser 449: (Nat Occurring Pest Bioregul) 278–292; Chem Abstr 1991, 115:494

Murach M, Hiller K, Franke P, Hintche R (1975) Bayogenin – ein Sapogenin in *Solidago canadensis* L. 2 Mitteilung: Zur Kenntnis der Saponine der Gattung *Solidago*. Pharmazie 30:619–620

Murashige T, Skoog F (1962) A revised medium for rapid growth and bio-assays with tobacco tissue cultures. Physiol Plant 15:473–497

Penso G (1983) Index plantarum medicinalium totius mundi corumque synonymorum. OEMF, Milano, pp 893–894

Pepeljnjak S, Kustrak D, Vukusic I (1997) Investigation of the antimycotic activities of *Solidago virgaurea* and *Solidago gigantea* extracts. 45th Annu Congr Med Plant Res, Regensburg, Germany, Abstr K 18

Pietta P, Gardana C, Mauri P, Zecca L (1991) High-performance liquid chromatographic analysis of flavonol glycosides of *Solidago virgaurea*. J Chromatogr 558:296–301

Poetsch F, Pauli GF, Nahrstedt A (1997) Specific access to the structures acid diesters – an example from *Solidago* sp. 45th Annu Congr Med Plant Res, Regensburg, Germany, Abstr C 22

Polos E, Polos E, Varadi G, Magoss S, Szabo M, Schadt I (1993) Insecticidal plant extracts. Hung Teljes HU 62, 435; Chem Abstr 1993, 119:133474e

Racz-Kotilla E, Majer M, Racz G (1977) Actiunea hipotensiva si sedativa a extracte – lor de *Solidago virgaurea* L. Note Botanice 13:1–5

Rebele F (1989) Lead concentration in leaves of *Solidago canadensis* near a lead battery plant in Berlin, Germany. Verh-Ges Oekol 18:437–442; Chem Abstr 1990, 112:41650m

Reznicek G, Jurenitsch J, Michl G, Haslinger E (1989a) The first structurally confirmed saponin from *Solidago gigantea*: structure elucidation by modern NMR techniques. Tetrahedron Lett 30:4097–4100

Reznicek G, Kubelka W, Jurenitsch J, Michl G, Halinger E (1989b) Structure elucidation of a saponin from *Solidago gigantea*. Bull Magn Reson 11:402

Reznicek G, Jurenitsch J, Kubelka W, Michl G, Korhammer S, Haslinger E (1990) Isolation and structure of the four main saponins from *Solidago gigantea* var *serotina*. Liebigs Ann Chem (10)989–994

Reznicek G, Jurenitsch J, Plausun M, Korhammer S, Haslinger E, Hiller K, Kubelka W (1991) Four major saponins from *Solidago candensis*. Phytochemistry 30:1629–1633

Reznicek G, Jurenitsch J, Freiler M, Korhammer S, Haslinger E, Hiller K, Kubelka W (1992) Isolation and structure elucidation of further new saponins from *Solidago canadensis*. Planta Med 58:94–98

Rostański K (1971) *Solidago* L. In: Pawtowski, Jasiewicz (eds) Flora Polska, vol 12. Państwowe Wydawnictwo Naukowe, Warszawa Kraków, pp 116–121

Rote Liste (1996) Das Fertigarzneimittelverzeichnis der Verbände der pharmazeutischen Industrie. Editio Cantor, Aulendorf/Württ

Sakamoto S, Tsuchiya N, Kuroyanagi M, Ueno A (1994) Biotransformation of germacrone by suspension cultured cells. Phytochemistry 35:1215–1219

Saraswathy A, Sarada A, Purushothaman KK (1996) Goldenoic acid from *Solidago sempervirens* Linn. Indian J Chem, Sec B Org Chem Incl Med Chem 34B:1107; Chem Abstr 1996, 124:25634b

Sawicka T, Drozd J, Prosińska J, Borkowski B (1994) Aktywność immunostymulacyjna niektórych fenolokwasów i związków estrowych. Herba Pol 40:31–34

Schilcher H (1965) Qualitätsprüfung von Handelsdrogen 1 Mitt: Herba *virgaureae* und Herba *Solidaginis*. Dtsch Apoth Ztg 105:681–683

Schilcher H (1984) Pflanzliche Urologika. Dtsch Apoth Ztg 124:2429–2436

Schilcher H (1987) Pflanzliche Diuretika. Z Phytother 8:141–147; Urologe (B) 27:215–222

Schilcher H, Bornschein U (1986) Goldrutenkraut. Dtsch Apoth Ztg 126:1377–1380

Schilcher H, Rau H (1988) Nachweis der aquaretischen Wirkung von Birkenbläter-und Goldrutenkrautauszügen im Tierversuch. Urologe (B) 28:274–280

Schilcher H, May P, Sökeland J (1988) Phytotherapie in der Urologie. Urologe (B) 28:265–271

Scoggan HJ (1979) The flora of Canada, part 4. National Museum of Canada, Ottawa, pp 1601–1613

Shimomura H, Sashida Y, Yoshinari K (1989) Phenolic glucosides from the heartwood of *Prunus grayana*. Phytochemistry 28:1499–1502

Shoji A, Horiguchi S, Endo G (1993) Frequency of positive reaction to pollinic allergens (common goldenrod, mugwort, timothy grass) in scratch tests on lead exposed workers. Rodo Kagaku 69:39–44 (Jpn); Chem Abstr 1993, 118:207093h

Skrzypczakowa L (1961) Flawonoidy w zielu nawłoci pospolitej – *Solidago virgaurea* L. I Wyodrębnienie i identyfikacja 3 – rutozydu kwercetyny (rutyny). Acta Pol Pharm 18:39–48

Skrzypczakowa L (1962a) Flawonoidy w zielu nawłoci pospolitej – *Solidago virgaurea* L. II Wyodrębnienie i identyfikacja dalszych pochodnych flawonolowych. Acta Pol Pharm 19:481–490

Skrzypczakowa L (1962b) Związki wielofenolowe w zielu nawłoci pospolitej – *Solidago virgaurea*. Dissertation, Medical Academy Poznań, Poland

Skrzypczak L, Ellnain-Wojtaszek M (1981) 2-β-D-glucopyronosyloxybenzyl-4-β-D glucopyronosyloxy-2-hydroxy-3-methoxybenzoate of *Solidago* species (Asteraceae). Pol J Chem 55:683–685

Strehl E, Schneider W, Elstner EF (1995) Inhibition of dihydrofolate reductase activity by alcoholic extracts from *Fraxinus excelsior*, *Populus tremula* and *Solidago virgaurea*. Arznem-Forsch 45:172–173

Sziszkin BK (1959) Flora, vol 25. USSR Academy of Sciences USSR, Moskow, pp 31–50

Tada E, Uchida M, Funayama T (1989) Extraction of brassinolide – active substances. Jpn Koai Tokkyo Koho JP 01, 117, 899; Chem Abstr 1989, 111:229277d

Tamas M (1986) Cercetari chemotaxonomice la genul *Solidago*. Contrib Bot Cluj-Napoca, pp 109–113

Tamas M, Rosca M (1988) Saponins from species of *Solidago* growing in Romania. Farmacia (Bucharest) 36:161171 (Rom); Chem Abstr 1990, 112:115804p

Vaz SM, Saiki M, Vasconcellos MBA, Sertie JAA (1995) Neutron activation analysis of medicinal plant extracts. J Radioanal Nucl Chem 195:185–193; Chem Abstr 1995, 123:123329e

Weiss RF (1991) Herbal medicine. Beaconsfield, Beaconsfield Publishers, UK pp 241–243

Wesołowska M, Skrzypczak L, Budzianowskij (1997a) In vitro cultures of the genus *Solidago* L. Polish Confer Section IAPTC, Cracow, Abstr, p 100 (in Polish)

Wesołowska M, Skrzypczak L, Budzianowskij J, Kamińska D (1997b) In vitro cultures of the genus *Solidago* L. 45th Annu Congr Med Plant Res, Regensburg, Germany, Abstr L 20

Weyerstahl P, Marschall H, Christiansen C, Kalemba D, Góra J (1993) Constituents of the essential oil of *Solidago canadensis* (goldenrod) from Poland – a correction. Planta Med 59:281–282

Zenkteler M (ed) (1984) Hodowla komórek i tkanek roślinnych. Państwowe Wydawnictwo Naukowé, Warszawa

XXIV *Zanthoxylum* Species: In Vitro Culture, Aroma Emanation, and the Production of Secondary Metabolites

K. IMAIZUMI

1 Introduction

1.1 Distribution and Importance

The genus *Zanthoxylum* L. (*Xanthoxylum* L.) belongs to the family Rutaceae, and has about 200 species (Tsukamoto 1989). These species are deciduous or evergreen woody plants (shrubs or trees). The leaf type is alternate phyllotaxis and compound. The flower is very small, and white or greenish yellow. About 15 species are distributed in the temperate regions of East Asia and North America. These species are dioecious plants and have thorns on their branches. The main medicinal species are *Z. ailanthoides* Seib. et Zucc., *Z. armatum* DC., *Z. nitidum* (Roxb.) DC., and *Z. piperitum* DC.

In Japan, there are three endemic species; *Z. piperitum* DC. (Japanese name, Sanshoo), and *Z. planispinum* Sieb. (Japanese name, Fuyuzanshoo) and one horticultural species, *Z. piperitum* var. f. *inerme* Makino (Japanese name, Asakura-sanshoo). *Z. piperitum* DC. and *Z. piperitum* var. f. *inerme* Makino are deciduous shrubs, *Z. planispinum* Sieb. is an evergreen shrub. Furthermore, *Z. piperitum* var. f. *inerme* Makino is a thornless variant. *Z. piperitum* (Fig. 1) is especially important in Japan, because it is used as a typical Japanese spice and a natural drug which has been used since ancient times. Their fresh leaves and fruits are useful as aromatic foods. Also, dried fruit and pericarp powder has been utilized as spice, and mature fruits and oil from the seeds as natural drugs as aromatic stomachics, antidotes and insecticides (Perry 1980).

1.2 Conventional Practices for Propagation, Production of Alkaloids/Medicinal Components

1.2.1 Propagation

Almost all *Zanthoxylum* species have been propagated by seeds. For example, *Z. piperitum* is planted from nursery stock which is cultivated from seeds and

PL Gakuen Women's Junior College, Botanical Institute, 2172-1, Shindou, Tondabayashi, Osaka 584-0091, Japan

Biotechnology in Agriculture and Forestry, Vol. 43
Medicinal and Aromatic Plants XI (ed. by Y.P.S. Bajaj)

Fig. 1. Photograph of *Zanthoxylum piperitum* var. f. *inerme* Makino (Japanese name, Asakura-sanshoo)

seedlings for 1 year. In Japan, the suitable season for seed harvesting is in early September, and the seedling season is in late February. The harvested seeds are stored in a dark and cool space without drying, because dried seeds lose their germination ability.

1.2.2 Production of Medicinal Components and Their Importance

Some of the alkaloids and medicinal compounds found in *Zanthoxylum* species are nitidine, dictamine, pinene, etc. (Fig. 2).

Zanthoxylum ailanthoides Sieb. et Zucc. Alkaloidal components were examined from the wood, bark and root. The tertiary furoquinoline-type alkaloids, dictamnine, and skimmianine, the quaternary benzophenanthridine-type base, nitidine chloride, and a water-soluble quaternary aporphine-type base, nitidine, laurifoline chloride were isolated (Ishii and Komaki 1956, 1966).

Zanthoxylum armatum DC. Some regard *Z. planispium* Sieb. et Zucc. as belonging to *Z. armatum* DC. The dictamnine and skimmianine isolated from the roots contain the alkaloids found in the wood of *Z. ailanthoides* Sieb. et Zucc. (Tomita and Ishii 1959). In this species, the reported constituents of

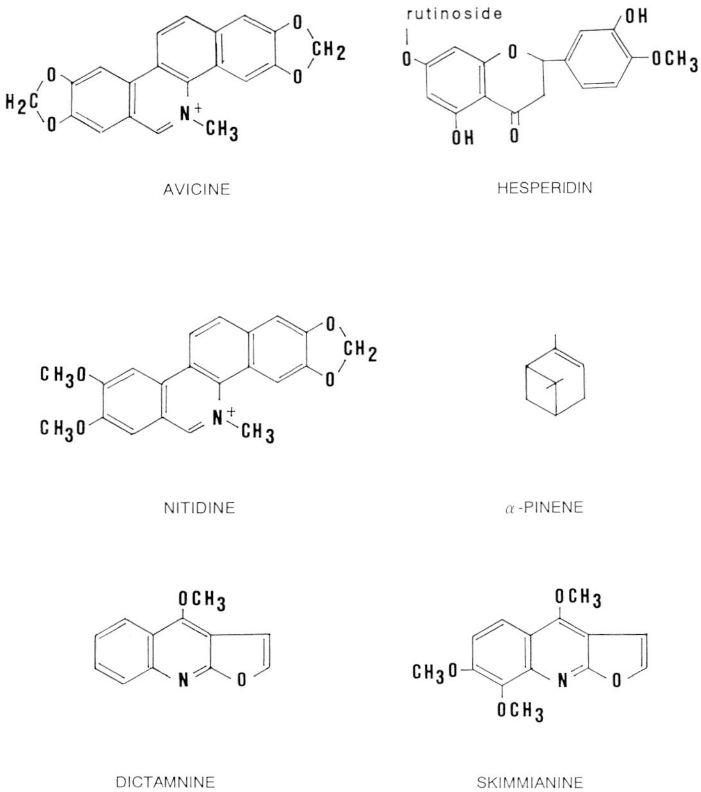

Fig. 2. The structures of some secondary metabolites in *Zanthoxylum* species

essential oils from the seeds are 1-α-phellandren, linaloll, and other sesquiterpenes (Simonsen and Rau 1922).

Zanthoxylum avicennae. A benzophenanthridine alkaloid, avicine, has been isolated from root bark. Avicine is related to nitidine in the same way that sanguinarine is related to chelerythrine. Hesperidin has been obtained from the leaves, and a mixture of hesperidin, diosmin, and avicennin, from the root bark of this plant (Arthur et al. 1959b).

Zanthoxylum nitidum (Roxb.) DC. This species has diverse medicinal uses in different regions of South East Asia. This plant is a woody climber with yellow roots and small white flowers. An alkaloid oxynitidine, and a salt of a related quaternary alkaloid, nitidine, have been isolated from methanol extracts of root bark and root wood (Arthur et al. 1959a).

Two alkaloids of nitidine and fagaronine are found in *Fagara* and *Zanthoxylum* species. These alkaloids are of the benzophenanthridine type and are very important compounds. They show excellent activitiy in the p-388 leukemia system (tumor cells of mice), and are the subject of intensive investigations (Messmer et al. 1972).

Zanthoxylum piperitum DC. From the root and stem of *Z. piperitum* and from the root of *Z. piperitum* var. *inerme*, a quanternary, aporphine-type alkaloid, magnoflorine, was detected and small amounts of tertiary bases were also found (Tomita and Ishii 1957).

Further, as other components without alkaloids were investigated, unsaturated aliphatic acid amides and essential oils (α-sanshool and β-sanshool) were found in the bark (Crombie and Taylar 1957).

1.2.3 Medicinal Components and Their Demand in the World Market

Zanthoxylum piperitum is a native of China, Japan, and Korea, and is also cultivated. In Japan, fresh leaves are mainly used as flavoring, and in other countries the leaves are used as an antidote. The sap of young leaves is applied to insect stings and cat bites. The bark is a stomachic and febrifuge. Styes may be cured by swallowing the fruits. In some areas of Asia, *Zanthoxylum* species (*Z. nitidum* DC.) are employed to treat sunstroke, cholera, and as an anthelmintic for children (Tanaka and Odashima 1939). The pungent seeds will stupefy fish (Groff et al. 1925). The juice of the root is utilized to treat scrofula, contusions, snake bite, and as an expectorant (How 1956).

In Taiwan, a decoction of leafy branches may be a cooling disinfectant that is gargled for inflammation of the throat. The roots are indicated as sudorific, febrifuge, and as an emmenagogue. The leaves are a medication for dropsy and colic, and are stimulating in baths, also vesicant in cataplasms (Perry 1980).

2 In Vitro Approaches

Our report (Imaizumi et al. 1993) appears to be the only research on tissue cultures of *Zanthoxylum* species published to date. This report analyzes the aromas of different leaves of plant lines derived from the meristem culture of Asakura-sanshoo (*Zanthoxylum piperitum* var. f. *inerme* Makino) from the parent plant (tissue cultute origin) and from seedlings. The total aroma contents were also compared, showing higher aroma emanation for the culture lines than for control lines during five subcultures.

2.1 Establishment of Tissue Cultures

Cultures were initiated from actively growing shoot tips collected from trees growing outdoors or in a greenhouse, and from meristem tips dissected from

Fig. 3. Shoot formation from Asakura-sanshoo callus on modified MS medium

buds on dormant or actively growing shoots. After removing the leaves, the material was surface sterilized by successive immersion in 70% alcohol for 30s and 3–5% sodium hypochlorite for 5min, followed by three rinses in sterile distilled water.

The age of the parent Asakura-sanshoo was estimated as being 7 to 10 years. Meristem tissues cut from the axillary and apical buds of sterilized shoot segments were incubated on modified MS medium (Murashige and Skoog 1962) with 3% sucrose and 0.9% agar containing 1mM α-naphthaleneacetic acid (NAA) and 1mM 6-benzylaminopurine (BAP) under light conditions (2000lx).

After 2 months, approximately 70% of the cultures had formed shoots on the same medium (Fig. 3). For root formation, the grown shoots were put on a two-layer medium consisting of an upper layer of modified MS agar and a lower layer of vermiculite containing only water. All experiments were conducted at 25°C. Rooted shoots of 5 to 10cm height were transplanted to pots and grown in a greenhouse (Fig. 4).

2.2 Extraction of Alkaloids/Medicinal Compounds

The leaves were extracted with excess volumes of hot methanol for 20h. The extract was concentrated, and left for 3 days. The green solid which was left

Fig. 4. A well-developed *Zanthoxylum piperitum* var. f. *inerme* Makino plant 6 months after transfer to garden soil

was extracted with light petroleum (b.p. 60–80 °C) and then with methanol. The pale brown residue was recrystallized twice with aqueous pyridine.

Milled root bark or wood bark was extracted with excess volumes of hot methanol for 25 h. The extracts were concentrated to 1/40 volume, then left for a few days. The crystalline mixture was collected, then extracted with boiling water. The residue (0.8 g) contained crude alkaloid compounds.

2.3 Commercial Aspects/Exploitation

Asakura-sanshoo (*Zanthoxylum piperitum* DC. f. *inerme* Makino) is a thornless variant of Japanese pepper (*Z. piperitum* DC.). In Japan, young leaves and fruits possessing a pleasant odor and strongly pungent taste are widely used as a spicy vegetable, and mature fruits are used as natural drugs in aromatic stomachics, antidotes and insecticides.

It has been reported that the major aroma components of the leaves are monoterpenoids (Kusumoto et al. 1968; Sakai et al. 1970), and the main components of the pungent taste substances in the stem, bark and mature seeds are acid amides (Yasuda et al. 1982). In these experiments, the essential oils extracted with organic solvents or steam distillation were analyzed by gas chromatography (GC). However, these extraction procedures were tedious and time-consuming. Instead of these methods, head-space GC (HS/GC) methods have been developed to directly analyze some aromatic plants (Chialva et al. 1982).

We tried an aroma analysis of different leaves from plant lines derived from a shoot-tip culture of Asakura-sanshoo, from the parent plant and from seedlings. The total aroma contents were also compared for the culture lines showing higher aroma emanation than the control lines during five subcultures.

2.3.1 Chemicals and Aroma Standards

α-Pinene, β-pinene, myrcene, limonene, 1,8-cineole and citronellal were purchased from Sigma (St. Louis, MO, USA). Sabinene was purchased from Extra Synthese (Geray, France) and β-caryophellen from Tokyo Kasei (Tokyo, Japan). These standards were dissolved in chloroform (Wako, Osaka, Japan). Standard solutions for aroma were used to obtain the GC retention time and calibration curve for quantification.

2.3.2 Head-Space Gas Chromatographic (HS/GC) Analysis

A Shimadzu GC-14A gas chromatograph coupled with an HSS-2B head-space sampler (Shimadzu, Kyoto, Japan) were used for the aroma analysis of Japanese pepper. Twenty ml vials containing 0.15 g of fresh leaves were retained at 40 °C for 30 min, and 0.8 ml of head-space gas in the vial was injected automatically to the GC equipment.

The GC conditions for the aroma analysis were as follows: column, Shimadzu CBP1-W25-300 wide-bore capillary type (25 m × 0.53 mm, 3 μm thickness); initial temperature 50 °C for 5 min; temperature gradient, 5 °C/min from 50 to 200 °C; flow rate of helium gas, 10.5 ml/min. A flame ionization detector (FID) and C-R4A Chromatopac integrator were used.

2.3.3 Identification of the Aroma Components

The aroma components from fresh leaves were identified by Shimadzu QP-1000A gas chromatography/mass spectrometry (GC-MS) with an HSS-2B sampler. The operating conditions were as follows: excitation, electron ioniza-

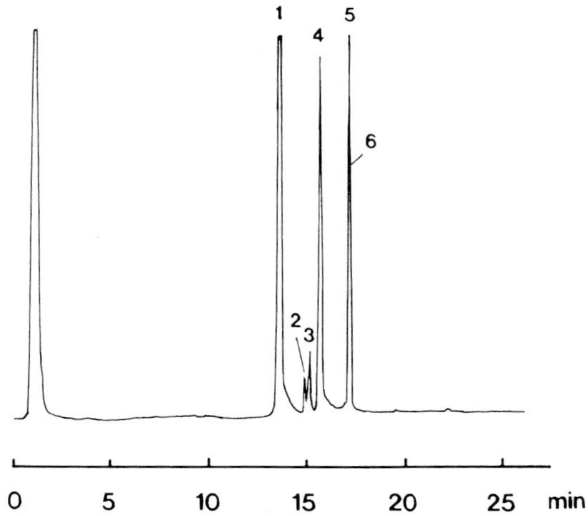

Fig. 5. Aroma components from fresh leaves of *Zanthoxylum piperitum* var. f. *inerme* Makino (Asakura-sanshoo) detected by the HS/GC method. Conditions for analysis: column, Shimadzu CBP1-W25-300 wide-bore capillary type (0.53 mm × 25 m); initial temperature, 50 °C for 5 min; temperature gradient, 5 °C/min from 50 °C to 200 °C; carrier gas, helium (flow rate, 10.5 ml/min); monitor, flame ionization detector (FID). Peaks: *1* α-pinene; *2* sabinene; *3* β-pinene; *4* myrcene; *5* limonene; *6* unknown

tion (EI); ionizing energy, 70 eV; separator temperature, 270 °C; ion source temperature, 250 °C; trapping current, 60 μA; and scanning range, m/z 33–500. The HS/GC conditions were the same as those just described.

Figure 5 shows a typical HS/GC gas chromatogram of the aromatic constituents from fresh leaves of Asakura-sanshoo, which were identified by GC-MS. The major aroma components were α-pinene, myrcene and limonene, and the minor components were β-pinene and sabinene. A quantitative analysis of the major aroma components in the fresh leaves was performed by HS/GC.

Standard solutions of α-pinene, sabinene, β-pinene, myrcene, limonene, citronellal and β-caryophellen were prepared to quantify the aroma of Asakura-sanshoo by using the external standard method. Using different injection volumes, standard curves were produced for each aroma component.

2.3.4 Comparison of the Aroma Levels in Fresh Leaves

Table 1 shows the average concentration of aroma substances in the fresh leaves from the parent plant, seedlings, and shoot-tip culture lines. The average leaf length of the shoot-tip culture lines was similar to that of the seedlings, while the parent plant leaves on average were smaller than the shoot-tip culture lines and seedlings. However, the average concentrations of aroma components from the shoot-tip culture lines were markedly different from those of seedlings (Table 1). The α-pinene level in fresh leaves from the seedlings was approximately twofold higher than that from the shoot-tip cultures and parent plant. With the shoot-tip culture lines and parent plant, the levels of myrcene and limonene were higher than those of the seedlings. Therefore, the pattern of aroma components in the culture lines was more similar to that of the parent than to the seedlings.

In order to select culture lines with high aroma emanation, we compared the total aroma levels calculated as the sum of α-pinene, myrcene, and limonene for every shoot-tip culture strain of Asakura-sanshoo. Figure 6 shows a histogram of the total aroma content of fresh leaves from the parent plant and shoot-tip culture lines. There was no significant difference in distribution between the parent plant and shoot-tip culture lines possessing a comparatively low level of total aroma. However, some of the shoot-tip culture strains exhibited a high level of total aroma compared to that of the parent plant. When strains Nos. 56 and 36 with an extremely high level of total aroma

Table 1. Average concentration of aroma components from fresh leaves of *Zanthoxylum piperitum*. (Imaizumi et al. 1993)

Plant group	No. of samples	Aroma content (μg/g fresh weight)		
		α-Pinene Mean ± SD	Myrcene Mean ± SD	Limonene Mean ± SD
Parent plant	29	1.39 ± 1.18	0.48 ± 0.53	1.91 ± 2.43
Seedlings	49	3.65 ± 2.40	0.34 ± 0.33	0.66 ± 0.70
Shoot-tip culture lines	116	1.93 ± 2.76	0.39 ± 0.57	1.54 ± 2.72

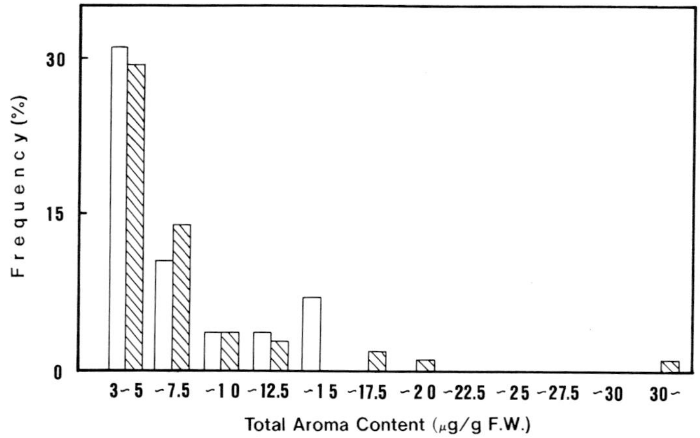

Fig. 6. Distribution of the total aroma contents from fresh leaves of Asakura-sanshoo. The sum of the α-pinene, myrcene and limonene levels was used as the total aroma content. *Open bars* show the percentage of the total number of leaf samples from the parent plant (N = 29), and *hatched bars* that from the shoot-tip culture lines (N = 116)

were examined, it was found that the α-pinene, myrcene, and limonene levels had all increased as a whole. This finding is different from the fact that the seedlings showed an increase in α-pinene level only.

The total aroma levels of almost all the shoot-tip culture lines were similar to that of the parent plant. This means that the hereditary characteristics of almost all the shoot-tip culture lines were preserved from the parent plant to some extent. Therefore, it has not been ascertained at this stage whether the few shoot-tip culture strains with a high level of total aroma were variants or mutants.

Tamura et al. (1984) examined the effects of shoot-tip culture on the secondary metabolite production of *Stevia rebaudiana*, and did not find a qualitative difference in the production of sweet diterpene glucosides such as stevioside and rebaudioside between the parent plants and shoot-tip cultures. They concluded that the shoot-tip culture strains of *Stevia* had inherited not only the shape but also the production of the secondary metabolites from their parents. In the case of Asakura-sanshoo, the leaf length and shape of the shoot-tip culture lines were similar to those of the young seedlings, while the aroma production was more similar to that of the parent plants than to the seedlings.

Our results seem to imply that genetic variation with respect to the aroma content could be avoided in clonal propagation by the shoot meristem culture, except for a few variants. Accordingly, clonal propagation of Asakura-sanshoo by the shoot-tip culture technique can be applied not only to the genetic preservation of a selected superior strain, but also to the mass propagation of clonal plants with uniform and better characteristics.

2.3.5 Selection of the Culture Lines with a High Aroma Level

The two culture lines of Asakura-sanshoo which exhibited a high aroma level in Fig. 6 were subcultured and separated into several culture lines. The aroma level of fresh leaves from each of the separated lines was then measured. Figure 7 shows a histogram of the total aroma levels of these selected culture lines, indicating that the total aroma level of fresh leaves from a parent plant was less than 50 μg/g fresh weight. Several ×66 lines derived from No. 36 with high aroma emanation had a total aroma level of 50 to 100 μg/g fresh weight, while some of the ×75 lines derived from No. 56 showed an even higher total aroma level of more than 100 μg/g fresh weight.

One morphological characteristic of the strains with an elevated aroma level is that their stems produced many branches. On the other hand, the strains with a low aroma level did not branch as much, and showed well-developed woody stems. Branching may correlate with the secondary metabolism of aroma productivity. In terms of the correlation of physiological activity to secondary metabolism, Nishioka (1988) reported that the elevation of physiological activity due to shoot-tip culture of a medicinal plant markedly promoted the productivity of pharmacological substances.

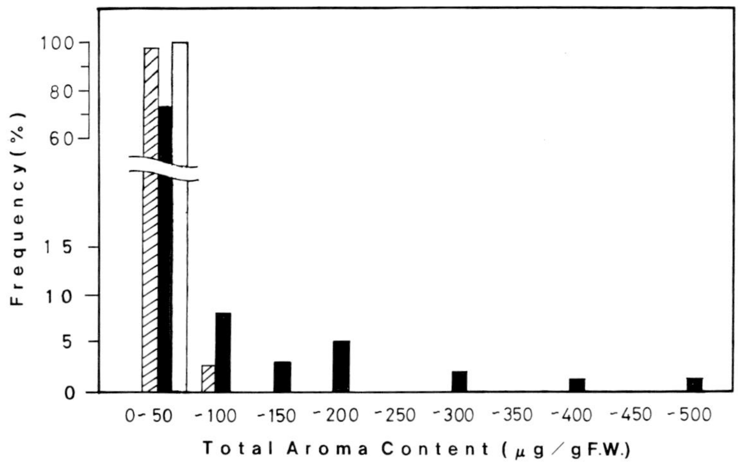

Fig. 7. Distribution of the total aroma contents from fresh leaves of Asakura-sanshoo. *Hatched bars* show the percentage of the total number of ×66 shoot-tip culture lines (N = 41), *closed bars* that of ×75 shoot-tip culture lines (N = 60), and *open bars* that of non-selected shoot-tip culture lines (N = 17)

In order to obtain culture lines with a high aroma level, the selected culture lines were further selected during several subcultures. The change in total aroma levels of the ×75 lines with comparatively high aroma content during five subcultures was examined. Each line was cultured without selection until the third transplantation. The lines with high aroma level were then selected at the fourth transplantation. One of the culture lines with a low aroma level at the first stage of transplantation was used as a control for those selected. Until the third transplantation, the total aroma level of both culture lines did not change significantly. After the fourth transplantation, however, the total aroma level of the selected culture lines had greatly increased, although that of the control line remained low.

Our experiments demonstrate that culture lines of Asakura-sanshoo with a high aroma level can be obtained by repeated in vitro selection after shoot-tip culture.

3 Conclusions and Prospects

The genus *Zanthoxylum* L. has about 200 species worldwide. In Asia, species are widely used as spices, vegetables and medicinal plants. Recently, some alkaloids with anti-tumor properties were isolated from *Zanthoxylum* species. Therefore, it is expected that the demand will increase for this medicinal plant.

The culture lines of Asakura-sanshoo (*Z. piperitum* var. f. *inerme* Makino) with a high aroma level can be obtained by repeated in vitro selection after shoot-tip culture. In the future, important medicinal materials such as anti-tumor substances may be produced from *Zanthoxylum* species by cell culture and other methods.

References

Arthur HR, Hui WH, Ng YL (1959a) An examination of the Rutaceae of Hong Kong, part 2. The alkaloids, nitidine and oxynitidine, from *Zanthoxylum nitidum*. J Chem Soc 1959:1840–1845

Arthur HR, Hui WH, Ng YL (1959b) An examination of the Rutaceae of Hong Kong, part 3. The alkaloid, avicine, from *Zanthoxylum avicennae*. J Chem Soc 1959:4007–4009

Arthur HR, Tam SW, Ng YL (1961) An examination of the Rutaceae of Hong Kong, part 7. The essential oils. J Chem Soc 1961:3551–3552

Chialva F, Gabri G, Liddle PHP, Ulian F (1982) Qualitative evaluation of aromatic herbs by direct head space $(GC)^2$ analysis. Applications of the method and comparison with the traditional analysis of essential oils. In: Margaris N, Koedam A, Vokou D (eds) Aromatic plants, basic and applied aspects. Nijhoff, Dordrecht, pp 183–195

Crombie L, Taylar JL (1957) Amides of vegetable oligine, part 8. The constitution and configuration of the sanshools. J Chem Soc 1957:2760–2766

Groff GW, Ding E, Groff EH (1925) An enumeration of the McClur collection of Hainan plants. Lingnaam Agar Rev 2:9–44

How FC (ed) (1956) Flora of Canton. South China Branch, Botanic Institute of the Academia Sinica, Canton, p 428

Imaizumi K, Totsuka Y, Oda S, Yamada Y (1993) Production of shoot-tip culture lines of Asakura-sanshoo (*Zanthoxylum piperitum* var. f. *inerme* Makino) with high aroma emanation. Biosci Biotechnol Biochem 57(4):591–593

Ishii H, Komaki T (1956) Studies on the alkaloids of rutaceous plants. 4. Alkaloids of *Xanthoxylum ailanthoides* Sieb. et Zucc. (*Fagara ailanthoides* (Seib. et Zucc.) Engel.).(1). The alkaloidal content of roots. J Pharm Soc Jpn 78:1441–1443

Ishii H, Komaki T (1966) Studies on the alkaloids of rutaceous plants. 18. Alkaloids of *Xanthoxylum ailanthoides* Sieb. et Zucc. (*Fagara ailanthoides* (Seib. et Zucc.) Engel.).(2). The alkaloidal content of roots. J Pharm Soc Jpn 86:631–633

Kusomoto S, Ohsuka A, Kotake M, Sakai T (1968) Constituents of leaf oil form Japanese pepper. Bull Chem Soc Jpn 41:1950–1953

Messmer WM, Tin-Wa M, Fong HHS, Bevelle C, Farnsworth NR, Abraham DJ, Trojanek J (1972) Fagaronine a new tumor inhibitor isolated from *Fagara zanthoxyloides* Lam (Rutaceae). J Pharm Sci 61(11):1858–1859

Murashige T, Skoog F (1962) A revised medium for rapid growth and bio-assays with tobacco tissue cultures. Physiol Plant 15:473–497

Nishioka I (1988) Clonal multiplication of medicinal plant by tissue culture. Shoyakugaku Zasshi 42:1–11 (in Japanese)

Perry LM (ed) (1980) Medical plants of East and Southeast Asia. MIT Press, Cambridge, MA

Sakai T, Yoshihra K, Hirose Y (1970) A comparative study of the constituents of volatile oils of *Zanthoxylum*. Bull Chem Soc Jpn 43:484–487

Simonsen JL, Rau MG (1922) The constituents of some Indian essential oils, part 5. The essential oil from the seeds of *Zanthoxylum alatum* Roxb. Indian For Rec 9:133–134

Tamura Y, Nakamura Y, Fukui H, Tabata M (1984) Comparison of *Stevia* plants grown from seeds, cuttings and stem-tip cultures for growth and sweet diterpene glucosides. Plant Cell Rep 3:180–182

Tanaka T, Odashima K (1939) Material for the study of vegetable resources of Hainan Island. Hortic Trop 9:1–31

Tomita M, Ishii H (1957) Studies on the alkaloids of rutaceous plants. 2. Alkaloids of *Xanthoxylum piperitum* (Linn.) DC. and *Xanthoxylum piperitum* (Linn.) DC. var. *inerme* Makino.(1). J Pharm Soc Jpn 77:810–812

Tomita M, Ishii H (1958) Studies on the alkaloids of rutaceous plants. 4. Alkaloids of *Xanthoxylum ailanthoides* Sieb, and Zucc. (*Fagara ailanthoides* (Sieb & Zucc.) Engel.). J Pharm Soc Jpn 78:1441–1443

Tomita M, Ishii H (1959) Studies on the alkaloids of rutaceous plants. 6. Alkaloids of *Xanthoxylum planispinum* Sieb. and Zucc. J Pharm Soc Jpn 79:1228–1230

Tsukamoto Y (ed) (1989) The grand dictionary of horticulture, vol 4. Shogakukan, Tokyo

Yasuda I, Takeya K, Itokawa H (1982) Distribution of unsaturated aliphatic acid amides in Japanese *Zanthoxylum* species. Phytochemistry 21:1295–1298

Subject Index

adventitious buds 3
aflatoxin 14, 254
Agrobacterium rhizogenes 31, 34, 157
A. tumefaciens 21, 34
agropine 145, 267
alder buckthorn 157
alkaloids 43, 136, 290, 338
aloe-emodin 161
amoorastatin 17
anagalline 1
anagallinone 2, 8
Anagallis arvensis 1–10
analgesia 261
anesthetic 354
anthocyanins 6, 177
anthocyanidins 158, 190
anthoquinones 157
anthracenone 223
antiasthma 106
antibacterial 160, 207, 244, 256
antibiotic 106, 337
anticancer 106
antidote 407
antifertility 100
antifungal 106, 160, 199, 244, 256, 393
antiinflammatory 296, 338, 355, 393
antiirradiation 393
antileprosy 106
antimalarial 18, 159
antimicrobial 15, 225, 338
antimitotic 271
antineoplastic 338
antiparasitic 11
antipyretic 11
antirachitic 364
antirheomatic 393
antiseptic 11
antispasmodic 274
antitumor 18, 138, 160, 244, 254, 261
antitussiv 261, 274
aphrodisiac 109
apigenin 274
apoatropine 350
apophyoscine 352
aporohines 292
aroma compound 410
aroma emanation 404
artificial seed 29
asthma 109
Asteraceae 261, 397
astragalin 112, 128
avicine 406
Azadirachta indica 11–44
azadirachtin 16, 30
azadirone 17

baldrinal 144
barbatene 219
barberine 337
bayogenin 391
belladonine 355
benzophenanthridine 136
betulinic acid 199
bioreactor studies 166
biosynthetic studies 283
biotransformation 310, 316, 331
birth control 15
bisaboline 219
bitter lettuce 261
buckthorn 223–242

caffeic acid 392
calcinogenic plant 364
calcinosis 364
calcitriol 368
callus 23, 65, k163, 265, 276, 372
callus differentiation 25
campesterol 71, 97, 314
camphene 275
cancer 238
carcinogenic 14
caryophylanes 249
castasterone 314
cat bites 408
cell culture 342
cell immobilization 373
cell suspension 165
Centranthus species 42–56
charcoal 24, 115
chelerythrine 139

chemotherapy 238
chlorogenic acid 392
cholesterol 86
choline 275
chrysophanol 161
clerodane 219
cocaine 354
commercial products 19
common pimpernel 1–10
cork tree 337–349
Costus speciosus 57–77
cough mixtures 262
coumarins 18
cryopreservation 88, 89, 95
Cuphea species 78–84
cytotoxic avtivity 254

dauricine 290
delphinidin 190
detergents 8
diarrhea 290
differentiation 7, 25
Dioscorea balcanica 85–104
D. caucasia 85
Dioscoreacea 86
diosgenin 57, 85
diterpenes 277
diterpenoids 15
diuretic 1, 385
Drosera species 105–135, 113
droserone 110

ectomycorrhizae 245
elicitors 72
embryogenesis 121
embryoids 327
emodin 159
emphysemia 262
enzootic calcinosis 364
epilepsy 79
essential oils 44
esters 180
euphol 16

Fabaceae 310
Fagara zanthoxyloides 136–156
fagaronine 136, 138
fatty acids 78
fever 137
flavonoids 110, 127, 338, 385
Frangula alnus 163
F. rupestris 166
Frangula species 157–176
fruiting bodies 244
fungicidal 1, 8
furanic labdane 277
furoquinoline 143

gastroenteritis 137
gedunin 18
gemmae 111
genetic transformation 34
geraniol 332
germanicol 263
goldenrod 384–403
gonorrhea 137
gossypin 112, 128
gossypitrin 112
gracillin 60
growth regulators 296, 329
Guttiferae 199

hairy roots 48, 145
hasubanans 293
hesperidin 406
HPLC 30, 49, 71, 147, 161, 190, 202, 235, 279, 302, 317, 344
hun horse 355
Hyacinthus orientalis 177–198
hydrophobia 1
hygrine 353
hypericin 200
Hypericum patulum 199–212
hyperin 128
hypocotyl culture 3
hyoscyamine 352

ileum 61
immobilized cells 332, 340
insect stings 407
insecticidal 11
insecticides 410
inulin 392
in vitro approaches 20, 181, 200, 228, 339, 372
in vitro culture 1, 42, 44, 57, 63, 78, 86, 145, 157, 162, 177, 199, 223, 261, 264, 274, 275, 290, 310, 311, 326, 350, 394, 404
in vitro propagation 181
ixerin 267

Jamesoniella automnalis 213–222
jamesionellides 219
japanese pepper 409
Jungermanniaceae 213

kaemferol 33
Karwinskia 223–242
knoghtalbinol 360

Lactarius 243–260
Lactuca virosa 261–273
lactucin 261, 263
lactucopicrin 261
Lamnaceae 274
leaf culture 3